MATHÉMATIQUES
&
APPLICATIONS

Directeurs de la collection :
G. Allaire et J. Garnier

69

For other titles published in this series, go to www.springer.com/series/2966

MATHÉMATIQUES & APPLICATIONS
Comité de Lecture 2008–2011/Editorial Board 2008–2011

RÉMI ABGRALL
INRIAet Mathématiques, Univ. Bordeaux 1, FR
abgrall@math.u-bordeaux.fr

GRÉGOIRE ALLAIRE
CMAP, École Polytechnique, Palaiseau, FR
gregoire.allaire@polytechnique.edu

MICHEL BENAÏM
Mathématiques, Univ. de Neuchâtel, CH
michel.benaim@unine.ch

OLIVIER CATONI
Proba. et Mod. Aléatoires, Univ. Paris 6, FR
catoni@ccr.jussieu.fr

THIERRY COLIN
Mathématiques, Univ. Bordeaux 1, FR
colin@math.u-bordeaux1.fr

MARIE-CHRISTINE COSTA
UMA, ENSTA, Paris, FR
marie-christine.costa@ensta.fr

ARNAUD DEBUSSCHE
ENS Cachan, Antenne de Bretagne
Avenue Robert Schumann,
35170 Bruz, FR
arnaud.debussche@bretagne.ens-cachan.fr

JACQUES DEMONGEOT
TIMC, IMAG, Univ. Grenoble I, FR
jacques.demongeot@imag.fr

NICOLE EL KAROUI
CMAP, École Polytechnique, Palaiseau, FR
elkaroui@cmapx.polytechnique.fr

JOSSELIN GARNIER
Proba. et Mod. Aléatoires, Univ. Paris 6 et 7, FR
garnier@math.jussieu.fr

STÉPHANE GAUBERT
INRIA, Saclay, Îles-de-France, Orsay et
CMAP, École Polytechnique, Palaiseau, FR
stephane.gaubert@inria.fr

CLAUDE LE BRIS
CERMICS, ENPC et INRIA
Marne la Vallée, FR
lebris@cermics.enpc.fr

CLAUDE LOBRY
INRA, INRIA, Sophia-Antipolis et
Analyse Systèmes et Biométrie
Montpellier, FR
lobrinria@wanadoo.fr

LAURENT MICLO
Analyse, Topologie et Proba., Univ. Provence, FR
miclo@cmi.univ-mrs.fr

FELIX OTTO
Institute for Applied Mathematics
University of Bonn, DE
otto@iam.uni-bonn.de

VALÉRIE PERRIER
Mod.. et Calcul, ENSIMAG, Grenoble, FR
valerie.perrier@imag.fr

BERNARD PRUM
Statist. et Génome, CNRS, INRA, Univ. Evry, FR
bernard.prum@genopole.cnrs.fr

PHILIPPE ROBERT
INRIA, Domaine de Voluceau, Rocquencourt, FR
philippe.robert@inria.fr

PIERRE ROUCHON
Automatique et Systèmes, École Mines, Paris, FR
pierre.rouchon@ensmp.fr

ANNICK SARTENAER
Mathématiques, Univ. Namur, BE
annick.sartenaer@fundp.ac.be

ERIC SONNENDRÜCKER
IRMA, Strasbourg, FR
sonnen@math.u-strasbg.fr

SYLVAIN SORIN
Combinat. et Optimisation, Univ. Paris 6, FR
sorin@math.jussieu.fr

ALAIN TROUVÉ
CMLA, ENS Cachan, FR
trouve@cmla.ens-cachan.fr

CÉDRIC VILLANI
UMPA, ENS Lyon, FR
cedric.villani@umpa.ens-lyon.fr

ENRIQUE ZUAZUA
Basque Center for Applied
Mathematics, Bilbao, Basque, ES
enrique.zuazua@uam.es

Directeurs de la collection :
G. ALLAIRE et J. GARNIER

Instructions aux auteurs :
Les textes ou projets peuvent être soumis directement à l'un des membres du comité de lecture avec copie à G. ALLAIRE OU J. GARNIER. Les manuscrits devront être remis à l'Éditeur sous format LaTeX2e (cf. ftp://ftp.springer.de/pub/tex/latex/svmonot1/).

Daniele Antonio Di Pietro · Alexandre Ern

Mathematical Aspects of Discontinuous Galerkin Methods

Daniele Antonio Di Pietro
IFP Energies nouvelles
Department of Applied Mathematics
1 & 4, avenue du Bois Préau
92852 Rueil-Malmaison
France
dipietrd@ifpen.fr

Alexandre Ern
Université Paris Est
CERMICS, Ecole des Ponts ParisTech
6 & 8, avenue Blaise Pascal
77455 Marne la Vallé cedex 2
France
ern@cermics.enpc.fr

ISSN 1154-483X
ISBN 978-3-642-22979-4   e-ISBN 978-3-642-22980-0
DOI 10.1007/978-3-642-22980-0
Springer Heidelberg Dordrecht Lodnon New York

Library of Congress Control Number: 2011940956

Mathematics Subject Classification (2010): 65N30, 65M60, 65N08, 65M08, 65N12,
                                            65N15, 65M12, 65M15, 76M10, 76M12

© Springer-Verlag Berlin Heidelberg 2012
This work is subject to copyright. All rights are reserved, whether the whole or part of the material is concerned, specifically the rights of translation, reprinting, reuse of illustrations, recitation, broadcasting, reproduction on microfilm or in any other way, and storage in data banks. Duplication of this publication or parts thereof is permitted only under the provisions of the German Copyright Law of September 9, 1965, in its current version, and permission for use must always be obtained from Springer. Violations are liable to prosecution under the German Copyright Law.

The use of general descriptive names, registered names, trademarks, etc. in this publication does not imply, even in the absence of a specific statement, that such names are exempt from the relevant protective laws and regulations and therefore free for general use.

Cover design: deblik, Berlin

Printed on acid-free paper

Springer is part of Springer Science + Business Media (www.springer.com)

Preface

Discontinuous Galerkin (dG) methods can be viewed as finite element methods allowing for discontinuities in the discrete trial and test spaces. Localizing test functions to single mesh elements and introducing numerical fluxes at interfaces, they can also be viewed as finite volume methods in which the approximate solution is represented on each mesh element by a polynomial function and not only by a constant function. From a practical viewpoint, working with discontinuous discrete spaces leads to compact discretization stencils and, at the same time, offers a substantial amount of flexibility, making the approach appealing for multi-domain and multi-physics simulations. Moreover, basic conservation principles can be incorporated into the method. Applications of dG methods cover a vast realm in engineering sciences. Examples can be found in the conference proceedings edited by Cockburn, Karniadakis, and Shu [106] and in recent special volumes of leading journals, e.g., those edited by Cockburn and Shu [114] and by Dawson [120]. There is also an increasing number of open source libraries implementing dG methods. A non-exhaustive list includes deal.II [27], Dune [36], FEniCS [251], freeFEM [118], libmesh [213], and Life [262].

A Brief Historical Perspective

Although dG methods have existed in various forms for more than 30 years, they have experienced a vigorous development only over the last decade, as illustrated in Fig. 1.

The first dG method to approximate first-order PDEs has been introduced by Reed and Hill in 1973 [268] in the context of steady neutron transport, while the first analysis for steady first-order PDEs was performed by Lesaint and Raviart in 1974 [227–229]. The error estimate was improved by Johnson and Pitkäranta in 1986 [204] who established an order of convergence in the L^2-norm of $(k + \frac{1}{2})$ if polynomials of degree k are used and the exact solution is smooth enough. Few years later, the method was further developed by Caussignac and Touzani [84,85] to approximate the three-dimensional boundary-layer equations for incompressible steady fluid flows. At around the same time, dG methods were extended to time-dependent hyperbolic PDEs by Chavent and Cockburn [86] using the forward Euler scheme for time discretization together with limiters. The order of accuracy was improved by Cockburn and Shu [110, 111] using

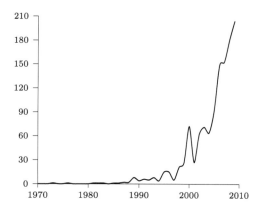

Fig. 1: Yearly number of entries with the keyword "discontinuous Galerkin" in the MathSciNet database

explicit Runge–Kutta schemes for time discretization, while a convergence proof to the entropy solution was obtained by Jaffré, Johnson, and Szepessy [200]. Extensions are discussed in a series of papers by Cockburn, Shu, and coworkers; see, e.g., [99, 108, 113].

For PDEs with diffusion, dG methods originated from the work of Nitsche on boundary-penalty methods in the early 1970s [248, 249] and the use of Interior Penalty (IP) techniques to weakly enforce continuity conditions imposed on the solution or its derivatives across interfaces, as in the work of Babuška [20], Babuška and Zlámal [24], Douglas and Dupont [134], Baker [25], Wheeler [306], and Arnold [14]. This latter work constitutes a milestone in the development of IP dG methods. In the late 1990s, following an approach more closely related to hyperbolic problems, dG methods were formulated using numerical fluxes by considering the mixed formulation of the diffusion term. Examples include the work of Bassi and Rebay [34] on the compressible Navier–Stokes equations and that of Cockburn and Shu [112] on convection-diffusion systems, leading to a new thrust in the development of dG methods. A unified analysis of dG methods for the Poisson problem can be found in the work of Arnold, Brezzi, Cockburn, and Marini [16], while a unified analysis encompassing both diffusive and first-order PDEs in the framework of Friedrichs' systems has been derived by Ern and Guermond [142–144].

Goals

The goal of this book is to provide the reader with the basic mathematical concepts to design and analyze dG methods for various model problems, starting at an introductory level and further elaborating on more advanced topics. Since the focus is on mathematical aspects of dG methods, linear model problems involving first-order PDEs and PDEs with diffusion play a central role.

Nonlinear problems, e.g., the incompressible Navier–Stokes equations and nonlinear conservation laws, are also treated, but in less detail for the latter, since very few mathematical results are yet available. This book also covers basic facts concerning the practical aspects of dG methods related to implementation; a more detailed treatment can be found in the recent textbook by Hesthaven and Warburton [192].

Some of the topics covered in this book stem from (very) recent work by the authors, e.g., on discrete functional analysis and convergence to minimal regularity solutions for PDEs with diffusion [131], the above-mentioned work with Guermond on Friedrichs' systems, and joint work with Vohralík on diffusive flux reconstruction and a posteriori error estimates for elliptic PDEs [147, 149, 151]. Another salient feature of this book is to bridge, as much as possible, the gap between dG methods and finite volume methods, where many important theoretical advances have been accomplished over the last decade (see, e.g., the work of Eymard, Gallouët, and Herbin [156, 157, 159]). In particular, we present so-called cell-centered Galerkin methods for elliptic PDEs with one degree of freedom per mesh cell, following recent work by Di Pietro and coworkers [5, 6, 128]. For the approximation of unsteady problems, we focus on the method of lines, whereby dG space semidiscretization is combined with various schemes to march in time. One important goal is to show how the stability of dG schemes interacts with the stability of the temporal scheme. This is reflected, for instance, in our covering the analysis based on energy estimates for explicit Runge–Kutta dG methods applied to first-order, linear PDEs, following the recent work of Burman, Ern, and Fernández [66]. For a description and analysis of space-time dG methods as an alternative to the method of lines, we refer the reader to the textbook by Thomée [294]. Finally, we consider a rather general setting for the underlying meshes so as to exploit, as much as possible, the flexibility offered by the dG setting.

Outline

Chapter 1 introduces basic concepts to formulate and analyze dG methods. In this chapter, we present:

(a) Two important results to assert the well posedness of linear model problems, namely the so-called Banach–Nečas–Babuška Theorem and the Lax–Milgram Lemma.

(b) The basic ingredients related to meshes and polynomials to build discrete functional spaces and, in particular, broken polynomial spaces.

(c) The three key properties for the convergence analysis of dG methods in the context of nonconforming finite elements, namely discrete stability, consistency, and boundedness.

(d) The basic analysis tools, in particular inverse and trace inequalities needed to assert discrete stability and boundedness, together with optimal polynomial approximation results, thereby leading to the concept of admissible mesh sequences.

In this book, we focus on mesh refinement as the main parameter to achieve convergence. Convergence analysis using, e.g., high-degree polynomials is possible by further tracking the dependency of the inverse and trace inequalities on the underlying polynomial degree and, in some cases, modifying accordingly the penalty strategy in the dG method to achieve discrete stability. Important tools in this direction can be found in the work of Babuška and Dorr [22], Babuška, Szabó, and Katz [23], Szabó and Babuška [288] and Schwab [275], and, in the context of dG methods, in the recent textbook of Hesthaven and Warburton [192].

The remaining chapters of this book are organized into three parts, each comprising two chapters. Part I, which comprises Chaps. 2 and 3, deals with scalar first-order PDEs. Chapter 2 focuses on the steady advection-reaction equation as a simple model problem. Therein, we present some mathematical tools to achieve a well-posed formulation at the continuous level and show how these ideas are built into the design of dG methods. Two methods are analyzed thoroughly, which correspond in the finite volume terminology to the use of centered and upwind fluxes.

Chapter 3 deals with unsteady first-order PDEs. Within the method of lines, we consider a dG method for space semidiscretization combined with an explicit scheme to march in time. We first present a detailed analysis in the linear case whereby the upwind dG method of Chap. 2 is used for space semidiscretization. Following the seminal work of Levy and Tadmor [231], the stability analysis hinges on energy estimates, and not on the more classical approach based on regions of absolute stability for explicit Runge–Kutta (RK) schemes. In particular, we show how the stability provided by the upwind dG scheme can compensate the anti-dissipative nature of the explicit time-marching scheme. We also show how this stability can be used to achieve quasi-optimal error estimates for smooth solutions. We present results for the forward Euler method combined with a finite volume scheme as an example of low-order approximation and then move on to explicit two- and three-stage RK schemes combined with a high-order dG approximation. The second part of Chap. 3 deals with nonlinear conservation laws. Taking inspiration from the linear case, we first design dG methods for space semidiscretization using the concept of numerical fluxes and provide various classical examples thereof, including Godunov, Rusanov, Lax–Friedrichs, and Roe fluxes. Then, we consider explicit RK schemes for time discretization and discuss Strong Stability-Preserving (SSP) RK schemes following the ideas of Gottlieb, Shu, and Tadmor [175, 176]. Finally, we discuss the use of limiters in the framework of one-dimensional total variation analysis and possible extensions to the multidimensional case.

Part II, which comprises Chaps. 4 and 5, addresses scalar, second-order PDEs. Chapter 4 covers various model problems with diffusion. We first present a heuristic derivation and a convergence analysis to smooth solutions for a purely diffusive problem approximated by the Symmetric Interior Penalty

(SIP) method of Arnold [14]. Then, we introduce the central concept of discrete gradients and present some important applications, including the link with the mixed dG approach and the local formulation of the discrete problem using numerical fluxes. Hybrid mixed dG methods are also briefly discussed. Focusing next on heterogeneous diffusion problems, we analyze a modification of the SIP method, the Symmetric Weighted Interior Penalty (SWIP) method, using weighted averages in the consistency term and harmonic means of the diffusion coefficient at interfaces in the penalty term. The SWIP method is then combined with the upwind dG method of Chap. 2 to approximate diffusion-advection-reaction problems. We examine singularly perturbed regimes due to dominant advection, and include in the analysis the case where the diffusion actually vanishes in some parts of the domain. Finally, as a simple example of time-dependent problem with diffusion, we consider the heat equation approximated by the SIP method in space and implicit time-marching schemes (backward Euler and BDF2).

Chapter 5 covers some additional topics related to diffusive PDEs, and, for the sake of simplicity, the scope is limited to purely diffusive problems. First, we present discrete functional analysis tools in broken polynomial spaces, namely discrete Sobolev embeddings, a discrete Rellich–Kondrachov compactness result, and a weak asymptotic consistency result for discrete gradients. As an example of application, we prove the convergence, as the meshsize goes to zero, of the sequence of discrete SIP solutions to minimal regularity solutions. These ideas are instrumental in Chap. 6 when analyzing the convergence of dG approximations for the steady incompressible Navier–Stokes equations. Then, we present some possible variations on symmetry and penalty for the SIP method. A further topic concerns the so-called cell-centered Galerkin methods which use a single degree of freedom per mesh cell and a suitable discrete gradient reconstruction, thereby providing a closer connection between dG and finite volume methods for elliptic PDEs. The last two topics covered in Chap. 5 are local postprocessing of the dG solution and a posteriori error estimates. One salient feature of local postprocessing is the possibility to reconstruct locally an accurate and conservative diffusive flux. An important application is contaminant transport where the diffusive flux resulting from a Darcy flow model is used as advection velocity. In turn, a posteriori error estimates provide fully computable error upper bounds that can be used to certify the accuracy of the simulation and to adapt the mesh. These last two topics can be nicely combined together since local postprocessing provides an efficient tool for deriving constant-free a posteriori error estimates.

Part III, which comprises Chaps. 6 and 7, deals with systems of PDEs. Chapter 6 is devoted to incompressible flows. Focusing first on the steady Stokes equations, we examine how the divergence-free constraint on the velocity field can be tackled using dG methods. We detail the analysis of equal-order approximations using both discontinuous velocities and pressures, whereby pressure jumps need to be penalized, and then briefly discuss alternative formulations avoiding the need for pressure jump penalty. The next step is the discretization of the nonlinear convection term in the momentum equation. To this purpose, we derive a discrete trilinear form that leads to the correct kinetic energy balance, using the so-called Temam's device to handle the fact that discrete velocities are only weakly divergence-free. Moreover, since the model problem is now nonlinear, the convergence analysis is performed under minimal regularity assumptions on the exact solution and without any smallness assumption on the data using the discrete functional analysis tools of Chap. 5. Finally, we discuss the approximation of the unsteady Navier–Stokes equations in the context of pressure-correction methods.

Chapter 7 presents a unified approach for the design and analysis of dG methods based on the class of symmetric positive systems of first-order PDEs introduced by Friedrichs [163]. Focusing first on the steady case, we review some examples of Friedrichs' systems and derive the main mathematical tools for asserting well posedness at the continuous level. Using these tools, we derive and analyze dG methods, and, in doing so, we follow a similar path of ideas to that undertaken in Chap. 2. Then, we consider more specifically the setting of two-field Friedrichs' systems and highlight the common ideas with the mixed dG approximation of elliptic PDEs. Finally, we consider unsteady Friedrichs' systems with explicit RK schemes in time and then specialize the setting to two-field Friedrichs' systems related to linear wave propagation, thereby addressing energy conservation and the possibility to accommodate local time stepping in the context of leap-frog schemes.

Appendix A covers practical implementation aspects of dG methods, focusing on matrix assembly and choice of local bases for which we discuss selection criteria and present various examples including both nodal and modal basis functions.

A bibliography comprising about 300 entries closes the manuscript. The amount of literature on dG methods is so vast that this bibliography is by no means exhaustive. We hope that the selected entries provide the reader with additional reading paths to examine more deeply the topics covered herein and to explore new ones. We mention, in particular, the recent textbooks by Hesthaven and Warburton [192], Kanschat [205], and Rivière [269].

Audience

This book is primarily geared to graduate students and researchers in applied mathematics and numerical analysis. It can be valuable also to graduate students and researchers in engineering sciences who are interested in further understanding the mathematical aspects that underlie the construction of dG methods, since these aspects are often important to formulate such methods when faced with new challenging applications. The reader is assumed to be familiar with conforming finite elements including weak formulations of model problems and error analysis (as presented, e.g., in the textbooks of Braess [49], Brenner and Scott [54], Ciarlet [92], or Ern and Guermond [141]) and to have some acquaintance with the basic PDEs in engineering and applied sciences. Special care has been devoted to making the material as much self-contained as possible. The general level of the book is best suited for a graduate-level course which can be built by drawing on some of the present chapters. The material is actually an elaboration on the lecture notes by the authors for a graduate course on dG methods at University Pierre et Marie Curie.

Acknowledgments

We are grateful to several colleagues for their constructive remarks that helped improve the manuscript. In particular, our warmest thanks go to Léo Agélas (IFP Énergies nouvelles), Erik Burman (University of Sussex), Miguel Fernández (INRIA), Jean-Luc Guermond (Texas A&M University), Rémi Joubaud (University Paris-Est), Olivier Le Maître (LIMSI), Serge Piperno (University Paris-Est), Quang Huy Tran (IFP Énergies nouvelles), and Martin Vohralík (University Pierre et Marie Curie). We are also thankful to Rémi Abgrall for offering us the opportunity to publish this book.

Contents

1 Basic Concepts ... 1
 1.1 Well-Posedness for Linear Model Problems ... 2
  1.1.1 The Banach-Nečas-Babuška Theorem ... 2
  1.1.2 The Lax–Milgram Lemma ... 3
  1.1.3 Lebesgue and Sobolev Spaces ... 4
 1.2 The Discrete Setting ... 7
  1.2.1 The Domain Ω ... 7
  1.2.2 Meshes ... 8
  1.2.3 Mesh Faces, Averages, and Jumps ... 9
  1.2.4 Broken Polynomial Spaces ... 12
  1.2.5 Broken Sobolev Spaces ... 13
  1.2.6 The Function Space $H(\mathrm{div};\Omega)$ and Its Broken Version ... 16
 1.3 Abstract Nonconforming Error Analysis ... 18
  1.3.1 The Discrete Problem ... 18
  1.3.2 Discrete Stability ... 19
  1.3.3 Consistency ... 21
  1.3.4 Boundedness ... 21
  1.3.5 Error Estimate ... 22
 1.4 Admissible Mesh Sequences ... 22
  1.4.1 Shape and Contact Regularity ... 23
  1.4.2 Geometric Properties ... 24
  1.4.3 Inverse and Trace Inequalities ... 26
  1.4.4 Polynomial Approximation ... 30
  1.4.5 The One-Dimensional Case ... 34

I Scalar First-Order PDEs ... 35

2 Steady Advection-Reaction ... 37
 2.1 The Continuous Setting ... 38
  2.1.1 Assumptions on the Data ... 38
  2.1.2 The Graph Space ... 39
  2.1.3 Traces in the Graph Space ... 40

XIII

	2.1.4	Weak Formulation and Well-Posedness	42
	2.1.5	Proof of Main Results	43
	2.1.6	Nonhomogeneous Boundary Condition	46
2.2	Centered Fluxes		47
	2.2.1	Heuristic Derivation	50
	2.2.2	Error Estimates	53
	2.2.3	Numerical Fluxes	55
2.3	Upwinding		56
	2.3.1	Tightened Stability Using Penalties	57
	2.3.2	Error Estimates Based on Coercivity	58
	2.3.3	Error Estimates Based on Inf-Sup Stability	61
	2.3.4	Numerical Fluxes	65

3 Unsteady First-Order PDEs   67

3.1	Unsteady Advection-Reaction		68
	3.1.1	The Continuous Setting	69
	3.1.2	Space Semidiscretization	72
	3.1.3	Time Discretization	74
	3.1.4	Main Convergence Results	78
	3.1.5	Analysis of Forward Euler and Finite Volume Schemes	83
	3.1.6	Analysis of Explicit RK2 Schemes	89
3.2	Nonlinear Conservation Laws		98
	3.2.1	The Continuous Setting	99
	3.2.2	Numerical Fluxes for Space Semidiscretization	100
	3.2.3	Time Discretization	106
	3.2.4	Limiters	108

II   Scalar Second-Order PDEs   117

4 PDEs with Diffusion   119

4.1	Pure Diffusion: The Continuous Setting		120
	4.1.1	Weak Formulation and Well-Posedness	120
	4.1.2	Potential and Diffusive Flux	121
4.2	Symmetric Interior Penalty		122
	4.2.1	Heuristic Derivation	122
	4.2.2	Other Boundary Conditions	126
	4.2.3	Basic Energy-Error Estimate	128
	4.2.4	L^2-Norm Error Estimate	133
	4.2.5	Analysis for Low-Regularity Solutions	134
4.3	Liftings and Discrete Gradients		137
	4.3.1	Liftings: Definition and Stability	138
	4.3.2	Discrete Gradients: Definition and Stability	140
	4.3.3	Reformulation of the SIP Bilinear Form	140
	4.3.4	Numerical Fluxes	142

	4.4	Mixed dG Methods	143
		4.4.1 The SIP Method As a Mixed dG Method	143
		4.4.2 Numerical Fluxes	145
		4.4.3 Hybrid Mixed dG Methods	148
	4.5	Heterogeneous Diffusion	150
		4.5.1 The Continuous Setting	151
		4.5.2 Discretization	153
		4.5.3 Error Estimates for Smooth Solutions	156
		4.5.4 Error Estimates for Low-Regularity Solutions	160
		4.5.5 Numerical Fluxes	162
		4.5.6 Anisotropy	163
	4.6	Diffusion-Advection-Reaction	163
		4.6.1 The Continuous Setting	163
		4.6.2 Discretization	165
		4.6.3 Error Estimates	166
		4.6.4 Locally Vanishing Diffusion	171
	4.7	An Unsteady Example: The Heat Equation	176
		4.7.1 The Continuous Setting	176
		4.7.2 Discretization	177
		4.7.3 Error Estimates	179
		4.7.4 BDF2 Time Discretization	182
		4.7.5 Improved $C^0(L^2(\Omega))$-Error Estimate	184
5	**Additional Topics on Pure Diffusion**		**187**
	5.1	Discrete Functional Analysis	188
		5.1.1 The BV Space and the $\|\cdot\|_{\mathrm{dG},p}$-Norms	188
		5.1.2 Discrete Sobolev Embeddings	190
		5.1.3 Discrete Compactness	193
	5.2	Convergence to Minimal Regularity Solutions	195
		5.2.1 Consistency Revisited	195
		5.2.2 Convergence	197
	5.3	Variations on Symmetry and Penalty	198
		5.3.1 Variations on Symmetry	199
		5.3.2 Variations on Penalty	203
		5.3.3 Synopsis	206
	5.4	Cell-Centered Galerkin	207
	5.5	Local Postprocessing	211
		5.5.1 Local Residuals	211
		5.5.2 Potential Reconstruction	214
		5.5.3 Diffusive Flux Reconstruction by Prescription	217
		5.5.4 Diffusive Flux Reconstruction by Solving Local Problems	222
	5.6	A Posteriori Error Estimates	227
		5.6.1 Overview	228
		5.6.2 Energy-Norm Error Upper Bounds	229
		5.6.3 Error Lower Bounds	236

III Systems — 239

6 Incompressible Flows — 241
- 6.1 Steady Stokes Flows … 243
  - 6.1.1 The Continuous Setting … 243
  - 6.1.2 Equal-Order Discontinuous Velocities and Pressures … 248
  - 6.1.3 Convergence to Smooth Solutions … 255
  - 6.1.4 Convergence to Minimal Regularity Solutions … 260
  - 6.1.5 Formulations Without Pressure Stabilization … 266
- 6.2 Steady Navier–Stokes Flows … 267
  - 6.2.1 The Continuous Setting … 267
  - 6.2.2 The Discrete Setting … 271
  - 6.2.3 Convergence Analysis … 276
  - 6.2.4 A Conservative Formulation … 280
- 6.3 The Unsteady Case … 283
  - 6.3.1 The Continuous Setting … 284
  - 6.3.2 The Projection Method … 284

7 Friedrichs' Systems — 293
- 7.1 Basic Ingredients and Examples … 294
  - 7.1.1 Basic Ingredients … 294
  - 7.1.2 Examples … 296
- 7.2 The Continuous Setting … 301
  - 7.2.1 Friedrichs' Operators … 302
  - 7.2.2 The Cone Formalism … 304
  - 7.2.3 The Boundary Operator M … 306
  - 7.2.4 The Well-Posedness Result … 308
  - 7.2.5 Examples … 309
- 7.3 Discretization … 312
  - 7.3.1 Assumptions on the Data and the Exact Solution … 312
  - 7.3.2 Design of the Discrete Bilinear Form … 314
  - 7.3.3 Convergence Analysis … 318
  - 7.3.4 Numerical Fluxes … 320
  - 7.3.5 Examples … 320
- 7.4 Two-Field Systems … 322
  - 7.4.1 The Continuous Setting … 323
  - 7.4.2 Discretization … 324
  - 7.4.3 Partial Coercivity … 329
- 7.5 The Unsteady Case … 331
  - 7.5.1 Explicit Runge–Kutta Schemes … 331
  - 7.5.2 Explicit Leap-Frog Schemes for Two-Field Systems … 334

Appendix A: Implementation — 343
- A.1 Matrix Assembly … 343
  - A.1.1 Basic Notation … 343
  - A.1.2 Mass Matrix … 344
  - A.1.3 Stiffness Matrix … 344

	A.2	Choice of Basis Functions	347
	A.2.1	Requirements	347
	A.2.2	Jacobi and Legendre Polynomials	349
	A.2.3	Nodal Basis Functions	350
	A.2.4	Modal Basis Functions	352

Bibliography **355**

Author Index **375**

Index **381**

Chapter 1
Basic Concepts

This chapter introduces the basic concepts to build discontinuous Galerkin (dG) approximations for the model problems examined in the subsequent chapters. In Sect. 1.1, we present two important results to assert the well-posedness of linear model problems, namely the so-called Banach–Nečas–Babuška (BNB) Theorem, which provides necessary and sufficient conditions for well-posedness, and the Lax–Milgram Lemma, which hinges on coercivity in a Hilbertian setting and provides sufficient conditions for well-posedness. We also state some basic results on Lebesgue and Sobolev spaces. In Sect. 1.2, we describe the main ideas to build finite-dimensional function spaces in the dG setting. The two ingredients are discretizing the domain Ω over which the model problem is posed using a mesh and then choosing a local functional behavior within each mesh element. For simplicity, we focus on a polynomial behavior, thereby leading to so-called broken polynomial spaces. We also introduce important concepts to be used extensively in this book, including mesh faces, jump and average operators, broken Sobolev spaces, and a broken gradient operator. In Sect. 1.3, we outline the key ingredients in the error analysis of nonconforming finite element methods. The error estimates are derived in the spirit of the Second Strang Lemma using discrete stability, (strong) consistency, and boundedness. The advantage of this approach is to deliver error estimates and (quasi-)optimal convergence rates for smooth solutions. This framework for error analysis is frequently used in what follows. Yet, it is not the only tool for analyzing the convergence of dG approximations. In Chaps. 5 and 6, we consider an alternative approach in the context of PDEs with diffusion based on a compactness argument and a different notion of consistency. This approach allows us to prove convergence (without delivering error estimates) with minimal regularity assumptions on the exact solution. Finally, in Sect. 1.4, we present technical, yet important, tools to analyze the convergence of dG methods as the meshsize goes to zero (the so-called h-convergence). A crucial issue is then to ensure that some key properties of the mesh hold uniformly in this limit, thereby leading to the important concept of admissible mesh sequences.

1.1 Well-Posedness for Linear Model Problems

Let X and Y be two Banach spaces equipped with their respective norms $\|\cdot\|_X$ and $\|\cdot\|_Y$ and assume that Y is reflexive. In many applications, X and Y are actually Hilbert spaces. We recall that $\mathcal{L}(X, Y)$ is the vector space spanned by bounded linear operators from X to Y, and that this space is equipped with the usual norm

$$\|A\|_{\mathcal{L}(X,Y)} := \sup_{v \in X \setminus \{0\}} \frac{\|Av\|_Y}{\|v\|_X} \qquad \forall A \in \mathcal{L}(X, Y).$$

We are interested in the abstract linear model problem

$$\text{Find } u \in X \text{ s.t. } a(u, w) = \langle f, w \rangle_{Y', Y} \text{ for all } w \in Y, \tag{1.1}$$

where $a \in \mathcal{L}(X \times Y, \mathbb{R})$ is a bounded bilinear form, $f \in Y' := \mathcal{L}(Y, \mathbb{R})$ is a bounded linear form, and $\langle \cdot, \cdot \rangle_{Y', Y}$ denotes the duality pairing between Y' and Y. Alternatively, it is possible to introduce the bounded linear operator $A \in \mathcal{L}(X, Y')$ such that

$$\langle Av, w \rangle_{Y', Y} := a(v, w) \qquad \forall (v, w) \in X \times Y, \tag{1.2}$$

and to consider the following problem:

$$\text{Find } u \in X \text{ s.t. } Au = f \text{ in } Y'. \tag{1.3}$$

Problems (1.1) and (1.3) are equivalent, that is, u solves (1.1) if and only if u solves (1.3).

Problem (1.1), or equivalently (1.3), is said to be *well-posed* if it admits one and only one solution $u \in X$. The well-posedness of problem (1.3) amounts to A being an isomorphism. In Banach spaces, if $A \in \mathcal{L}(X, Y')$ is an isomorphism, then A^{-1} is bounded, that is, $\|A^{-1}\|_{\mathcal{L}(Y', X)} \leq C$ (see, e.g., Ern and Guermond [141, Remark A.37]). As a result, the unique solution $u \in X$ satisfies the a priori estimate

$$\|u\|_X = \|A^{-1}f\|_X \leq C\|f\|_{Y'}.$$

1.1.1 The Banach–Nečas–Babuška Theorem

The key result for asserting the well-posedness of (1.1), or equivalently (1.3), is the so-called Banach–Nečas–Babuška (BNB) Theorem. We stress that this result provides *necessary and sufficient* conditions for well-posedness.

Theorem 1.1 (Banach–Nečas–Babuška (BNB)). *Let X be a Banach space and let Y be a reflexive Banach space. Let $a \in \mathcal{L}(X \times Y, \mathbb{R})$ and let $f \in Y'$. Then, problem (1.1) is well-posed if and only if:*

(i) *There is $C_{\text{sta}} > 0$ such that*

$$\forall v \in X, \qquad C_{\text{sta}} \|v\|_X \leq \sup_{w \in Y \setminus \{0\}} \frac{a(v, w)}{\|w\|_Y}, \tag{1.4}$$

(ii) *For all $w \in Y$,*
$$(\forall v \in X, \; a(v,w) = 0) \implies (w = 0). \tag{1.5}$$

Equivalently, the bounded linear operator $A \in \mathcal{L}(X,Y')$ defined by (1.2) is an isomorphism if and only if:

(i) *There is $C_{\text{sta}} > 0$ such that*
$$\forall v \in X, \quad C_{\text{sta}} \|v\|_X \le \|Av\|_{Y'}, \tag{1.6}$$

(ii) *For all $w \in Y$,*
$$(\forall v \in X, \; \langle Av, w\rangle_{Y',Y} = 0) \implies (w = 0). \tag{1.7}$$

Moreover, the following a priori estimate holds true:
$$\|u\|_X \le \frac{1}{C_{\text{sta}}} \|f\|_{Y'}.$$

Condition (1.4) is often called an inf-sup condition since it is equivalent to
$$C_{\text{sta}} \le \inf_{v \in X \setminus \{0\}} \sup_{w \in Y \setminus \{0\}} \frac{a(v,w)}{\|v\|_X \|w\|_Y}.$$

Furthermore, owing to the reflexivity of Y and introducing the adjoint operator $A^t \in \mathcal{L}(Y, X')$ such that, for all $(v,w) \in X \times Y$,
$$\langle A^t w, v \rangle_{X',X} = \langle Av, w \rangle_{Y',Y},$$
condition (1.7) means that, for all $w \in Y$, $A^t w = 0$ in X' implies $w = 0$, or equivalently that A^t is injective. Moreover, a classical result (see, e.g., [141, Lemma A.39]) is that condition (1.6) means that A is injective and that the range of A is closed.

Remark 1.2 (Name of Theorem 1.1). The terminology, proposed in [141], indicates that, from a functional analysis point of view, this theorem hinges on two key results of Banach, the Open Mapping Theorem and the Closed Range Theorem, while the theorem in the form stated above was derived by Nečas [243] and Babuška [21].

1.1.2 The Lax–Milgram Lemma

A simpler, yet less general, condition to assert the well-posedness of (1.1), or equivalently (1.3), is provided by the Lax–Milgram Lemma [222]. In this setting, X is a Hilbert space, $Y = X$, and the following coercivity property is invoked.

Definition 1.3 (Coercivity). Let X be a Hilbert space and let $a \in \mathcal{L}(X \times X, \mathbb{R})$. We say that the bilinear form a is *coercive* on X if there is $C_{\text{sta}} > 0$ such that
$$\forall v \in X, \quad C_{\text{sta}} \|v\|_X^2 \le a(v,v).$$

Equivalently, we say that the bounded linear operator $A \in \mathcal{L}(X, X')$ defined by (1.2) is *coercive* if there is $C_{\text{sta}} > 0$ such that

$$\forall v \in X, \qquad C_{\text{sta}} \|v\|_X^2 \leq \langle Av, v \rangle_{X', X}.$$

We can now state the Lax–Milgram Lemma. We stress that this result only provides *sufficient* conditions for well-posedness.

Lemma 1.4 (Lax–Milgram). *Let X be a Hilbert space, let $a \in \mathcal{L}(X \times X, \mathbb{R})$, and let $f \in X'$. Then, problem (1.1) is well-posed if the bilinear form a is coercive on X. Equivalently, problem (1.3) is well-posed if the linear operator $A \in \mathcal{L}(X, X')$ is coercive. Moreover, the following a priori estimate holds true:*

$$\|u\|_X \leq \frac{1}{C_{\text{sta}}} \|f\|_{X'}.$$

Proof. Let us verify that if a is coercive, conditions (1.4) and (1.5) hold true. To prove (1.4), we observe that, for all $v \in X \setminus \{0\}$,

$$C_{\text{sta}} \|v\|_X \leq \frac{a(v, v)}{\|v\|_X} \leq \sup_{w \in X \setminus \{0\}} \frac{a(v, w)}{\|w\|_X},$$

and that (1.4) trivially holds true if $v = 0$. To prove (1.5), let $w \in X$ be such that $a(v, w) = 0$, for all $v \in X$. Then, picking $v = w$ yields by coercivity that $\|w\|_X = 0$, i.e., $w = 0$. □

Remark 1.5 (Lax–Milgram Lemma and Hilbert spaces). Coercivity is essentially a Hilbertian property. Precisely, if X is a Banach space, then X can be equipped with a Hilbert structure with the same topology if and only if there is a coercive operator in $\mathcal{L}(X, X')$; see, e.g., [141, Proposition A.49].

1.1.3 Lebesgue and Sobolev Spaces

In practice, the model problem (1.1) corresponds to a PDE posed over a domain $\Omega \subset \mathbb{R}^d$ with space dimension $d \geq 1$. The domain Ω is a bounded, connected, open subset of $\mathbb{R}^d$ with Lipschitz boundary $\partial \Omega$. The spaces X and Y in (1.1) are then function spaces spanned by functions defined over Ω. For simplicity, we consider scalar-valued functions; the case of vector-valued functions can be treated similarly.

In this section, we briefly present two important classes of function spaces to be used in what follows, namely Lebesgue and Sobolev spaces. We only state the basic properties of such spaces, and we refer the reader to Evans [153, Chap. 5] or Brézis [55, Chaps. 8 and 9] for further background. A thorough presentation can also be found in the textbook of Adams [4].

1.1.3.1 Lebesgue Spaces

We consider functions $v : \Omega \to \mathbb{R}$ that are Lebesgue measurable and we denote by $\int_\Omega v$ the (Lebesgue) integral of v over Ω. Let $1 \le p \le \infty$ be a real number. We set

$$\|v\|_{L^p(\Omega)} := \left(\int_\Omega |v|^p \right)^{1/p} \qquad 1 \le p < \infty,$$

and

$$\begin{aligned}\|v\|_{L^\infty(\Omega)} &:= \sup\operatorname{ess}\{|v(x)| \text{ for a.e. } x \in \Omega\} \\ &= \inf\{M > 0 \mid |v(x)| \le M \text{ for a.e. } x \in \Omega\}.\end{aligned}$$

In either case, we define the *Lebesgue space*

$$L^p(\Omega) := \{v \text{ Lebesgue measurable} \mid \|v\|_{L^p(\Omega)} < \infty\}.$$

Equipped with the norm $\|\cdot\|_{L^p(\Omega)}$, $L^p(\Omega)$ is a Banach space for all $1 \le p \le \infty$ (see Evans [153, p. 249] or Brézis [55, p. 150]). Moreover, for all $1 \le p < \infty$, the space $C_0^\infty(\Omega)$ spanned by infinitely differentiable functions with compact support in Ω is dense in $L^p(\Omega)$. In the particular case $p = 2$, $L^2(\Omega)$ is a (real) Hilbert space when equipped with the scalar product

$$(v, w)_{L^2(\Omega)} := \int_\Omega vw.$$

A useful tool in Lebesgue spaces is *Hölder's inequality* which states that, for all $1 \le p, q \le \infty$ such that $1/p + 1/q = 1$, all $v \in L^p(\Omega)$, and all $w \in L^q(\Omega)$, there holds $vw \in L^1(\Omega)$ and

$$\int_\Omega vw \le \|v\|_{L^p(\Omega)} \|w\|_{L^q(\Omega)}.$$

The particular case $p = q = 2$ yields the Cauchy–Schwarz inequality, namely, for all $v, w \in L^2(\Omega)$, $vw \in L^1(\Omega)$ and

$$(v, w)_{L^2(\Omega)} \le \|v\|_{L^2(\Omega)} \|w\|_{L^2(\Omega)}.$$

1.1.3.2 Sobolev Spaces

On the Cartesian basis of $\mathbb{R}^d$ with coordinates $(x_1, \ldots, x_d)$, the symbol ∂_i with $i \in \{1, \ldots, d\}$ denotes the distributional partial derivative with respect to x_i. For a d-uple $\alpha \in \mathbb{N}^d$, $\partial^\alpha v$ denotes the distributional derivative $\partial_1^{\alpha_1} \ldots \partial_d^{\alpha_d} v$ of v, with the convention that $\partial^{(0,\ldots,0)} v = v$. For any real number $1 \le p \le \infty$, we define, for all $\xi \in \mathbb{R}^d$ with components $(\xi_1, \ldots, \xi_d)$ in the Cartesian basis of $\mathbb{R}^d$, the norm

$$|\xi|_{\ell^p} := \left(\sum_{i=1}^d |\xi_i|^p \right)^{1/p} \qquad 1 \le p < \infty,$$

and $|\xi|_{\ell^\infty} := \max_{1 \leq i \leq d} |\xi_i|$. The index is dropped for the Euclidean norm obtained with $p = 2$.

Let $m \geq 0$ be an integer and let $1 \leq p \leq \infty$ be a real number. We define the *Sobolev space*

$$W^{m,p}(\Omega) := \{v \in L^p(\Omega) \mid \forall \alpha \in A_d^m,\ \partial^\alpha v \in L^p(\Omega)\},$$

where

$$A_d^m := \{\alpha \in \mathbb{N}^d \mid |\alpha|_{\ell^1} \leq m\}. \tag{1.8}$$

Thus, $W^{m,p}(\Omega)$ is spanned by functions with derivatives of global order up to m in $L^p(\Omega)$. In particular, $W^{0,p}(\Omega) = L^p(\Omega)$. The Sobolev space $W^{m,p}(\Omega)$ is a Banach space when equipped with the norm

$$\|v\|_{W^{m,p}(\Omega)} := \left(\sum_{\alpha \in A_d^m} \|\partial^\alpha v\|_{L^p(\Omega)}^p \right)^{1/p} \qquad 1 \leq p < \infty,$$

and $\|v\|_{W^{m,\infty}(\Omega)} := \max_{\alpha \in A_d^m} \|\partial^\alpha v\|_{L^\infty(\Omega)}$. We also consider the seminorm $|\cdot|_{W^{m,p}(\Omega)}$ by restricting the above definitions to d-uples in the set $\overline{A}_d^m := \{\alpha \in \mathbb{N}^d \mid |\alpha|_{\ell^1} = m\}$, that is, by keeping only the derivatives of global order equal to m.

For $p = 2$, we use the notation $H^m(\Omega) := W^{m,2}(\Omega)$, so that

$$H^m(\Omega) = \{v \in L^2(\Omega) \mid \forall \alpha \in A_d^m,\ \partial^\alpha v \in L^2(\Omega)\}.$$

$H^m(\Omega)$ is a Hilbert space when equipped with the scalar product

$$(v, w)_{H^m(\Omega)} := \sum_{\alpha \in A_d^m} (\partial^\alpha v, \partial^\alpha w)_{L^2(\Omega)},$$

leading to the norm and seminorm

$$\|v\|_{H^m(\Omega)} := \left(\sum_{\alpha \in A_d^m} \|\partial^\alpha v\|_{L^2(\Omega)}^2 \right)^{1/2}, \qquad |v|_{H^m(\Omega)} := \left(\sum_{\alpha \in \overline{A}_d^m} \|\partial^\alpha v\|_{L^2(\Omega)}^2 \right)^{1/2}.$$

To allow for a more compact notation in the case $m = 1$, we consider the gradient $\nabla v = (\partial_1 v, \ldots, \partial_d v)^t$ with values in $\mathbb{R}^d$. The norm on $W^{1,p}(\Omega)$ becomes

$$\|v\|_{W^{1,p}(\Omega)} = \left(\|v\|_{L^p(\Omega)}^p + \|\nabla v\|_{[L^p(\Omega)]^d}^p \right)^{1/p} \qquad 1 \leq p < \infty,$$

with

$$\|\nabla v\|_{[L^p(\Omega)]^d} := \left(\int_\Omega |\nabla v|_{\ell^p}^p \right)^{1/p} = \left(\int_\Omega \sum_{i=1}^d |\partial_i v|^p \right)^{1/p}.$$

In the case $p = 2$, we obtain

$$(v, w)_{H^1(\Omega)} = (v, w)_{L^2(\Omega)} + (\nabla v, \nabla w)_{[L^2(\Omega)]^d}.$$

Boundary values of functions in the Sobolev space $W^{1,p}(\Omega)$ can be given a meaning (at least) in $L^p(\partial\Omega)$. More precisely (see, e.g., Brenner and Scott [54, Chap. 1]), for all $1 \leq p \leq \infty$, there is C such that

$$\|v\|_{L^p(\partial\Omega)} \leq C \|v\|_{L^p(\Omega)}^{1-1/p} \|v\|_{W^{1,p}(\Omega)}^{1/p} \qquad \forall v \in W^{1,p}(\Omega). \tag{1.9}$$

In particular, for $p = 2$, we obtain

$$\|v\|_{L^2(\partial\Omega)} \leq C \|v\|_{L^2(\Omega)}^{1/2} \|v\|_{H^1(\Omega)}^{1/2} \qquad \forall v \in H^1(\Omega). \tag{1.10}$$

The bounds (1.9) and (1.10) are called *continuous trace inequalities*.

Finally, at some instances, we consider Hilbert Sobolev spaces $H^s(\Omega)$ where the exponent s is a *positive real number*. We refer the reader, e.g., to Ern and Guermond [141, p. 484] for the definition of such spaces. In what follows, we use the fact that functions in $H^{1/2+\epsilon}(\Omega)$, $\epsilon > 0$, admit a trace in $L^2(\partial\Omega)$.

1.2 The Discrete Setting

In this section, we present the main ingredients to build finite-dimensional function spaces to approximate the model problem (1.1) using dG methods. The construction of such spaces hinges on discretizing the domain Ω (over which the PDE is posed) using a mesh and choosing a local functional behavior (e.g., polynomial) within each mesh element. This leads to broken polynomial spaces. We also introduce broken Sobolev spaces and a broken gradient operator. Finally, we briefly discuss the function space $H(\mathrm{div};\Omega)$ and its broken version; such spaces are particularly relevant in the context of PDEs with diffusion.

1.2.1 The Domain Ω

To simplify the presentation, we focus, throughout this book, on polyhedra.

Definition 1.6 (Polyhedron in $\mathbb{R}^d$)**.** We say that the set P is a *polyhedron* in $\mathbb{R}^d$ if P is an open, connected, bounded subset of $\mathbb{R}^d$ such that its boundary ∂P is a finite union of parts of hyperplanes, say $\{H_i\}_{1 \leq i \leq n_P}$. Moreover, for all $1 \leq i \leq n_P$, at each point in the interior of $\partial P \cap H_i$, the set P is assumed to lie on only one side of its boundary.

Assumption 1.7 (Domain Ω)**.** *The domain Ω is a polyhedron in $\mathbb{R}^d$.*

The advantage of Assumption 1.7 is that polyhedra can be exactly covered by a mesh consisting of polyhedral elements. PDEs posed over domains with curved boundary can also be approximated by dG methods using, e.g., isoparametric finite elements to build the mesh near curved boundaries as described, e.g., by Ciarlet [92, p. 224] and Brenner and Scott [54, p. 117].

Definition 1.8 (Boundary and outward normal)**.** The *boundary* of Ω is denoted by $\partial\Omega$ and its (unit) *outward normal*, which is defined a.e. on $\partial\Omega$, by n.

1.2.2 Meshes

The first step is to discretize the domain Ω using a mesh. Various types of meshes can be considered. We examine first the most familiar case, that of simplicial meshes. Such meshes should be familiar to the reader since they are one of the key ingredients to build continuous finite element spaces.

Definition 1.9 (Simplex). Given a family $\{a_0, \ldots, a_d\}$ of $(d+1)$ points in $\mathbb{R}^d$ such that the vectors $\{a_1 - a_0, \ldots, a_d - a_0\}$ are linearly independent, the interior of the convex hull of $\{a_0, \ldots, a_d\}$ is called a non-degenerate *simplex* of $\mathbb{R}^d$, and the points $\{a_0, \ldots, a_d\}$ are called its *vertices*.

By its definition, a non-degenerate simplex is an open subset of $\mathbb{R}^d$. In dimension 1, a non-degenerate simplex is an interval, in dimension 2 a triangle, and in dimension 3 a tetrahedron. The unit simplex of $\mathbb{R}^d$ is the set (cf. Fig. 1.1)

$$S_d := \{(x_1, \ldots, x_d) \in \mathbb{R}^d \mid x_i > 0 \ \forall i \in \{1, \ldots, d\} \text{ and } x_1 + \ldots + x_d < 1\}.$$

Any non-degenerate simplex of $\mathbb{R}^d$ is the image of the unit simplex by a bijective affine transformation of $\mathbb{R}^d$.

Definition 1.10 (Simplex faces). Let S be a non-degenerate simplex with vertices $\{a_0, \ldots, a_d\}$. For each $i \in \{0, \ldots, d\}$, the convex hull of $\{a_0, \ldots, a_d\} \setminus \{a_i\}$ is called a *face* of the simplex S.

Thus, a non-degenerate simplex has $(d+1)$ faces, and, by construction, a simplex face is a closed subset of $\mathbb{R}^d$. A simplex face has zero d-dimensional Hausdorff measure, but positive $(d-1)$-dimensional Hausdorff measure. In dimension 2, a simplex face is also called an edge, while in dimension 1, a simplex face is a point and its 0-dimensional Hausdorff measure is conventionally set to 1.

Definition 1.11 (Simplicial mesh). A *simplicial mesh* $\mathcal{T}$ of the domain Ω is a finite collection of disjoint non-degenerate simplices $\mathcal{T} = \{T\}$ forming a partition of Ω,

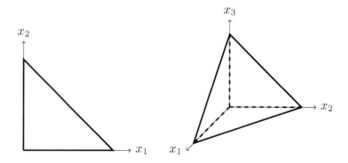

Fig. 1.1: Unit simplex in two (*left*) and three (*right*) space dimensions

1.2. The Discrete Setting

$$\overline{\Omega} = \bigcup_{T \in \mathcal{T}} \overline{T}. \tag{1.11}$$

Each $T \in \mathcal{T}$ is called a *mesh element*.

While simplicial meshes are quite convenient in the context of continuous finite elements, dG methods more easily accommodate general meshes.

Definition 1.12 (General mesh). A *general mesh* $\mathcal{T}$ of the domain Ω is a finite collection of disjoint polyhedra $\mathcal{T} = \{T\}$ forming a partition of Ω as in (1.11). Each $T \in \mathcal{T}$ is called a *mesh element*.

Obviously, a simplicial mesh is just a particular case of a general mesh.

Definition 1.13 (Element diameter, meshsize). Let $\mathcal{T}$ be a (general) mesh of the domain Ω. For all $T \in \mathcal{T}$, h_T denotes the *diameter* of T, and the *meshsize* is defined as the real number

$$h := \max_{T \in \mathcal{T}} h_T.$$

We use the notation $\mathcal{T}_h$ for a mesh $\mathcal{T}$ with meshsize h.

Definition 1.14 (Element outward normal). Let $\mathcal{T}_h$ be a mesh of the domain Ω and let $T \in \mathcal{T}_h$. We define n_T a.e. on ∂T as the (unit) *outward normal* to T.

Faces of a *single* polyhedral mesh element are not needed in what follows, and we leave them undefined to avoid confusion with the important concept of mesh faces introduced in Sect. 1.2.3. The key difference is that mesh faces depend on the way neighboring mesh elements come into contact.

Remark 1.15 (General hexahedra). In the present setting, general hexahedra in $\mathbb{R}^3$ cannot be mesh elements since their faces are not parts of (hyper)planes (since four distinct points do not generally belong to the same plane). One possibility is to approximate general hexahedra by subdividing nonplanar faces into two or four triangles.

1.2.3 Mesh Faces, Averages, and Jumps

The concepts of mesh faces, averages, and jumps play a central role in the design and analysis of dG methods.

Definition 1.16 (Mesh faces). Let $\mathcal{T}_h$ be a mesh of the domain Ω. We say that a (closed) subset F of $\overline{\Omega}$ is a *mesh face* if F has positive $(d-1)$-dimensional Hausdorff measure (in dimension 1, this means that F is nonempty) and if either one of the two following conditions is satisfied:

(i) There are distinct mesh elements T_1 and T_2 such that $F = \partial T_1 \cap \partial T_2$; in such a case, F is called an *interface*.

(ii) There is $T \in \mathcal{T}_h$ such that $F = \partial T \cap \partial \Omega$; in such a case, F is called a *boundary face*.

Interfaces are collected in the set $\mathcal{F}_h^i$, and boundary faces are collected in the set $\mathcal{F}_h^b$. Henceforth, we set
$$\mathcal{F}_h := \mathcal{F}_h^i \cup \mathcal{F}_h^b.$$
Moreover, for any mesh element $T \in \mathcal{T}_h$, the set
$$\mathcal{F}_T := \{F \in \mathcal{F}_h \mid F \subset \partial T\}$$
collects the mesh faces composing the boundary of T. The maximum number of mesh faces composing the boundary of mesh elements is denoted by
$$N_\partial := \max_{T \in \mathcal{T}_h} \operatorname{card}(\mathcal{F}_T). \tag{1.12}$$
Finally, for any mesh face $F \in \mathcal{F}_h$, we define the set
$$\mathcal{T}_F := \{T \in \mathcal{T}_h \mid F \subset \partial T\}, \tag{1.13}$$
and observe that $\mathcal{T}_F$ consists of two mesh elements if $F \in \mathcal{F}_h^i$ and of one mesh element if $F \in \mathcal{F}_h^b$.

Figure 1.2 depicts an interface between two mesh elements belonging to a simplicial mesh (left) or to a general mesh (right). We observe that in the case of simplicial meshes, interfaces are always parts of hyperplanes, but this is not necessarily the case for general meshes containing nonconvex polyhedra. We now define averages and jumps across interfaces of piecewise smooth functions; cf. Fig. 1.3 for a one-dimensional illustration.

Definition 1.17 (Interface averages and jumps). *Let v be a scalar-valued function defined on Ω and assume that v is smooth enough to admit on all $F \in \mathcal{F}_h^i$ a possibly two-valued trace. This means that, for all $T \in \mathcal{T}_h$, the restriction $v|_T$ of v to the open set T can be defined up to the boundary ∂T. Then, for all $F \in \mathcal{F}_h^i$ and a.e. $x \in F$, the* average *of v is defined as*
$$\{\!\{v\}\!\}_F(x) := \frac{1}{2}\Big(v|_{T_1}(x) + v|_{T_2}(x)\Big),$$

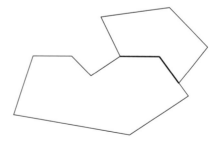

Fig. 1.2: Examples of interface for a simplicial mesh (*left*) and a general mesh (*right*)

1.2. The Discrete Setting

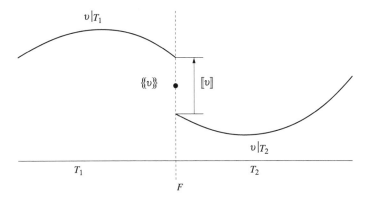

Fig. 1.3: One-dimensional example of average and jump operators; the face reduces to a point separating two adjacent intervals

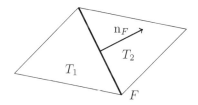

Fig. 1.4: Notation for an interface

and the *jump* of v as

$$[\![v]\!]_F(x) := v|_{T_1}(x) - v|_{T_2}(x).$$

When v is vector-valued, the above average and jump operators act componentwise on the function v. Whenever no confusion can arise, the subscript F and the variable x are omitted, and we simply write $\{\!\{v\}\!\}$ and $[\![v]\!]$.

Definition 1.18 (Face normals). For all $F \in \mathcal{F}_h$ and a.e. $x \in F$, we define the (unit) *normal* n_F to F at x as

(i) n_{T_1}, the unit normal to F at x pointing from T_1 to T_2 if $F \in \mathcal{F}_h^i$ with $F = \partial T_1 \cap \partial T_2$; the orientation of n_F is arbitrary depending on the choice of T_1 and T_2, but kept fixed in what follows. See Fig. 1.4.

(ii) n, the unit outward normal to Ω at x if $F \in \mathcal{F}_h^b$.

Remark 1.19 (Boundary averages and jumps). Averages and jumps can also be defined for boundary faces (this is particularly useful when discretizing PDEs with diffusion as in Chaps. 4–6). One possible definition is to set for a.e. $x \in F$ with $F \in \mathcal{F}_h^b$, $F = \partial T \cap \partial \Omega$, $\{\!\{v\}\!\}_F(x) = [\![v]\!]_F(x) := v|_T(x)$.

Remark 1.20 (Alternative definition of jumps). An alternative definition of jumps consists in setting for all $F \in \mathcal{F}_h^i$ with $F = \partial T_1 \cap \partial T_2$,

$$[\![v]\!]_* := v|_{T_1} \mathbf{n}_{T_1} + v|_{T_2} \mathbf{n}_{T_2},$$

so that $[\![v]\!]_* = \mathbf{n}_F [\![v]\!]$. This alternative definition allows T_1 and T_2 to play symmetric roles; however, the jump of a scalar-valued function is a vector-valued function.

1.2.4 Broken Polynomial Spaces

After having built a mesh of the domain Ω, the second step in the construction of discrete function spaces consists in choosing a certain functional behavior within each mesh element. For the sake of simplicity, we restrict ourselves to polynomial functions; more general cases can also be accommodated (see, e.g., Yuan and Shu [309]). The resulting spaces, consisting of piecewise polynomial functions, are termed *broken polynomial spaces*.

1.2.4.1 The Polynomial Space $\mathbb{P}_d^k$

Let $k \geq 0$ be an integer. Recalling the set A_d^k defined by (1.8), we define the space of polynomials of d variables, of total degree at most k, as

$$\mathbb{P}_d^k := \left\{ p : \mathbb{R}^d \ni x \mapsto p(x) \in \mathbb{R} \mid \exists (\gamma_\alpha)_{\alpha \in A_d^k} \in \mathbb{R}^{\operatorname{card}(A_d^k)} \text{ s.t. } p(x) = \sum_{\alpha \in A_d^k} \gamma_\alpha x^\alpha \right\},$$

with the convention that, for $x = (x_1, \ldots, x_d) \in \mathbb{R}^d$, $x^\alpha := \prod_{i=1}^d x_i^{\alpha_i}$. The dimension of the vector space $\mathbb{P}_d^k$ is

$$\dim(\mathbb{P}_d^k) = \operatorname{card}(A_d^k) = \binom{k+d}{k} = \frac{(k+d)!}{k!d!}. \tag{1.14}$$

The first few values of $\dim(\mathbb{P}_d^k)$ are listed in Table 1.1.

1.2.4.2 The Broken Polynomial Space $\mathbb{P}_d^k(\mathcal{T}_h)$

In what follows, we often consider the broken polynomial space

$$\mathbb{P}_d^k(\mathcal{T}_h) := \left\{ v \in L^2(\Omega) \mid \forall T \in \mathcal{T}_h,\ v|_T \in \mathbb{P}_d^k(T) \right\}, \tag{1.15}$$

Table 1.1: Dimension of the polynomial space $\mathbb{P}_d^k$ for $d \in \{1, 2, 3\}$ and $k \in \{0, 1, 2, 3\}$

k	d = 1	d = 2	d = 3
0	1	1	1
1	2	3	4
2	3	6	10
3	4	10	20

1.2. The Discrete Setting

where $\mathbb{P}_d^k(T)$ is spanned by the restriction to T of polynomials in $\mathbb{P}_d^k$. It is clear that
$$\dim(\mathbb{P}_d^k(\mathcal{T}_h)) = \mathrm{card}(\mathcal{T}_h) \times \dim(\mathbb{P}_d^k),$$
since the restriction of a function $v \in \mathbb{P}_d^k(\mathcal{T}_h)$ to each mesh element can be chosen independently of its restriction to other elements.

1.2.4.3 Other Broken Polynomial Spaces

It is possible to consider other broken polynomial spaces. Such spaces are encountered, e.g., when defining local bases using a reference element, e.g., nodal-based local bases associated with quadratures (see Gassner, Lörcher, Munz, and Hesthaven [164]). Further motivations for considering other broken polynomial spaces include, among others, bubble-stabilization techniques (see Burman and Stamm [70]) and inf-sup stable discretizations for incompressible flows (see Toselli [296]).

A relevant example is the space of polynomials of d variables, of degree at most k in each variable, namely

$$\mathbb{Q}_d^k := \left\{ p : \mathbb{R}^d \ni x \mapsto p(x) \in \mathbb{R} \mid \exists (\gamma_\alpha)_{\alpha \in B_d^k} \in \mathbb{R}^{\mathrm{card}(B_d^k)} \text{ s.t. } p(x) = \sum_{\alpha \in B_d^k} \gamma_\alpha x^\alpha \right\},$$

where B_d^k denotes the set of d-uples of ∞-norm smaller than or equal to k,

$$B_d^k := \left\{ \alpha \in \mathbb{N}^d \mid |\alpha|_{\ell^\infty} \leq k \right\}.$$

The dimension of the vector space $\mathbb{Q}_d^k$ is

$$\dim(\mathbb{Q}_d^k) = \mathrm{card}(B_d^k) = (k+1)^d.$$

The first few values of $\dim(\mathbb{Q}_d^k)$ are listed in Table 1.2.

1.2.5 Broken Sobolev Spaces

In this section, we introduce broken Sobolev spaces and a broken gradient operator. We also state the main properties of broken Sobolev spaces to be used in what follows.

Table 1.2: Dimension of the polynomial space $\mathbb{Q}_d^k$ for $d \in \{1,2,3\}$ and $k \in \{0,1,2,3\}$

k	$d=1$	$d=2$	$d=3$
0	1	1	1
1	2	4	8
2	3	9	27
3	4	16	64

Let $\mathcal{T}_h$ be a mesh of the domain Ω. For any mesh element $T \in \mathcal{T}_h$, the Sobolev spaces $H^m(T)$ and $W^{m,p}(T)$ can be defined as above by replacing Ω by T. We then define the *broken Sobolev spaces*

$$H^m(\mathcal{T}_h) := \left\{ v \in L^2(\Omega) \mid \forall T \in \mathcal{T}_h,\ v|_T \in H^m(T) \right\}, \quad (1.16)$$

$$W^{m,p}(\mathcal{T}_h) := \left\{ v \in L^p(\Omega) \mid \forall T \in \mathcal{T}_h,\ v|_T \in W^{m,p}(T) \right\}, \quad (1.17)$$

where $m \geq 0$ is an integer and $1 \leq p \leq \infty$ a real number.

In the context of broken Sobolev spaces, the continuous trace inequality (1.9) can be used to infer that, for all $v \in W^{1,p}(\mathcal{T}_h)$ and all $T \in \mathcal{T}_h$,

$$\|v\|_{L^p(\partial T)} \leq C \|v\|_{L^p(T)}^{1-1/p} \|v\|_{W^{1,p}(T)}^{1/p}, \quad (1.18)$$

while for $p = 2$, we obtain, for all $v \in H^1(\mathcal{T}_h)$ and all $T \in \mathcal{T}_h$,

$$\|v\|_{L^2(\partial T)} \leq C \|v\|_{L^2(T)}^{1/2} \|v\|_{H^1(T)}^{1/2}. \quad (1.19)$$

In what follows, it is implicitly understood that expressions such as $\|v\|_{L^2(\partial T)}$ (or such as $\|v\|_{L^2(F)}$ for a mesh face $F \in \mathcal{F}_T$ of a given mesh element $T \in \mathcal{T}_h$) should be evaluated using the restriction of v to T. A different version of the continuous trace inequality (1.19) is presented in Lemma 1.49 below. Continuous trace inequalities such as (1.18) or (1.19) are important in the context of dG methods to give a meaning to the traces of the exact solution or of its (normal) gradient on mesh faces.

It is natural to define a broken gradient operator acting on the broken Sobolev space $W^{1,p}(\mathcal{T}_h)$. In particular, this operator also acts on broken polynomial spaces.

Definition 1.21 (Broken gradient). *The broken gradient* $\nabla_h : W^{1,p}(\mathcal{T}_h) \to [L^p(\Omega)]^d$ *is defined such that, for all* $v \in W^{1,p}(\mathcal{T}_h)$,

$$\forall T \in \mathcal{T}_h, \qquad (\nabla_h v)|_T := \nabla(v|_T). \quad (1.20)$$

In what follows, we drop the index h in the broken gradient when this operator appears inside an integral over a fixed mesh element $T \in \mathcal{T}_h$.

It is important to observe that the usual Sobolev spaces are subspaces of the broken Sobolev spaces, and that on the usual Sobolev spaces, the broken gradient coincides with the distributional gradient. For completeness, we detail the proof of this result.

Lemma 1.22 (Broken gradient on usual Sobolev spaces). *Let* $m \geq 0$ *and let* $1 \leq p \leq \infty$. *There holds* $W^{m,p}(\Omega) \subset W^{m,p}(\mathcal{T}_h)$. *Moreover, for all* $v \in W^{1,p}(\Omega)$, $\nabla_h v = \nabla v$ *in* $[L^p(\Omega)]^d$.

Proof. It is sufficient to prove the inclusion for $m = 1$. Let $v \in W^{1,p}(\Omega)$. We first observe that $\nabla(v|_T) = (\nabla v)|_T$ for all $T \in \mathcal{T}_h$. Indeed, for all $\Phi \in [C_0^\infty(T)]^d$,

1.2. The Discrete Setting

the extension of Φ by zero to Ω, say $E\Phi$, is in $[C_0^\infty(\Omega)]^d$, so that

$$\int_T \nabla(v|_T)\cdot\Phi = -\int_T v(\nabla\cdot\Phi) = -\int_\Omega v(\nabla\cdot(E\Phi))$$
$$= \int_\Omega \nabla v \cdot E\Phi = \int_T (\nabla v)|_T\cdot\Phi.$$

Since Φ is arbitrary, this implies $\nabla(v|_T) = (\nabla v)|_T$ and, since $T \in \mathcal{T}_h$ is arbitrary, we infer $\nabla_h v = \nabla v$. This equality also shows that $v \in W^{1,p}(\mathcal{T}_h)$. $\square$

The reverse inclusion of Lemma 1.22 does not hold true in general (except obviously for $m = 0$). The reason is that functions in the broken Sobolev space $W^{1,p}(\mathcal{T}_h)$ can have nonzero jumps across interfaces, while functions in the usual Sobolev space $W^{1,p}(\Omega)$ have zero jumps across interfaces. We now give a precise statement of this important result.

Lemma 1.23 (Characterization of $W^{1,p}(\Omega)$). *Let $1 \leq p \leq \infty$. A function $v \in W^{1,p}(\mathcal{T}_h)$ belongs to $W^{1,p}(\Omega)$ if and only if*

$$[\![v]\!] = 0 \qquad \forall F \in \mathcal{F}_h^i. \tag{1.21}$$

Proof. The proof is again based on a distributional argument. Let $v \in W^{1,p}(\mathcal{T}_h)$. Then, for all $\Phi \in [C_0^\infty(\Omega)]^d$, we observe integrating by parts elementwise that

$$\int_\Omega \nabla_h v\cdot\Phi = \sum_{T\in\mathcal{T}_h}\int_T \nabla(v|_T)\cdot\Phi = -\sum_{T\in\mathcal{T}_h}\int_T v(\nabla\cdot\Phi) + \sum_{T\in\mathcal{T}_h}\int_{\partial T} v|_T(\Phi\cdot n_T)$$
$$= -\int_\Omega v(\nabla\cdot\Phi) + \sum_{F\in\mathcal{F}_h^i}\int_F (\Phi\cdot n_F)[\![v]\!], \tag{1.22}$$

since Φ is continuous across interfaces and vanishes on boundary faces. Assume first that (1.21) holds true. Then, (1.22) implies

$$\int_\Omega \nabla_h v\cdot\Phi = -\int_\Omega v(\nabla\cdot\Phi) \qquad \forall \Phi \in [C_0^\infty(\Omega)]^d,$$

meaning that $\nabla v = \nabla_h v$ in $[L^p(\Omega)]^d$. Hence, $v \in W^{1,p}(\Omega)$. Conversely, if $v \in W^{1,p}(\Omega)$, $\nabla v = \nabla_h v$ in $[L^p(\Omega)]^d$ owing to Lemma 1.22, so that (1.22) now implies

$$\sum_{F\in\mathcal{F}_h^i}\int_F (\Phi\cdot n_F)[\![v]\!] = \int_\Omega \nabla_h v\cdot\Phi + \int_\Omega v(\nabla\cdot\Phi)$$
$$= \int_\Omega \nabla_h v\cdot\Phi - \int_\Omega \nabla v\cdot\Phi = 0,$$

whence we infer (1.21) by choosing the support of Φ intersecting a single interface and since Φ is arbitrary. $\square$

1.2.6 The Function Space $H(\mathrm{div};\Omega)$ and Its Broken Version

In the context of PDEs with diffusion, the vector-valued field $\sigma = -\nabla u$ can be interpreted as the diffusive flux; here, u solves, e.g., the Poisson problem presented in Sect. 4.1. From a physical viewpoint, it is expected that the normal component of the diffusive flux does not jump across interfaces. From a mathematical viewpoint, the diffusive flux belongs to the function space

$$H(\mathrm{div};\Omega) := \{\tau \in [L^2(\Omega)]^d \mid \nabla \cdot \tau \in L^2(\Omega)\}. \tag{1.23}$$

It is therefore important to specify the meaning of the normal component on mesh faces of functions in $H(\mathrm{div};\Omega)$.

For all $T \in \mathcal{T}_h$, we define the function space $H(\mathrm{div};T)$ by replacing Ω by T in (1.23). We then introduce the broken space

$$H(\mathrm{div};\mathcal{T}_h) := \{\tau \in [L^2(\Omega)]^d \mid \forall T \in \mathcal{T}_h,\ \tau|_T \in H(\mathrm{div};T)\},$$

and the *broken divergence operator* $\nabla_h \cdot : H(\mathrm{div};\mathcal{T}_h) \to L^2(\Omega)$ such that, for all $\tau \in H(\mathrm{div};\mathcal{T}_h)$,

$$\forall T \in \mathcal{T}_h, \qquad (\nabla_h \cdot \tau)|_T := \nabla \cdot (\tau|_T).$$

Proceeding as in the proof of Lemma 1.22, we can verify that, if $\tau \in H(\mathrm{div};\Omega)$, then $\tau \in H(\mathrm{div};\mathcal{T}_h)$ and $\nabla_h \cdot \tau = \nabla \cdot \tau$.

Since the diffusive flux σ belongs to $H(\mathrm{div};\Omega)$, we infer that, for all $T \in \mathcal{T}_h$, $\sigma_T := \sigma|_T$ belongs to $H(\mathrm{div};T)$. This property allows us to give a weak meaning to the normal component $\sigma_{\partial T} := \sigma_T \cdot n_T$ on ∂T; cf. Remark 1.26 below for further insight. However, in the context of dG methods, we would like to give a meaning to the restriction $\sigma_F := \sigma_{\partial T}|_F$ independently on any face $F \in \mathcal{F}_T$, and additionally, it is convenient to assert $\sigma_F \in L^1(F)$. The fact that $\sigma_T \in H(\mathrm{div};T)$ does not provide enough regularity to assert this, as reflected by the counterexample in Remark 1.25 below. A suitable assumption to achieve $\sigma_F \in L^1(F)$ is

$$\sigma \in [W^{1,1}(\mathcal{T}_h)]^d.$$

Indeed, owing to the trace inequality (1.18), we infer, for all $T \in \mathcal{T}_h$, $\sigma_{\partial T} \in L^1(\partial T)$ so that, for all $F \in \mathcal{F}_T$, $\sigma_F \in L^1(F)$.

We can now characterize functions in $H(\mathrm{div};\Omega) \cap [W^{1,1}(\mathcal{T}_h)]^d$ using the jump of the normal component across interfaces.

Lemma 1.24 (Characterization of $H(\mathrm{div};\Omega)$). *A function $\tau \in H(\mathrm{div};\mathcal{T}_h) \cap [W^{1,1}(\mathcal{T}_h)]^d$ belongs to $H(\mathrm{div};\Omega)$ if and only if*

$$[\![\tau]\!] \cdot n_F = 0 \qquad \forall F \in \mathcal{F}_h^i. \tag{1.24}$$

Proof. The proof is similar to that of Lemma 1.23. Let $\tau \in H(\mathrm{div};\mathcal{T}_h) \cap [W^{1,1}(\mathcal{T}_h)]^d$ and let $\varphi \in C_0^\infty(\Omega)$. Integrating by parts on each mesh element and accounting for the fact that φ is smooth inside Ω and vanishes on $\partial\Omega$, we

1.2. The Discrete Setting

obtain

$$\int_\Omega \tau\cdot\nabla\varphi = \sum_{T\in\mathcal{T}_h}\int_T \tau\cdot\nabla\varphi = -\sum_{T\in\mathcal{T}_h}\int_T (\nabla\cdot\tau)\varphi + \sum_{T\in\mathcal{T}_h}\int_{\partial T}(\tau\cdot\mathbf{n}_T)\varphi$$

$$= -\int_\Omega (\nabla_h\cdot\tau)\varphi + \sum_{F\in\mathcal{F}_h^i}\int_F [\![\tau]\!]\cdot\mathbf{n}_F\varphi,$$

where we have used the fact that $\tau\cdot\mathbf{n}_T \in L^1(\partial T)$ to write a boundary integral and break it into face integrals. Hence, if (1.24) holds true,

$$\int_\Omega \tau\cdot\nabla\varphi = -\int_\Omega (\nabla_h\cdot\tau)\varphi \qquad \forall\varphi\in C_0^\infty(\Omega),$$

implying that $\nabla\cdot\tau = \nabla_h\cdot\tau \in L^2(\Omega)$, so that $\tau\in H(\mathrm{div};\Omega)$. Conversely, if $\tau\in H(\mathrm{div};\Omega)$, since $\nabla_h\cdot\tau = \nabla\cdot\tau$, the above identity yields

$$\sum_{F\in\mathcal{F}_h^i}\int_F [\![\tau]\!]\cdot\mathbf{n}_F\varphi = 0,$$

whence (1.24) is obtained by choosing the support of φ intersecting a single interface and since φ is arbitrary. $\square$

Remark 1.25 (Counter-example for $\sigma_F \in L^1(F)$). Following an idea by Carstensen and Peterseim [78], we consider the triangle

$$T = \{(x_1, x_2) \in \mathbb{R}^2 \mid 0 < x_1 < 1,\ 0 < x_2 < x_1\}.$$

For all $\epsilon > 0$, letting $r^2 = x_1^2 + x_2^2$, we define the vector field

$$\sigma_\epsilon = (\epsilon + r^2)^{-1}(x_2, -x_1)^t.$$

A direct calculation shows that σ_ϵ is divergence-free and, since $0 \le r^2 \le 1$,

$$\|\sigma_\epsilon\|_{[L^2(T)]^d}^2 = \int_0^{\pi/4}\int_0^{\cos(\theta)^{-1}} \frac{r^3}{(\epsilon+r^2)^2}\,\mathrm{d}r\,\mathrm{d}\theta$$

$$\le \int_0^{\pi/4}\int_0^{\sqrt{2}} \frac{r}{(\epsilon+r^2)}\,\mathrm{d}r\,\mathrm{d}\theta = \frac{\pi}{8}\ln(1+2\epsilon^{-1}).$$

Hence, letting $\tilde{\sigma}_\epsilon = \ln(1+2\epsilon^{-1})^{-1/2}\sigma_\epsilon$, we infer that $\|\tilde{\sigma}_\epsilon\|_{[L^2(T)]^d}$ is uniformly bounded in ϵ. Moreover, considering the face $F = \{(x_1,x_2) \in \partial T \mid x_1 = x_2\}$ with normal $\mathbf{n}_F = (-2^{-1/2}, 2^{-1/2})^t$, we obtain

$$\|\tilde{\sigma}_\epsilon\cdot\mathbf{n}_F\|_{L^1(F)} = \ln(1+2\epsilon^{-1})^{-1/2}\int_0^{\sqrt{2}} \frac{r}{\epsilon+r^2}\,\mathrm{d}r = \frac{1}{2}\ln(1+2\epsilon^{-1})^{1/2},$$

which grows unboundedly as $\epsilon \to 0^+$.

Remark 1.26 (Weak meaning of $\sigma_T \cdot n_T$ on ∂T). Let $T \in \mathcal{T}_h$ and let $\sigma_T \in H(\text{div}; T)$. Let $H^{1/2}(\partial T)$ be the vector space spanned by the traces on ∂T of functions in $H^1(T)$. Then, the normal component $\sigma_{\partial T} := \sigma_T \cdot n_T$ can be defined in the dual space $H^{-1/2}(\partial T) := (H^{1/2}(\partial T))'$ in such a way that, for all $g \in H^{1/2}(\partial T)$,

$$\langle \sigma_{\partial T}, g \rangle = \int_T \sigma_T \cdot \nabla \hat{g} + \int_T (\nabla \cdot \sigma_T) \hat{g}, \tag{1.25}$$

where $\hat{g} \in H^1(T)$ is such that its trace on ∂T is equal to g (the right-hand side of (1.25) is independent of the choice of $\hat{g}$). Consider now a face $F \in \mathcal{F}_T$ and let $H_{00}^{1/2}(F)$ be spanned by functions defined on F whose extension by zero to ∂T is in $H^{1/2}(\partial T)$. Then, the restriction $\sigma_F := \sigma_{\partial T}|_F$ can be given a meaning in the dual space $H^{-1/2}(F) := (H_{00}^{1/2}(F))'$ in such a way that, for all $g \in H_{00}^{1/2}(F)$,

$$\langle \sigma_F, g \rangle = \int_T \sigma_T \cdot \nabla \hat{g} + \int_T (\nabla \cdot \sigma_T) \hat{g},$$

where $\hat{g} \in H^1(T)$ is such that its trace on ∂T is equal to g on F and to zero elsewhere. However, this definition is of little use in the context of dG methods where the normal component σ_F has to act on polynomials on F which do not generally belong to $H_{00}^{1/2}(F)$ (unless they vanish on ∂F).

1.3 Abstract Nonconforming Error Analysis

The goal of this section is to present the key ingredients for the error analysis when approximating the linear model problem (1.1) by dG methods. We assume that (1.1) is well-posed; cf. Sect. 1.1. The error analysis presented in this section is derived in the spirit of Strang's Second Lemma [285] (see also Ern and Guermond [141, Sect. 2.3]). The three ingredients are (1) Discrete stability, (2) (Strong) consistency, and (3) Boundedness.

1.3.1 The Discrete Problem

Let $V_h \subset L^2(\Omega)$ denote a finite-dimensional function space; typically, V_h is a broken polynomial space (cf. Sect. 1.2.4). We are interested in the discrete problem

$$\text{Find } u_h \in V_h \text{ s.t. } a_h(u_h, w_h) = l_h(w_h) \text{ for all } w_h \in V_h, \tag{1.26}$$

with discrete bilinear form a_h defined (so far) only on $V_h \times V_h$ and discrete linear form l_h defined on V_h. We observe that we consider the so-called standard Galerkin approximation where the discrete trial and test spaces coincide. Moreover, since functions in V_h can be discontinuous across mesh elements, $V_h \not\subset X$ and $V_h \not\subset Y$ in general; cf., e.g., Lemma 1.23. In the terminology of finite elements, we say that the approximation is *nonconforming*.

1.3. Abstract Nonconforming Error Analysis

Alternatively, it is possible to introduce the discrete (linear) operator $A_h : V_h \to V_h$ such that, for all $v_h, w_h \in V_h$,

$$(A_h v_h, w_h)_{L^2(\Omega)} := a_h(v_h, w_h), \qquad (1.27)$$

and the discrete function $L_h \in V_h$ such that, for all $w_h \in V_h$, $(L_h, w_h)_{L^2(\Omega)} = l_h(w_h)$. This leads to the following problem (obviously equivalent to (1.26)):

$$\text{Find } u_h \in V_h \text{ s.t. } A_h u_h = L_h \text{ in } V_h. \qquad (1.28)$$

In what follows, we are often concerned with model problems where $Y \hookrightarrow L^2(\Omega)$ with dense and continuous injection. Identifying $L^2(\Omega)$ with its topological dual space $L^2(\Omega)'$ by means of the Riesz–Fréchet representation theorem, we are thus in the situation where

$$Y \hookrightarrow L^2(\Omega) \equiv L^2(\Omega)' \hookrightarrow Y',$$

with dense and continuous injections. For simplicity (cf. Remark 1.27 for further discussion), we often assume that the datum f is in $L^2(\Omega)$, so that the right-hand side of the model problem (1.1) becomes $(f, w)_{L^2(\Omega)}$, while the right-hand sides of the discrete problems (1.26) and (1.28) become, respectively,

$$l_h(w_h) = (f, w_h)_{L^2(\Omega)}, \qquad L_h = \pi_h f.$$

Here, π_h denotes the $L^2(\Omega)$-*orthogonal projection* onto V_h, that is, $\pi_h : L^2(\Omega) \to V_h$ is defined so that, for all $v \in L^2(\Omega)$, $\pi_h v \in V_h$ with

$$(\pi_h v, y_h)_{L^2(\Omega)} = (v, y_h)_{L^2(\Omega)} \qquad \forall y_h \in V_h. \qquad (1.29)$$

We observe that the restriction of $\pi_h v$ to a given mesh element $T \in \mathcal{T}_h$ can be computed independently from other mesh elements. For instance, if $V_h = \mathbb{P}^k_d(\mathcal{T}_h)$, we obtain that, for all $T \in \mathcal{T}_h$, $\pi_h v|_T \in \mathbb{P}^k_d(T)$ is such that

$$(\pi_h v|_T, \xi)_{L^2(T)} = (v, \xi)_{L^2(T)} \qquad \forall \xi \in \mathbb{P}^k_d(T).$$

We refer the reader to Sect. A.1.2 for further insight.

Remark 1.27 (Rough right-hand side). The assumption $f \in L^2(\Omega)$ is convenient so as to define the right-hand side of (1.26) using the $L^2(\Omega)$-scalar product. For rough right-hand sides, i.e., $f \in Y'$ but $f \notin L^2(\Omega)$, the right-hand side of (1.26) needs to be modified since the quantity $\langle f, w_h \rangle_{Y',Y}$ is in general not defined. One possibility is to consider the right-hand side $\langle f, \mathcal{I}_h w_h \rangle_{Y',Y}$ for some smoothing linear operator $\mathcal{I}_h : V_h \to V_h \cap Y$ (cf. Remark 4.9 for an example in the context of diffusive PDEs).

1.3.2 Discrete Stability

To formulate discrete stability, we introduce a norm, say $\|\cdot\|$, defined (at least) on V_h.

Definition 1.28 (Discrete stability). We say that the discrete bilinear form a_h enjoys *discrete stability* on V_h if there is $C_{\text{sta}} > 0$ such that

$$\forall v_h \in V_h, \quad C_{\text{sta}} \|v_h\| \leq \sup_{w_h \in V_h \setminus \{0\}} \frac{a_h(v_h, w_h)}{\|w_h\|}. \tag{1.30}$$

Remark 1.29 (h-dependency). In Definition 1.28, C_{sta} can depend on the meshsize h. In view of convergence analysis, it is important to ensure that C_{sta} be independent of h.

Property (1.30) is referred to as a discrete inf-sup condition since it is equivalent to

$$C_{\text{sta}} \leq \inf_{v_h \in V_h \setminus \{0\}} \sup_{w_h \in V_h \setminus \{0\}} \frac{a_h(v_h, w_h)}{\|v_h\| \|w_h\|}.$$

An important fact is that (1.30) is a necessary and sufficient condition for discrete well-posedness.

Lemma 1.30 (Discrete well-posedness). *The discrete problem (1.26), or equivalently (1.28), is well-posed if and only if the discrete inf-sup condition (1.30) holds true.*

Proof. Condition (1.30) is the discrete counterpart of condition (1.4) in the BNB Theorem. Hence, owing to this theorem, discrete well-posedness implies (1.30). Conversely, to prove that (1.30) implies discrete well-posedness, we first observe that (1.30) implies that the discrete operator A_h defined by (1.27) is injective. Indeed, $A_h v_h = 0$ yields $a_h(v_h, w_h) = 0$, for all $w_h \in V_h$. Hence, $v_h = 0$ by (1.30). In finite dimension, this implies that A_h is surjective. Hence, A_h is bijective. □

We observe that discrete well-posedness is equivalent to only one condition, namely (1.30), while two conditions appear in the continuous case. This is because injectivity is equivalent to bijectivity when the test and trial spaces have the same finite dimension.

A sufficient, and often easy to verify, condition for discrete stability is coercivity. This property can be stated as follows: There is $C_{\text{sta}} > 0$ such that

$$\forall v_h \in V_h, \quad C_{\text{sta}} \|v_h\|^2 \leq a_h(v_h, v_h). \tag{1.31}$$

Discrete coercivity implies the discrete inf-sup condition (1.30) since, for all $v_h \in V_h \setminus \{0\}$,

$$C_{\text{sta}} \|v_h\| \leq \frac{a_h(v_h, v_h)}{\|v_h\|} \leq \sup_{w_h \in V_h \setminus \{0\}} \frac{a_h(v_h, w_h)}{\|w_h\|}.$$

Property (1.31) is the discrete counterpart of that invoked in the Lax–Milgram Lemma.

1.3.3 Consistency

For the time being, we consider a rather strong form of consistency, namely that the exact solution u satisfies the discrete equations in (1.26). To formulate consistency, it is thus necessary to plug the exact solution into the first argument of the discrete bilinear form a_h, and this may not be possible in general since the discrete bilinear form a_h is so far defined on $V_h \times V_h$ only. Therefore, we assume that there is a subspace $X_* \subset X$ such that the exact solution u belongs to X_* and such that the discrete bilinear form a_h can be extended to $X_* \times V_h$ (it is not possible in general to extend a_h to $X \times V_h$). Consistency can now be formulated as follows.

Definition 1.31 (Consistency). We say that the discrete problem (1.26) is *consistent* if for the exact solution $u \in X_*$,

$$a_h(u, w_h) = l_h(w_h) \qquad \forall w_h \in V_h. \tag{1.32}$$

Remark 1.32 (Galerkin orthogonality). Consistency is equivalent to the usual *Galerkin orthogonality property* often considered in the context of finite element methods. Indeed, (1.32) holds true if and only if

$$a_h(u - u_h, w_h) = 0 \qquad \forall w_h \in V_h.$$

1.3.4 Boundedness

The last ingredient in the error analysis is boundedness. We introduce the vector space

$$X_{*h} := X_* + V_h,$$

and observe that the *approximation error* $(u - u_h)$ belongs to this space. We aim at measuring the approximation error using the discrete stability norm $\|\cdot\|$. Therefore, we assume in what follows that this norm can be extended to the space X_{*h}. In the present setting, we want to assert boundedness in the product space $X_{*h} \times V_h$, and not just in $V_h \times V_h$. It turns out that in most situations, it is not possible to assert boundedness using only the discrete stability norm $\|\cdot\|$. This is the reason why we introduce a second norm, say $\|\cdot\|_*$.

Definition 1.33 (Boundedness). We say that the discrete bilinear form a_h is *bounded* in $X_{*h} \times V_h$ if there is C_{bnd} such that

$$\forall (v, w_h) \in X_{*h} \times V_h, \qquad |a_h(v, w_h)| \le C_{\mathrm{bnd}} \|v\|_* \|w_h\|,$$

for a norm $\|\cdot\|_*$ defined on X_{*h} and such that, for all $v \in X_{*h}$, $\|v\| \le \|v\|_*$.

Remark 1.34 (h-dependency). In Definition 1.33, C_{bnd} can depend on the meshsize h. As mentioned above, in view of convergence analysis, it is important to ensure that C_{bnd} be independent of h.

1.3.5 Error Estimate

We can now state the main result of this section.

Theorem 1.35 (Abstract error estimate). *Let u solve (1.1) with $f \in L^2(\Omega)$. Let u_h solve (1.26). Let $X_* \subset X$ and assume that $u \in X_*$. Set $X_{*h} = X_* + V_h$ and assume that the discrete bilinear form a_h can be extended to $X_{*h} \times V_h$. Let $\|\cdot\|$ and $\|\cdot\|_*$ be two norms defined on X_{*h} and such that, for all $v \in X_{*h}$, $\|v\| \leq \|v\|_*$. Assume discrete stability, consistency, and boundedness. Then, the following error estimate holds true:*

$$\|u - u_h\| \leq C \inf_{y_h \in V_h} \|u - y_h\|_*, \tag{1.33}$$

with $C = 1 + C_{\text{sta}}^{-1} C_{\text{bnd}}$.

Proof. Let $y_h \in V_h$. Owing to discrete stability and consistency,

$$\|u_h - y_h\| \leq C_{\text{sta}}^{-1} \sup_{w_h \in V_h \setminus \{0\}} \frac{a_h(u_h - y_h, w_h)}{\|w_h\|} = C_{\text{sta}}^{-1} \sup_{w_h \in V_h \setminus \{0\}} \frac{a_h(u - y_h, w_h)}{\|w_h\|}.$$

Hence, owing to boundedness,

$$\|u_h - y_h\| \leq C_{\text{sta}}^{-1} C_{\text{bnd}} \|u - y_h\|_*.$$

Estimate (1.33) then results from the triangle inequality, the fact that $\|u - y_h\| \leq \|u - y_h\|_*$, and that y_h is arbitrary in V_h. $\square$

1.4 Admissible Mesh Sequences

The goal of this section is to derive some technical, yet important, tools to analyze the convergence of dG methods as the meshsize goes to zero. We consider a mesh sequence

$$\mathcal{T}_\mathcal{H} := (\mathcal{T}_h)_{h \in \mathcal{H}},$$

where $\mathcal{H}$ denotes a countable subset of $\mathbb{R}_{>0} := \{x \in \mathbb{R} \mid x > 0\}$ having 0 as only accumulation point. Our analysis tools are, on the one hand, inverse and trace inequalities that are instrumental to assert discrete stability and boundedness uniformly in h and, on the other hand, optimal polynomial approximation properties so as to infer from error estimates of the form (1.33) h-convergence rates for the approximation error whenever the exact solution is smooth enough.

In Sects. 1.4.1–1.4.4, we consider the case $d \geq 2$. We first introduce the concept of shape- and contact-regular mesh sequences, which is sufficient to derive inverse and trace inequalities, and then we combine it with an additional requirement on optimal polynomial approximation properties, leading to the concept of admissible mesh sequences. Finally, in Sect. 1.4.5, we deal with the case $d = 1$, where the requirements on admissible mesh sequences are much simpler.

1.4.1 Shape and Contact Regularity

A useful concept encountered in the context of conforming finite element methods is that of matching simplicial meshes.

Definition 1.36 (Matching simplicial mesh). We say that $\mathcal{T}_h$ is a *matching simplicial mesh* if it is a simplicial mesh and if for any $T \in \mathcal{T}_h$ with vertices $\{a_0, \ldots, a_d\}$, the set $\partial T \cap \partial T'$ for any $T' \in \mathcal{T}_h$, $T' \neq T$, is the convex hull of a (possibly empty) subset of $\{a_0, \ldots, a_d\}$.

For instance, in dimension 2, the set $\partial T \cap \partial T'$ for two distinct elements of a matching simplicial mesh is either empty, or a common vertex, or a common edge of the two elements. We now turn to the matching simplicial submesh of a general mesh.

Definition 1.37 (Matching simplicial submesh). Let $\mathcal{T}_h$ be a general mesh. We say that $\mathfrak{S}_h$ is a *matching simplicial submesh* of $\mathcal{T}_h$ if

(i) $\mathfrak{S}_h$ is a matching simplicial mesh,

(ii) For all $T' \in \mathfrak{S}_h$, there is only one $T \in \mathcal{T}_h$ such that $T' \subset T$,

(iii) For all $F' \in \mathfrak{F}_h$, the set collecting the mesh faces of $\mathfrak{S}_h$, there is at most one $F \in \mathcal{F}_h$ such that $F' \subset F$.

The simplices in $\mathfrak{S}_h$ are called *subelements*, and the mesh faces in $\mathfrak{F}_h$ are called *subfaces*. We set, for all $T \in \mathcal{T}_h$,

$$\mathfrak{S}_T := \{T' \in \mathfrak{S}_h \mid T' \subset T\},$$
$$\mathfrak{F}_T := \{F' \in \mathfrak{F}_h \mid F' \subset \partial T\}.$$

We also set, for all $F \in \mathcal{F}_h$,

$$\mathfrak{F}_F := \{F' \in \mathfrak{F}_h \mid F' \subset F\}.$$

Figure 1.5 illustrates the matching simplicial submesh for two polygonal mesh elements, say T_1 and T_2, that come into contact. The triangular subelements composing the sets $\mathfrak{S}_{T_1}$ and $\mathfrak{S}_{T_2}$ are indicated by dashed lines. We observe that the mesh face $F = \partial T_1 \cap \partial T_2$ (highlighted in bold) is not a part of a hyperplane and that the set $\mathfrak{F}_F$ contains two subfaces.

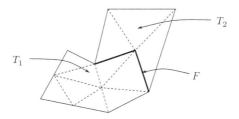

Fig. 1.5: Two polygonal mesh elements that come into contact with corresponding subelements indicated by *dashed lines* and interface indicated in *bold*

Definition 1.38 (Shape and contact regularity). We say that the mesh sequence $\mathcal{T}_\mathcal{H}$ is *shape- and contact-regular* if for all $h \in \mathcal{H}$, $\mathcal{T}_h$ admits a matching simplicial submesh $\mathfrak{S}_h$ such that

(i) The mesh sequence $\mathfrak{S}_\mathcal{H}$ is shape-regular in the usual sense of Ciarlet [92], meaning that there is a parameter $\varrho_1 > 0$, independent of h, such that, for all $T' \in \mathfrak{S}_h$,
$$\varrho_1 h_{T'} \leq r_{T'},$$
where $h_{T'}$ is the diameter of T' and $r_{T'}$ the radius of the largest ball inscribed in T',

(ii) There is a parameter $\varrho_2 > 0$, independent of h, such that, for all $T \in \mathcal{T}_h$ and for all $T' \in \mathfrak{S}_T$,
$$\varrho_2 h_T \leq h_{T'}.$$

Henceforth, the parameters ϱ_1 and ϱ_2 are called the *mesh regularity parameters* and are collectively denoted by the symbol ϱ. Finally, if $\mathcal{T}_h$ is itself matching and simplicial, then $\mathfrak{S}_h = \mathcal{T}_h$ and the only requirement is shape-regularity with parameter $\varrho_1 > 0$ independent of h.

As elaborated in Sect. 1.4.2, the two conditions in Definition 1.38 allow one to control the shape of the elements in $\mathcal{T}_h$ and the way these elements come into contact. The idea of considering a matching simplicial submesh has been proposed, e.g., by Brenner [51] to derive generalized Poincaré–Friedrichs inequalities in broken Sobolev spaces. More recently, in the context of dG methods, a matching simplicial submesh has been considered by Buffa and Ortner [61] for nonlinear minimization problems and by Ern and Vohralík [151] for a posteriori error estimates in the context of PDEs with diffusion.

Remark 1.39 (Anisotropic meshes). Definition 1.38 implies that the mesh is isotropic in the sense that, for all $T \in \mathcal{T}_h$, the d-dimensional measure $|T|_d$ is uniformly equivalent to h_T^d. In applications featuring sharp layers, anisotropic meshes can be advantageous. We refer the reader, e.g., to van der Vegt and van der Ven [297], Sun and Wheeler [287], Georgoulis [166], Georgoulis, Hall, and Houston [167], and Leicht and Hartmann [226] for various aspects of dG methods on anisotropic meshes.

1.4.2 Geometric Properties

This section collects some useful geometric properties of shape- and contact-regular mesh sequences. The first result is a uniform bound on $\operatorname{card}(\mathfrak{S}_T)$.

Lemma 1.40 (Bound on $\operatorname{card}(\mathfrak{S}_T)$). *Let $\mathcal{T}_\mathcal{H}$ be a shape- and contact-regular mesh sequence. Then, for all $h \in \mathcal{H}$ and all $T \in \mathcal{T}_h$, $\operatorname{card}(\mathfrak{S}_T)$ is bounded uniformly in h.*

1.4. Admissible Mesh Sequences

Proof. Let $|\cdot|_d$ denote the d-dimensional Hausdorff measure and let $\mathfrak{B}_d$ be the unit ball in $\mathbb{R}^d$. Then,

$$h_T^d \geq |T|_d = \sum_{T' \in \mathfrak{S}_T} |T'|_d \geq \sum_{T' \in \mathfrak{S}_T} |\mathfrak{B}_d|_d r_{T'}^d \geq \sum_{T' \in \mathfrak{S}_T} |\mathfrak{B}_d|_d \varrho_1^d h_{T'}^d$$

$$\geq \sum_{T' \in \mathfrak{S}_T} |\mathfrak{B}_d|_d \varrho_1^d \varrho_2^d h_T^d \geq |\mathfrak{B}_d|_d \varrho_1^d \varrho_2^d \operatorname{card}(\mathfrak{S}_T) h_T^d,$$

yielding the assertion. □

Our next result is a uniform bound on $\operatorname{card}(\mathcal{F}_T)$, $\operatorname{card}(\mathfrak{F}_T)$, the parameter N_∂ defined by (1.12), and $\operatorname{card}(\mathfrak{F}_F)$.

Lemma 1.41 (Bound on $\operatorname{card}(\mathcal{F}_T)$, $\operatorname{card}(\mathfrak{F}_T)$, N_∂, and $\operatorname{card}(\mathfrak{F}_F)$). *Let $\mathcal{T}_\mathcal{H}$ be a shape- and contact-regular mesh sequence with parameters ϱ. Then, for all $h \in \mathcal{H}$ and all $T \in \mathcal{T}_h$, $\operatorname{card}(\mathcal{F}_T)$, $\operatorname{card}(\mathfrak{F}_T)$, and N_∂ are bounded uniformly in h, while, for all $F \in \mathcal{F}_h$, $\operatorname{card}(\mathfrak{F}_F)$ is bounded uniformly in h.*

Proof. We observe that

$$\operatorname{card}(\mathcal{F}_T) \leq \operatorname{card}(\mathfrak{F}_T) \leq (d+1)\operatorname{card}(\mathfrak{S}_T),$$

so that the assertion on $\operatorname{card}(\mathcal{F}_T)$ and $\operatorname{card}(\mathfrak{F}_T)$ follows from Lemma 1.40. The bound on N_∂ results from its definition (1.12) and the bound on $\operatorname{card}(\mathcal{F}_T)$. Finally, to bound $\operatorname{card}(\mathfrak{F}_F)$ for all $F \in \mathcal{F}_h$, we pick $T \in \mathcal{T}_h$ such that $F \in \mathcal{F}_T$ and observe that $\operatorname{card}(\mathfrak{F}_F) \leq (d+1)\operatorname{card}(\mathfrak{S}_T)$, so that the bound on $\operatorname{card}(\mathfrak{F}_F)$ results from the bound on $\operatorname{card}(\mathfrak{S}_T)$. □

Our next result is a lower bound on the diameter of mesh faces.

Lemma 1.42 (Lower bound on face diameters). *Let $\mathcal{T}_\mathcal{H}$ be a shape- and contact-regular mesh sequence with parameters ϱ. Then, for all $h \in \mathcal{H}$, all $T \in \mathcal{T}_h$, and all $F \in \mathcal{F}_T$,*

$$\delta_F \geq \varrho_1 \varrho_2 h_T, \tag{1.34}$$

where δ_F denotes the diameter of F.

Proof. Let $T \in \mathcal{T}_h$ and let $F \in \mathcal{F}_T$. Then, we pick $F' \in \mathfrak{F}_F$ and denote by $T' \in \mathfrak{S}_T$ the simplex to which the subface F' belongs. We obtain

$$\delta_F \geq \delta_{F'} \geq r_{T'} \geq \varrho_1 h_{T'} \geq \varrho_1 \varrho_2 h_T,$$

yielding the assertion. □

A direct consequence of Lemma 1.42 is a comparison result on the diameter of neighboring elements.

Lemma 1.43 (Diameter comparison for neighboring elements). *Let $\mathcal{T}_\mathcal{H}$ be a shape- and contact-regular mesh sequence with parameters ϱ. Then, for all $h \in \mathcal{H}$ and all $T, T' \in \mathcal{T}_h$ sharing a face F, there holds*

$$\min(h_T, h_{T'}) \geq \varrho_1 \varrho_2 \max(h_T, h_{T'}). \tag{1.35}$$

Proof. Since F is a common face to both T and T', owing to (1.34),

$$\varrho_1 \varrho_2 \max(h_T, h_{T'}) \leq \delta_F \leq \min(h_T, h_{T'}),$$

yielding (1.35). □

1.4.3 Inverse and Trace Inequalities

Inverse and trace inequalities are useful tools to analyze dG methods. For simplicity, we derive these inequalities on the broken polynomial space $\mathbb{P}_d^k(\mathcal{T}_h)$ defined by (1.15); other broken polynomial spaces can be considered.

We begin with the following inverse inequality that delivers a local upper bound on the gradient of discrete functions.

Lemma 1.44 (Inverse inequality). *Let $\mathcal{T}_\mathcal{H}$ be a shape- and contact-regular mesh sequence with parameters ϱ. Then, for all $h \in \mathcal{H}$, all $v_h \in \mathbb{P}_d^k(\mathcal{T}_h)$, and all $T \in \mathcal{T}_h$,*

$$\|\nabla v_h\|_{[L^2(T)]^d} \leq C_{\text{inv}} h_T^{-1} \|v_h\|_{L^2(T)}, \tag{1.36}$$

where C_{inv} only depends on ϱ, d, and k.

Proof. Let $v_h \in \mathbb{P}_d^k(\mathcal{T}_h)$ and let $T \in \mathcal{T}_h$. For all $T' \in \mathfrak{S}_T$, the restriction $v_h|_{T'}$ is in $\mathbb{P}_d^k(T')$. Hence, owing to the usual inverse inequality on simplices (see Brenner and Scott [54, Sect. 4.5] or Ern and Guermond [141, Sect. 1.7]),

$$\|\nabla v_h\|_{[L^2(T')]^d} \leq C_{\text{inv,s}} h_{T'}^{-1} \|v_h\|_{L^2(T')},$$

where $C_{\text{inv,s}}$ only depends on ϱ_1, d, and k. Using point (ii) in Definition 1.38 yields

$$\|\nabla v_h\|_{[L^2(T')]^d} \leq \varrho_2^{-1} C_{\text{inv,s}} h_T^{-1} \|v_h\|_{L^2(T')}.$$

Squaring this inequality and summing over $T' \in \mathfrak{S}_T$ yields (1.36). □

Remark 1.45 (Nature of (1.36)). The inverse inequality (1.36) is local to mesh elements. As such, it depends on the shape of the mesh elements but not on the way mesh elements come into contact.

We now turn to the following discrete trace inequality that delivers an upper bound on the face values of discrete functions.

1.4. Admissible Mesh Sequences

Lemma 1.46 (Discrete trace inequality). *Let $\mathcal{T}_\mathcal{H}$ be a shape- and contact-regular mesh sequence with parameters ϱ. Then, for all $h \in \mathcal{H}$, all $v_h \in \mathbb{P}_d^k(\mathcal{T}_h)$, all $T \in \mathcal{T}_h$, and all $F \in \mathcal{F}_T$,*

$$h_T^{1/2} \|v_h\|_{L^2(F)} \leq C_{\mathrm{tr}} \|v_h\|_{L^2(T)}, \tag{1.37}$$

where C_{tr} only depends on ϱ, d, and k.

Proof. Let $v_h \in \mathbb{P}_d^k(\mathcal{T}_h)$, let $T \in \mathcal{T}_h$, and let $F \in \mathcal{F}_T$. We first assume that $\mathcal{T}_h$ is a matching simplicial mesh. Let $\widehat{T}$ be the unit simplex of $\mathbb{R}^d$ and let F_T be the bijective affine map such that $F_T(\widehat{T}) = T$. Let $\widehat{F}$ be any face of $\widehat{T}$. Since the unit sphere in $\mathbb{P}_d^k(\widehat{T})$ for the $L^2(\widehat{T})$-norm is a compact set, there is $\widehat{C}_{d,k}(\widehat{F})$, only depending on d, k, and $\widehat{F}$, such that, for all $\widehat{v} \in \mathbb{P}_d^k(\widehat{T})$,

$$\|\widehat{v}\|_{L^2(\widehat{F})} \leq \widehat{C}_{d,k}(\widehat{F}) \|\widehat{v}\|_{L^2(\widehat{T})}.$$

Applying the above inequality to the function $\widehat{v} = v_h|_T \circ F_T^{-1}$ which is in $\mathbb{P}_d^k(\widehat{T})$, we infer

$$|F|_{d-1}^{-1/2} \|v_h\|_{L^2(F)} \leq \widehat{C}_{d,k} |T|_d^{-1/2} \|v_h\|_{L^2(T)},$$

where $\widehat{C}_{d,k} := \max_{\widehat{F} \in \mathcal{F}_{\widehat{T}}} \widehat{C}_{d,k}(\widehat{F})$. Moreover, we observe that

$$\frac{|T|_d}{|F|_{d-1}} = \frac{1}{d} h_{T,F} \geq \frac{1}{d} r_T \geq \frac{1}{d} \varrho_1 h_T, \tag{1.38}$$

where $h_{T,F}$ denotes the distance of the vertex opposite to F to that face and r_T the radius of the largest ball inscribed in T. As a result,

$$h_T^{1/2} \|v_h\|_{L^2(F)} \leq C_{\mathrm{tr,s}} \|v_h\|_{L^2(T)}, \tag{1.39}$$

where $C_{\mathrm{tr,s}} := d^{1/2} \varrho_1^{-1/2} \widehat{C}_{d,k}$ only depends on ϱ_1, d, and k. We now consider the case of general meshes. For each $F' \in \mathfrak{F}_F$, let T' denote the simplex in $\mathfrak{S}_T$ of which F' is a face. Since the restriction $v_h|_{T'}$ is in $\mathbb{P}_d^k(T')$, the discrete trace inequality (1.39) yields

$$h_{T'}^{1/2} \|v_h\|_{L^2(F')} \leq C_{\mathrm{tr,s}} \|v_h\|_{L^2(T')} \leq C_{\mathrm{tr,s}} \|v_h\|_{L^2(T)}.$$

Squaring this inequality and summing over $F' \in \mathfrak{F}_F$ yields

$$\left(\sum_{F' \in \mathfrak{F}_F} h_{T'} \|v_h\|_{L^2(F')}^2 \right)^{1/2} \leq C_{\mathrm{tr,s}} \operatorname{card}(\mathfrak{F}_F)^{1/2} \|v_h\|_{L^2(T)},$$

whence the assertion follows, since $h_{T'} \geq \varrho_2 h_T$ and $\operatorname{card}(\mathfrak{F}_F)$ is bounded uniformly owing to Lemma 1.41. □

Remark 1.47 (Variant of (1.37)). Summing over $F \in \mathcal{F}_T$, we infer from (1.37) and the Cauchy–Schwarz inequality that

$$h_T^{1/2}\|v_h\|_{L^2(\partial T)} \leq C_{\mathrm{tr}} N_\partial^{1/2} \|v_h\|_{L^2(T)}, \qquad (1.40)$$

and we recall from Lemma 1.41 that N_∂ is bounded uniformly in h.

Remark 1.48 (k-dependency). When working with high-degree polynomials, it is important to determine the dependency of C_{inv} and C_{tr} on the polynomial degree k. This turns out to be a delicate question, and precise answers are available only in specific cases. Concerning the discrete trace inequality (1.37), it is proven by Warburton and Hesthaven [304] that on $\mathbb{P}_d^k(\mathcal{T}_h)$, C_{tr} scales as $\sqrt{k(k+d)}$. Moreover, on $\mathbb{Q}_d^k(\mathcal{T}_h)$ with the mesh elements being affine images of the unit hypercube in $\mathbb{R}^d$, one-dimensional results can be exploited using tensor-product polynomials, yielding that C_{tr} scales as $\sqrt{k(k+1)}$; see Canuto and Quarteroni [75], Bernardi and Maday [42], and Schwab [275]. One difficulty concerns the behavior of polynomials near the end points of an interval, and this can be dealt with by considering weighted norms in (1.37); see, e.g., Melenk and Wohlmuth [235]. Concerning the inverse inequality (1.36), C_{inv} scales as k^2 on triangles and parallelograms; see, e.g., Schwab [275].

We also need the following continuous trace inequality, which delivers an upper bound on the face values of functions in the broken Sobolev space $H^1(\mathcal{T}_h)$. We present here a simple proof inspired by Monk and Süli [238] and Carstensen and Funken [77] (see also Stephansen [283, Lemma 3.12]).

Lemma 1.49 (Continuous trace inequality). *Let $\mathcal{T}_\mathcal{H}$ be a shape- and contact-regular mesh sequence. Then, for all $h \in \mathcal{H}$, all $v \in H^1(\mathcal{T}_h)$, all $T \in \mathcal{T}_h$, and all $F \in \mathcal{F}_T$,*

$$\|v\|_{L^2(F)}^2 \leq C_{\mathrm{cti}} (2\|\nabla v\|_{[L^2(T)]^d} + d h_T^{-1} \|v\|_{L^2(T)}) \|v\|_{L^2(T)}, \qquad (1.41)$$

with $C_{\mathrm{cti}} := \varrho_1^{-1}$ if $\mathcal{T}_h$ is matching and simplicial, while $C_{\mathrm{cti}} := (1+d)(\varrho_1 \varrho_2)^{-1}$ otherwise.

Proof. Let $v \in H^1(\mathcal{T}_h)$, let $T \in \mathcal{T}_h$, and let $F \in \mathcal{F}_T$. Assume first that T is a simplex and consider the $\mathbb{R}^d$-valued function

$$\sigma_F = \frac{|F|_{d-1}}{d|T|_d}(x - a_F),$$

where a_F is the vertex of T opposite to F; cf. Fig. 1.6. The normal component of σ_F is constant and equal to one on F, and it vanishes on all the remaining faces in $\mathcal{F}_T$. (The function σ_F is proportional to the lowest-order Raviart–Thomas–Nédélec shape function in T; see, e.g., Brezzi and Fortin [57, p. 116] or Ern and Guermond [141, Sect. 1.2.7] and also cf. Sect. 5.5.3.) Owing to the divergence theorem,

$$\|v\|_{L^2(F)}^2 = \int_F |v|^2 = \int_{\partial T} |v|^2(\sigma_F \cdot n_T) = \int_T \nabla \cdot (|v|^2 \sigma_F)$$
$$= \int_T 2 v \sigma_F \cdot \nabla v + \int_T |v|^2(\nabla \cdot \sigma_F).$$

1.4. Admissible Mesh Sequences

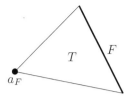

Fig. 1.6: Notation for the proof of Lemma 1.49

Since
$$\|\sigma_F\|_{[L^\infty(T)]^d} \leq \frac{|F|_{d-1} h_T}{d|T|_d}, \qquad \nabla \cdot \sigma_F = \frac{|F|_{d-1}}{|T|_d},$$
we infer using the Cauchy–Schwarz inequality that
$$\|v\|_{L^2(F)}^2 \leq \frac{|F|_{d-1} h_T}{d|T|_d} \left(2\|\nabla v\|_{[L^2(T)]^d} + dh_T^{-1}\|v\|_{L^2(T)}\right) \|v\|_{L^2(T)}. \qquad (1.42)$$

Using (1.38) yields (1.41) with $C_{\text{cti}} = \varrho_1^{-1}$ on matching simplicial meshes. Consider now the case where T belongs to a general mesh. For each $F' \in \mathfrak{F}_F$, let T' denote the simplex in $\mathfrak{S}_T$ of which F' is a face. Applying the continuous trace inequality using F' and T' yields
$$\|v\|_{L^2(F')}^2 \leq \varrho_1^{-1} \left(2\|\nabla v\|_{[L^2(T')]^d} + dh_{T'}^{-1}\|v\|_{L^2(T')}\right) \|v\|_{L^2(T')}.$$

Since $h_{T'} \geq \varrho_2 h_T$ and $\varrho_2 \leq 1$, we infer
$$\|v\|_{L^2(F')}^2 \leq (\varrho_1 \varrho_2)^{-1} \left(2\|\nabla v\|_{[L^2(T')]^d} + dh_T^{-1}\|v\|_{L^2(T')}\right) \|v\|_{L^2(T')}.$$

Summing over $F' \in \mathfrak{F}_F$ and using the Cauchy–Schwarz inequality yields (1.41) since each $T' \in \mathfrak{S}_T$ appears at most $(d+1)$ times in the summation. $\square$

We close this section with some useful inverse and trace inequalities in a non-Hilbertian setting. Our first result allows us to compare the $\|\cdot\|_{L^p(T)}$- and $\|\cdot\|_{L^q(T)}$-norms. The proof is skipped since it hinges again on the simplicial submesh and the corresponding inverse inequality on simplices.

Lemma 1.50 (Comparison of $\|\cdot\|_{L^p(T)}$- and $\|\cdot\|_{L^q(T)}$-norms). *Let $\mathcal{T}_\mathcal{H}$ be a shape- and contact-regular mesh sequence with parameters ϱ. Let $1 \leq p, q \leq \infty$ be two real numbers. Then, for all $h \in \mathcal{H}$, all $v_h \in \mathbb{P}_d^k(\mathcal{T}_h)$, and all $T \in \mathcal{T}_h$,*
$$\|v_h\|_{L^p(T)} \leq C_{\text{inv},p,q} h_T^{d(1/p - 1/q)} \|v_h\|_{L^q(T)}, \qquad (1.43)$$
where $C_{\text{inv},p,q}$ only depends on ϱ, d, k, p, and q.

Remark 1.51 (Dependency on p and q). The quantity $C_{\text{inv},p,q}$ in (1.43) can be uniformly bounded in p and q. We first observe that owing to Hölder's inequality, $C_{\text{inv},p,q} = 1$ if $p < q$. Additionally, it is shown by Verfürth [299] that on simplices, for $p > 2$ and $q = 2$, $C_{\text{inv},p,2} \leq C_{d,k}^{1-2/p}$ and for $p = 2$ and $q < 2$, $C_{\text{inv},2,q} \leq C_{d,k}^{2/q-1}$, where $C_{d,k} := ((2k+2)(4k+2)^{d-1})^{1/2}$. Hence, the largest value for $C_{\text{inv},p,q}$ is obtained in the case where $p > 2 > q$, and observing that $|1 - 2/p| \leq 1$, $|1 - 2/q| \leq 1$, and $C_{d,k} \geq 1$, we infer that the quantity $C_{\text{inv},p,q}$ can always be bounded by $C_{d,k}^2$. A uniform upper bound in p and q can also be derived on general meshes by summing over the simplicial subelements.

Our second and last result is a non-Hilbertian version of the discrete trace inequality (1.37).

Lemma 1.52 (Discrete trace inequality in $L^p(F)$). *Let $\mathcal{T}_\mathcal{H}$ be a shape- and contact-regular mesh sequence with parameters ϱ. Let $1 \leq p \leq \infty$ be a real number. Then, for all $h \in \mathcal{H}$, all $v_h \in \mathbb{P}_d^k(\mathcal{T}_h)$, all $T \in \mathcal{T}_h$, and all $F \in \mathcal{F}_T$,*

$$h_T^{1/p} \|v_h\|_{L^p(F)} \leq C_{\text{tr},p} \|v_h\|_{L^p(T)}, \qquad (1.44)$$

where $C_{\text{tr},p}$ only depends on ϱ, d, k, and p.

Proof. We combine the discrete trace inequality (1.37) with the inverse inequality (1.43) (in F) to infer

$$\begin{aligned} h_T^{1/p} \|v_h\|_{L^p(F)} &\leq C_{\text{inv},p,2} h_T^{1/p} \delta_F^{(d-1)(1/p-1/2)} \|v_h\|_{L^2(F)} \\ &\leq C_{\text{inv},p,2} C_{\text{tr}} h_T^{1/p-1/2} \delta_F^{(d-1)(1/p-1/2)} \|v_h\|_{L^2(T)} \\ &\leq C_{\text{inv},p,2} C_{\text{tr}} C_{\text{inv},2,p} h_T^{1/p-1/2} \delta_F^{(d-1)(1/p-1/2)} h_T^{d(1/2-1/p)} \|v_h\|_{L^p(T)}. \end{aligned}$$

The assertion follows by observing that δ_F in uniformly equivalent to h_T. □

Remark 1.53 (Dependency on p). The quantity $C_{\text{tr},p}$ in (1.44) can be uniformly bounded in p. This is a direct consequence of the above proof and Remark 1.51.

1.4.4 Polynomial Approximation

To infer from estimate (1.33) a convergence rate in h for the *approximation error* $(u - u_h)$ measured in the $\|\cdot\|$-norm when the exact solution u is smooth enough, we need to estimate the right-hand side given by

$$\inf_{y_h \in V_h} \|u - y_h\|_*,$$

when V_h is typically the broken polynomial space $\mathbb{P}_d^k(\mathcal{T}_h)$ defined by (1.15); other broken polynomial spaces can be considered. Since $u_h \in V_h$, we infer from (1.33) that

$$\inf_{y_h \in V_h} \|u - y_h\| \leq \|u - u_h\| \leq C \inf_{y_h \in V_h} \|u - y_h\|_*. \qquad (1.45)$$

Definition 1.54 (Optimality, quasi-optimality, and suboptimality of the error estimate). We say that the error estimate (1.45) is

1.4. Admissible Mesh Sequences

(i) *Optimal* if $\|\cdot\| = \|\cdot\|_*$,

(ii) *Quasi-optimal* if the two norms are different, but the lower and upper bounds in (1.45) converge, for smooth u, at the same convergence rate as $h \to 0$,

(iii) *Suboptimal* if the upper bound converges at a slower rate than the lower bound.

The analysis of the upper bound $\inf_{y_h \in V_h} \|u - y_h\|_*$ depends on the polynomial approximation properties that can be achieved in the broken polynomial space V_h.

Definition 1.55 (Optimal polynomial approximation). We say that the mesh sequence $\mathcal{T}_\mathcal{H}$ has *optimal polynomial approximation* properties if, for all $h \in \mathcal{H}$, all $T \in \mathcal{T}_h$, and all polynomial degree k, there is a linear interpolation operator $\mathcal{I}_T^k : L^2(T) \to \mathbb{P}_d^k(T)$ such that, for all $s \in \{0, \ldots, k+1\}$ and all $v \in H^s(T)$, there holds

$$|v - \mathcal{I}_T^k v|_{H^m(T)} \leq C_{\text{app}} h_T^{s-m} |v|_{H^s(T)} \qquad \forall m \in \{0, \ldots, s\}, \tag{1.46}$$

where C_{app} is independent of both T and h.

Remark 1.56 (Nature of (1.46)). As for the inverse inequality (1.36), the optimal polynomial approximation property (1.46) is local to mesh elements. As such, it depends on the shape of the mesh elements, but not on the way mesh elements come into contact.

Definition 1.57 (Admissible mesh sequences). We say that the mesh sequence $\mathcal{T}_\mathcal{H}$ is *admissible* if it is shape- and contact-regular and if it has optimal polynomial approximation properties.

In what follows, we often consider the L^2-orthogonal projection onto the broken polynomial space $\mathbb{P}_d^k(\mathcal{T}_h)$. One reason is that its definition (cf. (1.29)) is very simple, even on general meshes.

Lemma 1.58 (Optimality of L^2-orthogonal projection). *Let $\mathcal{T}_\mathcal{H}$ be an admissible mesh sequence. Let π_h be the L^2-orthogonal projection onto $\mathbb{P}_d^k(\mathcal{T}_h)$. Then, for all $s \in \{0, \ldots, k+1\}$ and all $v \in H^s(T)$, there holds*

$$|v - \pi_h v|_{H^m(T)} \leq C'_{\text{app}} h_T^{s-m} |v|_{H^s(T)} \qquad \forall m \in \{0, \ldots, s\}, \tag{1.47}$$

where C'_{app} is independent of both T and h.

Proof. For $m = 0$, we obtain by definition of the L^2-orthogonal projection,

$$\|v - \pi_h v\|_{L^2(T)} \leq \|v - \mathcal{I}_T^k v\|_{L^2(T)} \leq C_{\text{app}} h_T^s |v|_{H^s(T)}.$$

For $m \geq 1$, we use m times the inverse inequality (1.36) together with the triangle inequality to infer

$$\begin{aligned}|v - \pi_h v|_{H^m(T)} &\leq |v - \mathcal{I}_T^k v|_{H^m(T)} + |\mathcal{I}_T^k v - \pi_h v|_{H^m(T)} \\ &\leq |v - \mathcal{I}_T^k v|_{H^m(T)} + C' h_T^{-m} \|\mathcal{I}_T^k v - \pi_h v\|_{L^2(T)} \\ &\leq |v - \mathcal{I}_T^k v|_{H^m(T)} + 2C' h_T^{-m} \|v - \mathcal{I}_T^k v\|_{L^2(T)},\end{aligned}$$

where C' has the same dependencies as C'_{app}, whence (1.47) owing to (1.46). $\square$

In the analysis of dG methods, we often need to bound polynomial approximation errors on mesh faces. The following result is a direct consequence of (1.47) and of the continuous trace inequality in Lemma 1.49.

Lemma 1.59 (Polynomial approximation on mesh faces). *Under the hypotheses of Lemma 1.58, assume additionally that $s \geq 1$. Then, for all $h \in \mathcal{H}$, all $T \in \mathcal{T}_h$, and all $F \in \mathcal{F}_T$, there holds*

$$\|v - \pi_h v\|_{L^2(F)} \leq C''_{\text{app}} h_T^{s-1/2} |v|_{H^s(T)},$$

and if $s \geq 2$,

$$\|\nabla(v - \pi_h v)|_T \cdot \mathbf{n}_T\|_{L^2(F)} \leq C'''_{\text{app}} h_T^{s-3/2} |v|_{H^s(T)},$$

where C''_{app} and C'''_{app} are independent of both T and h.

Lemmata 1.58 and 1.59 are instrumental in deriving convergence rates, as the meshsize goes to zero, for the approximation error owing to the error estimate (1.33) which yields

$$\|u - u_h\| \leq \inf_{y_h \in V_h} \|u - y_h\|_* \leq \|u - \pi_h u\|_*.$$

On general meshes, asserting optimal polynomial approximation is a delicate question since this property depends on the shape of mesh elements. In practice, meshes are generated by successive refinements of an initial mesh, and the shape of mesh elements depends on the refinement procedure. It is convenient to identify sufficient conditions on the mesh sequence $\mathcal{T}_{\mathcal{H}}$ to assert optimal polynomial approximation in broken polynomial spaces. One approach is based on the star-shaped property with respect to a ball.

Definition 1.60 (Star-shaped property with respect to a ball). We say that a polyhedron P is *star-shaped with respect to a ball* if there is a ball $\mathfrak{B}_P \subset P$ such that, for all $x \in P$, the convex hull of $\{x\} \cup \mathfrak{B}_P$ is included in $\overline{P}$.

Figure 1.7 displays two polyhedra. The one on the left is star-shaped with respect to the ball indicated in black. Instead, the one on the right is not star-shaped with respect to any ball.

1.4. Admissible Mesh Sequences

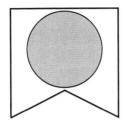 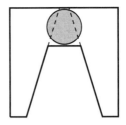

Fig. 1.7: Example (*left*) and counter-example (*right*) of a polyhedron which is star-shaped with respect to a ball

Lemma 1.61 (Mesh sequences with star-shaped property). *Let $\mathcal{T}_{\mathcal{H}}$ be a shape- and contact-regular mesh sequence. Assume that, for all $h \in \mathcal{H}$ and all $T \in \mathcal{T}_h$, the mesh element T is star-shaped with respect to a ball with uniformly comparable diameter with respect to h_T. Then, the mesh sequence $\mathcal{T}_{\mathcal{H}}$ is admissible.*

Proof. Optimal polynomial approximation is proven by Brenner and Scott [54, Chap. 4] using averaged Taylor polynomials. □

Another sufficient condition ensuring optimal polynomial approximation, but somewhat less general than the star-shaped property, is that of finitely shaped mesh sequences. A simple example is that of shape- and contact-regular mesh sequences whose elements are either simplices or parallelotopes in $\mathbb{R}^d$.

Lemma 1.62 (Finitely shaped mesh sequences). *Let $\mathcal{T}_{\mathcal{H}}$ be a shape- and contact-regular mesh sequence. Assume that $\mathcal{T}_{\mathcal{H}}$ is finitely shaped in the sense that there is a finite set $\widehat{\mathcal{R}} = \{\widehat{T}\}$ whose elements are reference polyhedra in $\mathbb{R}^d$ and such that, for all $h \in \mathcal{H}$, each $T \in \mathcal{T}_h$ is the image of a reference polyhedron in $\widehat{\mathcal{R}}$ by an affine bijective map F_T. Then, the mesh sequence $\mathcal{T}_{\mathcal{H}}$ is admissible.*

Proof. The proof is sketched since it uses classical finite element techniques. Let $h \in \mathcal{H}$ and let $T \in \mathcal{T}_h$. Let $v \in H^s(T)$. Since T is such that $T = F_T(\widehat{T})$ for some $\widehat{T} \in \widehat{\mathcal{R}}$, we set $\widehat{v} = v \circ F_T$ and observe that $\widehat{v} \in H^s(\widehat{T})$. Let $k \geq 0$ and let $s \in \{0, \ldots, k+1\}$. Owing to the Deny–Lions Lemma (see Deny and Lions [123] or Ern and Guermond [141, Lemma B.67]), we infer

$$|\widehat{v} - \pi_{\widehat{T}}^k \widehat{v}|_{H^m(\widehat{T})} \leq C_{\widehat{T}} |\widehat{v}|_{H^s(\widehat{T})} \qquad \forall m \in \{0, \ldots, s\},$$

where $\pi_{\widehat{T}}^k$ is the L^2-orthogonal projection onto $\mathbb{P}_d^k(\widehat{T})$. Since T contains a ball with diameter comparable to h_T, transforming back to T yields

$$|v - \mathcal{I}_T^k v|_{H^m(T)} \leq C'_{\widehat{T}} h_T^{s-m} |v|_{H^s(T)},$$

where $\mathcal{I}_T^k v = (\pi_{\widehat{T}}^k \widehat{v}) \circ F_T^{-1}$. The assertion follows by taking the maximal value of $C'_{\widehat{T}}$ for all $\widehat{T} \in \widehat{\mathcal{R}}$, and this yields a bounded quantity since the set $\widehat{\mathcal{R}}$ is finite. □

Remark 1.63 (Role of finitely-shaped assumption). As reflected in the above proof, the assumption that the mesh sequence is finitely shaped allows us to derive a uniform upper bound on the quantities $C'_{\widehat{T}}$ resulting from the application of the Deny–Lions Lemma in each mesh element.

Remark 1.64 (Quadrangular meshes). In dimension 2, it is possible to consider more general quadrangular meshes, where the elements are generated using bilinear mappings from the reference unit square in $\mathbb{R}^2$. Then, under the regularity conditions derived by Girault and Raviart [170, Sect. A.2], optimal polynomial approximation is achieved using polynomials in $\mathbb{Q}_d^k$.

1.4.5 The One-Dimensional Case

The situation is much simpler in dimension 1 where all the mesh elements are intervals which can come into contact only through their endpoints. Thus, shape and contact regularity are void concepts. Moreover, optimal polynomial approximation properties can be classically asserted. However, we need the counterpart of Lemma 1.43 to compare the diameter of neighboring mesh intervals. This is the only requirement for admissibility of mesh sequences.

Definition 1.65 (Admissible mesh sequences, $d = 1$). We say that the mesh sequence $\mathcal{T}_\mathcal{H}$ is *admissible* if there is $\varrho > 0$ such that, for all $h \in \mathcal{H}$ and all $T, T' \in \mathcal{T}_h$ with $\overline{T} \cap \overline{T}'$ nonempty,

$$\min(h_T, h_{T'}) \geq \varrho \max(h_T, h_{T'}).$$

The important properties derived in Sects. 1.4.3 and 1.4.4 are available on admissible mesh sequences in dimension 1, namely:

(a) The inverse inequality (1.36) holds true.

(b) The discrete trace inequality (1.37) holds true (recalling that, in dimension 1, face integrals reduce to pointwise evaluation) and the continuous trace inequality (1.41) also holds true (recalling that the 0-dimensional Hausdorff measure of a face is conventionally set to 1).

(c) The optimal polynomial approximation property (1.46) holds true, together with the optimal bounds on the L^2-projection (cf. Lemmata 1.58 and 1.59).

Part I

Scalar First-Order PDEs

Chapter 2
Steady Advection-Reaction

The steady advection-reaction equation with homogeneous inflow boundary condition

$$\beta \cdot \nabla u + \mu u = f \quad \text{in } \Omega, \tag{2.1a}$$
$$u = 0 \quad \text{on } \partial\Omega^-, \tag{2.1b}$$

is one of the simplest model problems based on a linear, scalar, steady first-order PDE. Here, the unknown function u is scalar-valued and represents, e.g., a solute concentration; β is the $\mathbb{R}^d$-valued advective velocity, μ the reaction coefficient, f the source term, and $\partial\Omega^-$ denotes the *inflow* part of the boundary of Ω, namely

$$\partial\Omega^- := \{x \in \partial\Omega \mid \beta(x) \cdot n(x) < 0\}. \tag{2.2}$$

The main goal of this chapter is to design and analyze dG methods to approximate the model problem (2.1). The mathematical analysis of this problem entails some subtleties. Since dG methods are essentially tailored to approximate PDEs in an L^2-setting where discrete stability is enhanced by suitable least-squares penalties, the most natural weak formulation at the continuous level is that based on the concept of graph space. A further question concerns the mathematical sense of the boundary condition (2.1b) for functions belonging to the graph space. To this purpose, we introduce reasonably mild, sufficient conditions on the advective field β to achieve a well-posed weak formulation. We formulate the boundary condition (2.1b) weakly in the continuous problem since this is the way boundary conditions are enforced in dG methods. Then, we present a step-by-step derivation of suitable dG bilinear forms that match the discrete stability, consistency, and boundedness properties outlined in Sect. 1.3 for nonconforming finite element error analysis. We also discuss an alternative viewpoint using local (elementwise) problems and numerical fluxes. Two dG methods are analyzed, resulting from the use of so-called centered or upwind fluxes.

D.A. Di Pietro and A. Ern, *Mathematical Aspects of Discontinuous Galerkin Methods*, Mathématiques et Applications 69, DOI 10.1007/978-3-642-22980-0_2,
© Springer-Verlag Berlin Heidelberg 2012

We observe that it is possible to consider instead of (2.1a) the PDE

$$\nabla\cdot(\beta u) + \tilde{\mu} u = f \quad \text{in } \Omega, \tag{2.3}$$

with the reaction coefficient $\tilde{\mu} := \mu - \nabla\cdot\beta$ so that, at least formally, $\nabla\cdot(\beta u) + \tilde{\mu} u = \beta\cdot\nabla u + \mu u$. The advection operator in (2.1a) is in *nonconservative* form, whereas, in (2.3), it is in *conservative* form. In the present framework of graph spaces, both forms are equivalent regarding the well-posedness of the weak formulation and the design and analysis of the dG approximation. Therefore, we focus on (2.1a), and occasionally indicate equivalent reformulations associated with (2.3).

2.1 The Continuous Setting

The purpose of this section is to specify the assumptions on the data for the model problem (2.1), to formulate this problem in weak form, and to show that it is well-posed. Most of the material is drawn from Ern and Guermond [142]; we also refer the reader to the earlier work of Bardos [28].

2.1.1 Assumptions on the Data

Concerning the data μ and β, we assume that

$$\mu \in L^\infty(\Omega), \qquad \beta \in [\text{Lip}(\Omega)]^d, \tag{2.4}$$

where $\text{Lip}(\Omega)$ denotes the space spanned by Lipschitz continuous functions, that is, $v \in \text{Lip}(\Omega)$ means that there is L_v such that, for all $x, y \in \Omega$, $|v(x) - v(y)| \leq L_v|x-y|$ where $|x-y|$ denotes the Euclidean norm of $(x-y)$ in $\mathbb{R}^d$. The quantity L_v is called the *Lipschitz module* of v. Since $\beta \in [\text{Lip}(\Omega)]^d$, there holds (see, e.g., Brenner and Scott [54, Chap. 1] for further insight)

$$\beta \in [W^{1,\infty}(\Omega)]^d$$

with $\|\nabla\beta_i\|_{[L^\infty(\Omega)]^d} \leq L_{\beta_i}$ for all $i \in \{1,\ldots,d\}$, $(\beta_1,\ldots,\beta_d)$ being the components of β in the Cartesian basis of $\mathbb{R}^d$. In what follows, we set

$$L_\beta := \max_{1\leq i \leq d} L_{\beta_i}. \tag{2.5}$$

The assumption on the regularity of β can be weakened (at least, a bound on $\|\beta\|_{[L^\infty(\Omega)]^d}$ and on $\|\nabla\cdot\beta\|_{L^\infty(\Omega)}$ is needed), but the fact that all the components of β are Lipschitz continuous functions simplifies the presentation.

In addition to (2.4), we assume that there is a real number $\mu_0 > 0$ such that

$$\Lambda := \mu - \frac{1}{2}\nabla\cdot\beta \geq \mu_0 \quad \text{a.e. in } \Omega. \tag{2.6}$$

Concerning the source term f, we assume that

$$f \in L^2(\Omega).$$

2.1. The Continuous Setting

Finally, we recall that Ω is a polyhedron in $\mathbb{R}^d$ (cf. Definition 1.6). This assumption is solely made to facilitate the meshing of Ω and is not used in the continuous setting.

In what follows, we consider a reference time τ_c and a reference velocity β_c defined as

$$\tau_c := \{\max(\|\mu\|_{L^\infty(\Omega)}, L_\beta)\}^{-1}, \qquad \beta_c := \|\beta\|_{[L^\infty(\Omega)]^d}. \tag{2.7}$$

Since μ and L_β scale as the reciprocal of a time, τ_c can be interpreted as the (fastest) time scale in the problem. Moreover, β_c represents the maximum velocity. We observe that τ_c is finite since $\|\mu\|_{L^\infty(\Omega)} = L_\beta = 0$ implies $\Lambda = 0$ which contradicts (2.6). We keep track of the parameters τ_c and β_c in the convergence analysis of dG approximations. This allows us to work with norms consisting of terms having the same physical dimension. Keeping track of these parameters is also useful when dealing with singularly perturbed regimes. For simplicity, the reader can assume that both parameters are of order unity and discard them in what follows.

2.1.2 The Graph Space

Our first goal is to specify the functional space in which the solution to the model problem (2.1) is sought. Let $C_0^\infty(\Omega)$ denote the space of infinitely differentiable functions with compact support in Ω and recall that this space is dense in $L^2(\Omega)$. For a function $v \in L^2(\Omega)$, the statement $\beta \cdot \nabla v \in L^2(\Omega)$ means that the linear form

$$C_0^\infty(\Omega) \ni \varphi \longmapsto -\int_\Omega v \nabla \cdot (\beta \varphi) \in \mathbb{R}$$

is bounded in $L^2(\Omega)$, that is, there is C_v such that

$$\forall \varphi \in C_0^\infty(\Omega), \qquad \int_\Omega v \nabla \cdot (\beta \varphi) \leq C_v \|\varphi\|_{L^2(\Omega)}.$$

The function $\beta \cdot \nabla v$ is then defined as the function representing this linear form in $L^2(\Omega)$ by means of the Riesz–Fréchet theorem.

Definition 2.1 (Graph space). *The graph space is defined as*

$$V := \{v \in L^2(\Omega) \mid \beta \cdot \nabla v \in L^2(\Omega)\}, \tag{2.8}$$

and is equipped with the natural scalar product: For all $v, w \in V$,

$$(v, w)_V := (v, w)_{L^2(\Omega)} + (\beta \cdot \nabla v, \beta \cdot \nabla w)_{L^2(\Omega)}, \tag{2.9}$$

and the associated graph norm $\|v\|_V = (v,v)_V^{1/2}$.

Proposition 2.2 (Hilbertian structure of graph space). *The graph space V defined by (2.8) and equipped with the scalar product (2.9) is a Hilbert space.*

Proof. Let $(v_n)_{n\in\mathbb{N}}$ be a Cauchy sequence in V. Then, $(v_n)_{n\in\mathbb{N}}$ and $(\beta\cdot\nabla v_n)_{n\in\mathbb{N}}$ are Cauchy sequences in $L^2(\Omega)$. Let v and w be their respective limits in $L^2(\Omega)$. Let $\varphi \in C_0^\infty(\Omega)$. Then, by definition, we obtain, for all $n \in \mathbb{N}$,

$$\int_\Omega v_n \nabla\cdot(\beta\varphi) = -\int_\Omega (\beta\cdot\nabla v_n)\varphi,$$

so that

$$\int_\Omega v\nabla\cdot(\beta\varphi) \leftarrow \int_\Omega v_n\nabla\cdot(\beta\varphi) = -\int_\Omega (\beta\cdot\nabla v_n)\varphi \to -\int_\Omega w\varphi.$$

This means that $v \in V$ with $\beta\cdot\nabla v = w$. □

Remark 2.3 (Conservative form (2.3)). When working with (2.3), the natural graph space is defined as

$$\tilde{V} := \left\{v \in L^2(\Omega) \mid \nabla\cdot(\beta v) \in L^2(\Omega)\right\},$$

the second assertion meaning that the linear form

$$C_0^\infty(\Omega) \ni \varphi \longmapsto -\int_\Omega v(\beta\cdot\nabla\varphi) \in \mathbb{R}$$

is bounded in $L^2(\Omega)$. Since

$$-\int_\Omega v(\beta\cdot\nabla\varphi) = -\int_\Omega v\nabla\cdot(\beta\varphi) + \int_\Omega (\nabla\cdot\beta)v\varphi,$$

and the last term is always bounded by $\|\nabla\cdot\beta\|_{L^\infty(\Omega)}\|v\|_{L^2(\Omega)}\|\varphi\|_{L^2(\Omega)}$, it is clear that $\tilde{V} = V$. Hence, all the developments presented herein can be applied to the conservative form (2.3).

Remark 2.4 (Gelfand triple). Since $C_0^\infty(\Omega) \subset V \subset L^2(\Omega)$ and $C_0^\infty(\Omega)$ is dense in $L^2(\Omega)$, V is dense in $L^2(\Omega)$. Hence, denoting by V' the topological dual space of V (spanned by the continuous linear forms on V), we are in the situation where

$$V \hookrightarrow L^2(\Omega) \equiv L^2(\Omega)' \hookrightarrow V',$$

with dense and continuous injections. In the literature, the triple $\{V, L^2(\Omega), V'\}$ is sometimes referred to as a Gelfand triple.

2.1.3 Traces in the Graph Space

The next step is to specify mathematically the meaning of the boundary condition (2.1b). To this purpose, we need to investigate the trace on $\partial\Omega$ of functions in the graph space V. Our aim is to give a meaning to such traces in the space

$$L^2(|\beta\cdot n|; \partial\Omega) := \left\{v \text{ is measurable on } \partial\Omega \mid \int_{\partial\Omega} |\beta\cdot n|v^2 < \infty\right\}. \tag{2.10}$$

2.1. The Continuous Setting

Recalling definition (2.2) of the inflow boundary, we also define the *outflow boundary* as
$$\partial\Omega^+ := \{x \in \partial\Omega \mid \beta(x)\cdot\mathrm{n}(x) > 0\},$$
and following [142], we assume that the inflow and outflow boundaries are well-separated, namely
$$\mathrm{dist}(\partial\Omega^-,\partial\Omega^+) := \min_{(x,y)\in\partial\Omega^-\times\partial\Omega^+} |x-y| > 0.$$

The following result is very important since it allows us to define traces of functions belonging to the graph space and to use an integration by parts formula. The proof is postponed to Sect. 2.1.5.

Lemma 2.5 (Traces and integration by parts). *In the above framework, the trace operator*
$$\gamma : C^0(\overline{\Omega}) \ni v \longmapsto \gamma(v) := v|_{\partial\Omega} \in L^2(|\beta\cdot\mathrm{n}|;\partial\Omega)$$
extends continuously to V, meaning that there is C_γ such that, for all $v \in V$,
$$\|\gamma(v)\|_{L^2(|\beta\cdot\mathrm{n}|;\partial\Omega)} \le C_\gamma \|v\|_V.$$
Moreover, the following integration by parts formula holds true: For all $v, w \in V$,
$$\int_\Omega [(\beta\cdot\nabla v)w + (\beta\cdot\nabla w)v + (\nabla\cdot\beta)vw] = \int_{\partial\Omega} (\beta\cdot\mathrm{n})\gamma(v)\gamma(w).$$

To alleviate the notation, we omit henceforth the trace operator γ when writing boundary integrals, so that the above integration by parts formula becomes, for all $v, w \in V$,
$$\int_\Omega [(\beta\cdot\nabla v)w + (\beta\cdot\nabla w)v + (\nabla\cdot\beta)vw] = \int_{\partial\Omega} (\beta\cdot\mathrm{n})vw. \tag{2.11}$$

Remark 2.6 (Counter-example for inflow/outflow separation). The separation assumption on inflow and outflow boundaries cannot be circumvented if we wish to work with traces in $L^2(|\beta\cdot\mathrm{n}|;\partial\Omega)$. Consider for instance the triangular domain
$$\Omega = \{(x_1, x_2) \in \mathbb{R}^2 \mid 0 < x_2 < 1 \text{ s.t. } |x_1| < x_2\},$$
and the field $\beta = (1,0)^t$; cf. Fig. 2.1. Then, the function $u(x_1, x_2) = x_2^\alpha$ is in V provided $\alpha > -1$, but it has a trace in $L^2(|\beta\cdot\mathrm{n}|;\partial\Omega)$ only if $\alpha > -1/2$. Indeed, a direct calculation shows that
$$\int_\Omega u^2 \,\mathrm{d}x_1\,\mathrm{d}x_2 = \int_0^1 x_2^{2\alpha} \left(\int_{-x_2}^{x_2} \mathrm{d}x_1 \right) \mathrm{d}x_2 = \int_0^1 2x_2^{2\alpha+1} \,\mathrm{d}x_2,$$
which is finite if and only if $2\alpha + 1 > -1$, that is, $\alpha > -1$. Hence, under this condition, $u \in L^2(\Omega)$, and since $\beta\cdot\nabla u = 0$, we infer $u \in V$. Furthermore, $\partial\Omega^- = \{(x_1, x_2) \in \mathbb{R}^2 \mid x_2 = -x_1, x_1 \in (-1, 0)\}$, and a direct calculation yields
$$\int_{\partial\Omega^-} |\beta\cdot\mathrm{n}|u^2 = \int_0^1 x_2^{2\alpha} \,\mathrm{d}x_2,$$
which is finite if and only if $2\alpha > -1$, that is, $\alpha > -1/2$. The same condition is obtained for the integral over $\partial\Omega^+$.

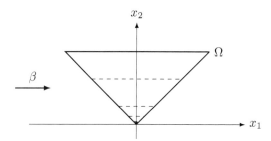

Fig. 2.1: Counter-example for inflow/outflow separation: constant advection field β, triangular domain Ω, and, in *dashed lines*, contour lines of u

2.1.4 Weak Formulation and Well-Posedness

The goal of this section is to derive a well-posed weak formulation of the model problem (2.1).

For a real number x, we define its positive and negative parts respectively as

$$x^\oplus := \frac{1}{2}(|x|+x), \qquad x^\ominus := \frac{1}{2}(|x|-x). \tag{2.12}$$

We observe that both quantities are, by definition, nonnegative. We introduce the following bilinear form: For all $v, w \in V$,

$$a(v,w) := \int_\Omega \mu v w + \int_\Omega (\beta \cdot \nabla v) w + \int_{\partial\Omega} (\beta \cdot n)^\ominus v w. \tag{2.13}$$

This bilinear form is bounded in $V \times V$ owing to Lemma 2.5. Precisely, for all $v, w \in V$, the Cauchy–Schwarz inequality yields

$$|a(v,w)| \leq (1+\|\mu\|^2_{L^\infty(\Omega)})^{1/2} \|v\|_V \|w\|_{L^2(\Omega)} + C_\gamma^2 \|v\|_V \|w\|_V.$$

Using the graph space V and the bilinear form a, the model problem (2.1) can be cast into the weak form

$$\text{Find } u \in V \text{ s.t. } a(u,w) = \int_\Omega fw \text{ for all } w \in V. \tag{2.14}$$

This problem turns out to be well-posed (cf. Theorem 2.9 and Sect. 2.1.5). Before addressing this, we examine in which sense does a solution to (2.14) solve the original problem (2.1). In particular, we observe that the boundary condition is weakly enforced in (2.14).

Proposition 2.7 (Characterization of the solution to (2.14)). *Assume that $u \in V$ solves (2.14). Then,*

$$\beta \cdot \nabla u + \mu u = f \quad \text{a.e. in } \Omega, \tag{2.15}$$

$$u = 0 \quad \text{a.e. in } \partial\Omega^-. \tag{2.16}$$

Proof. Taking $w = \varphi \in C_0^\infty(\Omega)$ in (2.14) yields
$$\int_\Omega (\mu u + \beta \cdot \nabla u - f)\varphi = 0,$$
whence (2.15) follows from the density of $C_0^\infty(\Omega)$ in $L^2(\Omega)$. Using (2.15) in (2.14) implies that, for all $w \in V$,
$$\int_{\partial\Omega} (\beta \cdot n)^\ominus uw = 0.$$
Using $w = u$ as test function yields $\int_{\partial\Omega} (\beta \cdot n)^\ominus u^2 = 0$, whence (2.16). $\square$

An important (yet, not sufficient) ingredient for the well-posedness of the weak problem (2.14) is the L^2-coercivity of a in the graph space V.

Lemma 2.8 (L^2-coercivity of a). *The bilinear form a defined by (2.13) is L^2-coercive on V, namely,*
$$\forall v \in V, \qquad a(v,v) \geq \mu_0 \|v\|^2_{L^2(\Omega)} + \int_{\partial\Omega} \frac{1}{2}|\beta \cdot n|v^2. \qquad (2.17)$$

Proof. This is a straightforward consequence of assumption (2.6) and the integration by parts formula (2.11) since, for all $v \in V$,
$$a(v,v) = \int_\Omega \left(\mu - \frac{1}{2}\nabla\cdot\beta\right)v^2 + \int_{\partial\Omega} \frac{1}{2}(\beta\cdot n)v^2 + \int_{\partial\Omega} (\beta\cdot n)^\ominus v^2$$
$$= \int_\Omega \Lambda v^2 + \int_{\partial\Omega} \frac{1}{2}|\beta\cdot n|v^2 \geq \mu_0 \|v\|^2_{L^2(\Omega)} + \int_{\partial\Omega} \frac{1}{2}|\beta\cdot n|v^2,$$
completing the proof. $\square$

A consequence of Lemma 2.8 is that the weak problem (2.14) admits at most one solution. We are now in a position to state the main result of this section. The proof of existence is postponed to Sect. 2.1.5.

Theorem 2.9 (Well-posedness). *Problem (2.14) is well-posed.*

Remark 2.10 (Conservative form (2.3)). When working with (2.3), the bilinear form a can be rewritten as
$$a(v,w) = \int_\Omega \tilde{\mu} vw + \int_\Omega \nabla\cdot(\beta v)w + \int_{\partial\Omega} (\beta\cdot n)^\ominus vw.$$

2.1.5 Proof of Main Results

In this section, we prove Lemma 2.5 and Theorem 2.9.

Proof of Lemma 2.5. Owing to the separation assumption on inflow and outflow boundaries, there are two functions ψ^- and ψ^+ in $C^\infty(\overline{\Omega})$ such that

$$\psi^- + \psi^+ \equiv 1 \text{ in } \overline{\Omega}, \quad \psi^-|_{\partial\Omega^+} = 0, \quad \psi^+|_{\partial\Omega^-} = 0. \tag{2.18}$$

Let $v \in C^\infty(\overline{\Omega})$. Observe that

$$\int_{\partial\Omega} v^2 |\beta \cdot n| = \int_{\partial\Omega} v^2(\psi^- + \psi^+)|\beta \cdot n| = \int_{\partial\Omega^-} v^2 \psi^- |\beta \cdot n| + \int_{\partial\Omega^+} v^2 \psi^+ |\beta \cdot n|$$

$$= \int_{\partial\Omega} (\psi^+ - \psi^-) v^2 (\beta \cdot n) = \int_\Omega \nabla \cdot (v^2 (\psi^+ - \psi^-) \beta)$$

$$= \int_\Omega \nabla \cdot ((\psi^+ - \psi^-)\beta) v^2 + \int_\Omega 2(\psi^+ - \psi^-)(\beta \cdot \nabla v) v$$

$$\leq C_\gamma^2 \|v\|_V^2,$$

with $C_\gamma^2 = \|\nabla \cdot ((\psi^+ - \psi^-)\beta)\|_{L^\infty(\Omega)} + \|\psi^+ - \psi^-\|_{L^\infty(\Omega)}$. Hence, for all $v \in C^\infty(\overline{\Omega})$,

$$\|v\|_{L^2(|\beta \cdot n|; \partial\Omega)} \leq C_\gamma \|v\|_V.$$

Let now $v \in V$. Since the space $C^\infty(\overline{\Omega})$ is dense in V (see Jensen [201, Theorem 4, p. 21]), there is a sequence $(v_n)_{n\in\mathbb{N}}$ in $C^\infty(\overline{\Omega})$ converging to v in V. The inequality

$$\|\gamma(v_n)\|_{L^2(|\beta \cdot n|; \partial\Omega)} \leq C_\gamma \|v_n\|_V,$$

implies that $(\gamma(v_n))_{n\in\mathbb{N}}$ is a Cauchy sequence in $L^2(|\beta \cdot n|; \partial\Omega)$. Letting $\gamma(v)$ be its limit, it is clear that this defines a continuous extension of the trace operator γ to V. Finally, the integration by parts formula (2.11) is also proven by a density argument since it holds true on $C^\infty(\overline{\Omega})$ and all the terms pass to the limit in V. This concludes the proof. □

Proof of Theorem 2.9. The proof proceeds in four steps.
(i) We first prove that a variant of (2.14), where the boundary condition is strongly enforced, is well-posed. Specifically, letting $V_0 = \{v \in V \mid v|_{\partial\Omega^-} = 0\}$ and using $L^2(\Omega)$ as test space, we consider the problem:

$$\text{Find } u \in V_0 \text{ s.t. } a_0(u, w) = \int_\Omega fw \text{ for all } w \in L^2(\Omega), \tag{2.19}$$

with the bilinear form

$$a_0(v, w) = \int_\Omega \mu vw + \int_\Omega (\beta \cdot \nabla v) w. \tag{2.20}$$

To prove that (2.19) is well-posed, we use the BNB Theorem (cf. Theorem 1.1) with spaces $X = V_0$ and $Y = L^2(\Omega)$. Since V_0 is, by construction, closed in V owing to Lemma 2.5, V_0 is a Hilbert space. Moreover, $L^2(\Omega)$ is reflexive, and the right-hand side in (2.19) is a bounded linear form in $L^2(\Omega)$. It remains to

2.1. The Continuous Setting

verify conditions (1.4) and (1.5) of the BNB Theorem.
(ii) Proof of condition (1.4). Let $v \in V_0$ and set

$$\mathbb{S} = \sup_{w \in L^2(\Omega) \setminus \{0\}} \frac{a_0(v, w)}{\|w\|_{L^2(\Omega)}}.$$

Proceeding as in the proof of Lemma 2.8, we observe that, for all $v \in V_0$,

$$a_0(v, v) = \int_\Omega \left(\mu - \frac{1}{2}\nabla \cdot \beta\right) v^2 + \int_{\partial\Omega} \frac{1}{2}(\beta \cdot \mathrm{n})v^2$$

$$\geq \mu_0 \|v\|^2_{L^2(\Omega)} + \int_{\partial\Omega} \frac{1}{2}(\beta \cdot \mathrm{n})v^2$$

$$= \mu_0 \|v\|^2_{L^2(\Omega)} + \int_{\partial\Omega^+} \frac{1}{2}(\beta \cdot \mathrm{n})v^2 \geq \mu_0 \|v\|^2_{L^2(\Omega)},$$

since $v|_{\partial\Omega^-} = 0$. As a result, for $v \neq 0$,

$$\|v\|^2_{L^2(\Omega)} \leq \mu_0^{-1} a_0(v, v) \leq \mu_0^{-1} \frac{a_0(v, v)}{\|v\|_{L^2(\Omega)}} \|v\|_{L^2(\Omega)} \leq \mu_0^{-1} \mathbb{S} \|v\|_{L^2(\Omega)},$$

whence $\|v\|_{L^2(\Omega)} \leq \mu_0^{-1} \mathbb{S}$ for all $v \in V_0$. Moreover,

$$\|\beta \cdot \nabla v\|_{L^2(\Omega)} = \sup_{w \in L^2(\Omega) \setminus \{0\}} \frac{\int_\Omega (\beta \cdot \nabla v) w}{\|w\|_{L^2(\Omega)}} = \sup_{w \in L^2(\Omega) \setminus \{0\}} \frac{a_0(v, w) - \int_\Omega \mu v w}{\|w\|_{L^2(\Omega)}}$$

$$\leq \mathbb{S} + \|\mu\|_{L^\infty(\Omega)} \|v\|_{L^2(\Omega)} \leq (1 + \mu_0^{-1} \|\mu\|_{L^\infty(\Omega)}) \mathbb{S}.$$

Collecting these bounds yields

$$\|v\|_V^2 \leq (\mu_0^{-2} + (1 + \mu_0^{-1} \|\mu\|_{L^\infty(\Omega)})^2) \mathbb{S}^2,$$

whence we infer condition (1.4).
(iii) Proof of condition (1.5). Let $w \in L^2(\Omega)$ be such that $a_0(v, w) = 0$ for all $v \in V_0$. Since $C_0^\infty(\Omega) \subset V_0$, a distributional argument yields $\mu w - \nabla \cdot (\beta w) = 0$ in Ω. Hence, $\beta \cdot \nabla w = (\mu - \nabla \cdot \beta) w \in L^2(\Omega)$, so that $w \in V$. Combining this expression for $\beta \cdot \nabla w$ with the integration by parts formula (2.11), we infer that, for all $v \in V_0$,

$$\int_{\partial\Omega} (\beta \cdot \mathrm{n}) v w = \int_\Omega [(\beta \cdot \nabla v) w + (\beta \cdot \nabla w) v + (\nabla \cdot \beta) v w]$$

$$= a_0(v, w) - \int_\Omega (\mu - \nabla \cdot \beta) v w + \int_\Omega (\beta \cdot \nabla w) v = a_0(v, w) = 0.$$

Taking $v = \psi^+ w$ with ψ^+ defined by (2.18) (so that $v \in V_0$) yields $\int_{\partial\Omega} (\beta \cdot \mathrm{n})^\oplus w^2 = 0$ so that $w|_{\partial\Omega^+} = 0$. Finally, since $\mu w - \nabla \cdot (\beta w) = 0$, we observe that

$$0 = \int_\Omega [\mu w^2 - \nabla \cdot (\beta w) w] = \int_\Omega \left(\mu - \frac{1}{2} \nabla \cdot \beta\right) w^2 - \int_{\partial\Omega} \frac{1}{2} (\beta \cdot \mathrm{n}) w^2 \geq \mu_0 \|w\|^2_{L^2(\Omega)},$$

where we have used the fact that $\int_{\partial\Omega}(\beta\cdot n)w^2 \leq 0$ since $w|_{\partial\Omega^+} = 0$. The above inequality implies that $w = 0$, thereby completing the proof of condition (1.5).
(iv) Let us finally prove that (2.14) is well-posed. Existence of a solution results from the fact that u, the unique solution to (2.19), solves (2.14) since $u \in V_0$ implies that, for all $w \in V$, $a(u,w) = a_0(u,w)$. Moreover, uniqueness of the solution has already been deduced from the L^2-coercivity of a on V. □

2.1.6 Nonhomogeneous Boundary Condition

In this section, we consider the nonhomogeneous boundary condition

$$u = g \quad \text{on } \partial\Omega^-.$$

We extend the boundary datum g to $\partial\Omega$ by setting it to zero outside $\partial\Omega^-$ and we assume that

$$g \in L^2(|\beta\cdot n|; \partial\Omega).$$

The model problem in weak form now becomes:

$$\text{Find } u \in V \text{ s.t. } a(u,w) = \int_\Omega fw + \int_{\partial\Omega} (\beta\cdot n)^\ominus gw \text{ for all } w \in V. \qquad (2.21)$$

The key result to investigate nonhomogeneous boundary conditions is the surjectivity of the trace operator $\gamma : V \to L^2(|\beta\cdot n|; \partial\Omega)$ identified in Lemma 2.5.

Lemma 2.11 (Surjectivity of traces). *For all $g \in L^2(|\beta\cdot n|; \partial\Omega)$, there is $u_g \in V$ such that $u_g = g$ a.e. in $\partial\Omega^- \cup \partial\Omega^+$. Moreover, there is C, only depending on Ω and β, such that $\|u_g\|_V \leq C\|g\|_{L^2(|\beta\cdot n|;\partial\Omega)}$.*

Proof. Let $g \in L^2(|\beta\cdot n|; \partial\Omega)$ and let $\psi_g : V \to \mathbb{R}$ be the linear map such that, for all $w \in V$,

$$\psi_g(w) = \int_{\partial\Omega} (\beta\cdot n)gw.$$

Using the Cauchy–Schwarz inequality and the fact that $\|w\|_{L^2(|\beta\cdot n|;\partial\Omega)} \leq C_\gamma \|w\|_V$ owing to Lemma 2.5, we infer that

$$|\psi_g(w)| \leq \|g\|_{L^2(|\beta\cdot n|;\partial\Omega)}\|w\|_{L^2(|\beta\cdot n|;\partial\Omega)}$$
$$\leq C_\gamma \|g\|_{L^2(|\beta\cdot n|;\partial\Omega)}\|w\|_V.$$

Hence, $\psi_g \in V'$ and $\|\psi_g\|_{V'} \leq C_\gamma \|g\|_{L^2(|\beta\cdot n|;\partial\Omega)}$. Owing to the Riesz–Fréchet representation theorem, there exists $z \in V$ such that, for all $w \in V$,

$$(z,w)_V = \int_\Omega zw + \int_\Omega (\beta\cdot\nabla z)(\beta\cdot\nabla w) = \int_{\partial\Omega} (\beta\cdot n)gw.$$

Set $u_g := \beta\cdot\nabla z \in L^2(\Omega)$ and let us verify that $u_g \in V$. Taking $w = \varphi \in C_0^\infty(\Omega)$ yields

$$\int_\Omega u_g(\beta\cdot\nabla\varphi) = -\int_\Omega z\varphi,$$

so that
$$\int_\Omega u_g \nabla\cdot(\beta\varphi) = -\int_\Omega z\varphi + \int_\Omega (\nabla\cdot\beta) u_g \varphi.$$
Since the right-hand side is a bounded linear form in $L^2(\Omega)$ acting on φ, we infer that $\beta\cdot\nabla u_g = z - (\nabla\cdot\beta)u_g \in L^2(\Omega)$, so that $u_g \in V$. Furthermore, it is easily seen that
$$\|u_g\|_V \leq C'\|z\|_V = C'\|\psi_g\|_{V'} \leq C\|g\|_{L^2(|\beta\cdot n|;\partial\Omega)},$$
where C' and C only depend on Ω and β. Moreover, for all $w \in V$, we obtain from the integration by parts formula (2.11) that
$$\int_{\partial\Omega} (\beta\cdot n) u_g w = \int_\Omega (\beta\cdot\nabla w) u_g + \int_\Omega (\beta\cdot\nabla u_g) w + \int_\Omega (\nabla\cdot\beta) u_g w$$
$$= \int_\Omega (\beta\cdot\nabla w)(\beta\cdot\nabla z) + \int_\Omega zw = \int_{\partial\Omega} (\beta\cdot n) g w.$$
The fact that $\int_{\partial\Omega}(\beta\cdot n)(u_g - g)w = 0$ for all $w \in V$ and the density of $C^\infty(\overline\Omega)$ in V imply that $u_g = g$ a.e. in $\partial\Omega^- \cup \partial\Omega^+$. □

Theorem 2.12 (Well-posedness). *Problem (2.21) is well-posed. Moreover, its unique solution $u \in V$ satisfies (2.15) and $u = g$ a.e. in $\partial\Omega^-$.*

Proof. Let u_g be given by Lemma 2.11. We consider the problem:
$$\text{Find } v \in V \text{ s.t. } a(v,w) = \int_\Omega fw - a_0(u_g, w) \text{ for all } w \in V, \qquad (2.22)$$
where the bilinear form a_0 is defined by (2.20). The map $V \ni w \longmapsto a_0(u_g, w) \in \mathbb{R}$ is bounded in $L^2(\Omega)$ since, for all $w \in L^2(\Omega)$,
$$|a_0(u_g, w)| \leq (1 + \|\mu\|^2_{L^\infty(\Omega)})^{1/2} \|u_g\|_V \|w\|_{L^2(\Omega)} \leq C\|g\|_{L^2(|\beta\cdot n|;\partial\Omega)} \|w\|_{L^2(\Omega)}.$$
Hence, by the Riesz–Fréchet representation theorem, the right-hand side of (2.22) can be written as $\int_\Omega \tilde f w$ for some $\tilde f \in L^2(\Omega)$. Owing to Theorem 2.9, problem (2.22) is therefore well-posed. Proceeding as in the proof of Proposition 2.7, we infer that the function $u = v + u_g$ satisfies $\mu u + \beta\cdot\nabla u = f$ in Ω, and, since $v = 0$ and $u_g = g$ on $\partial\Omega^-$, we obtain $u = g$ on $\partial\Omega^-$. This proves the existence of a solution to (2.21), while uniqueness results, as before, from the L^2-coercivity of a in V. □

2.2 Centered Fluxes

The goal of this section is to design and analyze the simplest dG method to approximate the model problem (2.14). Referring the reader to Sect. 1.3, the method is designed so as to be consistent, and a minimal discrete stability is ensured by L^2-coercivity. Using the terminology of Definition 1.54, the resulting error estimate turns out to be suboptimal. Alternatively, the method can be viewed as based on the use of centered fluxes. As in the previous section, we assume that the data μ and β satisfy (2.4) and (2.6).

We seek an approximate solution in the broken polynomial space $\mathbb{P}_d^k(\mathcal{T}_h)$ defined by (1.15); other choices are possible. We assume that $k \geq 1$ and that $\mathcal{T}_h$ belongs to an admissible mesh sequence. We set

$$V_h := \mathbb{P}_d^k(\mathcal{T}_h)$$

and consider the discrete problem:

$$\text{Find } u_h \in V_h \text{ s.t. } a_h(u_h, v_h) = \int_\Omega f v_h \text{ for all } v_h \in V_h,$$

for a discrete bilinear form a_h yet to be designed.

To analyze the method, we make a slightly more stringent regularity assumption on the exact solution u than just belonging to the graph space V. This assumption is needed to formulate the consistency of the method by directly plugging in the exact solution into the discrete bilinear form a_h. In particular, we need to consider the trace of the exact solution on each mesh face.

Assumption 2.13 (Regularity of exact solution and space V_*). *We assume that there is a partition $P_\Omega = \{\Omega_i\}_{1 \leq i \leq N_\Omega}$ of Ω into disjoint polyhedra such that, for the exact solution u,*

$$u \in V_* := V \cap H^1(P_\Omega).$$

In the spirit of Sect. 1.3, *we set* $V_{*h} := V_* + V_h$.

Owing to the trace inequality (1.18) with $p = 2$, Assumption 2.13 implies that, for all $T \in \mathcal{T}_h$, the restriction $u|_T$ has traces a.e. on each face $F \in \mathcal{F}_T$, and these traces belong to $L^2(F)$ (even if the mesh $\mathcal{T}_h$ is not fitted to P_Ω). Weaker regularity assumptions on u can be made; cf. Remark 2.16 below. We now examine the jumps of u across interfaces.

Lemma 2.14 (Jumps of u across interfaces). *The exact solution $u \in V_*$ is such that, for all $F \in \mathcal{F}_h^i$,*

$$(\beta \cdot n_F)[\![u]\!](x) = 0 \quad \text{for a.e. } x \in F. \tag{2.23}$$

Proof. Let $F \in \mathcal{F}_h^i$ with $F = \partial T_1 \cap \partial T_2$. This interface can be partitioned into a finite number of subsets $\{F_j\}_{1 \leq j \leq N_F}$ such that each F_j is shared by at most two elements of the partition P_Ω. For instance, in the situation depicted in Fig. 2.2, the interface F is partitioned into two subsets, one of which belongs to only one element of the partition, while the other belongs to two elements. We now prove that, a.e. on each F_j, (2.23) holds true. We consider the case where F_j is shared by two elements of the partition, say Ω_1 and Ω_2 (the case where F_j belongs to only one element is similar). Let $\varphi \in C_0^\infty(\Omega)$ with support intersecting only F_j and only Ω_1 and Ω_2; cf. Fig. 2.2. Since $\varphi \in C_0^\infty(\Omega)$ and $u \in V$, there holds

$$\int_\Omega \left\{ (\nabla \cdot \beta) u \varphi + (\beta \cdot \nabla u) \varphi + u(\beta \cdot \nabla \varphi) \right\} = 0.$$

2.2. Centered Fluxes

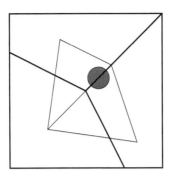

Fig. 2.2: Interface $F \in \mathcal{F}_h^i$ separating two triangles; the partition P_Ω in Assumption 2.13 consists of three polygons, and the support of the test function φ in the proof of Lemma 2.14 is represented by a *grey ball*

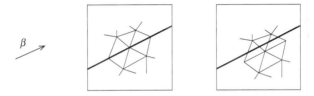

Fig. 2.3: Fitted (*left*) and unfitted (*right*) simplicial mesh; the partition P_Ω consists of two polygons, and the exact solution can jump across the *thick line*

Moreover, owing to the choice for the support of φ, the integral over Ω is the sum of the integrals over $T_1 \cap \Omega_1$ and $T_2 \cap \Omega_2$. On both of these sets, u is smooth enough to integrate by parts, yielding

$$0 = \int_\Omega \{\ldots\} = \int_{T_1 \cap \Omega_1} \{\ldots\} + \int_{T_2 \cap \Omega_2} \{\ldots\} = \int_{F_j} (\beta \cdot n_F)[\![u]\!]\varphi.$$

The assertion then follows from a density argument. $\square$

Remark 2.15 (Singularities of exact solution). Condition (2.23) does not say anything on the jumps of the exact solution across interfaces to which the advective velocity β is tangential. We also observe that Assumption 2.13 does not require the mesh to be fitted to solution singularities, that is, both situations depicted in Fig. 2.3 are admissible.

Remark 2.16 (Weaker regularity assumption). A weaker regularity assumption on the exact solution is $u \in V \cap H^{1/2+\epsilon}(P_\Omega)$, $\epsilon > 0$, since this assumption suffices to assert the existence of traces in $L^2(F)$ for all $F \in \mathcal{F}_h$. Another possibility is to assume that

$$u \in V \cap W^{1,1}(P_\Omega),$$

which yields the existence of traces in $L^1(F)$ for all $F \in \mathcal{F}_h$. This latter property is sufficient to prove Lemma 2.14; however, the norm introduced to assert boundedness in the convergence analysis of Sect. 2.2.2 needs to be modified following the ideas presented in Sect. 4.2.5.

Remark 2.17 (Nonhomogeneous boundary condition). When working with the nonhomogeneous boundary condition $u = g$ on $\partial\Omega^-$ (cf. Sect. 2.1.6), the discrete problem becomes

$$\text{Find } u_h \in V_h \text{ s.t. } a_h(u_h, v_h) = \int_\Omega f v_h + \int_{\partial\Omega} (\beta{\cdot}\mathrm{n})^\ominus g v_h \text{ for all } v_h \in V_h.$$

2.2.1 Heuristic Derivation

The main idea in the design of the discrete bilinear form a_h is to mimic at the discrete level the L^2-coercivity that holds at the continuous level (cf. (2.17)), while, at the same time, ensuring consistency. Our starting point is a discrete bilinear form $a_h^{(0)}$ simply derived from the exact bilinear form a by replacing the advective derivative $\beta{\cdot}\nabla$ by its broken counterpart $\beta{\cdot}\nabla_h$, namely, we define on $V_{*h} \times V_h$,

$$a_h^{(0)}(v, w_h) := \int_\Omega \left\{ \mu v w_h + (\beta{\cdot}\nabla_h v) w_h \right\} + \int_{\partial\Omega} (\beta{\cdot}\mathrm{n})^\ominus v w_h.$$

That $a_h^{(0)}$ yields consistency is clear since the exact solution satisfies (2.15) and (2.16).

Let us now focus on discrete coercivity. An important observation is that this property is not transferred from a to $a_h^{(0)}$. Indeed, integration by parts on each mesh element yields, for all $v_h \in V_h$,

$$\begin{aligned} a_h^{(0)}(v_h, v_h) &= \int_\Omega \left\{ \mu v_h^2 + (\beta{\cdot}\nabla_h v_h) v_h \right\} + \int_{\partial\Omega} (\beta{\cdot}\mathrm{n})^\ominus v_h^2 \\ &= \int_\Omega \mu v_h^2 + \sum_{T \in \mathcal{T}_h} \int_T (\beta{\cdot}\nabla v_h) v_h + \int_{\partial\Omega} (\beta{\cdot}\mathrm{n})^\ominus v_h^2 \\ &= \int_\Omega \Lambda v_h^2 + \sum_{T \in \mathcal{T}_h} \int_{\partial T} \frac{1}{2}(\beta{\cdot}\mathrm{n}_T) v_h^2 + \int_{\partial\Omega} (\beta{\cdot}\mathrm{n})^\ominus v_h^2, \end{aligned}$$

where we recall that $\Lambda = \mu - \frac{1}{2}\nabla{\cdot}\beta$ and that n_T denotes the outward normal to T on ∂T. The second term on the right-hand side can be reformulated as a sum over mesh faces. Indeed, exploiting the continuity of (the normal component of) β across interfaces leads to

$$\sum_{T \in \mathcal{T}_h} \int_{\partial T} \frac{1}{2}(\beta{\cdot}\mathrm{n}_T) v_h^2 = \sum_{F \in \mathcal{F}_h^i} \int_F \frac{1}{2}(\beta{\cdot}\mathrm{n}_F) [\![v_h^2]\!] + \sum_{F \in \mathcal{F}_h^b} \int_F \frac{1}{2}(\beta{\cdot}\mathrm{n}) v_h^2.$$

For all $F \in \mathcal{F}_h^i$ with $F = \partial T_1 \cap \partial T_2$, $v_i = v_h|_{T_i}$, $i \in \{1, 2\}$, there holds

$$\frac{1}{2}[\![v_h^2]\!] = \frac{1}{2}(v_1^2 - v_2^2) = \frac{1}{2}(v_1 - v_2)(v_1 + v_2) = [\![v_h]\!]\{\!\{v_h\}\!\}.$$

2.2. Centered Fluxes

As a result,

$$a_h^{(0)}(v_h, v_h) = \int_\Omega \Lambda v_h^2 + \sum_{F \in \mathcal{F}_h^i} \int_F (\beta \cdot n_F) [\![v_h]\!] \{\!\{v_h\}\!\}$$
$$+ \sum_{F \in \mathcal{F}_h^b} \int_F \frac{1}{2}(\beta \cdot n) v_h^2 + \int_{\partial\Omega} (\beta \cdot n)^\ominus v_h^2,$$

and combining the two rightmost terms, we arrive at

$$a_h^{(0)}(v_h, v_h) = \int_\Omega \Lambda v_h^2 + \sum_{F \in \mathcal{F}_h^i} \int_F (\beta \cdot n_F) [\![v_h]\!] \{\!\{v_h\}\!\} + \int_{\partial\Omega} \frac{1}{2} |\beta \cdot n| v_h^2.$$

The second term on the right-hand side, involving interfaces, has no sign a priori. Therefore, it must be removed, and this can be achieved while maintaining consistency if we set, for all $(v, w_h) \in V_{*h} \times V_h$,

$$a_h^{\mathrm{cf}}(v, w_h) := \int_\Omega \left\{ \mu v w_h + (\beta \cdot \nabla_h v) w_h \right\} + \int_{\partial\Omega} (\beta \cdot n)^\ominus v w_h$$
$$- \sum_{F \in \mathcal{F}_h^i} \int_F (\beta \cdot n_F) [\![v]\!] \{\!\{w_h\}\!\}, \qquad (2.24)$$

since $(\beta \cdot n_F)[\![u]\!] = 0$ for all $F \in \mathcal{F}_h^i$ owing to (2.23). The superscript indicates the use of centered fluxes, as detailed in Sect. 2.2.3.

We can now summarize the properties of the discrete bilinear form a_h^{cf} established so far. The coercivity of a_h^{cf} is expressed using the following norm defined on V_{*h}:

$$\|v\|_{\mathrm{cf}}^2 := \tau_\mathrm{c}^{-1} \|v\|_{L^2(\Omega)}^2 + \int_{\partial\Omega} \frac{1}{2} |\beta \cdot n| v^2, \qquad (2.25)$$

with the time scale τ_c defined by (2.7). We observe that $\|\cdot\|_{\mathrm{cf}}$ is indeed a norm since it controls the L^2-norm.

Lemma 2.18 (Consistency and discrete coercivity). *The discrete bilinear form a_h^{cf} defined by (2.24) satisfies the following properties:*

(i) Consistency, *namely for the exact solution $u \in V_*$,*

$$a_h^{\mathrm{cf}}(u, v_h) = \int_\Omega f v_h \qquad \forall v_h \in V_h,$$

(ii) Coercivity *on V_h with respect to the $\|\cdot\|_{\mathrm{cf}}$-norm, namely*

$$\forall v_h \in V_h, \qquad a_h^{\mathrm{cf}}(v_h, v_h) \geq C_{\mathrm{sta}} \|v_h\|_{\mathrm{cf}}^2,$$

with $C_{\mathrm{sta}} := \min(1, \tau_\mathrm{c} \mu_0)$.

Proof. Consistency has already been verified. Concerning coercivity, we observe that the above calculation yields, for all $v_h \in V_h$,

$$a_h^{\text{cf}}(v_h, v_h) = \int_\Omega \Lambda v_h^2 + \int_{\partial\Omega} \frac{1}{2}|\beta\cdot\mathbf{n}|v_h^2,$$

whence the assertion follows from assumption (2.6). □

Remark 2.19 (Definition of $\|\cdot\|_{\text{cf}}$). Defining the $\|\cdot\|_{\text{cf}}$-norm as

$$\|v\|_{\text{cf}}^2 := \mu_0 \|v\|_{L^2(\Omega)}^2 + \int_{\partial\Omega} \frac{1}{2}|\beta\cdot\mathbf{n}|v^2$$

yields coercivity in the simpler form $a_h^{\text{cf}}(v_h, v_h) \geq \|v_h\|_{\text{cf}}^2$. The present definition using the time scale τ_c instead of μ_0^{-1} is more convenient to examine the boundedness of the discrete bilinear form a_h^{cf}.

Before proceeding further, we record an equivalent expression of the discrete bilinear form a_h^{cf} obtained after integrating by parts the advective derivative in each mesh element. This expression is useful when introducing the notion of fluxes in Sect. 2.2.3 and when analyzing the dG method based on upwinding in Sect. 2.3.

Lemma 2.20 (Equivalent expression for a_h^{cf}). *For all $(v, w_h) \in V_{*h} \times V_h$, there holds*

$$a_h^{\text{cf}}(v, w_h) = \int_\Omega \left\{ (\mu - \nabla\cdot\beta)vw_h - v(\beta\cdot\nabla_h w_h) \right\} + \int_{\partial\Omega} (\beta\cdot\mathbf{n})^\oplus vw_h$$
$$+ \sum_{F \in \mathcal{F}_h^i} \int_F (\beta\cdot\mathbf{n}_F)\{v\}[\![w_h]\!]. \tag{2.26}$$

Proof. Integrating by parts in each mesh element the advective derivative in (2.24) leads to

$$a_h^{\text{cf}}(v, w_h) = \int_\Omega \left\{ (\mu - \nabla\cdot\beta)vw_h - v(\beta\cdot\nabla_h w_h) \right\} + \sum_{T \in \mathcal{T}_h} \int_{\partial T} (\beta\cdot\mathbf{n}_T)vw_h$$
$$+ \int_{\partial\Omega} (\beta\cdot\mathbf{n})^\ominus vw_h - \sum_{F \in \mathcal{F}_h^i} \int_F (\beta\cdot\mathbf{n}_F)[\![v]\!]\{w_h\}. \tag{2.27}$$

The third term on the right-hand side can be reformulated as a sum over mesh faces, namely

$$\sum_{T \in \mathcal{T}_h} \int_{\partial T} (\beta\cdot\mathbf{n}_T)vw_h = \sum_{F \in \mathcal{F}_h^i} \int_F (\beta\cdot\mathbf{n}_F)[\![vw_h]\!] + \sum_{F \in \mathcal{F}_h^b} \int_F (\beta\cdot\mathbf{n})vw_h,$$

2.2. Centered Fluxes

exploiting the continuity of the normal component of β across interfaces. For all $F \in \mathcal{F}_h^i$ with $F = \partial T_1 \cap \partial T_2$, $v_i = v|_{T_i}$, $w_i = w_h|_{T_i}$, $i \in \{1,2\}$, we observe that

$$[\![vw_h]\!] = v_1 w_1 - v_2 w_2$$
$$= \frac{1}{2}(v_1 - v_2)(w_1 + w_2) + \frac{1}{2}(v_1 + v_2)(w_1 - w_2)$$
$$= [\![v]\!]\{\!\{w_h\}\!\} + \{\!\{v\}\!\}[\![w_h]\!].$$

The expression (2.26) then results from the combination of the three rightmost in (2.27). □

Remark 2.21 (Conservative form (2.3)). When working with (2.3), the right-hand side of (2.24) can be rewritten as

$$\int_\Omega \left\{ \tilde{\mu} v w_h + \nabla_h \cdot (\beta v) w_h \right\} + \int_{\partial \Omega} (\beta \cdot n)^\ominus v w_h - \sum_{F \in \mathcal{F}_h^i} \int_F (\beta \cdot n_F)[\![v]\!]\{\!\{w_h\}\!\}.$$

2.2.2 Error Estimates

We consider the discrete problem:

$$\text{Find } u_h \in V_h \text{ s.t. } a_h^{\mathrm{cf}}(u_h, v_h) = \int_\Omega f v_h \text{ for all } v_h \in V_h. \qquad (2.28)$$

This problem is well-posed owing to the discrete coercivity of a_h^{cf} on V_h. Our goal is to estimate the approximation error $(u - u_h)$ in the $\|\cdot\|_{\mathrm{cf}}$-norm. The convergence analysis is performed in the spirit of Theorem 1.35. Owing to Lemma 2.18, it only remains to address the boundedness of the discrete bilinear form a_h^{cf}. To this purpose, we define on V_{*h} the norm

$$\|v\|_{\mathrm{cf},*}^2 = \|v\|_{\mathrm{cf}}^2 + \sum_{T \in \mathcal{T}_h} \tau_{\mathrm{c}} \|\beta \cdot \nabla v\|_{L^2(T)}^2 + \sum_{T \in \mathcal{T}_h} \tau_{\mathrm{c}} \beta_{\mathrm{c}}^2 h_T^{-1} \|v\|_{L^2(\partial T)}^2,$$

with time scale τ_{c} and reference velocity β_{c} defined by (2.7).

Lemma 2.22 (Boundedness). *There holds*

$$\forall (v, w_h) \in V_{*h} \times V_h, \qquad a_h^{\mathrm{cf}}(v, w_h) \leq C_{\mathrm{bnd}} \|v\|_{\mathrm{cf},*} \|w_h\|_{\mathrm{cf}},$$

with C_{bnd} independent of h and of the data μ and β.

Proof. Let $(v, w_h) \in V_{*h} \times V_h$. We bound the terms on the right-hand side of (2.24). Using the Cauchy–Schwarz inequality, it is clear that

$$\int_\Omega \left\{ \mu v w_h + (\beta \cdot \nabla_h v) w_h \right\} + \int_{\partial \Omega} (\beta \cdot n)^\ominus v w_h \leq 2 \|v\|_{\mathrm{cf},*} \|w_h\|_{\mathrm{cf}}.$$

To bound the last term in (2.24), we first use the Cauchy–Schwarz inequality to infer

$$\sum_{F\in\mathcal{F}_h^i}\int_F (\beta\cdot n_F)[\![v]\!]\{\!\{w_h\}\!\} \leq \left(\sum_{F\in\mathcal{F}_h^i}\frac{1}{2}\tau_c\beta_c^2\{\!\{h\}\!\}^{-1}\|[\![v]\!]\|_{L^2(F)}^2\right)^{1/2}$$

$$\times\left(\sum_{F\in\mathcal{F}_h^i}2\tau_c^{-1}\{\!\{h\}\!\}\|\{\!\{w_h\}\!\}\|_{L^2(F)}^2\right)^{1/2},$$

where for all $F\in\mathcal{F}_h^i$ with $F=\partial T_1\cap\partial T_2$, $\{\!\{h\}\!\}=\frac{1}{2}(h_{T_1}+h_{T_2})$. We also set $v_i=v|_{T_i}$ and $w_i=w_h|_{T_i}$, $i\in\{1,2\}$, and observe that $\frac{1}{2}[\![v]\!]^2\leq(v_1^2+v_2^2)$ and $2\{\!\{w_h\}\!\}^2\leq(w_1^2+w_2^2)$. Moreover, owing to Lemma 1.43,

$$C_\varrho^{-1}\max(h_{T_1},h_{T_2})\leq\{\!\{h\}\!\}\leq C_\varrho\min(h_{T_1},h_{T_2}),$$

where C_ϱ only depends on mesh regularity. As a result,

$$\sum_{F\in\mathcal{F}_h^i}\int_F(\beta\cdot n_F)[\![v]\!]\{\!\{w_h\}\!\}\leq C_\varrho\|v\|_{\mathrm{cf},*}\left(\sum_{T\in\mathcal{T}_h}\tau_c^{-1}h_T\|w_h\|_{L^2(\partial T)}^2\right)^{1/2},$$

To conclude, we use the discrete trace inequality (1.40) to bound the last factor on the right-hand side by $\tau_c^{-1/2}C_{\mathrm{tr}}N_\partial^{1/2}\|w_h\|_{L^2(\Omega)}$. □

A straightforward consequence of Theorem 1.35 is the following error estimate.

Theorem 2.23 (Error estimate). *Let u solve (2.14) and let u_h solve (2.28) where a_h^{cf} is defined by (2.24) and $V_h=\mathbb{P}_d^k(\mathcal{T}_h)$ with $k\geq 1$ and $\mathcal{T}_h$ belonging to an admissible mesh sequence. Then, there holds*

$$\|u-u_h\|_{\mathrm{cf}}\leq C\inf_{y_h\in V_h}\|u-y_h\|_{\mathrm{cf},*}, \tag{2.29}$$

with C independent of h and depending on the data only through the factor $\{\min(1,\tau_c\mu_0)\}^{-1}$.

To infer a convergence result from (2.29), we assume that the exact solution is smooth enough, take $y_h=\pi_h u$, the L^2-orthogonal projection of u onto V_h, in (2.29), and use Lemmata 1.58 and 1.59.

Corollary 2.24 (Convergence rate for smooth solutions). *Besides the hypotheses of Theorem 2.23, assume $u\in H^{k+1}(\Omega)$. Then, there holds*

$$\|u-u_h\|_{\mathrm{cf}}\leq C_u h^k, \tag{2.30}$$

with $C_u=C\|u\|_{H^{k+1}(\Omega)}$ and C independent of h and depending on the data only through the factor $\{\min(1,\tau_c\mu_0)\}^{-1}$.

2.2.3 Numerical Fluxes

At this stage, it is instructive to consider an alternative viewpoint based on numerical fluxes. Because we are working with broken polynomial spaces, the discrete problem (2.28) admits a local formulation obtained by considering an arbitrary mesh element $T \in \mathcal{T}_h$ and an arbitrary polynomial $\xi \in \mathbb{P}_d^k(T)$. For a set $S \subset \Omega$, we denote by χ_S its characteristic function, namely

$$\chi_S(x) = \begin{cases} 1 & \text{if } x \in S, \\ 0 & \text{otherwise.} \end{cases}$$

Then, using the test function $v_h = \xi \chi_T$ in the discrete problem (2.28), observing that

$$[\![\xi \chi_T]\!] = \epsilon_{T,F} \xi \qquad \text{with} \qquad \epsilon_{T,F} := n_T \cdot n_F,$$

and owing to the expression (2.26) for the discrete bilinear form a_h^{cf}, we infer

$$\int_T \left\{ (\mu - \nabla \cdot \beta) u_h \xi - u_h (\beta \cdot \nabla \xi) \right\} + \sum_{F \in \mathcal{F}_T} \epsilon_{T,F} \int_F \phi_F(u_h) \xi = \int_T f \xi, \qquad (2.31)$$

where the *numerical fluxes* $\phi_F(u_h)$ are given by

$$\phi_F(u_h) := \begin{cases} (\beta \cdot n_F) \{\!\{ u_h \}\!\} & \text{if } F \in \mathcal{F}_h^i, \\ (\beta \cdot n)^{\oplus} u_h & \text{if } F \in \mathcal{F}_h^b. \end{cases}$$

The numerical fluxes $\phi_F(u_h)$ are called *centered fluxes* because the average value of u_h is used on each $F \in \mathcal{F}_h^i$. Since these fluxes are *single-valued* and since for all $F \in \mathcal{F}_h^i$ with $F = \partial T_1 \cap \partial T_2$, $\epsilon_{T_1,F} + \epsilon_{T_2,F} = 0$, the local formulation (2.31) is *conservative* in the sense that whatever "flows" out of a mesh element through one of its faces "flows" into the neighboring element through that face. Finally, taking $\xi \equiv 1$ in (2.31) leads to the usual balance formulation encountered in finite volume methods, namely

$$\int_T (\mu - \nabla \cdot \beta) u_h + \sum_{F \in \mathcal{F}_T} \epsilon_{T,F} \int_F \phi_F(u_h) = \int_T f.$$

Remark 2.25 (Conservative form (2.3)). When working with (2.3), the local problems (2.31) can be rewritten as

$$\int_T \left\{ \tilde{\mu} u_h \xi - u_h (\beta \cdot \nabla \xi) \right\} + \sum_{F \in \mathcal{F}_T} \epsilon_{T,F} \int_F \phi_F(u_h) \xi = \int_T f \xi.$$

Observe that the numerical fluxes have not been modified.

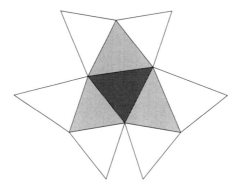

Fig. 2.4: Example of stencil of an element $T \in \mathcal{T}_h$ when $\mathcal{T}_h$ is a matching triangular mesh; the mesh element is highlighted in *dark grey*, and its three neighbors, which all belong to the stencil, are highlighted in *light grey*; the other triangles do not belong to the stencil

A useful concept in practical implementations is that of stencil (cf. Appendix A for further insight).

Definition 2.26 (Stencil). For a given element $T \in \mathcal{T}_h$, we define the elementary *stencil* $\mathcal{S}(a_h^{\mathrm{cf}}; T)$ associated with the bilinear form a_h^{cf} as

$$\mathcal{S}(a_h^{\mathrm{cf}}; T) := \left\{ T' \in \mathcal{T}_h \mid \exists q \in \mathbb{P}_d^k(T), \exists r \in \mathbb{P}_d^k(T'), a_h^{\mathrm{cf}}(q\chi_T, r\chi_{T'}) \neq 0 \right\},$$

where χ_T and $\chi_{T'}$ denote characteristic functions.

Owing to the local formulation (2.31), the *stencil* of a given element $T \in \mathcal{T}_h$ consists of T itself and its neighbors in the sense of faces. For instance, on a matching simplicial mesh, the stencil contains $(d+2)$ mesh elements; cf. Fig. 2.4 for a two-dimensional illustration.

2.3 Upwinding

The goal of this section is to strengthen the stability of the dG bilinear form so as to arrive at quasi-optimal error estimates in the sense of Definition 1.54. This goal is achieved by penalizing in a least-squares sense the interface jumps of the discrete solution. In terms of fluxes, this approach can be interpreted as upwinding. We keep assumptions (2.4) and (2.6) on the data μ and β as well as Assumption 2.13 on the regularity of the exact solution u, but the polynomial degree k is here such that $k \geq 0$. For $k = 0$, the dG method considered in this section coincides with a finite volume approximation with upwinding.

The idea of presenting dG methods with upwinding through a suitable penalty of interface jumps has been highlighted recently by Brezzi, Marini, and Süli [60]. Therein, a quasi-optimal error estimate on the norm comprising the L^2-error and the jumps is derived. As reflected in Sect. 2.3.2, this analysis

2.3. Upwinding

hinges on discrete coercivity to establish stability. To tighten the error estimate further by including an optimal bound on the advective derivative of the error, a discrete inf-sup condition is needed; this condition is established in Sect. 2.3.3 following the ideas of Johnson and Pitkäranta [204].

2.3.1 Tightened Stability Using Penalties

We consider the new bilinear form

$$a_h^{\text{upw}}(v_h, w_h) := a_h^{\text{cf}}(v_h, w_h) + s_h(v_h, w_h), \tag{2.32}$$

with the stabilization bilinear form

$$s_h(v_h, w_h) = \sum_{F \in \mathcal{F}_h^i} \int_F \frac{\eta}{2} |\beta \cdot n_F| \llbracket v_h \rrbracket \llbracket w_h \rrbracket, \tag{2.33}$$

where $\eta > 0$ is a user-dependent parameter. Specifically, using (2.24),

$$a_h^{\text{upw}}(v_h, w_h) := \int_\Omega \left\{ \mu v_h w_h + (\beta \cdot \nabla_h v_h) w_h \right\} + \int_{\partial \Omega} (\beta \cdot n)^\ominus v_h w_h \tag{2.34}$$
$$- \sum_{F \in \mathcal{F}_h^i} \int_F (\beta \cdot n_F) \llbracket v_h \rrbracket \{w_h\} + \sum_{F \in \mathcal{F}_h^i} \int_F \frac{\eta}{2} |\beta \cdot n_F| \llbracket v_h \rrbracket \llbracket w_h \rrbracket,$$

or, equivalently, using (2.26),

$$a_h^{\text{upw}}(v_h, w_h) = \int_\Omega \left\{ (\mu - \nabla \cdot \beta) v_h w_h - v_h (\beta \cdot \nabla_h w_h) \right\} + \int_{\partial \Omega} (\beta \cdot n)^\oplus v_h w_h \tag{2.35}$$
$$+ \sum_{F \in \mathcal{F}_h^i} \int_F (\beta \cdot n_F) \{v_h\} \llbracket w_h \rrbracket + \sum_{F \in \mathcal{F}_h^i} \int_F \frac{\eta}{2} |\beta \cdot n_F| \llbracket v_h \rrbracket \llbracket w_h \rrbracket.$$

We observe that the discrete bilinear forms a_h^{cf} and a_h^{upw} lead to the same stencil (cf. Definition 2.26). The numerical flux associated with the discrete bilinear form a_h^{upw} depends on the penalty parameter η (cf. Sect. 2.3.4). Choosing $\eta = 1$ is particularly interesting since it leads to the usual upwind fluxes in the context of finite volume schemes. More generally, the discrete bilinear form a_h^{upw} is henceforth referred to as the *upwind dG bilinear form*.

We consider the discrete problem:

$$\text{Find } u_h \in V_h \text{ s.t. } a_h^{\text{upw}}(u_h, v_h) = \int_\Omega f v_h \text{ for all } v_h \in V_h. \tag{2.36}$$

We first examine the consistency and discrete coercivity of the upwind dG bilinear form. Recalling definition (2.25) of the $\|\cdot\|_{\text{cf}}$-norm considered for centered fluxes, we now assert coercivity with respect to the following stronger norm, also defined on V_{*h}:

$$\|v\|_{\text{uwb}}^2 := \|v\|_{\text{cf}}^2 + \sum_{F \in \mathcal{F}_h^i} \int_F \frac{\eta}{2} |\beta \cdot n_F| \llbracket v \rrbracket^2. \tag{2.37}$$

Lemma 2.27 (Consistency and discrete coercivity). *The upwind dG bilinear form a_h^{upw} defined by (2.32) and (2.33) satisfies the following properties:*

(i) *Consistency, namely for the exact solution $u \in V_*$,*

$$a_h^{\mathrm{upw}}(u, v_h) = \int_\Omega f v_h \qquad \forall v_h \in V_h,$$

(ii) *Coercivity on V_h with respect to the $\|\cdot\|_{\mathrm{uw}\flat}$-norm, namely*

$$\forall v_h \in V_h, \qquad a_h^{\mathrm{upw}}(v_h, v_h) \geq C_{\mathrm{sta}} \|v_h\|_{\mathrm{uw}\flat}^2,$$

with $C_{\mathrm{sta}} = \min(1, \tau_c \mu_0)$ as in Lemma 2.18.

Proof. The consistency of a_h^{upw} results from the consistency of a_h^{cf} established in Lemma 2.18 and the fact that $(\beta \cdot n_F)\llbracket u \rrbracket$ vanishes across interfaces owing to (2.23). Furthermore, the coercivity of a_h^{upw} results from that of a_h^{cf} established in Lemma 2.18 and the fact that $s_h(v_h, v_h) = \sum_{F \in \mathcal{F}_h^i} \int_F \frac{\eta}{2} |\beta \cdot n_F| \llbracket v_h \rrbracket^2$. □

The discrete coercivity of a_h^{upw} on V_h implies the well-posedness of the discrete problem (2.36). In the following two subsections, we estimate the approximation error $(u - u_h)$, first using coercivity and then a tightened stability property hinging on a discrete inf-sup condition.

Remark 2.28 (Choice of penalty parameter). Assigning a large value to the penalty parameter η is not appropriate since this choice significantly reduces the accuracy of the discrete solution. This issue has been examined by Burman, Quarteroni, and Stamm [68] where it is shown (on matching simplicial meshes) that the dG solution solving (2.36) converges to the discrete solution produced by the standard continuous Galerkin approach as the penalty parameter tends to infinity. Furthermore, we refer the reader to Burman and Stamm [69] for a study on minimal stabilization for dG approximations of the two-dimensional advection-reaction equation. In particular, on matching triangulations with polynomial degree $k \geq 2$, a high-pass filter can be applied to the jumps before penalizing them. One motivation for minimal stabilization is achieving numerical fluxes with a moderate dependency on the penalty parameter.

2.3.2 Error Estimates Based on Coercivity

It turns out that the discrete coercivity norm $\|\cdot\|_{\mathrm{uw}\flat}$ defined by (2.37) is not strong enough to establish a boundedness result for the upwind dG bilinear form a_h^{upw} leading to quasi-optimal error estimates (this is why we denote this norm by $\|\cdot\|_{\mathrm{uw}\flat}$; a stronger norm, denoted by $\|\cdot\|_{\mathrm{uw}\sharp}$, is introduced in Sect. 2.3.3). To circumvent this difficulty, one idea consists in establishing boundedness by restricting the functions in the first argument of a_h^{upw} to those functions in V_{*h} that are L^2-orthogonal to the discrete space V_h, that is, to functions of the form $v - \pi_h v$ for $v \in V_*$ (recall that π_h denotes the L^2-orthogonal projection onto V_h). Functions of the form $v - \pi_h v$ are termed orthogonal subscales.

2.3. Upwinding

Definition 2.29 (Boundedness on orthogonal subscales)**.** We say that *boundedness on orthogonal subscales* holds true for a_h^{upw} (uniformly in h, μ, and β) if there is C_{bnd}, independent of h, μ, and β, such that, for all $(v, w_h) \in V_* \times V_h$,

$$|a_h^{\text{upw}}(v - \pi_h v, w_h)| \leq C_{\text{bnd}} \|v - \pi_h v\|_{\text{uwb},*} \|w_h\|_{\text{uwb}}, \qquad (2.38)$$

for a norm $\|\cdot\|_{\text{uwb},*}$ defined on V_{*h} such that, for all $v \in V_{*h}$, $\|v\|_{\text{uwb}} \leq \|v\|_{\text{uwb},*}$.

Lemma 2.30 (Boundedness on orthogonal subscales)**.** *Boundedness on orthogonal subscales holds true for the upwind dG bilinear form* a_h^{upw} *when defining on* V_{*h} *the norm*

$$\|v\|_{\text{uwb},*}^2 = \|v\|_{\text{uwb}}^2 + \sum_{T \in \mathcal{T}_h} \beta_c \|v\|_{L^2(\partial T)}^2.$$

Proof. Let $(v, w_h) \in V_* \times V_h$ and set $y = v - \pi_h v$. We consider the expression (2.26) for a_h^{cf} and bound the various terms on the right-hand side. First, it is clear using the Cauchy–Schwarz inequality that

$$\int_\Omega (\mu - \nabla\cdot\beta) y w_h + \int_{\partial\Omega} (\beta\cdot\mathbf{n})^\oplus y w_h \leq C_1 \|y\|_{\text{uwb}} \|w_h\|_{\text{uwb}},$$

with C_1 independent of h, μ, and β. Moreover, letting $\langle\beta\rangle_T$ denote the mean value of β on each $T \in \mathcal{T}_h$, we observe, on the one hand, that $\|\beta - \langle\beta\rangle_T\|_{L^\infty(T)} \leq L_\beta h_T \leq \tau_c^{-1} h_T$ since β is Lipschitz continuous and, on the other hand, that, for all $w_h \in V_h$, $\langle\beta\rangle_T \cdot \nabla w_h \in \mathbb{P}_d^{k-1}(T) \subset \mathbb{P}_d^k(T)$ so that

$$\forall T \in \mathcal{T}_h, \qquad \int_T y \langle\beta\rangle_T \cdot \nabla w_h = 0.$$

Hence, owing to the inverse inequality (1.36),

$$\int_\Omega y \beta \cdot \nabla_h w_h = \sum_{T \in \mathcal{T}_h} \int_T y \beta \cdot \nabla w_h = \sum_{T \in \mathcal{T}_h} \int_T y (\beta - \langle\beta\rangle_T) \cdot \nabla w_h$$

$$\leq \sum_{T \in \mathcal{T}_h} \|y\|_{L^2(T)} \tau_c^{-1} h_T \|\nabla w_h\|_{[L^2(T)]^d}$$

$$\leq \sum_{T \in \mathcal{T}_h} \|y\|_{L^2(T)} \tau_c^{-1} C_{\text{inv}} \|w_h\|_{L^2(T)}$$

$$\leq C_{\text{inv}} \|y\|_{\text{uwb}} \|w_h\|_{\text{uwb}}.$$

In addition, the Cauchy–Schwarz inequality yields

$$\sum_{F \in \mathcal{F}_h^i} \int_F (\beta\cdot\mathbf{n}_F)\{y\}[w_h] \leq \left(\sum_{F \in \mathcal{F}_h^i} \int_F 2\eta^{-1} |\beta\cdot\mathbf{n}_F| \{y\}^2 \right)^{1/2} \|w_h\|_{\text{uwb}}$$

$$\leq \left(\eta^{-1} \sum_{T \in \mathcal{T}_h} \beta_c \|y\|_{L^2(\partial T)}^2 \right)^{1/2} \|w_h\|_{\text{uwb}}.$$

Collecting the above bounds yields

$$a_h^{\mathrm{cf}}(y, w_h) \leq C_2 \|y\|_{\mathrm{uwb},*} \|w_h\|_{\mathrm{uwb}},$$

with C_2 independent of h, μ, and β. Finally, observing that

$$\sum_{F \in \mathcal{F}_h^i} \int_F \frac{\eta}{2} |\beta \cdot n_F| [\![y]\!] [\![w_h]\!] \leq \|y\|_{\mathrm{uwb}} \|w_h\|_{\mathrm{uwb}},$$

the desired bound is obtained. □

We can now state the main result of this section.

Theorem 2.31 (Error estimate). *Let u solve (2.14) and let u_h solve (2.36) where a_h^{upw} is defined by (2.34) and $V_h = \mathbb{P}_d^k(\mathcal{T}_h)$ with $k \geq 0$ and $\mathcal{T}_h$ belonging to an admissible mesh sequence. Then, there holds*

$$\|u - u_h\|_{\mathrm{uwb}} \leq C \|u - \pi_h u\|_{\mathrm{uwb},*}, \qquad (2.39)$$

with C independent of h and depending on the data only through the factor $\{\min(1, \tau_c \mu_0)\}^{-1}$.

Proof. The proof is similar to that of Theorem 1.35. Assume first that $\pi_h u \neq u_h$. Then, owing to discrete coercivity and consistency (cf. Lemma 2.27), we infer

$$\|u_h - \pi_h u\|_{\mathrm{uwb}} \leq C_{\mathrm{sta}}^{-1} \frac{a_h^{\mathrm{upw}}(u_h - \pi_h u, u_h - \pi_h u)}{\|u_h - \pi_h u\|_{\mathrm{uwb}}} = C_{\mathrm{sta}}^{-1} \frac{a_h^{\mathrm{upw}}(u - \pi_h u, u_h - \pi_h u)}{\|u_h - \pi_h u\|_{\mathrm{uwb}}},$$

recalling that $C_{\mathrm{sta}} = \min(1, \tau_c \mu_0)$. Hence, using boundedness on orthogonal subscales,

$$\|u_h - \pi_h u\|_{\mathrm{uwb}} \leq C_{\mathrm{sta}}^{-1} C_{\mathrm{bnd}} \|u - \pi_h u\|_{\mathrm{uwb},*},$$

and this inequality holds true also if $\pi_h u = u_h$. Estimate (2.39) then results from the triangle inequality and the fact that $\|u - \pi_h u\|_{\mathrm{uwb}} \leq \|u - \pi_h u\|_{\mathrm{uwb},*}$. □

To infer a convergence result from (2.39), we assume that the exact solution is smooth enough and use Lemmata 1.58 and 1.59.

Corollary 2.32 (Convergence rate for smooth solutions). *Besides the hypotheses of Theorem 2.31, assume $u \in H^{k+1}(\Omega)$. Then, there holds*

$$\|u - u_h\|_{\mathrm{uwb}} \leq C_u h^{k+1/2}, \qquad (2.40)$$

with $C_u = C \|u\|_{H^{k+1}(\Omega)}$ and C independent of h and depending on the data only through the factor $\{\min(1, \tau_c \mu_0)\}^{-1}$.

Estimate (2.40) yields the convergence of the dG approximation for any polynomial degree $k \geq 0$. The result is quasi-optimal for the boundary and jump terms that dominate the stability norm $\|\cdot\|_{\mathrm{uwb}}$, but suboptimal by a factor $h^{1/2}$ for the L^2-norm. The L^2-norm error estimate turns out to be sharp, as confirmed by

2.3. Upwinding

Peterson [258] on specific meshes. However, one drawback of the present error estimate is that it does not convey information on the behavior of first-order derivatives, and more specifically on the advective derivative. This shortcoming is remedied in the next subsection upon strengthening the discrete stability result.

Remark 2.33 (On the error estimate). The error estimate in Theorem 2.31 is less sharp than that derived in Theorem 1.35 since the infimum on the right-hand side is replaced by the projection error $u - \pi_h u$. However, this bound is sufficient to infer quasi-optimal convergence rates for smooth solutions (cf. Corollary 2.32).

2.3.3 Error Estimates Based on Inf-Sup Stability

The convergence analysis is now performed in the spirit of Theorem 1.35. Consistency has already been established, and it remains to address discrete stability and boundedness. We augment the $\|\cdot\|_{\mathrm{uw}\flat}$-norm defined by (2.37) by adding a scaled advective derivative and define the stronger norm

$$\|v\|_{\mathrm{uw}\sharp}^2 := \|v\|_{\mathrm{uw}\flat}^2 + \sum_{T \in \mathcal{T}_h} \beta_c^{-1} h_T \|\beta \cdot \nabla v\|_{L^2(T)}^2.$$

We are now in a position to state the main stability result regarding dG methods with upwinding. The proof, which elaborates on the ideas of Johnson and Pitkäranta [204] using locally the scaled advective derivative as a test function, follows the recent presentation by Ern and Guermond [142]. To simplify the arguments while keeping track of the dependency of the estimates on μ and β, we assume that

$$h \leq \beta_c \tau_c. \tag{2.41}$$

Remark 2.34 (Meaning of assumption (2.41)). Recalling (2.7), (2.41) yields

$$\max(h\|\mu\|_{L^\infty(\Omega)}\|\beta\|_{[L^\infty(\Omega)]^d}^{-1}, hL_\beta\|\beta\|_{[L^\infty(\Omega)]^d}^{-1}) \leq 1.$$

The quantity $h\|\mu\|_{L^\infty(\Omega)}\|\beta\|_{[L^\infty(\Omega)]^d}^{-1}$ is termed the local Damköhler number (the term local indicates here the use of the meshsize as reference length), and assuming that this quantity is bounded (e.g., by 1) means that we are not concerned with dominant reactions, but instead with tempered or mild reactions. Furthermore, assuming that $hL_\beta\|\beta\|_{[L^\infty(\Omega)]^d}^{-1} \leq 1$ means that the meshsize resolves the spatial variations of the advective velocity.

Lemma 2.35 (Discrete inf-sup condition). *There is $C'_{\mathrm{sta}} > 0$, independent of h, μ, and β, such that*

$$\forall v_h \in V_h, \qquad C'_{\mathrm{sta}} C_{\mathrm{sta}} \|v_h\|_{\mathrm{uw}\sharp} \leq \sup_{w_h \in V_h \setminus \{0\}} \frac{a_h^{\mathrm{upw}}(v_h, w_h)}{\|w_h\|_{\mathrm{uw}\sharp}},$$

with $C_{\mathrm{sta}} = \min(1, \tau_c \mu_0)$ as in Lemma 2.27.

Proof. Let $v_h \in V_h \setminus \{0\}$ and set $\mathbb{S} = \sup_{w_h \in V_h \setminus \{0\}} \frac{a_h^{\mathrm{upw}}(v_h, w_h)}{\|w_h\|_{\mathrm{uw}\sharp}}$. Lemma 2.27 implies that

$$C_{\mathrm{sta}} \|v_h\|_{\mathrm{uw}\flat}^2 \le a_h^{\mathrm{upw}}(v_h, v_h) = \frac{a_h^{\mathrm{upw}}(v_h, v_h)}{\|v_h\|_{\mathrm{uw}\sharp}} \|v_h\|_{\mathrm{uw}\sharp} \le \mathbb{S} \|v_h\|_{\mathrm{uw}\sharp}. \qquad (2.42)$$

It remains to bound the contribution of the advective derivative in the expression for $\|v_h\|_{\mathrm{uw}\sharp}$. To this purpose, we consider the function $w_h \in V_h$ such that, for all $T \in \mathcal{T}_h$, $w_h|_T = \beta_c^{-1} h_T \langle \beta \rangle_T \cdot \nabla v_h$, where we recall that $\langle \beta \rangle_T$ denotes the mean value of β over T. To alleviate the notation, we abbreviate as $a \lesssim b$ the inequality $a \le Cb$ with positive C independent of h, μ, and β. The dependency on the penalty parameter η is not tracked since in practice this parameter can be taken equal to one.

(i) Let us bound $\|w_h\|_{\mathrm{uw}\sharp}$ by $\|v_h\|_{\mathrm{uw}\sharp}$. We first observe that

$$\sum_{T \in \mathcal{T}_h} \beta_c h_T^{-1} \|w_h\|_{L^2(T)}^2 \lesssim \|v_h\|_{\mathrm{uw}\sharp}^2. \qquad (2.43)$$

Indeed, using the triangle inequality, the regularity of β, and the inverse inequality (1.36), we obtain

$$\sum_{T \in \mathcal{T}_h} \beta_c h_T^{-1} \|w_h\|_{L^2(T)}^2 = \sum_{T \in \mathcal{T}_h} \beta_c^{-1} h_T \|\langle \beta \rangle_T \cdot \nabla v_h\|_{L^2(T)}^2$$

$$\le 2 \sum_{T \in \mathcal{T}_h} \beta_c^{-1} \left(h_T \|\beta \cdot \nabla v_h\|_{L^2(T)}^2 + h_T \|(\beta - \langle \beta \rangle_T) \cdot \nabla v_h\|_{L^2(T)}^2 \right)$$

$$\lesssim \sum_{T \in \mathcal{T}_h} \beta_c^{-1} h_T \|\beta \cdot \nabla v_h\|_{L^2(T)}^2 + \sum_{T \in \mathcal{T}_h} \tau_c^{-2} \beta_c^{-1} h_T \|v_h\|_{L^2(T)}^2$$

$$\le \sum_{T \in \mathcal{T}_h} \beta_c^{-1} h_T \|\beta \cdot \nabla v_h\|_{L^2(T)}^2 + \sum_{T \in \mathcal{T}_h} \tau_c^{-1} \|v_h\|_{L^2(T)}^2 \lesssim \|v_h\|_{\mathrm{uw}\sharp}^2,$$

where we have used (2.41) to simplify the last term. A first consequence of (2.43) is that using the discrete trace inequality (1.37), we infer

$$\int_{\partial \Omega} \frac{1}{2} |\beta \cdot \mathbf{n}| w_h^2 + \sum_{F \in \mathcal{F}_h^i} \int_F \frac{\eta}{2} |\beta \cdot \mathbf{n}_F| [\![w_h]\!]^2 \lesssim \sum_{T \in \mathcal{T}_h} \beta_c h_T^{-1} \|w_h\|_{L^2(T)}^2 \lesssim \|v_h\|_{\mathrm{uw}\sharp}^2.$$

Moreover, the inverse inequality (1.36) yields $\|w_h\|_{L^2(\Omega)} \lesssim \|v_h\|_{L^2(\Omega)}$ and

$$\sum_{T \in \mathcal{T}_h} \beta_c^{-1} h_T \|\beta \cdot \nabla w_h\|_{L^2(T)}^2 \lesssim \sum_{T \in \mathcal{T}_h} \beta_c h_T^{-1} \|w_h\|_{L^2(T)}^2 \lesssim \|v_h\|_{\mathrm{uw}\sharp}^2.$$

Collecting these bounds, we infer

$$\|w_h\|_{\mathrm{uw}\sharp} \lesssim \|v_h\|_{\mathrm{uw}\sharp}. \qquad (2.44)$$

2.3. Upwinding

(ii) Now, using the definition (2.32) and (2.33) of a_h^{upw} and the expression (2.24) of a_h^{cf}, we obtain

$$\sum_{T \in \mathcal{T}_h} \beta_c^{-1} h_T \|\beta \cdot \nabla v_h\|_{L^2(T)}^2 = a_h^{\text{upw}}(v_h, w_h) - \int_\Omega \mu v_h w_h$$

$$+ \sum_{T \in \mathcal{T}_h} \beta_c^{-1} h_T \int_T (\beta \cdot \nabla v_h)(\beta - \langle \beta \rangle_T) \cdot \nabla v_h$$

$$- \int_{\partial\Omega} (\beta \cdot n)^\ominus v_h w_h + \sum_{F \in \mathcal{F}_h^i} \int_F (\beta \cdot n_F) [\![v_h]\!] \{\!\{w_h\}\!\}$$

$$- \sum_{F \in \mathcal{F}_h^i} \int_F \frac{\eta}{2} |\beta \cdot n_F| [\![v_h]\!] [\![w_h]\!] = \mathfrak{T}_1 + \ldots + \mathfrak{T}_6,$$

and we estimate the terms $\mathfrak{T}_1, \ldots, \mathfrak{T}_6$ on the right-hand side. Owing to (2.44),

$$|\mathfrak{T}_1| = |a_h^{\text{upw}}(v_h, w_h)| = \frac{|a_h^{\text{upw}}(v_h, w_h)|}{\|w_h\|_{\text{uw}\sharp}} \|w_h\|_{\text{uw}\sharp} \leq \mathfrak{S}\|w_h\|_{\text{uw}\sharp} \lesssim \mathfrak{S}\|v_h\|_{\text{uw}\sharp}.$$

Moreover, using the Cauchy–Schwarz inequality and the definition of the $\|\cdot\|_{\text{uw}\flat}$-norm yields

$$|\mathfrak{T}_2| + |\mathfrak{T}_4| + |\mathfrak{T}_6| \lesssim \|v_h\|_{\text{uw}\flat} \|w_h\|_{\text{uw}\flat} \lesssim \|v_h\|_{\text{uw}\flat} \|v_h\|_{\text{uw}\sharp}.$$

Using again the Cauchy–Schwarz inequality, together with the discrete trace inequality (1.37), leads to

$$|\mathfrak{T}_5| \leq \left(\sum_{F \in \mathcal{F}_h^i} \int_F \frac{\eta}{2} |\beta \cdot n_F| [\![v_h]\!]^2 \right)^{1/2} \left(\sum_{F \in \mathcal{F}_h^i} \int_F \frac{2}{\eta} |\beta \cdot n_F| \{\!\{w_h\}\!\}^2 \right)^{1/2}$$

$$\lesssim \|v_h\|_{\text{uw}\flat} \left(\sum_{T \in \mathcal{T}_h} \beta_c h_T^{-1} \|w_h\|_{L^2(T)}^2 \right)^{1/2},$$

so that, owing to (2.43),

$$|\mathfrak{T}_5| \lesssim \|v_h\|_{\text{uw}\flat} \|v_h\|_{\text{uw}\sharp}.$$

Finally, to bound $\mathfrak{T}_3$, we use the Cauchy–Schwarz inequality, the inverse inequality (1.36), and assumption (2.41) to obtain

$$|\mathfrak{T}_3| \lesssim \left(\sum_{T \in \mathcal{T}_h} \beta_c^{-1} h_T \|\beta \cdot \nabla v_h\|_{L^2(T)}^2 \right)^{1/2} \left(\sum_{T \in \mathcal{T}_h} \beta_c^{-1} h_T \tau_c^{-2} \|v_h\|_{L^2(T)}^2 \right)^{1/2}$$

$$\lesssim \left(\sum_{T \in \mathcal{T}_h} \beta_c^{-1} h_T \|\beta \cdot \nabla v_h\|_{L^2(T)}^2 \right)^{1/2} \|v_h\|_{\text{uw}\flat}.$$

Using Young's inequality in the form $ab \leq \gamma a^2 + (4\gamma)^{-1} b^2$ where $\gamma > 0$ can be chosen as small as desired, we infer

$$|\mathfrak{T}_3| - \frac{1}{2}\sum_{T\in\mathcal{T}_h} \beta_c^{-1} h_T \|\beta\cdot\nabla v_h\|_{L^2(T)}^2 \lesssim \|v_h\|_{\text{uw}\flat}^2.$$

Collecting the above bounds leads to

$$\sum_{T\in\mathcal{T}_h} \beta_c^{-1} h_T \|\beta\cdot\nabla v_h\|_{L^2(T)}^2 \lesssim \mathbb{S}\|v_h\|_{\text{uw}\sharp} + \|v_h\|_{\text{uw}\flat}\|v_h\|_{\text{uw}\sharp} + \|v_h\|_{\text{uw}\flat}^2.$$

(iii) Combining the above result with (2.42) and since $C_{\text{sta}} \leq 1$ yields

$$C_{\text{sta}}\|v_h\|_{\text{uw}\sharp}^2 \lesssim \mathbb{S}\|v_h\|_{\text{uw}\sharp} + C_{\text{sta}}\|v_h\|_{\text{uw}\flat}\|v_h\|_{\text{uw}\sharp}.$$

Using Young's inequality for the last term together with (2.42) yields

$$C_{\text{sta}}\|v_h\|_{\text{uw}\sharp}^2 \lesssim \mathbb{S}\|v_h\|_{\text{uw}\sharp} + C_{\text{sta}}\|v_h\|_{\text{uw}\flat}^2 \lesssim \mathbb{S}\|v_h\|_{\text{uw}\sharp},$$

whence the assertion. $\square$

To formulate a boundedness result, we define the following norm:

$$\|v\|_{\text{uw}\sharp,*}^2 := \|v\|_{\text{uw}\sharp}^2 + \sum_{T\in\mathcal{T}_h} \beta_c \left(h_T^{-1}\|v\|_{L^2(T)}^2 + \|v\|_{L^2(\partial T)}^2\right).$$

Lemma 2.36 (Boundedness). *There holds*

$$\forall (v, w_h) \in V_{*h} \times V_h, \qquad |a_h^{\text{upw}}(v, w_h)| \leq C_{\text{bnd}} \|v\|_{\text{uw}\sharp,*} \|w_h\|_{\text{uw}\sharp},$$

with C_{bnd} independent of h, μ, and β.

Proof. We proceed as in the proof of Lemma 2.30. The only difference lies in bounding the contribution of the advective derivative. Here, using the Cauchy–Schwarz inequality, we infer $\int_\Omega v(\beta\cdot\nabla_h w_h) \leq \|v\|_{\text{uw}\sharp,*}\|w_h\|_{\text{uw}\sharp}$. $\square$

A straightforward consequence of Theorem 1.35 is the following error estimate.

Theorem 2.37 (Error estimate). *Let u solve (2.14) and let u_h solve (2.36) where a_h^{upw} is defined by (2.34) and $V_h = \mathbb{P}_d^k(\mathcal{T}_h)$ with $k \geq 0$ and $\mathcal{T}_h$ belonging to an admissible mesh sequence. Then, there holds*

$$\|u - u_h\|_{\text{uw}\sharp} \leq C \inf_{y_h \in V_h} \|u - y_h\|_{\text{uw}\sharp,*}, \qquad (2.45)$$

with C independent of h and depending on the data only through the factor $\{\min(1, \tau_c\mu_0)\}^{-1}$.

To infer a convergence result from (2.45), we assume, as in Sect. 2.3.2, that the exact solution is smooth enough and use Lemmata 1.58 and 1.59.

2.3. Upwinding

Corollary 2.38 (Convergence rate for smooth solutions). *Besides the hypotheses of Theorem 2.37, assume that $u \in H^{k+1}(\Omega)$. Then, there holds*

$$\|u - u_h\|_{\text{uw}\sharp} \leq C_u h^{k+1/2}, \qquad (2.46)$$

with $C_u = C\|u\|_{H^{k+1}(\Omega)}$ and C independent of h and depending on the data only through the factor $\{\min(1, \tau_c \mu_0)\}^{-1}$.

Estimate (2.46) improves (2.40) since it provides an optimal convergence estimate for the scaled advective derivative.

2.3.4 Numerical Fluxes

To conclude this section, we examine how the additional penalty term on the interface jumps modifies the numerical fluxes. Proceeding as in Sect. 2.2.3, we obtain the following local formulation: For all $T \in \mathcal{T}_h$ and all $\xi \in \mathbb{P}_d^k(T)$,

$$\int_T \left\{ (\mu - \nabla \cdot \beta) u_h \xi - u_h (\beta \cdot \nabla \xi) \right\} + \sum_{F \in \mathcal{F}_T} \epsilon_{T,F} \int_F \phi_F(u_h) \xi = \int_T f\xi, \qquad (2.47)$$

where the numerical fluxes now take the form

$$\phi_F(u_h) = \begin{cases} \beta \cdot n_F \{\!\{u_h\}\!\} + \frac{1}{2} \eta |\beta \cdot n_F| [\![u_h]\!] & \text{if } F \in \mathcal{F}_h^i, \\ (\beta \cdot n)^{\oplus} u_h & \text{if } F \in \mathcal{F}_h^b. \end{cases}$$

The choice $\eta = 1$ leads to the so-called *upwind fluxes*

$$\phi_F(u_h) = \begin{cases} \beta \cdot n_F u_h^{\uparrow} & \text{if } F \in \mathcal{F}_h^i, \\ (\beta \cdot n)^{\oplus} u_h & \text{if } F \in \mathcal{F}_h^b, \end{cases}$$

where $u_h^{\uparrow} = u_h|_{T_1}$ if $\beta \cdot n_F > 0$ and $u_h^{\uparrow} = u_h|_{T_2}$ otherwise (recall that $F = \partial T_1 \cap \partial T_2$ and that n_F points from T_1 toward T_2). The upwind fluxes can also be written as

$$\phi_F(u_h) = \begin{cases} (\beta \cdot n_F)^{\oplus} u_h|_{T_1} - (\beta \cdot n_F)^{\ominus} u_h|_{T_2} & \text{if } F \in \mathcal{F}_h^i, \\ (\beta \cdot n)^{\oplus} u_h & \text{if } F \in \mathcal{F}_h^b. \end{cases}$$

Remark 2.39 (Nonhomogeneous boundary condition). When working with the nonhomogeneous boundary condition $u = g$ on $\partial\Omega^-$ (cf. Sect. 2.1.6), the local formulation can still be written in the form (2.47), and the boundary datum g incorporated in the numerical flux. This yields

$$\phi_F(u_h) = \begin{cases} (\beta \cdot n_F)^{\oplus} u_h|_{T_1} - (\beta \cdot n_F)^{\ominus} u_h|_{T_2} & \text{if } F \in \mathcal{F}_h^i, \\ (\beta \cdot n)^{\oplus} u_h - (\beta \cdot n_F)^{\ominus} g & \text{if } F \in \mathcal{F}_h^b. \end{cases}$$

Chapter 3
Unsteady First-Order PDEs

This chapter is devoted to the approximation of unsteady, scalar, first-order PDEs. We focus on the so-called *method of lines* in which the evolution problem is first semidiscretized in space yielding a system of coupled ordinary differential equations (ODEs) which is then discretized in time. Specifically, we consider space semidiscretization by a dG method together with an explicit Runge–Kutta (RK) scheme for time discretization. Space-time dG methods provide an alternative to the method of lines. Such methods are not covered herein, and we refer the reader to the textbook by Thomée [294].

In Sect. 3.1, we consider linear PDEs and examine the unsteady version of the advection-reaction equation studied in Chap. 2. Namely, for a given finite time $t_F > 0$, we consider the time evolution problem:

$$\partial_t u + \beta \cdot \nabla u + \mu u = f \quad \text{in } \Omega \times (0, t_F), \tag{3.1a}$$

$$u = 0 \quad \text{on } \partial\Omega^- \times (0, t_F), \tag{3.1b}$$

$$u(\cdot, t = 0) = u_0 \quad \text{in } \Omega, \tag{3.1c}$$

with advective velocity β, reaction coefficient μ, source term f, and initial datum u_0. We recall that $\partial\Omega^- := \{x \in \partial\Omega \mid \beta(x) \cdot n(x) < 0\}$ denotes the inflow part of the boundary $\partial\Omega$. Space semidiscretization is achieved using the upwind dG method introduced in Sect. 2.3 for the steady case. Then, we analyze some explicit RK schemes for time discretization. We consider the forward Euler method combined with a finite volume scheme in space as an example of low-order approximation and then explicit two- and three-stage RK schemes combined with a dG scheme with polynomial degree $k \geq 1$ in space as examples of higher-order approximations. In all cases, we derive quasi-optimal error estimates for smooth solutions in the natural energy norm comprising the L^2-norm and the jumps. These results are achieved under Courant–Friedrichs–Lewy (CFL) conditions restricting the size of the time step in terms of the advective velocity and the meshsize. Two points in the analysis deserve comments. First, as pointed out in the seminal work of Levy and Tadmor [231], the classical

D.A. Di Pietro and A. Ern, *Mathematical Aspects of Discontinuous Galerkin Methods*, Mathématiques et Applications 69, DOI 10.1007/978-3-642-22980-0_3, © Springer-Verlag Berlin Heidelberg 2012

stability analysis hinging on the eigenvalues of the space semidiscrete linear operator and absolute stability regions for the RK schemes can be misleading in the case of nonnormal operators such as semidiscrete first-order PDEs. Following the recent work by Burman, Ern, and Fernández [66], our stability analysis instead hinges on energy (that is, L^2-norm) estimates in physical space. The main idea is that the energy production resulting from the explicit nature of the RK scheme can be compensated under a CFL condition by the dissipativity of the dG scheme. Second, as in the steady case, the boundedness of the space semidiscrete operator is needed to derive quasi-optimal error estimates (and not only stability estimates).

In Sect. 3.2, we consider nonlinear conservation laws of the form

$$\partial_t u + \nabla\cdot\mathfrak{f}(u) = 0.$$

The unknown u is a scalar-valued function and the flux function $\mathfrak{f}$ is $\mathbb{R}^d$-valued. In the linear case where $f(u) = \beta u$ for some $\mathbb{R}^d$-valued field β, we recover the unsteady advection-reaction equation (3.1a) with $\mu = \nabla\cdot\beta$ and zero source term. Nonlinear conservation laws, and hyperbolic systems of nonlinear conservation laws, are of interest in many contexts, ranging from compressible fluid dynamics to gas dynamics and trafic modeling. For the mathematical theory of nonlinear conservation laws (and systems thereof), we refer the reader to the textbooks by Godlewski and Raviart [172], LeFloch [225], and LeVeque [230]. The application of dG methods to such problems has been propelled by the pioneering works of Cockburn, Shu, and coworkers [99, 108, 110, 111, 113]. Taking inspiration from finite volume schemes and the local formulation of dG methods using numerical fluxes, the dG method for space semidiscretization is formulated using suitably designed numerical fluxes. Classical examples, discussed in the textbooks by LeVeque [230] and Toro [295], include Godunov, Rusanov, Lax–Friedrichs, and Roe fluxes. Then, an explicit RK scheme can be used for time discretization. In particular, we consider Strong Stability-Preserving (SSP) RK schemes following the ideas of Gottlieb, Shu, and Tadmor [175, 176]; see also Spiteri and Ruuth [281]. Finally, we discuss the use of limiters, i.e., devices which help preventing the appearance of spurious oscillations when approximating rough solutions. Although dG discretizations of hyperbolic conservation laws have been used for over two decades, their theoretical analysis is far from being accomplished. Part of the material in Sect. 3.2 is based on heuristics, and only in some cases a rigorous mathematical justification is available (often in space dimension $d = 1$ only).

3.1 Unsteady Advection-Reaction

The goal of this section is to analyze an approximation of the linear time evolution problem (3.1) using the upwind dG method for space semidiscretization together with an explicit RK scheme for time discretization. Most of the results presented in this section are drawn from [66].

3.1.1 The Continuous Setting

Herein, we introduce some basic notation for space-time functions, specify the assumptions on the data and on the regularity of the exact solution, and present an energy estimate for the latter.

3.1.1.1 Notation for Space-Time Functions

Let ψ be a function defined on the space-time cylinder $\Omega \times (0, t_F)$. Then, it is convenient to consider ψ as a function of the time variable with values in a Hilbert space, say V, spanned by functions of the space variable, in such a way that
$$\psi : (0, t_F) \ni t \longmapsto \psi(t) \equiv \psi(\cdot, t) \in V.$$

In what follows, for an integer $l \geq 0$, we consider the space
$$C^l(V) := C^l([0, t_F]; V)$$

spanned by V-valued functions that are l times continuously differentiable in the interval $[0, t_F]$. For $\psi \in C^l(V)$, we denote by $d_t^m \psi$, $m \in \{0, \ldots, l\}$, the time derivative of ψ of order m with the convention that $d_t^0 \psi = \psi$; clearly $d_t^m \psi \in C^{l-m}(V)$. The space $C^0(V)$ is a Banach space when equipped with the norm
$$\|\psi\|_{C^0(V)} := \max_{t \in [0, t_F]} \|\psi(t)\|_V,$$

and the space $C^l(V)$ is a Banach space when equipped with the norm
$$\|\psi\|_{C^l(V)} := \max_{0 \leq m \leq l} \|d_t^m \psi\|_{C^0(V)}.$$

3.1.1.2 Assumptions on the Data and the Exact Solution

We consider the time evolution problem (3.1) with source term
$$f \in C^0(L^2(\Omega)).$$

We assume for simplicity that the data μ and β are *time-independent* and that they satisfy the regularity assumption (2.4), namely
$$\mu \in L^\infty(\Omega), \qquad \beta \in [\text{Lip}(\Omega)]^d.$$

However, we no longer assume (2.6) so that
$$\Lambda = \mu - \frac{1}{2} \nabla \cdot \beta \in L^\infty(\Omega)$$

does not necessarily take positive values.

As in the steady case, we consider the reference time τ_c and the reference velocity β_c such that (cf. (2.7))
$$\tau_c = \{\max(\|\mu\|_{L^\infty(\Omega)}, L_\beta)\}^{-1}, \qquad \beta_c = \|\beta\|_{[L^\infty(\Omega)]^d},$$

where L_β is the Lipschitz module of β (cf. (2.5)). In the absence of assumption (2.6) providing a uniform positive lower bound on $\Lambda = \mu - \frac{1}{2}\nabla\cdot\beta$, it is possible that $\|\mu\|_{L^\infty(\Omega)} = L_\beta = 0$ (this corresponds to the case of constant advective velocity and no reaction), yielding $\tau_c = \infty$. We introduce an additional time scale
$$\tau_* := \min(t_F, \tau_c),$$
which is always finite. Moreover, in what follows, we use the quantity τ_c with the convention that expressions involving τ_c^{-1} are evaluated as zero if $\tau_c = \infty$.

An important point is that we focus on *smooth solutions*, that is, we assume
$$u \in C^0(H^1(\Omega)) \cap C^1(L^2(\Omega)). \tag{3.2}$$

If such a solution exists, it is unique; this is a straightforward consequence of the energy estimate derived below in Lemma 3.2. The regularity assumption (3.2) implies that the PDE (3.1a) and the boundary condition (3.1b) hold up to the initial time $t = 0$. Assumption (3.2) also implies that $u_0 \in H^1(\Omega)$ with $u_0|_{\partial\Omega^-} = 0$; the initial datum is said to be *well-prepared*. Introducing the bounded operator $A \in \mathcal{L}(H^1(\Omega), L^2(\Omega))$ such that for all $v \in H^1(\Omega)$, $Av := \mu v + \beta\cdot\nabla v$, the PDE (3.1a) can be rewritten as
$$d_t u(t) + Au(t) = f(t) \qquad \forall t \in [0, t_F]. \tag{3.3}$$

More general mathematical settings for the unsteady advection equation allowing for rough solutions have been investigated, e.g., by Ambrosio [10], Boyer [48], and Crippa [117].

Remark 3.1 (Using $H^1(\Omega)$ instead of the graph space). At the continuous level, it is also possible to assume for the exact solution that $u \in C^0(V)$ instead of $u \in C^0(H^1(\Omega))$, where V is the graph space introduced for the steady advection-reaction equation and defined by (2.8).

3.1.1.3 Energy Estimate on the Exact Solution

At any time $t \in [0, t_F]$, we define the energy of the exact solution as the quantity $\|u(t)\|^2_{L^2(\Omega)}$. We now establish a basic energy estimate on the exact solution under the regularity assumption (3.2).

Lemma 3.2 (Energy estimate). *Let $u \in C^0(H^1(\Omega)) \cap C^1(L^2(\Omega))$ solve (3.1). Then, introducing the time scale $\varsigma := (t_F^{-1} + 2\|\Lambda\|_{L^\infty(\Omega)})^{-1}$, there holds*
$$\|u(t)\|^2_{L^2(\Omega)} \leq e^{t/\varsigma}(\|u_0\|^2_{L^2(\Omega)} + \varsigma t_F \|f\|^2_{C^0(L^2(\Omega))}) \qquad \forall t \in [0, t_F]. \tag{3.4}$$

Proof. For all $t \in [0, t_F]$, we take the L^2-scalar product of (3.3) with $u(t)$ yielding
$$\frac{1}{2}d_t\|u(t)\|^2_{L^2(\Omega)} + (Au(t), u(t))_{L^2(\Omega)} = (f(t), u(t))_{L^2(\Omega)}.$$

3.1. Unsteady Advection-Reaction

Integrating by parts and accounting for the boundary condition (3.1b) yields

$$(Au(t), u(t))_{L^2(\Omega)} = \int_\Omega \left(\mu - \frac{1}{2}\nabla\cdot\beta\right) u(t)^2 + \int_{\partial\Omega^+} \frac{1}{2}(\beta\cdot\mathrm{n}) u(t)^2$$
$$= \int_\Omega \Lambda u(t)^2 + \int_{\partial\Omega^+} \frac{1}{2}(\beta\cdot\mathrm{n}) u(t)^2 \geq \int_\Omega \Lambda u(t)^2,$$

where $\partial\Omega^+ := \{x \in \partial\Omega \mid \beta(x)\cdot\mathrm{n}(x) > 0\}$ denotes the outflow part of the boundary. Hence,

$$\frac{1}{2} d_t \|u(t)\|_{L^2(\Omega)}^2 = (f(t), u(t))_{L^2(\Omega)} - (Au(t), u(t))_{L^2(\Omega)}$$
$$\leq (f(t), u(t))_{L^2(\Omega)} + \|\Lambda\|_{L^\infty(\Omega)} \|u(t)\|_{L^2(\Omega)}^2. \qquad (3.5)$$

Using the Cauchy–Schwarz followed by Young's inequality for the first term on the right-hand side leads to

$$\frac{1}{2} d_t \|u(t)\|_{L^2(\Omega)}^2 \leq \|f(t)\|_{L^2(\Omega)} \|u(t)\|_{L^2(\Omega)} + \|\Lambda\|_{L^\infty(\Omega)} \|u(t)\|_{L^2(\Omega)}^2$$
$$\leq \frac{t_\mathrm{F}}{2} \|f(t)\|_{L^2(\Omega)}^2 + \frac{1}{2t_\mathrm{F}} \|u(t)\|_{L^2(\Omega)}^2 + \|\Lambda\|_{L^\infty(\Omega)} \|u(t)\|_{L^2(\Omega)}^2$$
$$= \frac{t_\mathrm{F}}{2} \|f(t)\|_{L^2(\Omega)}^2 + \frac{1}{2\varsigma} \|u(t)\|_{L^2(\Omega)}^2. \qquad (3.6)$$

Setting $\varphi(t) = \|u(t)\|_{L^2(\Omega)}^2$ and $\alpha = t_\mathrm{F} \|f\|_{C^0(L^2(\Omega))}^2$, we obtain

$$d_t \varphi(t) \leq \alpha + \varsigma^{-1} \varphi(t).$$

Hence, $d_t(e^{-t/\varsigma} \varphi(t)) = e^{-t/\varsigma}(d_t\varphi(t) - \varsigma^{-1}\varphi(t)) \leq e^{-t/\varsigma}\alpha$, so that integrating in time, we arrive at

$$\varphi(t) \leq e^{t/\varsigma} \varphi(0) + \alpha \int_0^t e^{(t-s)/\varsigma} \, ds \leq e^{t/\varsigma}(\varphi(0) + \varsigma\alpha),$$

whence the assertion follows. $\qquad \square$

Remark 3.3 (Gronwall's Lemma and long-time behavior). The argument by which a bound on $\|u(t)\|_{L^2(\Omega)}^2$ is inferred from an estimate of the form (3.6) is often referred to as Gronwall's Lemma. This approach is not suitable for long-time stability estimates (that is, for $t_\mathrm{F} \gg \|\Lambda\|_{L^\infty(\Omega)}^{-1}$) because it leads to the exponential factor $e^{t/\varsigma}$ in (3.4) which becomes extremely large in this situation. When interested in the long-time behavior of the solution, an assumption on Λ is needed to obtain sharper energy estimates, for instance assumption (2.6) (i.e., $\Lambda \geq \mu_0 > 0$ a.e. in Ω) as in the steady case. Then, (3.5) can be replaced by

$$\frac{1}{2} d_t \|u(t)\|_{L^2(\Omega)}^2 + \mu_0 \|u(t)\|_{L^2(\Omega)}^2 \leq (f(t), u(t))_{L^2(\Omega)}.$$

Using the Cauchy–Schwarz inequality followed by Young's inequality on the right-hand side leads to

$$\frac{1}{2}d_t\|u(t)\|_{L^2(\Omega)}^2 \leq \frac{1}{4\mu_0}\|f(t)\|_{L^2(\Omega)}^2,$$

whence we infer a (much) sharper energy estimate than (3.4).

3.1.2 Space Semidiscretization

Space semidiscretization is achieved by means of the upwind dG method introduced in Sect. 2.3 in the context of the steady advection-reaction equation. Thus, the discrete space V_h is taken to be the broken polynomial space $\mathbb{P}_d^k(\mathcal{T}_h)$ defined by (1.15) with polynomial degree $k \geq 0$ and $\mathcal{T}_h$ belonging to an admissible mesh sequence. We consider quasi-uniform mesh sequences, meaning that there is C such that, for all $h \in \mathcal{H}$,

$$\max_{T \in \mathcal{T}_h} h_T \leq C \min_{T \in \mathcal{T}_h} h_T.$$

It is possible to relax this assumption; cf. Remark 3.6 below for further discussion. As in the steady case (cf. Sect. 2.3.3), we also make the mild assumption that

$$h \leq \beta_c \tau_*,$$

recalling that the assumption $h \leq \beta_c \tau_c$ means that the local Damköhler number is not too large (thereby avoiding strong reaction regimes) and that the meshsize resolves the spatial variations of the advective velocity, while the assumption $h \leq \beta_c t_F$ means that a particle advected at speed β_c crosses at least one mesh element over the time interval $(0, t_F)$.

Instead of working with discrete bilinear forms, we consider discrete (differential) operators to allow for a more compact notation. In what follows, we often manipulate functions of the form $(u(t) - v_h)$ where $t \in [0, t_F]$ and $v_h \in V_h$. Recalling that $u \in C^0(H^1(\Omega))$ owing to assumption (3.2), the function $(u(t) - v_h)$ belongs to the space

$$V_{*h} := H^1(\Omega) + V_h.$$

We define the discrete operator $A_h^{\mathrm{upw}} : V_{*h} \to V_h$ such that for all $(v, w_h) \in V_{*h} \times V_h$,

$$(A_h^{\mathrm{upw}} v, w_h)_{L^2(\Omega)} = \int_\Omega \mu v w_h + \int_\Omega (\beta \cdot \nabla_h v) w_h + \int_{\partial\Omega} (\beta \cdot n)^\ominus v w_h \qquad (3.7)$$
$$- \sum_{F \in \mathcal{F}_h^i} \int_F (\beta \cdot n_F)[\![v]\!]\{\!\{w_h\}\!\} + \sum_{F \in \mathcal{F}_h^i} \int_F \frac{1}{2}|\beta \cdot n_F|[\![v]\!][\![w_h]\!].$$

In terms of the discrete bilinear forms considered in Chap. 2, this yields

$$(A_h^{\mathrm{upw}} v, w_h)_{L^2(\Omega)} = a_h^{\mathrm{upw}}(v, w_h) = a_h^{\mathrm{cf}}(v, w_h) + s_h(v, w_h),$$

where a_h^{upw} is the upwind dG bilinear form (cf. (2.34)), a_h^{cf} the dG bilinear form associated with centered fluxes (cf. (2.24)), and s_h the stabilization bilinear form

3.1. Unsteady Advection-Reaction

defined by (2.33). The penalty parameter η in the stabilization bilinear form s_h has been set to 1 for simplicity.

The discrete operator A_h^{upw} can be used to formulate the space semidiscrete problem in the form (compare with (3.3))

$$d_t u_h(t) + A_h^{\text{upw}} u_h(t) = f_h(t) \qquad \forall t \in [0, t_F], \tag{3.8}$$

with initial condition $u_h(0) = \pi_h u_0$ and source term

$$f_h(t) = \pi_h f(t) \qquad \forall t \in [0, t_F],$$

where π_h denotes the L^2-orthogonal projection onto V_h. Choosing a basis for V_h, the space semidiscrete evolution problem (3.8) can be transformed into a system of coupled ODEs for the time-dependent components of $u_h(t)$ on the selected basis. Written in component form, (3.8) leads to the appearance of the block-diagonal mass matrix on its left-hand side; cf. Sect. A.1.2.

Two important properties of the discrete operator A_h^{upw} are consistency and discrete dissipation. These two properties are the counterpart of the consistency and discrete coercivity established in the steady case for the discrete bilinear form a_h^{upw} (cf. Lemma 2.27).

Lemma 3.4 (Consistency and discrete dissipation for A_h^{upw}). *The discrete operator A_h^{upw} satisfies the following properties:*

(i) *Consistency: For the exact solution $u \in C^0(H^1(\Omega)) \cap C^1(L^2(\Omega))$,*

$$\pi_h d_t u(t) + A_h^{\text{upw}} u(t) = f_h(t) \qquad \forall t \in [0, t_F]. \tag{3.9}$$

(ii) *Discrete dissipation: For all $v_h \in V_h$,*

$$(A_h^{\text{upw}} v_h, v_h)_{L^2(\Omega)} = |v_h|_\beta^2 + (\Lambda v_h, v_h)_{L^2(\Omega)}, \tag{3.10}$$

where we have defined on V_{*h} the seminorm

$$|v|_\beta^2 := \int_{\partial \Omega} \frac{1}{2} |\beta \cdot n| v^2 + \sum_{F \in \mathcal{F}_h^i} \int_F \frac{1}{2} |\beta \cdot n_F| [\![v]\!]^2. \tag{3.11}$$

Proof. (i) Let $w_h \in V_h$. Taking the L^2-scalar product of (3.3) with w_h implies for all $t \in [0, t_F]$,

$$(d_t u(t), w_h)_{L^2(\Omega)} + (Au(t), w_h)_{L^2(\Omega)} = (f(t), w_h)_{L^2(\Omega)}.$$

Since $u \in C^0(H^1(\Omega))$ and accounting for the boundary condition $u(t)|_{\partial \Omega^-} = 0$ and the fact that $[\![u(t)]\!] = 0$ for all $F \in \mathcal{F}_h^i$, the expression (3.7) yields

$$(A_h^{\text{upw}} u(t), w_h)_{L^2(\Omega)} = \int_\Omega (\mu u(t) + \beta \cdot \nabla u(t), w_h)_{L^2(\Omega)} = (Au(t), w_h)_{L^2(\Omega)}.$$

Moreover, owing to the definition of π_h, $(f(t), w_h)_{L^2(\Omega)} = (\pi_h f(t), w_h)_{L^2(\Omega)}$ and similarly for $d_t u(t)$. Hence,

$$(\pi_h d_t u(t), w_h)_{L^2(\Omega)} + (A_h^{\text{upw}} u(t), w_h)_{L^2(\Omega)} = (f_h(t), w_h)_{L^2(\Omega)},$$

thereby proving (3.9).
(ii) To prove (3.10), we observe that, for all $v_h \in V_h$, integrating by parts on each mesh element the advective derivative yields, similarly to the steady case,

$$\begin{aligned}(A_h^{\text{upw}} v_h, v_h)_{L^2(\Omega)} &= \int_\Omega \left(\mu - \frac{1}{2}\nabla\cdot\beta\right) v_h^2 + \int_{\partial\Omega} \frac{1}{2}(\beta\cdot\mathbf{n}) v_h^2 + \int_{\partial\Omega} (\beta\cdot\mathbf{n})^\ominus v_h^2 \\ &\quad + \sum_{F \in \mathcal{F}_h^i} \int_F \frac{1}{2}|\beta\cdot\mathbf{n}_F|[\![v_h]\!]^2 \\ &= \int_\Omega \Lambda v_h^2 + \int_{\partial\Omega} \frac{1}{2}|\beta\cdot\mathbf{n}|v_h^2 + \sum_{F \in \mathcal{F}_h^i} \int_F \frac{1}{2}|\beta\cdot\mathbf{n}_F|[\![v_h]\!]^2,\end{aligned}$$

and the assertion follows from the definition of the $|\cdot|_\beta$-seminorm. □

Remark 3.5 (Discrete dissipation). We slightly abuse the terminology when referring to (3.10) as discrete dissipation since the term $(\Lambda v_h, v_h)_{L^2(\Omega)}$ has no a priori sign. The important point, however, is that the term $|v_h|_\beta^2$ is nonnegative. This latter term plays a crucial role in the stability analysis of explicit RK schemes.

3.1.3 Time Discretization

Let δt be the time step, taken to be constant for simplicity and such that $t_F = N\delta t$ where N is an integer. For $n \in \{0, \ldots, N\}$, we define the discrete times $t^n := n\delta t$. A superscript n indicates the value of a function at the discrete time $n\delta t$, so that, e.g., $u^n = u(t^n)$ and $f^n = f(t^n)$. We make the mild assumption that

$$\delta t \leq \tau_*,$$

and observe that the inequality $\delta t \leq t_F$ is trivial, while the inequality $\delta t \leq \tau_c$ means that the time step resolves the reference time τ_c.

The simplest scheme to discretize in time is the *forward* (also called *explicit*) *Euler scheme* which takes the form

$$u_h^{n+1} = u_h^n - \delta t A_h^{\text{upw}} u_h^n + \delta t f_h^n, \qquad (3.12)$$

with the initial condition $u_h^0 = \pi_h u_0$. The forward Euler scheme can be derived by replacing the time derivative by a forward finite difference in (3.8); indeed, (3.12) can be equivalently rewritten as

$$\frac{u_h^{n+1} - u_h^n}{\delta t} + A_h^{\text{upw}} u_h^n = f_h^n.$$

To improve the accuracy of time discretization, one possibility is to consider explicit *Runge–Kutta* (RK) schemes. Such schemes are one-step methods where,

3.1. Unsteady Advection-Reaction

at each time step, starting from u_h^n, s stages, $s \geq 1$, are performed to compute u_h^{n+1}. Explicit RK schemes can be formulated in various forms. Herein, we consider two approaches. The first one takes the form

$$k_i = -A_h^{\text{upw}}\left(u_h^n + \delta t \sum_{j=1}^{s} a_{ij} k_j\right) + f_h(t^n + c_i \delta t) \qquad \forall i \in \{1, \ldots, s\}, \tag{3.13a}$$

$$u_h^{n+1} = u_h^n + \delta t \sum_{i=1}^{s} b_i k_i. \tag{3.13b}$$

Here, $(a_{ij})_{1 \leq i,j \leq s}$ are real numbers, $(b_i)_{1 \leq i \leq s}$ are real numbers such that $\sum_{i=1}^{s} b_i = 1$, and $(c_i)_{1 \leq i \leq s}$ are real numbers in $[0,1]$ such that $c_i = \sum_{j=1}^{s} a_{ij}$ for all $i \in \{1, \ldots, s\}$. These quantities are usually collected in the so-called *Butcher's array*

$$\begin{bmatrix} c_1 & a_{11} & \cdots & a_{1s} \\ \vdots & \vdots & & \vdots \\ c_s & a_{s1} & \cdots & a_{ss} \\ \hline & b_1 & \cdots & b_s \end{bmatrix}.$$

The scheme is explicit whenever $a_{ij} = 0$ for all $j \geq i$. We observe that explicit schemes still require the inversion of the mass matrix at each of the s stages (3.13a). In the context of dG methods, mass matrices are block-diagonal and with a suitable choice of basis functions can be made diagonal; cf. Sects. A.1.2 and A.2.

The *forward Euler* scheme is actually a one-stage RK method for which

$$\begin{bmatrix} 0 & 0 \\ \hline & 1 \end{bmatrix} \qquad \begin{cases} k_1 = -A_h^{\text{upw}} u_h^n + f_h^n \\ u_h^{n+1} = u_h^n + \delta t k_1 \end{cases}$$

Two examples of two-stage RK schemes are, on the one hand, the two-stage *Runge scheme* (also called the *improved forward Euler scheme*) for which

$$\begin{bmatrix} 0 & 0 & 0 \\ 1/2 & 1/2 & 0 \\ \hline & 0 & 1 \end{bmatrix} \qquad \begin{cases} k_1 = -A_h^{\text{upw}} u_h^n + f_h^n \\ k_2 = -A_h^{\text{upw}}(u_h^n + \tfrac{1}{2}\delta t k_1) + f_h^{n+1/2} \\ u_h^{n+1} = u_h^n + \delta t k_2 \end{cases} \tag{3.14}$$

with the notation $f_h^{n+1/2} = f_h(t^n + \tfrac{1}{2}\delta t)$ and, on the other hand, the *two-stage Heun scheme* for which

$$\begin{bmatrix} 0 & 0 & 0 \\ 1 & 1 & 0 \\ \hline & 1/2 & 1/2 \end{bmatrix} \qquad \begin{cases} k_1 = -A_h^{\text{upw}} u_h^n + f_h^n \\ k_2 = -A_h^{\text{upw}}(u_h^n + \delta t k_1) + f_h^{n+1} \\ u_h^{n+1} = u_h^n + \delta t \tfrac{1}{2}(k_1 + k_2) \end{cases} \tag{3.15}$$

In the absence of external forcing ($f = 0$) and since the discrete operator A_h^{upw} is linear, the schemes (3.14) and (3.15) produce the same result for u_h^{n+1}. Indeed, straightforward algebra shows that both (3.14) and (3.15) yield

$$u_h^{n+1} = u_h^n - \delta t A_h^{\text{upw}} u_h^n + \tfrac{1}{2}\delta t^2 (A_h^{\text{upw}})^2 u_h^n.$$

On the right-hand side, we recognize a second-order Taylor expansion in time at t^n where the time derivatives have been substituted using the space semidiscrete scheme (3.8) without forcing (that is, $d_t u(t^n) = -A_h^{\text{upw}} u(t^n)$ and $d_{tt} u(t^n) = (A_h^{\text{upw}})^2 u(t^n)$). Furthermore, an example of three-stage RK scheme is the *three-stage Heun* scheme for which

$$\begin{bmatrix} 0 & 0 & 0 & 0 \\ 1/3 & 1/3 & 0 & 0 \\ 2/3 & 0 & 2/3 & 0 \\ \hline & 1/4 & 0 & 3/4 \end{bmatrix} \quad \begin{cases} k_1 = -A_h^{\text{upw}} u_h^n + f_h^n, \\ k_2 = -A_h^{\text{upw}} (u_h^n + \tfrac{1}{3}\delta t k_1) + f_h^{n+1/3} \\ k_3 = -A_h^{\text{upw}} (u_h^n + \tfrac{2}{3}\delta t k_2) + f_h^{n+2/3} \\ u_h^{n+1} = u_h^n + \tfrac{1}{4}\delta t (k_1 + 3k_3) \end{cases} \quad (3.16)$$

with obvious notation for the source term. In the absence of external forcing ($f = 0$), straightforward algebra shows that (3.16) yields

$$u_h^{n+1} = u_h^n - \delta t A_h^{\text{upw}} u_h^n + \tfrac{1}{2}\delta t^2 (A_h^{\text{upw}})^2 u_h^n - \tfrac{1}{6}\delta t^3 (A_h^{\text{upw}})^3 u_h^n.$$

On the right-hand side, we recognize now a third-order Taylor expansion in time. Finally, an example of four-stage RK scheme is

$$\begin{bmatrix} 0 & 0 & 0 & 0 & 0 \\ 1/2 & 1/2 & 0 & 0 & 0 \\ 1/2 & 0 & 1/2 & 0 & 0 \\ 1 & 0 & 0 & 1 & 0 \\ \hline & 1/6 & 1/3 & 1/3 & 1/6 \end{bmatrix} \quad \begin{cases} k_1 = -A_h^{\text{upw}} u_h^n + f_h^n, \\ k_2 = -A_h^{\text{upw}} (u_h^n + \tfrac{1}{2}\delta t k_1) + f_h^{n+1/2} \\ k_3 = -A_h^{\text{upw}} (u_h^n + \tfrac{1}{2}\delta t k_2) + f_h^{n+1/2} \\ k_4 = -A_h^{\text{upw}} (u_h^n + \delta t k_3) + f_h^{n+1} \\ u_h^{n+1} = u_h^n + \tfrac{1}{6}\delta t (k_1 + 2k_2 + 2k_3 + k_4) \end{cases}$$

leading to a fourth-order Taylor expansion in time in the absence of external forcing. A consequence of these Taylor expansions is that the above s-stage RK schemes with $s \leq 4$ have formal order of accuracy equal to s. Unfortunately, it is not possible to devise an s-stage RK scheme with formal order of accuracy s for $s \geq 5$; this is possible using more than s steps (see Butcher [74]). In practice, values of $s \leq 4$ are most often considered.

An alternative formulation of RK schemes consists in introducing intermediate stages for the discrete solution instead of the intermediate increments k_i. Following Shu and Osher [279], we write in the absence of external forcing the s-stage explicit RK scheme in the form

$$u_h^{n,0} = u_h^n,$$

$$u_h^{n,i} = \sum_{j=0}^{i-1} \left(\alpha_{ij} u_h^{n,j} - \delta t \beta_{ij} A_h^{\text{upw}} u_h^{n,j} \right) \qquad \forall i \in \{1, \ldots, s\},$$

$$u_h^{n+1} = u_h^{n,s}.$$

3.1. Unsteady Advection-Reaction

The scheme is specified through the lower triangular matrices $(\alpha_{ij})_{1 \leq i \leq s, 0 \leq j \leq i-1}$ and $(\beta_{ij})_{1 \leq i \leq s, 0 \leq j \leq i-1}$. Two examples are the two-stage scheme for which

$$\alpha = \begin{bmatrix} 1 & \\ 1/2 & 1/2 \end{bmatrix}, \qquad \beta = \begin{bmatrix} 1 & \\ 0 & 1/2 \end{bmatrix},$$

yielding (eliminating $u_h^{n,0} = u_h^n$ and $u_h^{n,2} = u_h^{n+1}$ and observing that the diagonal of α and β corresponds to the coefficients $\alpha_{i,i-1}$ and $\beta_{i,i-1}$)

$$u_h^{n,1} = u_h^n - \delta t A_h^{\text{upw}} u_h^n, \tag{3.17a}$$
$$u_h^{n+1} = \tfrac{1}{2}(u_h^n + u_h^{n,1}) - \tfrac{1}{2}\delta t A_h^{\text{upw}} u_h^{n,1}, \tag{3.17b}$$

and the three-stage scheme for which

$$\alpha = \begin{bmatrix} 1 & & \\ 1/2 & 1/2 & \\ 1/3 & 1/3 & 1/3 \end{bmatrix}, \qquad \beta = \begin{bmatrix} 1 & & \\ 0 & 1/2 & \\ 0 & 0 & 1/3 \end{bmatrix},$$

yielding

$$u_h^{n,1} = u_h^n - \delta t A_h^{\text{upw}} u_h^n, \tag{3.18a}$$
$$u_h^{n,2} = \tfrac{1}{2}(u_h^n + u_h^{n,1}) - \tfrac{1}{2}\delta t A_h^{\text{upw}} u_h^{n,1}, \tag{3.18b}$$
$$u_h^{n+1} = \tfrac{1}{3}(u_h^n + u_h^{n,1} + u_h^{n,2}) - \tfrac{1}{3}\delta t A_h^{\text{upw}} u_h^{n,2}. \tag{3.18c}$$

Since A_h^{upw} is linear, (3.17) produces the same result for u_h^{n+1} as the two-stage Runge and Heun schemes (3.14) and (3.15), while (3.18) produces the same result for u_h^{n+1} as the three-stage Heun scheme (3.16). Moreover, the schemes (3.17) and (3.18) produce, as intermediate stages, Taylor expansions of increasing order in time. For instance, (3.18) can be equivalently rewritten as

$$u_h^{n,1} = u_h^n - \delta t A_h^{\text{upw}} u_h^n,$$
$$u_h^{n,2} = u_h^n - \delta t A_h^{\text{upw}} u_h^n + \tfrac{1}{2}\delta t^2 (A_h^{\text{upw}})^2 u_h^n,$$
$$u_h^{n+1} = u_h^n - \delta t A_h^{\text{upw}} u_h^n + \tfrac{1}{2}\delta t^2 (A_h^{\text{upw}})^2 u_h^n - \tfrac{1}{6}\delta t^3 (A_h^{\text{upw}})^3 u_h^n.$$

As a result, a natural way to account for external forcing ($f \neq 0$) is to set for the two-stage scheme (3.17),

$$u_h^{n,1} = u_h^n - \delta t A_h^{\text{upw}} u_h^n + \delta t f_h^n, \tag{3.19a}$$
$$u_h^{n+1} = \tfrac{1}{2}(u_h^n + u_h^{n,1}) - \tfrac{1}{2}\delta t A_h^{\text{upw}} u_h^{n,1} + \tfrac{1}{2}\delta t \psi_h^n, \tag{3.19b}$$

and to require that ψ_h^n provides, for smooth f, a second-order accurate approximation of $f_h^n + \delta t d_t f_h^n$ (cf. (3.24) below for the precise statement). For instance, the two-stage Runge scheme (3.14) corresponds to (3.19) with $\psi_h^n = 2 f_h^{n+1/2} - f_h^n$, while the two-stage Heun scheme corresponds to (3.19) with $\psi_h^n = f_h^{n+1}$. Similarly, for the three-stage scheme (3.17), one possibility is to set

$$u_h^{n,1} = u_h^n - \delta t A_h^{\text{upw}} u_h^n + \delta t f_h^n, \tag{3.20a}$$
$$u_h^{n,2} = \tfrac{1}{2}(u_h^n + u_h^{n,1}) - \tfrac{1}{2}\delta t A_h^{\text{upw}} u_h^{n,1} + \tfrac{1}{2}\delta t(f_h^n + \delta t d_t f_h^n), \tag{3.20b}$$
$$u_h^{n+1} = \tfrac{1}{3}(u_h^n + u_h^{n,1} + u_h^{n,2}) - \tfrac{1}{3}\delta t A_h^{\text{upw}} u_h^{n,2} + \tfrac{1}{3}\delta t \psi_h^n, \tag{3.20c}$$

and to require that ψ_h^n provides, for smooth f, a third-order accurate approximation of $f_h^n + \delta t d_t f_h^n + \frac{1}{2}\delta t^2 d_{tt} f_h^n$ (cf. (3.28) for the precise statement). The three-stage Heun scheme (3.16) fits this form; see [66]. Henceforth, (3.17)–(3.20) are termed *explicit RK2* and *explicit RK3* schemes, respectively.

When the discrete operator A_h^{upw} is linear, the two formulations of explicit RK methods, using either intermediate increments or intermediate stages for the discrete solution, are equivalent in the absence of external forcing (differences do appear in the handling of the external forcing). The situation is different in the nonlinear case, e.g., when the discrete operator A_h^{upw} results from the space semidiscretization of a nonlinear conservation law. Then, the second form, based on intermediate stages for the discrete solution, is more appropriate; cf. Sect. 3.2.3 for further discussion.

3.1.4 Main Convergence Results

In this section, we present convergence results for the forward Euler and the explicit RK2 and RK3 schemes for time discretization combined with the discrete operator A_h^{upw} resulting from the upwind dG method for space semidiscretization. In particular, we derive quasi-optimal error estimates for smooth solutions under CFL-type restrictions on the time step of the form

$$\delta t \leq \varrho \frac{h}{\beta_c}, \tag{3.21}$$

for some positive real number ϱ. For the forward Euler scheme, we only consider finite volume schemes (that is, dG schemes with polynomial degree $k = 0$) for space semidiscretization. The reason for this is that the restriction on the time step to achieve stability (or, in other words, to compensate the anti-dissipative nature of the forward Euler scheme) is too stringent whenever polynomials with degree $k \geq 1$ are employed because there is not enough dissipation produced by the space semidiscretization; cf. Remark 3.21 below for a detailed discussion. For explicit RK2 and RK3 schemes, we consider dG schemes with polynomial degree $k \geq 1$ for space semidiscretization. It is also possible to use finite volume schemes, but this is seldom considered in practice because, in such an approach, the error estimates are dominated by space approximation errors.

Herein, we only state the convergence results. Proofs for the forward Euler scheme and the explicit RK2 scheme are presented in Sects. 3.1.5 and 3.1.6, respectively, while the proof for the explicit RK3 scheme can be found in [66]. General convergence results for forward Euler and finite volume schemes have been obtained recently by Després [124,125] and by Merlet and Vovelle [236]. The convergence result for explicit RK2 and RK3 schemes was proven, using different arguments, by Zhang and Shu [310,311] for nonlinear conservation laws.

In what follows, we abbreviate as $a \lesssim b$ the inequality $a \leq Cb$ with positive C independent of h, δt, and the data f, μ, and β; the actual value of C can change at each occurrence.

Remark 3.6 (Mesh quasi-uniformity and CFL condition). It is possible to discard the assumption on mesh quasi-uniformity and to consider admissible mesh

3.1. Unsteady Advection-Reaction

sequences to build the broken polynomial spaces V_h. In this case, the local length scale in the CFL condition is no longer the largest mesh element diameter, but the smallest. It is also possible, with additional technicalities, to consider in the CFL condition the lowest ratio $\|\beta\|_{[L^\infty(T)]^d}^{-1} h_T$ over $T \in \mathcal{T}_h$.

3.1.4.1 Forward Euler and Finite Volume Schemes

We are interested in the forward Euler scheme (3.12),
$$u_h^{n+1} = u_h^n - \delta t A_h^{\text{upw}} u_h^n + \delta t f_h^n,$$
with initial condition $u_h^0 = \pi_h u_0$. We focus on finite volume schemes for space semidiscretization so that $V_h = \mathbb{P}_d^0(\mathcal{T}_h)$.

Theorem 3.7 (Convergence for forward Euler). *Assume $u \in C^0(H^1(\Omega)) \cap C^2(L^2(\Omega))$. Set $V_h = \mathbb{P}_d^0(\mathcal{T}_h)$. Assume the CFL condition (3.21) with $\varrho \leq \varrho^{\text{Eul}}$ for a suitable threshold ϱ^{Eul}, independent of h, δt, and the data f, μ, and β. Then, there holds*
$$\|u^N - u_h^N\|_{L^2(\Omega)} + \left(\sum_{m=0}^{N-1} \delta t |u^m - u_h^m|_\beta^2\right)^{1/2} \lesssim e^{C_{\text{sta}} t_F/\tau_*}(\chi_1 \delta t + \chi_2 h^{1/2}), \quad (3.22)$$
where $\chi_1 = t_F^{1/2} \tau_*^{1/2} \|d_t^2 u\|_{C^0(L^2(\Omega))}$ and $\chi_2 = t_F^{1/2} \beta_c^{1/2} \|u\|_{C^0(H^1(\Omega))}$, the $|\cdot|_\beta$-seminorm is defined by (3.11), and the quantity C_{sta} is independent of h, δt, and the data f, μ, and β.

Remark 3.8 (Convergence rate with CFL condition). Under the CFL condition (3.21), the first term on the right-hand side of the error estimate (3.22) converges as h, while the second one converges as $h^{1/2}$ and is, therefore, dominant as $h \to 0$.

Remark 3.9 (Long-time behavior). The exponential factor $e^{C_{\text{sta}} t_F/\tau_*}$ in the error estimate (3.22) means that the estimate deteriorates in long time, that is, whenever $t_F \gg \tau_c$ (recall that τ_c can be interpreted as the fastest time-scale in the time evolution problem (3.1)). In practice, it is reasonable to assume that τ_c is not too small with respect to t_F.

3.1.4.2 Explicit RK2 Schemes

We are interested in explicit RK2 schemes of the form (3.19) (for later use, we introduce a specific symbol for the intermediate stage)
$$w_h^n = u_h^n - \delta t A_h^{\text{upw}} u_h^n + \delta t f_h^n, \quad (3.23a)$$
$$u_h^{n+1} = \tfrac{1}{2}(u_h^n + w_h^n) - \tfrac{1}{2}\delta t A_h^{\text{upw}} w_h^n + \tfrac{1}{2}\delta t \psi_h^n, \quad (3.23b)$$
with initial condition $u_h^0 = \pi_h u_0$.

Theorem 3.10 (Convergence for RK2). *Assume $f \in C^2(L^2(\Omega))$ and that the source term ψ_h^n in (3.23b) is such that*

$$\|\psi_h^n - f_h^n - \delta t d_t f_h^n\|_{L^2(\Omega)} \lesssim \delta t^2 \|d_t^2 f(t)\|_{C^0(L^2(\Omega))}. \tag{3.24}$$

Assume $u \in C^3(L^2(\Omega)) \cap C^0(H^1(\Omega))$. Set $V_h = \mathbb{P}_d^k(\mathcal{T}_h)$ with $k \geq 1$.

(i) *In the case $k \geq 2$, assume the 4/3-CFL condition*

$$\delta t \leq \varrho' \tau_*^{-1/3} \left(\frac{h}{\beta_c}\right)^{4/3}, \tag{3.25}$$

for some positive real number ϱ'.

(ii) *In the case $k = 1$, assume the CFL condition (3.21), that is,*

$$\delta t \leq \varrho^{\mathrm{RK2}} \frac{h}{\beta_c},$$

for a threshold ϱ^{RK2} independent of h, δt, and the data f, μ, and β. Finally, assume $d_t^s u \in C^0(H^{k+1-s}(\Omega))$ for $s \in \{0, 1\}$. Then,

$$\|u^N - u_h^N\|_{L^2(\Omega)} + \left(\sum_{m=0}^{N-1} \delta t |u^m - u_h^m|_\beta^2\right)^{1/2} \lesssim e^{C_{\mathrm{sta}} t_{\mathrm{F}}/\tau_*} (\chi_1 \delta t^2 + \chi_2 h^{k+1/2}), \tag{3.26}$$

where

$$\chi_1 = t_{\mathrm{F}}^{1/2} \tau_*^{1/2} (\|d_t^2 f\|_{C^0(L^2(\Omega))} + \|d_t^3 u\|_{C^0(L^2(\Omega))}),$$
$$\chi_2 = t_{\mathrm{F}}^{1/2} \beta_c^{1/2} \sum_{s \in \{0,1\}} \beta_c^{-s} \|d_t^s u\|_{C^0(H^{k+1-s}(\Omega))},$$

and the quantity C_{sta} is independent of h, δt, and the data f, μ, and β.

Remark 3.11 (Convergence rate with 4/3-CFL condition). Under the 4/3-CFL condition, the convergence rate with respect to h of the upper bound in (3.26) is $h^{8/3} + h^{k+1/2}$. For $k = 2$, this yields $h^{8/3} + h^{5/2}$, so that the second contribution, related to space approximation, is slightly dominant, although in practice, the two contributions can be considered to be equilibrated. For $k \geq 3$, the first contribution, related to the time error, is dominant, so that equilibrating both contributions actually imposes a more stringent restriction on the time step than the 4/3-CFL condition.

Remark 3.12 (Upper bound on ϱ' in the 4/3-CFL condition). Although there is no specific upper bound on the value of parameter ϱ' in the 4/3-CFL condition in the statement of Theorem 3.10, the quantity C_{sta} in the error estimate (3.26) depends on this parameter. Hence, in practice, a small enough value should be considered.

3.1.4.3 Explicit RK3 Schemes

We are interested in explicit RK3 schemes of the form (3.20) (for later use, we introduce a specific symbol for the intermediates stages)

$$w_h^n = u_h^n - \delta t A_h^{\mathrm{upw}} u_h^n + \delta t f_h^n, \tag{3.27a}$$
$$y_h^n = \tfrac{1}{2}(u_h^n + w_h^n) - \tfrac{1}{2}\delta t A_h^{\mathrm{upw}} w_h^n + \tfrac{1}{2}\delta t(f_h^n + \delta t d_t f_h^n), \tag{3.27b}$$
$$u_h^{n+1} = \tfrac{1}{3}(u_h^n + w_h^n + y_h^n) - \tfrac{1}{3}\delta t A_h^{\mathrm{upw}} y_h^n + \tfrac{1}{3}\delta t \psi_h^n, \tag{3.27c}$$

with initial condition $u_h^0 = \pi_h u_0$.

Theorem 3.13 (Convergence for RK3). *Assume $f \in C^3(L^2(\Omega))$ and that the source term ψ_h^n in (3.27c) is such that*

$$\|\psi_h^n - f_h^n - \delta t d_t f_h^n - \tfrac{1}{2}\delta t^2 d_t^2 f_h^n\|_{L^2(\Omega)} \lesssim \delta t^3 \|d_t^3 f\|_{C^0(L^2(\Omega))}. \tag{3.28}$$

Assume $u \in C^4(L^2(\Omega)) \cap C^0(H^1(\Omega))$. Set $V_h = \mathbb{P}_d^k(\mathcal{T}_h)$ for $k \geq 1$. Assume the usual CFL condition (3.21), that is,

$$\delta t \leq \varrho^{\mathrm{RK3}} \frac{h}{\beta_{\mathrm{c}}},$$

for a threshold ϱ^{RK3} independent of h, δt, and the data f, μ, and β. Finally, assume $d_t^s u \in C^0(H^{k+1-s}(\Omega))$ for $s \in \{0,1,2\}$. Then,

$$\|u^N - u_h^N\|_{L^2(\Omega)} + \left(\sum_{m=0}^{N-1} \delta t |u^m - u_h^m|_\beta^2\right)^{1/2} \lesssim e^{C_{\mathrm{sta}} t_{\mathrm{F}}/\tau_*}(\chi_1 \delta t^3 + \chi_2 h^{k+1/2}), \tag{3.29}$$

where

$$\chi_1 = t_{\mathrm{F}}^{1/2} \tau_*^{1/2}(\|d_t^3 f\|_{C^0(L^2(\Omega))} + \|d_t^4 u\|_{C^0(L^2(\Omega))}),$$
$$\chi_2 = t_{\mathrm{F}}^{1/2} \beta_{\mathrm{c}}^{1/2} \sum_{s \in \{0,1,2\}} \beta_{\mathrm{c}}^{-s}\|d_t^s u\|_{C^0(H^{k+1-s}(\Omega))},$$

and the quantity C_{sta} is independent of h, δt, and the data f, μ, and β.

Remark 3.14 (Stability of explicit RK3 schemes). Explicit RK3 schemes enjoy stronger stability properties than explicit RK2 schemes. This is for instance reflected by the fact that explicit RK3 schemes are stable under the usual CFL condition (3.21) even in the absence of stabilization; see, e.g., Tadmor [289, Theorem 2]). Stability alone is, however, not sufficient to derive quasi-optimal error estimates. Indeed, in the absence of stabilization (that is, if centered fluxes are used), the source terms in the error equation cannot be controlled properly, and this leads to suboptimal error estimates of the form $\chi_1 \delta t^3 + \chi_2 h^k$ (compare with (3.29)). The stability of explicit RK3 schemes in the presence of stabilization has been studied recently in [66].

Remark 3.15 (Influence of polynomial degree). The threshold ϱ^{RK3} depends on the polynomial degree k, and the larger k, the lower ϱ^{RK3}. This dependency results from the fact that the actual value for ϱ^{RK3} depends on some boundedness properties of the discrete operator A_h^{upw}, and that these properties are established using inverse and discrete trace inequalities where the factors depend on the polynomial degree (cf. Remark 1.48).

3.1.4.4 Numerical Illustration for Rough Solutions

Numerical results illustrating the convergence rates predicted by Theorems 3.10 and 3.13 for smooth solutions are reported in [66] for polynomial degrees $k \in \{1, 2\}$ in the dG scheme. Numerical experiments indicate that the 4/3-CFL condition for explicit RK2 schemes and polynomial degree $k = 2$ appears to be sharp. Here, we focus instead on approximating rough solutions. We consider the unsteady advection-reaction equation (3.1) with (divergence-free) advective velocity $\beta = (x_2, -x_1)^{\mathrm{t}}$, reaction coefficient $\mu = 0$, source term $f = 0$, and circular domain $\Omega = \{x \in \mathbb{R}^2 \, : \, x_1^2 + x_2^2 \leq 1\}$ (a polygonal domain can be considered as well). The initial datum is

$$u_0(x) = \frac{1}{2}(\tanh(10^3(e^{-\phi(x)} - 0.5)) + 1),$$

with $\phi(x) = 10((x_1 - 0.3)^2 + (x_2 - 0.3)^2)$ so that the solution exhibits a sharp inner layer with thickness of order 0.001.

We consider a fixed uniform mesh with 256 elements along the boundary of Ω ($h \approx 0.025$). The inner layer is thus under-resolved. Approximate solutions obtained using explicit RK2 and RK3 schemes combined with polynomial degrees $k \in \{1, 2\}$ for the dG scheme are reported in Fig. 3.1. The results illustrate the fact that the present methods are able to avoid the global spreading of spurious oscillations, so that the discrete solution after a complete rotation still exhibits a sharp inner layer. In all cases, the largest possible time step has been used; the corresponding value of ϱ in the CFL condition is reported in the caption of Fig. 3.1. This figure also shows that the present schemes do not avoid spurious oscillations in the vicinity of the inner layer. To eliminate such oscillations, limiters can be employed. These techniques are discussed in Sect. 3.2.4 in the context of nonlinear conservation laws.

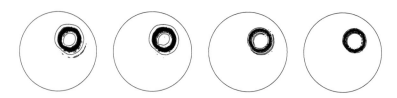

Fig. 3.1: Contour lines of discrete solution after a complete rotation of the initial datum. From *left* to *right*: RK2, $k = 1$, $\delta t = 0.2h$; RK3, $k = 1$, $\delta t = 0.26h$; RK2, $k = 2$, $\delta t = 0.12h$; RK3, $k = 2$, $\delta t = 0.15h$

3.1.5 Analysis of Forward Euler and Finite Volume Schemes

The goal of this section is to prove Theorem 3.7. Although this section is restricted to finite volume schemes, we recommend reading it before tackling the more elaborate analysis of Sect. 3.1.6.

3.1.5.1 The Error Equation

Define
$$\xi_h^n := u_h^n - \pi_h u^n, \qquad \xi_\pi^n := u^n - \pi_h u^n,$$
so that the approximation error at the discrete time t^n can be decomposed as
$$u^n - u_h^n = \xi_\pi^n - \xi_h^n.$$

The convergence analysis proceeds as follows. Since a bound on ξ_π^n can be inferred from polynomial approximation properties and the smoothness of u^n, an error upper bound is derived by first obtaining a bound on ξ_h^n in terms of ξ_π^n and then using the triangle inequality.

Our first step consists in identifying the error equation for the time evolution of ξ_h^n. The form of this equation is similar to the forward Euler scheme (3.12) with data depending on ξ_π^n and u.

Lemma 3.16 (Error equation). *Assume $u \in C^0(H^1(\Omega)) \cap C^2(L^2(\Omega))$. Then,*
$$\xi_h^{n+1} = \xi_h^n - \delta t A_h^{\text{upw}} \xi_h^n + \delta t \alpha_h^n, \qquad (3.30)$$
where
$$\alpha_h^n := A_h^{\text{upw}} \xi_\pi^n - \pi_h \theta^n, \qquad (3.31)$$
with $\theta^n := \delta t^{-1} \int_{t^n}^{t^{n+1}} (t^{n+1} - t) d_t^2 u(t) \, dt$.

Proof. A second-order Taylor expansion in time with integral remainder for the exact solution u yields
$$u^{n+1} = u^n + \delta t \, d_t u^n + \delta t \theta^n.$$

Projecting onto V_h and using the consistency property (3.9) at the discrete time t^n leads to
$$\pi_h u^{n+1} = \pi_h u^n + \delta t \pi_h d_t u^n + \delta t \pi_h \theta^n$$
$$= \pi_h u^n + \delta t (-A_h^{\text{upw}} u^n + f_h^n) + \delta t \pi_h \theta^n.$$

Subtracting this equation from (3.12), we infer that
$$\xi_h^{n+1} = \xi_h^n - \delta t A_h^{\text{upw}} (\xi_h^n - \xi_\pi^n) - \delta t \pi_h \theta^n,$$
whence the assertion. $\square$

3.1.5.2 Energy Identity

Our second step consists in deriving an energy identity for the scheme (3.30). This identity is obtained by taking the L^2-scalar product of (3.30) with ξ_h^n.

Lemma 3.17 (Energy identity). *There holds*

$$\frac{1}{2}\|\xi_h^{n+1}\|_{L^2(\Omega)}^2 - \frac{1}{2}\|\xi_h^n\|_{L^2(\Omega)}^2 + \delta t|\xi_h^n|_\beta^2 = \frac{1}{2}\|\xi_h^{n+1} - \xi_h^n\|_{L^2(\Omega)}^2$$
$$+ \delta t(\alpha_h^n, \xi_h^n)_{L^2(\Omega)} - \delta t(\Lambda\xi_h^n, \xi_h^n)_{L^2(\Omega)}. \quad (3.32)$$

Proof. Taking the L^2-scalar product of (3.30) with ξ_h^n and using discrete dissipation (cf. (3.10)) yields

$$(\xi_h^{n+1}, \xi_h^n)_{L^2(\Omega)} = \|\xi_h^n\|_{L^2(\Omega)}^2 - \delta t(A_h^{\mathrm{upw}}\xi_h^n, \xi_h^n)_{L^2(\Omega)} + \delta t(\alpha_h^n, \xi_h^n)_{L^2(\Omega)}$$
$$= \|\xi_h^n\|_{L^2(\Omega)}^2 - \delta t|\xi_h^n|_\beta^2 - \delta t(\Lambda\xi_h^n, \xi_h^n)_{L^2(\Omega)} + \delta t(\alpha_h^n, \xi_h^n)_{L^2(\Omega)}.$$

Using on the left-hand side the algebraic identity $ab = \frac{1}{2}a^2 + \frac{1}{2}b^2 - \frac{1}{2}(a-b)^2$ valid for arbitrary real numbers a and b leads to

$$\frac{1}{2}\|\xi_h^{n+1}\|_{L^2(\Omega)}^2 + \frac{1}{2}\|\xi_h^n\|_{L^2(\Omega)}^2 - \frac{1}{2}\|\xi_h^{n+1} - \xi_h^n\|_{L^2(\Omega)}^2 = \|\xi_h^n\|_{L^2(\Omega)}^2$$
$$- \delta t|\xi_h^n|_\beta^2 - \delta t(\Lambda\xi_h^n, \xi_h^n)_{L^2(\Omega)} + \delta t(\alpha_h^n, \xi_h^n)_{L^2(\Omega)},$$

whence the assertion. $\square$

3.1.5.3 Boundedness of A_h^{upw}

Before proceeding further, we need to obtain suitable bounds on the discrete operator A_h^{upw}. To this purpose, we consider the following norms: For all $v \in V_{*h}$,

$$\|v\|_{\mathrm{uwb}}^2 = \tau_c^{-1}\|v\|_{L^2(\Omega)}^2 + |v|_\beta^2,$$
$$\|v\|_{\mathrm{uwb},*}^2 = \|v\|_{\mathrm{uwb}}^2 + \sum_{T \in \mathcal{T}_h} \beta_c\|v\|_{L^2(\partial T)}^2,$$

where the $|\cdot|_\beta$-seminorm is defined by (3.11). The $\|\cdot\|_{\mathrm{uwb}}$- and $\|\cdot\|_{\mathrm{uwb},*}$-norms coincide with those considered in Sect. 2.3 in the steady case.

Lemma 3.18 (Boundedness of A_h^{upw}). *There holds*

(i) *For all $(v, w_h) \in V_{*h} \times V_h$,*

$$(A_h^{\mathrm{upw}}v, w_h)_{L^2(\Omega)} \leq C_{*,1}\|v\|_{\mathrm{uwb},*}\|w_h\|_{\mathrm{uwb}}, \quad (3.33)$$

(ii) *For all $v_h \in V_h$,*

$$\|A_h^{\mathrm{upw}}v_h\|_{L^2(\Omega)} \leq C_{*,2}\beta_c^{1/2}h^{-1/2}\|v_h\|_{\mathrm{uwb}}. \quad (3.34)$$

Here, $C_{*,1}$ and $C_{*,2}$ are independent of h, δt, and the data f, μ, and β.

3.1. Unsteady Advection-Reaction

Proof. (i) Proof of (3.33). Let $(v, w_h) \in V_{*h} \times V_h$. Using the expression (2.26) for a_h^{cf} (that is, integrating by parts the advective derivative in each mesh element in the expression (3.7) defining A_h^{upw}) and exploiting the fact that $\nabla_h w_h = 0$ since w_h is piecewise constant, we obtain

$$(A_h^{\mathrm{upw}} v, w_h)_{L^2(\Omega)} = \int_\Omega (\mu - \nabla\cdot\beta) v w_h + \int_{\partial\Omega} (\beta\cdot\mathrm{n})^\oplus v w_h$$
$$+ \sum_{F \in \mathcal{F}_h^i} \int_F (\beta\cdot\mathrm{n}_F) \{v\} [\![w_h]\!] + \sum_{F \in \mathcal{F}_h^i} \int_F \frac{1}{2} |\beta\cdot\mathrm{n}_F| [\![v]\!] [\![w_h]\!],$$

whence we infer (3.33) using the Cauchy–Schwarz inequality.

(ii) Proof of (3.34). Let $v_h \in V_h$. We use the expression (3.7) defining A_h^{upw} and observe that $\nabla_h v_h = 0$ since v_h is piecewise constant. As a result, for all $w_h \in V_h$,

$$(A_h^{\mathrm{upw}} v_h, w_h)_{L^2(\Omega)} = \int_\Omega \mu v_h w_h + \int_{\partial\Omega} (\beta\cdot\mathrm{n})^\ominus v_h w_h$$
$$- \sum_{F \in \mathcal{F}_h^i} \int_F (\beta\cdot\mathrm{n}_F) [\![v_h]\!] \{w_h\} + \sum_{F \in \mathcal{F}_h^i} \int_F \frac{1}{2} |\beta\cdot\mathrm{n}_F| [\![v_h]\!] [\![w_h]\!].$$

Using the Cauchy–Schwarz inequality, the discrete trace inequality (1.37) to bound $\{w_h\}$, and the fact that $\|A_h^{\mathrm{upw}} v_h\|_{L^2(\Omega)} = \sup_{w_h \in V_h \setminus \{0\}} \frac{(A_h^{\mathrm{upw}} v_h, w_h)_{L^2(\Omega)}}{\|w_h\|_{L^2(\Omega)}}$, we infer (3.34). □

Remark 3.19 (On the bounds (3.33) and (3.34)). The bound (3.33) is similar to that derived in Lemma 2.30 concerning boundedness on orthogonal subscales in the steady case. The present bound is slightly more general since $w_h \in V_h = \mathbb{P}_d^0(\mathcal{T}_h)$ implies $\nabla_h w_h = 0$, and this circumvents the boundedness argument on orthogonal subscales. Furthermore, the bound (3.34) is equivalent to the coercivity condition introduced by Levy and Tadmor in [231]. This property is specific to the finite volume setting involving piecewise constant functions (recall that we used $\nabla_h v_h = 0$ in the proof) and cannot be extended to high-order dG methods with polynomial degree $k \geq 1$. We also observe that the bound (3.34) cannot be inferred directly from (3.33). Indeed, using the discrete trace inequality (1.37), and since $h \leq \beta_c \tau_*$, we infer, for all $w_h \in V_h$,

$$\|w_h\|_{\mathrm{uw}\flat} \lesssim \beta_c^{1/2} h^{-1/2} \|w_h\|_{L^2(\Omega)},$$

so that (3.33) yields, for all $v_h \in V_h$,

$$\|A_h^{\mathrm{upw}} v_h\|_{L^2(\Omega)} = \sup_{w_h \in V_h \setminus \{0\}} \frac{(A_h^{\mathrm{upw}} v_h, w_h)_{L^2(\Omega)}}{\|w_h\|_{L^2(\Omega)}} \lesssim \beta_c^{1/2} h^{-1/2} \|v_h\|_{\mathrm{uw}\flat,*},$$

which is less sharp than (3.34) since $\|\cdot\|_{\mathrm{uw}\flat} \leq \|\cdot\|_{\mathrm{uw}\flat,*}$.

3.1.5.4 Stability

Our next step is to bound the three terms on the right-hand side of the energy identity (3.32) using the CFL condition (3.21) together with the positive term $\delta t |\xi_h^n|_\beta^2$ appearing on the left-hand side and which results from the dissipative properties of the dG discretization. The first term, namely $\|\xi_h^{n+1} - \xi_h^n\|_{L^2(\Omega)}^2$, results from the anti-dissipative nature of the forward Euler scheme. The second term, namely $\delta t(\alpha_h^n, \xi_h^n)_{L^2(\Omega)}$, contains the contribution of space approximation errors through the term $A_h^{\text{upw}} \xi_\pi^n$, and bounding this contribution also requires some care. Finally, bounding the third term, namely $\delta t(\Lambda \xi_h^n, \xi_h^n)_{L^2(\Omega)}$, is straightforward.

We can now tackle the main stability result of this section. The bound (3.33) on A_h^{upw} is crucial to estimate $\delta t(\alpha_h^n, \xi_h^n)_{L^2(\Omega)}$, while the bound (3.34) on A_h^{upw} is crucial to control the anti-dissipative term $\|\xi_h^{n+1} - \xi_h^n\|_{L^2(\Omega)}^2$.

Lemma 3.20 (Stability). *Assume* $u \in C^0(H^1(\Omega)) \cap C^2(L^2(\Omega))$. *Assume the CFL condition* (3.21) *with* $\varrho \le (2C_{*,2})^{-2}$, *that is,*

$$\delta t \le (2C_{*,2})^{-2} \frac{h}{\beta_c}. \tag{3.35}$$

Then, there is C_{sta}, *independent of* h, δt, *and* f, μ, *and* β, *such that*

$$\|\xi_h^{n+1}\|_{L^2(\Omega)}^2 - \|\xi_h^n\|_{L^2(\Omega)}^2 + \delta t |\xi_h^n|_\beta^2$$
$$\le C_{\text{sta}} \delta t (\|\xi_\pi^n\|_{\text{uwb},*}^2 + \tau_*^{-1} \|\xi_h^n\|_{L^2(\Omega)}^2 + \tau_* C_u^2 \delta t^2), \tag{3.36}$$

with $C_u := \|d_t^2 u\|_{C^0(L^2(\Omega))}$.

Proof. We start from the energy identity (3.32) and bound the three terms on the right-hand side, say $\mathfrak{T}_1$, $\mathfrak{T}_2$, and $\mathfrak{T}_3$.
(i) Bound on $\mathfrak{T}_3$. Owing to the Cauchy–Schwarz inequality and since $\|\Lambda\|_{L^\infty(\Omega)} \lesssim \tau_c^{-1}$, we obtain

$$|\mathfrak{T}_3| = \delta t |(\Lambda \xi_h^n, \xi_h^n)_{L^2(\Omega)}| \lesssim \delta t \tau_c^{-1} \|\xi_h^n\|_{L^2(\Omega)}^2. \tag{3.37}$$

(ii) Bound on $\mathfrak{T}_2$. Here, the bound (3.33) on A_h^{upw} is crucial. Using the definition (3.31) of α_h^n yields

$$|\mathfrak{T}_2| = \delta t |(\alpha_h^n, \xi_h^n)_{L^2(\Omega)}| \le \delta t |(A_h^{\text{upw}} \xi_\pi^n, \xi_h^n)_{L^2(\Omega)}| + \delta t |(\pi_h \theta^n, \xi_h^n)_{L^2(\Omega)}|.$$

For the first term on the right-hand side, owing to the bound (3.33) on A_h^{upw}, Young's inequality in the form $ab \le \gamma a^2 + (4\gamma)^{-1} b^2$ for arbitrary real numbers a and b and positive γ that can be chosen as small as needed, and the definition of the $\|\cdot\|_{\text{uwb}}$-norm, we infer

$$\delta t |(A_h^{\text{upw}} \xi_\pi^n, \xi_h^n)_{L^2(\Omega)}| \le \delta t C_{*,1} \|\xi_\pi^n\|_{\text{uwb},*} \|\xi_h^n\|_{\text{uwb}} \le \frac{1}{4} \delta t \|\xi_h^n\|_{\text{uwb}}^2 + C \delta t \|\xi_\pi^n\|_{\text{uwb},*}^2$$

$$= \frac{1}{4} \delta t (|\xi_h^n|_\beta^2 + \tau_c^{-1} \|\xi_h^n\|_{L^2(\Omega)}^2) + C \delta t \|\xi_\pi^n\|_{\text{uwb},*}^2$$

$$\le \frac{1}{4} \delta t |\xi_h^n|_\beta^2 + C \delta t (\|\xi_\pi^n\|_{\text{uwb},*}^2 + \tau_c^{-1} \|\xi_h^n\|_{L^2(\Omega)}^2).$$

3.1. Unsteady Advection-Reaction

Concerning the second term on the right-hand side, we first observe that owing to the Cauchy–Schwarz inequality,

$$\|\theta^n\|_{L^2(\Omega)}^2 = \delta t^{-2} \int_\Omega \left(\int_{t^n}^{t^{n+1}} (t^{n+1} - t) d_t^2 u(t) \, dt \right)^2 dx$$

$$\leq \delta t^{-2} \int_\Omega \left(\int_{t^n}^{t^{n+1}} (t^{n+1} - t)^2 \, dt \right) \left(\int_{t^n}^{t^{n+1}} |d_t^2 u(t)|^2 \, dt \right) dx$$

$$= \frac{1}{3}\delta t \int_\Omega \int_{t^n}^{t^{n+1}} |d_t^2 u(t)|^2 \, dt \, dx \leq \frac{1}{3}\delta t^2 \max_{t \in [t^n, t^{n+1}]} \|d_t^2 u\|_{L^2(\Omega)}^2,$$

so that
$$\|\theta^n\|_{L^2(\Omega)} \lesssim C_u \delta t. \tag{3.38}$$

Hence, using again the Cauchy–Schwarz inequality followed by Young's inequality yields

$$\delta t |(\pi_h \theta^n, \xi_h^n)_{L^2(\Omega)}| = \delta t |(\theta^n, \xi_h^n)_{L^2(\Omega)}| \leq \delta t \|\theta^n\|_{L^2(\Omega)} \|\xi_h^n\|_{L^2(\Omega)}$$

$$\leq 2\delta t \tau_* \|\theta^n\|_{L^2(\Omega)}^2 + 2\delta t \tau_*^{-1} \|\xi_h^n\|_{L^2(\Omega)}^2$$

$$\lesssim \delta t \tau_* C_u^2 \delta t^2 + \delta t \tau_*^{-1} \|\xi_h^n\|_{L^2(\Omega)}^2.$$

Collecting the two above bounds and recalling that $\tau_* \leq \tau_c$, we infer

$$|\mathfrak{T}_2| \leq \frac{1}{4}\delta t |\xi_h^n|_\beta^2 + C\delta t (\|\xi_\pi^n\|_{\text{uwb},*}^2 + \tau_*^{-1} \|\xi_h^n\|_{L^2(\Omega)}^2 + \tau_* C_u^2 \delta t^2). \tag{3.39}$$

(iii) Bound on $\mathfrak{T}_1$. Here, the bound (3.34) on A_h^{upw} is crucial. We first obtain using the error equation (3.30) that

$$|\mathfrak{T}_1| = \frac{1}{2}\|\xi_h^{n+1} - \xi_h^n\|_{L^2(\Omega)}^2 = \frac{1}{2}\delta t^2 \|A_h^{\text{upw}} \xi_h^n - \alpha_h^n\|_{L^2(\Omega)}^2$$

$$\leq \delta t^2 \|A_h^{\text{upw}} \xi_h^n\|_{L^2(\Omega)}^2 + \delta t^2 \|\alpha_h^n\|_{L^2(\Omega)}^2.$$

For the first term on the right-hand side, owing to (3.34) and the choice (3.35) for the CFL condition, we infer

$$\delta t^2 \|A_h^{\text{upw}} \xi_h^n\|_{L^2(\Omega)}^2 \leq \delta t^2 C_{*,2}^2 \beta_c h^{-1} \|\xi_h^n\|_{\text{uwb}}^2 \leq \frac{1}{4}\delta t \|\xi_h^n\|_{\text{uwb}}^2.$$

Hence, using the definition of the $\|\cdot\|_{\text{uwb}}$-norm,

$$\delta t^2 \|A_h^{\text{upw}} \xi_h^n\|_{L^2(\Omega)}^2 \leq \frac{1}{4}\delta t |\xi_h^n|_\beta^2 + \frac{1}{4}\delta t \tau_c^{-1} \|\xi_h^n\|_{L^2(\Omega)}^2.$$

We now consider the second term on the right-hand side, namely $\delta t^2 \|\alpha_h^n\|_{L^2(\Omega)}^2$. Proceeding as in Remark 3.19, we infer, for all $v \in V_{*h}$,

$$\|A_h^{\text{upw}} v\|_{L^2(\Omega)} \lesssim \beta_c^{1/2} h^{-1/2} \|v\|_{\text{uwb},*}.$$

Therefore, we obtain using the definition (3.31) of α_h^n, the triangle inequality, the above bound on $A_h^{\mathrm{upw}} v$ with $v = \xi_\pi^n$, the CFL condition (3.21), and the bound (3.38) on $\|\theta^n\|_{L^2(\Omega)}$,

$$\begin{aligned}\delta t^2 \|\alpha_h^n\|_{L^2(\Omega)}^2 &\lesssim \delta t^2 \|A_h^{\mathrm{upw}} \xi_\pi^n\|_{L^2(\Omega)}^2 + \delta t^2 \|\pi_h \theta^n\|_{L^2(\Omega)}^2 \\ &\lesssim \delta t^2 \beta_c h^{-1} \|\xi_\pi^n\|_{\mathrm{uwb},*}^2 + \delta t^2 \|\theta^n\|_{L^2(\Omega)}^2 \\ &\lesssim \delta t \|\xi_\pi^n\|_{\mathrm{uwb},*}^2 + \delta t^2 C_u^2 \delta t^2.\end{aligned}$$

Hence, since $\delta t \leq \tau_*$,

$$|\mathfrak{T}_1| \leq \frac{1}{4}\delta t |\xi_h^n|_\beta^2 + C\delta t(\|\xi_\pi^n\|_{\mathrm{uwb},*}^2 + \tau_c^{-1}\|\xi_h^n\|_{L^2(\Omega)}^2 + \tau_* C_u^2 \delta t^2). \qquad (3.40)$$

(iv) Collecting estimates (3.37), (3.39), and (3.40) yields

$$|\mathfrak{T}_1| + |\mathfrak{T}_2| + |\mathfrak{T}_3| \leq \frac{1}{2}\delta t |\xi_h^n|_\beta^2 + C\delta t(\|\xi_\pi^n\|_{\mathrm{uwb},*}^2 + \tau_*^{-1}\|\xi_h^n\|_{L^2(\Omega)}^2 + \tau_* C_u^2 \delta t^2),$$

since $\tau_* \leq \tau_c$, whence the assertion. $\square$

Remark 3.21 (High-order dG approximation). The above proof breaks down if a high-order dG method is used for space semidiscretization, that is, if $V_h = \mathbb{P}_d^k(\mathcal{T}_h)$ with polynomial degree $k \geq 1$. The difficulty lies in bounding the term $\|A_h^{\mathrm{upw}} \xi_h^n\|_{L^2(\Omega)}$ (cf. the above estimate on $\mathfrak{T}_1$) resulting from the anti-dissipative nature of the forward Euler scheme. Indeed, for $k \geq 1$, the bound (3.34) is no longer available because $\nabla_h v_h$ does not necessarily vanish for $v_h \in V_h$. Instead, using the inverse inequality (1.36) to control $\nabla_h v_h$, it is possible to prove that, for all $v_h \in V_h$,
$$\|A_h^{\mathrm{upw}} v_h\|_{L^2(\Omega)} \lesssim \beta_c h^{-1} \|v_h\|_{L^2(\Omega)}.$$
As a result, when bounding $\mathfrak{T}_1$, we obtain
$$\delta t^2 \|A_h^{\mathrm{upw}} \xi_h^n\|_{L^2(\Omega)}^2 \lesssim \delta t^2 \beta_c^2 h^{-2} \|\xi_h^n\|_{L^2(\Omega)}^2,$$
and a so-called 2-CFL condition (also called parabolic CFL condition) of the form $\delta t \lesssim \tau_*^{-1} \beta_c^{-2} h^2$ must be invoked to obtain
$$\delta t^2 \|A_h^{\mathrm{upw}} \xi_h^n\|_{L^2(\Omega)}^2 \lesssim \delta t \tau_*^{-1} \|\xi_h^n\|_{L^2(\Omega)}^2.$$
Unfortunately, the 2-CFL condition is too restrictive on the size of the time step to be used in practice.

3.1.5.5 Proof of Theorem 3.7

Letting, for all $n \in \{0, \ldots, N\}$, $a^n := \|\xi_h^n\|_{L^2(\Omega)}^2$ and $b^n := \delta t |\xi_h^n|_\beta^2$, we deduce from the stability estimate (3.36) that

$$a^{n+1} + b^n \leq (1+\gamma)a^n + d^n, \qquad (3.41)$$

3.1. Unsteady Advection-Reaction

with $\gamma := C_{\text{sta}} \delta t \tau_*^{-1}$ and $d^n := C_{\text{sta}} \delta t (\|\xi_\pi^n\|_{\text{uw}\flat,*}^2 + \tau_* C_u^2 \delta t^2)$. It is easily shown by induction that

$$a^{n+1} + \sum_{m=0}^{n} (1+\gamma)^{n-m} b^m \leq (1+\gamma)^{n+1} a^0 + \sum_{m=0}^{n} (1+\gamma)^{n-m} d^m.$$

We apply this estimate at $n = N - 1$ and simplify it by observing that $1 \leq (1+\gamma)^{n-m} \leq (1+\gamma)^N \leq e^{N\gamma} = e^{C_{\text{sta}} t_F / \tau_*}$. Moreover, because of the initial condition $u_h^0 = \pi_h u_0$, $\xi_h^0 = 0$ so that $a^0 = 0$. This yields

$$a^N + \sum_{m=0}^{N-1} b^m \leq e^{C_{\text{sta}} t_F / \tau_*} \sum_{m=0}^{N-1} d^m. \qquad (3.42)$$

Using the triangle inequality, recalling that $\xi_\pi^m = u^m - \pi_h u^m$, and since $|\xi_\pi^m|_\beta \leq \|\xi_\pi^m\|_{\text{uw}\flat,*}$, we infer

$$\|u^N - u_h^N\|_{L^2(\Omega)} + \left(\sum_{m=0}^{N-1} \delta t |u^m - u_h^m|_\beta^2 \right)^{1/2} \lesssim \left(e^{C_{\text{sta}} t_F / \tau_*} \sum_{m=0}^{N-1} d^m \right)^{1/2} + \|\xi_\pi^N\|_{L^2(\Omega)}.$$

Using the definition of the $\|\cdot\|_{\text{uw}\flat,*}$-norm, the polynomial approximation properties stated in Lemmata 1.58 and 1.59, and the assumption $h \leq \beta_c \tau_c$ leads to

$$\|\xi_\pi^m\|_{\text{uw}\flat,*}^2 \lesssim \beta_c h \|u^m\|_{H^1(\Omega)}^2.$$

As a result,

$$\sum_{m=0}^{N-1} d^m \lesssim \sum_{m=0}^{N-1} \delta t (\tau_* C_u^2 \delta t^2 + \|\xi_\pi^m\|_{\text{uw}\flat,*}^2)$$
$$\leq t_F \tau_* C_u^2 \delta t^2 + t_F \beta_c h \|u\|_{C^0(H^1(\Omega))}^2 = \chi_1^2 \delta t^2 + \chi_2^2 h.$$

Furthermore, since $h \leq \beta_c t_F$,

$$\|\xi_\pi^N\|_{L^2(\Omega)} \lesssim h \|u^N\|_{H^1(\Omega)} \leq (h t_F^{-1} \beta_c^{-1})^{1/2} \chi_2 h^{1/2} \leq \chi_2 h^{1/2}.$$

Collecting the above bounds yields (3.22), thereby completing the proof.

Remark 3.22 (Discrete Gronwall). The argument by which the bound (3.42) is inferred from an estimate of the form (3.41) is often called a discrete Gronwall Lemma. This is the argument leading to the exponential factor $e^{C_{\text{sta}} t_F / \tau_*}$ in the error estimate (3.22).

3.1.6 Analysis of Explicit RK2 Schemes

This section is devoted to the analysis of explicit RK2 schemes. The proofs follow a similar path to that deployed in Sect. 3.1.5 for the forward Euler scheme combined with a finite volume method, namely an energy identity is obtained from a suitable error equation, whence a stability estimate is inferred, leading finally to the error estimate using a discrete Gronwall Lemma. However, the arguments are more elaborate because of the higher-order nature of both temporal and spatial discretizations. As for the Euler scheme, there are typically two terms to bound in the energy identity, namely the contribution of the source terms in the error equation and the anti-dissipative term resulting from the explicit nature of the RK2 scheme. To bound the contribution of the source terms, the key argument is the boundedness on orthogonal subscales of the discrete operator A_h^{upw}. This property is the counterpart of that already considered in the steady case and is crucial to achieve quasi-optimal error estimates instead of suboptimal ones (cf. Remark 3.26 below for further discussion).

Moreover, bounding the anti-dissipative term is not straightforward since, contrary to the finite volume case examined in Sect. 3.1.5, the broken gradient of discrete functions does no longer vanish, so that the bound (3.34) on A_h^{upw} is no longer available. Two strategies are possible depending on the polynomial degree k used in the dG scheme. In the piecewise affine case ($k=1$), it is possible to exploit the fact that the broken gradient is piecewise constant so as to derive an additional property for A_h^{upw}, a so-called adjoint boundedness on orthogonal subscales (cf. Lemma 3.28 below), thereby leading to a stability estimate under the usual CFL condition. In the general case $k \geq 2$, one possibility is to exploit the fact that the anti-dissipative term is one order higher in time than with the forward Euler scheme (we need to bound a term of the form $\delta t^2 \|(A_h^{\mathrm{upw}})^2 \xi_h^n\|_{L^2(\Omega)}$ instead of a term of the form $\delta t \|A_h^{\mathrm{upw}} \xi_h^n\|_{L^2(\Omega)}$). This fact allows us to invoke a 4/3-CFL condition (instead of the 2-CFL condition mentioned in Remark 3.21).

3.1.6.1 The Error Equation

Define

$$\xi_h^n := u_h^n - \pi_h u^n, \qquad \zeta_h^n := w_h^n - \pi_h w^n,$$
$$\xi_\pi^n := u^n - \pi_h u^n, \qquad \zeta_\pi^n := w^n - \pi_h w^n,$$

with $w := u + \delta t d_t u$. Using these quantities, the errors can be written as

$$u^n - u_h^n = \xi_\pi^n - \xi_h^n, \qquad w^n - w_h^n = \zeta_\pi^n - \zeta_h^n.$$

Our first step is to identify the error equation governing the time evolution of ξ_h^n and ζ_h^n.

Lemma 3.23 (Error equation). *There holds*

$$\zeta_h^n = \xi_h^n - \delta t A_h^{\mathrm{upw}} \xi_h^n + \delta t \alpha_h^n, \tag{3.43a}$$
$$\xi_h^{n+1} = \tfrac{1}{2}(\xi_h^n + \zeta_h^n) - \tfrac{1}{2}\delta t A_h^{\mathrm{upw}} \zeta_h^n + \tfrac{1}{2}\delta t \beta_h^n, \tag{3.43b}$$

where

$$\alpha_h^n := A_h^{\mathrm{upw}} \xi_\pi^n, \qquad \beta_h^n := A_h^{\mathrm{upw}} \zeta_\pi^n - \pi_h \theta^n + \delta_h^n, \tag{3.44}$$

3.1. Unsteady Advection-Reaction

with

$$\delta_h^n := \psi_h^n - f_h^n - \delta t d_t f_h^n, \qquad \theta^n := \delta t^{-1} \int_{t^n}^{t^{n+1}} (t^{n+1} - t)^2 d_t^3 u(t)\,dt. \qquad (3.45)$$

Proof. Consistency at the discrete time t^n (cf. (3.9)) yields

$$\pi_h w^n = \pi_h u^n + \delta t \pi_h d_t u^n = \pi_h u^n - \delta t A_h^{\mathrm{upw}} u^n + \delta t f_h^n.$$

Subtracting this equation from (3.23a) yields (3.43a). To derive (3.43b), we observe that

$$u^{n+1} = u^n + \delta t d_t u^n + \tfrac{1}{2}\delta t^2 d_t^2 u^n + \tfrac{1}{2}\delta t \theta^n = w^n + \tfrac{1}{2}\delta t^2 d_t^2 u^n + \tfrac{1}{2}\delta t \theta^n,$$

so that projecting onto V_h yields

$$\begin{aligned}\pi_h u^{n+1} &= \pi_h w^n + \tfrac{1}{2}\delta t^2 \pi_h d_t^2 u^n + \tfrac{1}{2}\delta t \pi_h \theta^n \\ &= \tfrac{1}{2}(\pi_h u^n + \pi_h w^n) - \tfrac{1}{2}\delta t A_h^{\mathrm{upw}} u^n + \tfrac{1}{2}\delta t f_h^n + \tfrac{1}{2}\delta t^2 \pi_h d_t^2 u^n + \tfrac{1}{2}\delta t \pi_h \theta^n.\end{aligned}$$

Moreover,

$$\delta t \pi_h d_t^2 u^n = \delta t d_t(\pi_h d_t u^n) = -\delta t A_h^{\mathrm{upw}} d_t u^n + \delta t d_t f_h^n = -A_h^{\mathrm{upw}}(w^n - u^n) + \delta t d_t f_h^n,$$

whence

$$\pi_h u^{n+1} = \tfrac{1}{2}(\pi_h u^n + \pi_h w^n) - \tfrac{1}{2}\delta t A_h^{\mathrm{upw}} w^n + \tfrac{1}{2}\delta t(\pi_h \theta^n + f_h^n + \delta t d_t f_h^n).$$

Subtracting this equation from (3.23b) yields (3.43b). $\square$

3.1.6.2 Energy Identity

Our next step is to derive an energy identity for the scheme (3.43).

Lemma 3.24 (Energy identity). *There holds*

$$\begin{aligned}\|\xi_h^{n+1}\|_{L^2(\Omega)}^2 - \|\xi_h^n\|_{L^2(\Omega)}^2 + \delta t |\xi_h^n|_\beta^2 + \delta t |\zeta_h^n|_\beta^2 &= \|\xi_h^{n+1} - \zeta_h^n\|_{L^2(\Omega)}^2 \\ &\quad + \delta t(\alpha_h^n, \xi_h^n)_{L^2(\Omega)} + \delta t(\beta_h^n, \zeta_h^n)_{L^2(\Omega)} - \Lambda_h^n,\end{aligned} \qquad (3.46)$$

where $\Lambda_h^n := \delta t(\Lambda \xi_h^n, \xi_h^n)_{L^2(\Omega)} + \delta t(\Lambda \zeta_h^n, \zeta_h^n)_{L^2(\Omega)}$.

Proof. We multiply (3.43a) by ξ_h^n and (3.43b) by $2\zeta_h^n$ and sum both equations to obtain

$$\begin{aligned}2(\xi_h^{n+1}, \zeta_h^n)_{L^2(\Omega)} &= \|\zeta_h^n\|_{L^2(\Omega)}^2 + \|\xi_h^n\|_{L^2(\Omega)}^2 + \delta t(\alpha_h^n, \xi_h^n)_{L^2(\Omega)} + \delta t(\beta_h^n, \zeta_h^n)_{L^2(\Omega)} \\ &\quad - \delta t(A_h^{\mathrm{upw}} \xi_h^n, \xi_h^n)_{L^2(\Omega)} - \delta t(A_h^{\mathrm{upw}} \zeta_h^n, \zeta_h^n)_{L^2(\Omega)},\end{aligned}$$

so that, owing to discrete dissipation (cf. (3.10)),

$$\begin{aligned}2(\xi_h^{n+1}, \zeta_h^n)_{L^2(\Omega)} &= \|\zeta_h^n\|_{L^2(\Omega)}^2 + \|\xi_h^n\|_{L^2(\Omega)}^2 + \delta t(\alpha_h^n, \xi_h^n)_{L^2(\Omega)} + \delta t(\beta_h^n, \zeta_h^n)_{L^2(\Omega)} \\ &\quad - \delta t|\xi_h^n|_\beta^2 - \delta t|\zeta_h^n|_\beta^2 - \Lambda_h^n.\end{aligned}$$

Moreover,

$$2(\xi_h^{n+1}, \zeta_h^n)_{L^2(\Omega)} - \|\zeta_h^n\|_{L^2(\Omega)}^2 = \|\xi_h^{n+1}\|_{L^2(\Omega)}^2 - \|\xi_h^{n+1} - \zeta_h^n\|_{L^2(\Omega)}^2.$$

Rearranging terms yields the assertion. $\square$

3.1.6.3 Preliminary Stability Bound

We first bound the three rightmost terms in the energy identity (3.46). For the time being, we only need the usual CFL condition (3.21). We remark that the 4/3-CFL condition (3.25) implies the usual CFL condition (3.21) with $\varrho = (\varrho')^{3/4}$ since $\delta t \leq \tau_*$.

An important ingredient to derive our preliminary stability bound is the *boundedness on orthogonal subscales* for the discrete operator A_h^{upw}, namely, for all $(v, w_h) \in H^1(\Omega) \times V_h$,

$$(A_h^{\text{upw}}(v - \pi_h v), w_h)_{L^2(\Omega)} \lesssim \|v - \pi_h v\|_{\text{uwb},*} \|w_h\|_{\text{uwb}}. \tag{3.47}$$

This property is a direct consequence of Lemma 2.30. We recall that, for all $v \in V_{*h}$,

$$\|v\|_{\text{uwb}}^2 = \tau_c^{-1} \|v\|_{L^2(\Omega)}^2 + |v|_\beta^2,$$

$$\|v\|_{\text{uwb},*}^2 = \|v\|_{\text{uwb}}^2 + \sum_{T \in \mathcal{T}_h} \beta_c \|v\|_{L^2(\partial T)}^2.$$

We also need to consider the following norm: For all $v \in V_{*h}$,

$$\|v\|_{**}^2 = \|v\|_{\text{uwb},*}^2 + \beta_c h \|\nabla_h v\|_{[L^2(\Omega)]^d}^2.$$

Finally, for brevity of notation, letting $C_{fu} := \|d_t^2 f\|_{C^0(L^2(\Omega))} + \|d_t^3 u\|_{C^0(L^2(\Omega))}$, we introduce the quantity

$$E_h^n := \|\xi_\pi^n\|_{**} + \|\zeta_\pi^n\|_{**} + \tau_*^{1/2} C_{fu} \delta t^2 + \tau_*^{-1/2} \|\xi_h^n\|_{L^2(\Omega)},$$

which collects, in addition to $\tau_*^{-1/2} \|\xi_h^n\|_{L^2(\Omega)}$, the contributions of the space and time approximation errors.

Lemma 3.25 (Preliminary stability bound). *Assume $f \in C^2(L^2(\Omega))$ and $u \in C^3(L^2(\Omega)) \cap C^0(H^1(\Omega))$. Assume the CFL condition (3.21). Then, there is C_{sta}, independent of h, δt, and the data f, μ, and β, such that*

$$\|\xi_h^{n+1}\|_{L^2(\Omega)}^2 - \|\xi_h^n\|_{L^2(\Omega)}^2 + \tfrac{1}{2}\delta t |\xi_h^n|_\beta^2 + \tfrac{1}{2}\delta t |\zeta_h^n|_\beta^2 \leq \|\xi_h^{n+1} - \zeta_h^n\|_{L^2(\Omega)}^2 + C_{\text{sta}} \delta t (E_h^n)^2. \tag{3.48}$$

Proof. We proceed in four steps.
(i) Further bounds on A_h^{upw}. Let us first prove that

$$\|A_h^{\text{upw}} v\|_{L^2(\Omega)} \lesssim \beta_c^{1/2} h^{-1/2} \|v\|_{**} \qquad \forall v \in V_{*h}. \tag{3.49}$$

Let $w_h \in V_h$. We obtain using (3.7) together with the Cauchy–Schwarz inequality, the discrete trace inequality (1.37), and mesh quasi-uniformity,

$$(A_h^{\text{upw}} v, w_h)_{L^2(\Omega)} \leq (\tau_c^{-1} \|v\|_{L^2(\Omega)} + \beta_c \|\nabla_h v\|_{[L^2(\Omega)]^d}) \|w_h\|_{L^2(\Omega)}$$

$$+ |v|_\beta |w_h|_\beta + |v|_\beta \left(\sum_{F \in \mathcal{F}_h^i} \int_F 2|\beta \cdot n| \{\!\{w_h\}\!\}^2 \right)^{1/2}$$

$$\lesssim (\tau_c^{-1} \|v\|_{L^2(\Omega)} + \beta_c \|\nabla_h v\|_{[L^2(\Omega)]^d} + \beta_c^{1/2} h^{-1/2} |v|_\beta) \|w_h\|_{L^2(\Omega)}.$$

3.1. Unsteady Advection-Reaction

Owing to the definition of the $\|\cdot\|_{**}$-norm, and the fact that $\tau_c^{-1/2} \leq \beta_c^{1/2} h^{-1/2}$, we infer

$$(A_h^{\text{upw}} v, w_h)_{L^2(\Omega)} \lesssim \beta_c^{1/2} h^{-1/2} \|v\|_{**} \|w_h\|_{L^2(\Omega)}.$$

Since $\|A_h^{\text{upw}} v\|_{L^2(\Omega)} = \sup_{w_h \in V_h \setminus \{0\}} \frac{(A_h^{\text{upw}} v, w_h)_{L^2(\Omega)}}{\|w_h\|_{L^2(\Omega)}}$, this yields (3.49). Moreover, using inverse and trace inequalities and $h \leq \beta_c \tau_c$ yields

$$\|v_h\|_{**} \lesssim \beta_c^{1/2} h^{-1/2} \|v_h\|_{L^2(\Omega)} \qquad \forall v_h \in V_h,$$

so that for discrete arguments in (3.49), we obtain

$$\|A_h^{\text{upw}} v_h\|_{L^2(\Omega)} \lesssim \beta_c h^{-1} \|v_h\|_{L^2(\Omega)} \qquad \forall v_h \in V_h. \tag{3.50}$$

(ii) Bound on α_h^n and β_h^n. Using the definition (3.44) of α_h^n, the bound (3.49) on A_h^{upw}, and the usual CFL condition yields

$$\delta t^{1/2} \|\alpha_h^n\|_{L^2(\Omega)} = \delta t^{1/2} \|A_h^{\text{upw}} \xi_\pi^n\|_{L^2(\Omega)}$$
$$\lesssim \delta t^{1/2} \beta_c^{1/2} h^{-1/2} \|\xi_\pi^n\|_{**} \lesssim \|\xi_\pi^n\|_{**} \leq E_h^n. \tag{3.51}$$

Using the triangle inequality yields

$$\|\beta_h^n\|_{L^2(\Omega)} = \|A_h^{\text{upw}} \zeta_\pi^n - \pi_h \theta^n + \delta_h^n\|_{L^2(\Omega)}$$
$$\leq \|A_h^{\text{upw}} \zeta_\pi^n\|_{L^2(\Omega)} + \|\delta_h^n\|_{L^2(\Omega)} + \|\pi_h \theta^n\|_{L^2(\Omega)}.$$

The definition (3.45) of δ_h^n together with assumption (3.24) yield

$$\|\delta_h^n\|_{L^2(\Omega)} \lesssim \delta t^2 \|d_t^2 f(t)\|_{C^0(L^2(\Omega))},$$

while proceeding as in the proof of (3.38) to bound $\|\theta^n\|_{L^2(\Omega)}$ yields

$$\|\pi_h \theta^n\|_{L^2(\Omega)} \leq \|\theta^n\|_{L^2(\Omega)} \lesssim \delta t^2 \|d_t^3 u\|_{C^0(L^2(\Omega))}.$$

Finally, proceeding as above, we obtain $\delta t^{1/2} \|A_h^{\text{upw}} \zeta_\pi^n\|_{L^2(\Omega)} \lesssim \|\zeta_\pi^n\|_{**}$. As a result,

$$\delta t^{1/2} \|\beta_h^n\|_{L^2(\Omega)} \lesssim \|\zeta_\pi^n\|_{**} + \delta t^{1/2} C_{fu} \delta t^2 \leq E_h^n. \tag{3.52}$$

(iii) Bound on ζ_h^n and Λ_h^n. Using the error equation (3.43a), the triangle inequality, the bound (3.50) on A_h^{upw}, the usual CFL condition, and the bound (3.51) on α_h^n, we infer

$$\|\zeta_h^n\|_{L^2(\Omega)} \leq \|\xi_h^n\|_{L^2(\Omega)} + \delta t \|A_h^{\text{upw}} \xi_h^n\|_{L^2(\Omega)} + \delta t \|\alpha_h^n\|_{L^2(\Omega)}$$
$$\lesssim \|\xi_h^n\|_{L^2(\Omega)} + \delta t \beta_c h^{-1} \|\xi_h^n\|_{L^2(\Omega)} + \delta t^{1/2} \|\xi_\pi^n\|_{**}$$
$$\lesssim \|\xi_h^n\|_{L^2(\Omega)} + \delta t^{1/2} \|\xi_\pi^n\|_{**}.$$

Hence, since $\delta t \leq \tau_* \leq \tau_c$,

$$\delta t \tau_c^{-1} \|\zeta_h^n\|_{L^2(\Omega)}^2 \lesssim \delta t \tau_c^{-1} \|\xi_h^n\|_{L^2(\Omega)}^2 + \delta t \|\xi_\pi^n\|_{**}^2. \tag{3.53}$$

Using $\|\Lambda\|_{L^\infty(\Omega)} \lesssim \tau_c^{-1} \leq \tau_*^{-1}$, this yields
$$\Lambda_h^n \leq \delta t \tau_c^{-1} \|\xi_h^n\|_{L^2(\Omega)}^2 + \delta t \tau_c^{-1} \|\zeta_h^n\|_{L^2(\Omega)}^2 \lesssim \delta t (E_h^n)^2. \tag{3.54}$$

(iv) Bound on $\delta t (\alpha_h^n, \xi_h^n)_{L^2(\Omega)} + \delta t (\beta_h^n, \zeta_h^n)_{L^2(\Omega)}$. The bounds (3.51) and (3.52) are not sufficiently sharp to bound these quantities by means of the Cauchy–Schwarz inequality. Instead, we make use of boundedness on orthogonal subscales (cf. (3.47)) and the fact that $\|\cdot\|_{\mathrm{uwb},*} \leq \|\cdot\|_{**}$ to infer

$$\delta t (\alpha_h^n, \xi_h^n)_{L^2(\Omega)} = \delta t (A_h^{\mathrm{upw}} \xi_\pi^n, \xi_h^n)_{L^2(\Omega)} \lesssim \delta t \|\xi_\pi^n\|_{**} \|\xi_h^n\|_{\mathrm{uwb}}$$
$$\lesssim \delta t \|\xi_\pi^n\|_{**} (|\xi_h^n|_\beta + \tau_c^{-1/2} \|\xi_h^n\|_{L^2(\Omega)}).$$

Hence, using Young's inequality leads to

$$\delta t (\alpha_h^n, \xi_h^n)_{L^2(\Omega)} - \tfrac{1}{2} \delta t |\xi_h^n|_\beta^2 \lesssim \delta t \|\xi_\pi^n\|_{**}^2 + \delta t \tau_c^{-1} \|\xi_h^n\|_{L^2(\Omega)}^2 < \delta t (E_h^n)^2.$$

Similarly, using again (3.47) together with the above bounds on δ_h^n and $\pi_h \theta^n$ and the Cauchy–Schwarz inequality yields

$$\delta t (\beta_h^n, \zeta_h^n)_{L^2(\Omega)} \lesssim \delta t \|\zeta_\pi^n\|_{**} \|\zeta_h^n\|_{\mathrm{uwb}} + \delta t (\delta_h^n - \pi_h \theta^n, \zeta_h^n)_{L^2(\Omega)}$$
$$\lesssim \delta t \|\zeta_\pi^n\|_{**} (|\zeta_h^n|_\beta + \tau_c^{-1/2} \|\zeta_h^n\|_{L^2(\Omega)}) + C_{fu} \delta t^3 \|\zeta_h^n\|_{L^2(\Omega)},$$

so that

$$\delta t (\beta_h^n, \zeta_h^n)_{L^2(\Omega)} - \tfrac{1}{2} \delta t |\zeta_h^n|_\beta^2 \lesssim \delta t \|\zeta_\pi^n\|_{**}^2 + \delta t \tau_c^{-1} \|\zeta_h^n\|_{L^2(\Omega)}^2 + C_{fu} \delta t^3 \|\zeta_h^n\|_{L^2(\Omega)}.$$

Using the estimate (3.53) for the two rightmost terms together with Young's inequality leads to

$$\delta t (\beta_h^n, \zeta_h^n)_{L^2(\Omega)} - \tfrac{1}{2} \delta t |\zeta_h^n|_\beta^2 \lesssim \delta t (E_h^n)^2.$$

Inserting the above bounds on $\delta t (\alpha_h^n, \xi_h^n)_{L^2(\Omega)}$ and $\delta t (\beta_h^n, \zeta_h^n)_{L^2(\Omega)}$ together with the bound (3.54) in the energy identity (3.46) yields (3.48). $\square$

To turn (3.48) into a stability estimate, we need to bound the anti-dissipative term $\|\xi_h^{n+1} - \zeta_h^n\|_{L^2(\Omega)}^2$ appearing on the right-hand side of (3.48) and which results from the explicit nature of the RK2 scheme. There are two ways to bound this term depending on the polynomial degree used for the dG approximation. For $k \geq 2$, a 4/3-CFL condition is invoked. By proceeding differently for $k = 1$, this term can be controlled using only the usual CFL condition.

Remark 3.26 (Importance of the boundedness on orthogonal subscales). The boundedness on orthogonal subscales (3.47) is a stronger property than the bound (3.49) on A_h^{upw}. Indeed, (3.49) combined with the Cauchy–Schwarz inequality only yields

$$(A_h^{\mathrm{upw}}(v - \pi_h v), w_h)_{L^2(\Omega)} \lesssim \beta_c^{1/2} h^{-1/2} \|v - \pi_h v\|_{**} \|w_h\|_{L^2(\Omega)}.$$

The difficulty is not the presence of the $\|\cdot\|_{**}$-norm instead of the $\|\cdot\|_{\mathrm{uwb},*}$-norm (cf. (3.47)) since $\|\xi_\pi^n\|_{**}$ and $\|\xi_\pi^n\|_{\mathrm{uwb},*}$ exhibit the same convergence rate as

$h \to 0$ for smooth enough u^n, but the presence of the factor $h^{-1/2}$ which causes the loss of half a power of h and thus the suboptimality of the error estimate. In other words, the benefit of boundedness on orthogonal subscales is to remove the factor $h^{-1/2}$ by increasing the norm of w_h on the right-hand side ($\|\|w_h\|\|_{\text{uwb}}$ instead of $\|w_h\|_{L^2(\Omega)}$). The additional term $|w_h|_\beta$ is then controlled by the dissipative properties of the dG scheme.

3.1.6.4 The Case $k \geq 2$: 4/3-CFL Condition

Lemma 3.27 (Stability, $k \geq 2$). *Assume $f \in C^2(L^2(\Omega))$ and $u \in C^3(L^2(\Omega)) \cap C^0(H^1(\Omega))$. Assume the 4/3-CFL condition (3.25) for some positive real number ϱ', that is,*

$$\delta t \leq \varrho' \tau_*^{-1/3} \left(\frac{h}{\beta_c}\right)^{4/3}.$$

Then, there is C_{sta}, independent of h, δt, and the data f, μ, and β, such that

$$\|\xi_h^{n+1}\|_{L^2(\Omega)}^2 - \|\xi_h^n\|_{L^2(\Omega)}^2 + \tfrac{1}{2}\delta t |\xi_h^n|_\beta^2 + \tfrac{1}{2}\delta t |\zeta_h^n|_\beta^2 \leq C_{\text{sta}} \delta t (E_h^n)^2. \quad (3.55)$$

Proof. We first observe that the error equations (3.43) imply

$$\xi_h^{n+1} = \zeta_h^n - \tfrac{1}{2}\delta t A_h^{\text{upw}}(\zeta_h^n - \xi_h^n) + \tfrac{1}{2}\delta t(\beta_h^n - \alpha_h^n), \quad (3.56)$$

and using again (3.43a) to eliminate $(\zeta_h^n - \xi_h^n)$ yields

$$\xi_h^{n+1} - \zeta_h^n = \tfrac{1}{2}\delta t^2 (A_h^{\text{upw}})^2 \xi_h^n + \tfrac{1}{2}\delta t(\beta_h^n - \alpha_h^n - \delta t A_h^{\text{upw}} \alpha_h^n).$$

Let $\mathfrak{T}_1$ and $\mathfrak{T}_2$ denote the L^2-norm of the two terms on the right-hand side. We first bound $\mathfrak{T}_2$. Owing to the usual CFL condition (which is implied by the 4/3-CFL condition) and the bound (3.50),

$$\mathfrak{T}_2 \lesssim \delta t \|\alpha_h^n\|_{L^2(\Omega)} + \delta t \|\beta_h^n\|_{L^2(\Omega)},$$

so that owing to (3.51) and (3.52) and since $\delta t \leq \tau_*$, $\mathfrak{T}_2 \lesssim \delta t^{1/2} E_h^n$. Furthermore, to bound $\mathfrak{T}_1$, we apply (3.50) twice, yielding $\|(A_h^{\text{upw}})^2 \xi_h^n\|_{L^2(\Omega)} \lesssim \beta_c^2 h^{-2} \|\xi_h^n\|_{L^2(\Omega)}$, so that

$$\mathfrak{T}_1 \lesssim \delta t^2 \beta_c^2 h^{-2} \|\xi_h^n\|_{L^2(\Omega)} = (\delta t^{3/2} \beta_c^2 h^{-2} \tau_*^{1/2})(\delta t^{1/2} \tau_*^{-1/2}) \|\xi_h^n\|_{L^2(\Omega)}$$

$$\lesssim \delta t^{1/2} \tau_*^{-1/2} \|\xi_h^n\|_{L^2(\Omega)},$$

owing to the 4/3-CFL condition (3.25). Hence, $\|\xi_h^{n+1} - \zeta_h^n\|_{L^2(\Omega)}^2 \lesssim |\mathfrak{T}_1|^2 + |\mathfrak{T}_2|^2 \lesssim \delta t (E_h^n)^2$. Substituting this estimate into (3.48) yields (3.55). $\square$

3.1.6.5 The Piecewise Affine Case ($k = 1$): Usual CFL Condition

To bound the anti-dissipative term $\|\xi_h^{n+1} - \zeta_h^n\|_{L^2(\Omega)}^2$ in the piecewise affine case, we need some technical results. First, a close inspection at the proof of (3.49) shows that there holds, for all $v \in V_{*h}$,

$$\|A_h^{\text{upw}} v\|_{L^2(\Omega)} \leq \tau_c^{-1} \|v\|_{L^2(\Omega)} + \beta_c \|\nabla_h v\|_{[L^2(\Omega)]^d} + C_{*,3} \beta_c^{1/2} h^{-1/2} |v|_\beta, \quad (3.57)$$

where $C_{*,3}$ is independent of h, δt, and the data f, μ, and β. The bound (3.57) is actually valid for any polynomial degree. We also need an additional property of the discrete operator A_h^{upw}, this time specific to the case $k=1$.

Lemma 3.28 (Adjoint boundedness on orthogonal subscales). *Let π_h^0 denote the L^2-orthogonal projection onto $\mathbb{P}_d^0(\mathcal{T}_h)$ (recall that this space is spanned by piecewise constant functions on $\mathcal{T}_h$). Then, there holds, for all $v_h, w_h \in V_h = \mathbb{P}_d^1(\mathcal{T}_h)$,*

$$(A_h^{\mathrm{upw}} v_h, w_h - \pi_h^0 w_h)_{L^2(\Omega)} \leq C_{*,4}\beta_c^{1/2} h^{-1/2}\|v_h\|_{\mathrm{uwb}}\|w_h - \pi_h^0 w_h\|_{L^2(\Omega)}, \quad (3.58)$$

where $C_{*,4}$ is independent of h, δt, and the data f, μ, and β.

Proof. We start from the expression (3.7) observing that $w_h - \pi_h^0 w_h \in V_h$ and bound all the terms using the Cauchy–Schwarz inequality except the term $\int_\Omega (\beta \cdot \nabla_h v_h)(w_h - \pi_h^0 w_h)$ for which we exploit the fact that $\nabla_h v_h$ is piecewise constant since $k=1$ to obtain using an inverse inequality and the regularity of β that

$$\int_\Omega (\beta\cdot\nabla_h v_h)(w_h - \pi_h^0 w_h) = \int_\Omega ((\beta - \pi_h^0\beta)\cdot\nabla_h v_h)(w_h - \pi_h^0 w_h)$$
$$\leq L_\beta h C_{\mathrm{inv}} h^{-1} \|v_h\|_{L^2(\Omega)} \|w_h - \pi_h^0 w_h\|_{L^2(\Omega)}$$
$$\leq C_{\mathrm{inv}} \tau_c^{-1} \|v_h\|_{L^2(\Omega)} \|w_h - \pi_h^0 w_h\|_{L^2(\Omega)},$$

whence we infer (3.58). □

We now turn to our main stability result in the piecewise affine case. To formulate the CFL condition, we use in particular the quantity C'_{inv} such that, for all $v_h \in \mathbb{P}_d^1(\mathcal{T}_h)$,

$$\|\nabla_h v_h\|_{[L^2(\Omega)]^d} = \|\nabla_h(v_h - \pi_h^0 v_h)\|_{[L^2(\Omega)]^d} \leq C'_{\mathrm{inv}} h^{-1}\|v_h - \pi_h^0 v_h\|_{L^2(\Omega)}. \quad (3.59)$$

Property (3.59) results from the inverse inequality (1.36) and mesh quasi-uniformity (this property is actually valid for any polynomial degree).

Lemma 3.29 (Stability, $k=1$). *Let $V_h = \mathbb{P}_d^1(\mathcal{T}_h)$. Assume that $f \in C^2(L^2(\Omega))$ and that $u \in C^3(L^2(\Omega)) \cap C^0(H^1(\Omega))$. Assume the CFL condition (3.21) with $\varrho \leq \min\{\frac{1}{8}(C_{*,3})^{-2}, \frac{1}{2}(C'_{\mathrm{inv}} C_{*,4})^{-2/3}\}$, that is,*

$$\delta t \leq \min\left\{\tfrac{1}{8}(C_{*,3})^{-2}, \tfrac{1}{2}(C'_{\mathrm{inv}} C_{*,4})^{-2/3}\right\} \frac{h}{\beta_c}. \quad (3.60)$$

Then, there is C_{sta}, independent of h, δt, and the data f, μ, and β, such that

$$\|\xi_h^{n+1}\|_{L^2(\Omega)}^2 - \|\xi_h^n\|_{L^2(\Omega)}^2 + \tfrac{1}{8}\delta t|\xi_h^n|_\beta^2 + \tfrac{1}{8}\delta t|\zeta_h^n|_\beta^2 \leq \delta t (E_h^n)^2. \quad (3.61)$$

Proof. Setting $x_h^n := \xi_h^n - \zeta_h^n$, (3.56) yields

$$\xi_h^{n+1} - \zeta_h^n = \tfrac{1}{2}\delta t A_h^{\mathrm{upw}} x_h^n + \tfrac{1}{2}\delta t(\beta_h^n - \alpha_h^n).$$

3.1. Unsteady Advection-Reaction

Let $\mathfrak{T}_1$ and $\mathfrak{T}_2$ denote the L^2-norm of the two terms on the right-hand side. Using the above bounds on α_h^n and β_h^n yields $\mathfrak{T}_2 \lesssim \delta t^{1/2} E_h^n$. To bound $\mathfrak{T}_1$, we start from the bound (3.57) yielding

$$\mathfrak{T}_1 = \tfrac{1}{2}\delta t \|A_h^{\text{upw}} x_h^n\|_{L^2(\Omega)}$$
$$\leq \tfrac{1}{2}\delta t \beta_c \|\nabla_h x_h^n\|_{[L^2(\Omega)]^d} + \tfrac{1}{2} C_{*,3} \delta t \beta_c^{1/2} h^{-1/2} |x_h^n|_\beta + \tfrac{1}{2} \delta t \tau_c^{-1} \|x_h^n\|_{L^2(\Omega)}. \quad (3.62)$$

The main difficulty is to bound $\|\nabla_h x_h^n\|_{[L^2(\Omega)]^d}$. Owing to (3.59),

$$\|\nabla_h x_h^n\|_{[L^2(\Omega)]^d} \leq C'_{\text{inv}} h^{-1} \|y_h^n\|_{L^2(\Omega)}, \qquad y_h^n := x_h^n - \pi_h^0 x_h^n. \quad (3.63)$$

To bound $\|y_h^n\|_{L^2(\Omega)}$, we exploit the fact that, up to a non-essential perturbation, x_h^n is in the range of the discrete operator A_h^{upw} and we use adjoint boundedness on orthogonal subscales (cf. Lemma 3.28). Specifically, the error equation (3.43a) yields

$$x_h^n = \delta t A_h^{\text{upw}} \xi_h^n - \delta t \alpha_h^n.$$

Thus,

$$\|y_h^n\|_{L^2(\Omega)}^2 = (x_h^n, y_h^n)_{L^2(\Omega)} = \delta t (A_h^{\text{upw}} \xi_h^n, y_h^n)_{L^2(\Omega)} - \delta t (\alpha_h^n, y_h^n)_{L^2(\Omega)}.$$

Owing to Lemma 3.28,

$$\delta t |(A_h^{\text{upw}} \xi_h^n, y_h^n)_{L^2(\Omega)}| \leq C_{*,4} \delta t \beta_c^{1/2} h^{-1/2} (|\xi_h^n|_\beta + \tau_c^{-1/2} \|\xi_h^n\|_{L^2(\Omega)}) \|y_h^n\|_{L^2(\Omega)},$$

while, owing to the Cauchy–Schwarz inequality and the bound (3.51) on α_h^n,

$$\delta t |(\alpha_h^n, y_h^n)_{L^2(\Omega)}| \lesssim \delta t^{1/2} \|\xi_\pi^n\|_{**} \|y_h^n\|_{L^2(\Omega)}.$$

Hence, simplifying by $\|y_h^n\|_{L^2(\Omega)}$ leads to

$$\|y_h^n\|_{L^2(\Omega)} \leq C_{*,4} \delta t \beta_c^{1/2} h^{-1/2} (|\xi_h^n|_\beta + \tau_c^{-1/2} \|\xi_h^n\|_{L^2(\Omega)}) + C \delta t^{1/2} \|\xi_\pi^n\|_{**}$$
$$\leq C_{*,4} \beta_c^{1/2} h^{-1/2} \delta t |\xi_h^n|_\beta + C \delta t^{1/2} E_h^n.$$

Using (3.63) yields

$$\|\nabla_h x_h^n\|_{[L^2(\Omega)]^d} \leq C'_{\text{inv}} C_{*,4} \beta_c^{1/2} h^{-3/2} \delta t |\xi_h^n|_\beta + C h^{-1} \delta t^{1/2} E_h^n.$$

Plugging the above bound into (3.62), we infer

$$\mathfrak{T}_1 \leq \tfrac{1}{2} C'_{\text{inv}} C_{*,4} \beta_c^{3/2} h^{-3/2} \delta t^2 |\xi_h^n|_\beta + \tfrac{1}{2} C_{*,3} \delta t \beta_c^{1/2} h^{-1/2} |x_h^n|_\beta + C \delta t^{1/2} E_h^n,$$

where we have used the fact that

$$\delta t \tau_c^{-1} \|x_h^n\|_{L^2(\Omega)} \leq \delta t^{1/2} \tau_c^{-1/2} \|\xi_h^n\|_{L^2(\Omega)} + \delta t^{1/2} \tau_c^{-1/2} \|\zeta_h^n\|_{L^2(\Omega)}$$
$$\lesssim \delta t^{1/2} \|\xi_\pi^n\|_{**} + \delta t^{1/2} \tau_c^{-1/2} \|\xi_h^n\|_{L^2(\Omega)},$$

owing to the bound (3.53) on ζ_h^n. Collecting the bounds on $\mathfrak{T}_1$ and $\mathfrak{T}_2$ yields

$$\|\xi_h^{n+1} - \zeta_h^n\|_{L^2(\Omega)} \leq \tfrac{1}{2}C'_{\mathrm{inv}}C_{*,4}\beta_{\mathrm{c}}^{3/2}h^{-3/2}\delta t^2|\xi_h^n|_\beta + \tfrac{1}{2}C_{*,3}\delta t\beta_{\mathrm{c}}^{1/2}h^{-1/2}|\xi_h^n - \zeta_h^n|_\beta$$
$$+ C\delta t^{1/2}E_h^n.$$

Owing to the condition (3.60) on the parameter ϱ in the CFL condition, we obtain

$$\|\xi_h^{n+1} - \zeta_h^n\|_{L^2(\Omega)} \leq 2^{-5/2}\delta t^{1/2}(|\xi_h^n|_\beta + |\xi_h^n - \zeta_h^n|_\beta) + C\delta t^{1/2}E_h^n$$
$$\leq 2^{-3/2}\delta t^{1/2}(|\xi_h^n|_\beta + |\zeta_h^n|_\beta) + C\delta t^{1/2}E_h^n.$$

Squaring yields

$$\|\xi_h^{n+1} - \zeta_h^n\|_{L^2(\Omega)}^2 \leq \tfrac{3}{8}\delta t|\xi_h^n|_\beta^2 + \tfrac{3}{8}\delta t|\zeta_h^n|_\beta^2 + C\delta t(E_h^n)^2,$$

whence (3.61) easily follows. $\square$

3.1.6.6 Proof of Theorem 3.10

Starting from Lemmata 3.27 or 3.29 depending on the polynomial degree, we proceed as in the proof of Theorem 3.7 using the same techniques. The only difference lies in the use of the $\|\cdot\|_{**}$-norm instead of the $\|\cdot\|_{\mathrm{uwb},*}$-norm and in the additional presence of terms involving ζ_π^n. Using the definition of the $\|\cdot\|_{**}$-norm, Lemmata 1.58 and 1.59, and the assumption $h \leq \beta_{\mathrm{c}}\tau_{\mathrm{c}}$ leads to

$$\|\xi_\pi^m\|_{**}^2 \lesssim \beta_{\mathrm{c}}h^{2k+1}\|u^m\|_{H^{k+1}(\Omega)}^2.$$

Similarly, since $\zeta_\pi^m = w^m - \pi_h w^m$ with $w^m = u^m + \delta t d_t u^m$ and using the usual CFL condition leads to

$$\|\zeta_\pi^m\|_{**}^2 \lesssim \beta_{\mathrm{c}}(h^{2k+1}\|u^m\|_{H^{k+1}(\Omega)}^2 + \delta t^2 h^{2k-1}\|d_t u^m\|_{H^k(\Omega)}^2)$$
$$\lesssim \beta_{\mathrm{c}}h^{2k+1}(\|u^m\|_{H^{k+1}(\Omega)}^2 + \delta t^2 h^{-2}\|d_t u^m\|_{H^k(\Omega)}^2)$$
$$\lesssim \beta_{\mathrm{c}}h^{2k+1}(\|u^m\|_{H^{k+1}(\Omega)}^2 + \beta_{\mathrm{c}}^{-2}\|d_t u^m\|_{H^k(\Omega)}^2).$$

The conclusion is straightforward.

3.2 Nonlinear Conservation Laws

In this section, we investigate the approximation of scalar, nonlinear conservation laws, whereby space semidiscretization is achieved by dG methods formulated using numerical fluxes and time discretization by SSP-RK schemes. We also discuss the use of limiters to achieve tighter control of spurious oscillations when approximating rough solutions.

3.2.1 The Continuous Setting

We consider nonlinear conservation laws of the form

$$\partial_t u + \nabla \cdot \mathfrak{f}(u) = 0 \quad \text{in } \Omega \times (0, t_\text{F}). \tag{3.64}$$

The unknown u is a scalar-valued function. In many models, this function cannot take values in the whole real line, but only in a subset of *admissible states*, denoted by $\mathcal{U}$. For instance, when u represents a concentration, $\mathcal{U}$ is the set of nonnegative real numbers. The *flux function* $\mathfrak{f}$ is a $\mathbb{R}^d$-valued function of class C^1 defined on the set of admissible states $\mathcal{U}$. In what follows, we assume, without precluding generality, that $\mathcal{U}$ is an interval with $0 \in \mathcal{U}$. The flux function can sometimes also depend explicitly on the space variable x. For instance, in the linear case, we obtain $\mathfrak{f}(u) = \beta u$ where β denotes the $\mathbb{R}^d$-valued advective velocity.

We supplement problem (3.64) with the initial condition

$$u(\cdot, t = 0) = u_0 \quad \text{in } \Omega.$$

Enforcing a boundary condition on (3.64) is a subtle issue. We refer the reader to the pioneering work of Bardos, Le Roux, and Nédélec [29] and to the more recent work by Otto [254] and Vovelle [303]. Let g denote a $\mathcal{U}$-valued external state defined on $\partial\Omega \times (0, t_\text{F})$. Recalling that n denotes the outward normal on $\partial\Omega$, we define the normal flux function on the boundary as

$$\forall v \in \mathcal{U}, \qquad \mathfrak{f}_\text{n}(v) := \mathfrak{f}(v) \cdot \text{n}.$$

Then, on any point $(x, t) \in \partial\Omega \times (0, t_\text{F})$, we enforce the boundary condition in the form

$$\mathfrak{f}_\text{n}(u) = \Phi^\text{R}(\text{n}, u, g) := \begin{cases} \min_{w \in [u, g]} \mathfrak{f}_\text{n}(w) & \text{if } u \leq g, \\ \max_{w \in [g, u]} \mathfrak{f}_\text{n}(w) & \text{otherwise.} \end{cases} \tag{3.65}$$

This condition means that $u < g$ is possible only if, for all $w \in [u, g]$, $\mathfrak{f}_\text{n}(w) \geq \mathfrak{f}_\text{n}(u)$, and, similarly, $u > g$ is possible only if, for all $w \in [g, u]$, $\mathfrak{f}_\text{n}(w) \leq \mathfrak{f}_\text{n}(u)$; otherwise, there holds $u = g$. In the linear case where $\mathfrak{f}_\text{n}(v) = (\beta \cdot \text{n})v$, condition (3.65) is equivalent to enforcing $u = g$ whenever $\beta \cdot \text{n} < 0$, that is, on the inflow boundary. The peculiarity of the nonlinear case is that it is not possible to assert a priori the part of the boundary where the boundary condition $u = g$ is enforced. This is the reason why we define the external state on the whole boundary.

Remark 3.30 (Riemann problem at boundary). The quantity $\Phi^\text{R}(\text{n}, u, g)$ defined by (3.65) corresponds to the flux obtained from a one-dimensional Riemann problem with flux function $\mathfrak{f}_\text{n}$, left state u, and right state g. Thus, condition (3.65) means that a state $u \neq g$ at the boundary is possible only if the flux for the Riemann problem does not depend on g.

3.2.2 Numerical Fluxes for Space Semidiscretization

Let $\mathcal{T}_h$ belong to an admissible mesh sequence. For a given integer $k \geq 0$, we consider the broken polynomial space $V_h := \mathbb{P}_d^k(\mathcal{T}_h)$. Formally, we test the conservation law (3.64) with a discrete function $v_h \in V_h$ (observe that v_h is time-independent) and integrate by parts elementwise the divergence term to obtain

$$d_t \int_\Omega u v_h - \int_\Omega \mathfrak{f}(u) \cdot \nabla_h v_h + \sum_{T \in \mathcal{T}_h} \int_{\partial T} (\mathfrak{f}(u) \cdot n_T) v_h = 0.$$

Using the fact that $\mathfrak{f}(u) \cdot n_T$ takes opposite values across interfaces and expressing the third term in the left-hand side as a sum over mesh faces yields

$$d_t \int_\Omega u v_h - \int_\Omega \mathfrak{f}(u) \cdot \nabla_h v_h + \sum_{F \in \mathcal{F}_h} \int_F \mathfrak{f}_{n_F}(u) [\![v_h]\!] = 0,$$

where we have introduced the normal flux function

$$\mathfrak{f}_{n_F}(v) := \mathfrak{f}(v) \cdot n_F,$$

and we have set $[\![v_h]\!] = v_h$ on boundary faces. Accounting for the boundary condition (3.65), we obtain

$$d_t \int_\Omega u v_h - \int_\Omega \mathfrak{f}(u) \cdot \nabla_h v_h + \sum_{F \in \mathcal{F}_h^i} \int_F \mathfrak{f}_{n_F}(u) [\![v_h]\!] + \sum_{F \in \mathcal{F}_h^b} \int_F \Phi^R(n, u, g) v_h = 0.$$

The space semidiscrete solution $u_h \in V_h$ is not single-valued at interfaces so that the normal flux $\mathfrak{f}_{n_F}(u_h)$ is not well-defined. For all $F \in \mathcal{F}_h$, we define the so-called left and right states of u_h as

$$u_F^- := \begin{cases} u_h|_{T_1} & \text{if } F \in \mathcal{F}_h^i, \\ u_h|_T & \text{if } F \in \mathcal{F}_h^b, \end{cases} \qquad u_F^+ := \begin{cases} u_h|_{T_2} & \text{if } F \in \mathcal{F}_h^i, \\ g & \text{if } F \in \mathcal{F}_h^b, \end{cases}$$

with the usual notation $F = \partial T_1 \cap \partial T_2$ if $F \in \mathcal{F}_h^i$ and $F = \partial T \cap \partial \Omega$ if $F \in \mathcal{F}_h^b$. We observe that, on boundary faces, one of the values (here, u_F^+) is replaced by the external state g. The space semidiscrete problem then reads: For all $v_h \in V_h$,

$$d_t \int_\Omega u_h v_h - \int_\Omega \mathfrak{f}(u_h) \cdot \nabla_h v_h + \sum_{F \in \mathcal{F}_h} \int_F \Phi(n_F, u_F^-, u_F^+) [\![v_h]\!] = 0. \qquad (3.66)$$

This problem is formulated in terms of a *numerical flux*

$$\Phi : \mathbb{R}^d \times \mathcal{U} \times \mathcal{U} \to \mathbb{R}$$

which provides, with its first argument set to n_F, a single-valued approximation of the normal flux $\mathfrak{f}_{n_F}$ at the mesh face $F \in \mathcal{F}_h$. For simplicity, we assume that the boundary condition (3.65) is weakly enforced in (3.66) by setting, for all

3.2. Nonlinear Conservation Laws

$F \in \mathcal{F}_h^b$, $\Phi(\mathrm{n}_F, u_F^-, u_F^+) = \Phi^R(\mathrm{n}_F, u_F^-, g)$. Moreover, (3.66) is supplemented by the initial condition

$$u_h(\cdot, t = 0) = \pi_h u_0,$$

where π_h denotes the L^2-orthogonal projection onto V_h.

A first requirement on the numerical flux is to match the following consistency property.

Definition 3.31 (Consistency of numerical flux). We say that the numerical flux Φ is *consistent* if it is linear in its first argument and Lipschitz continuous with respect to its second and third argument and if it satisfies, for all $\mathrm{n} \in \mathbb{R}^d$,

$$\forall v \in \mathcal{U}, \qquad \Phi(\mathrm{n}, v, v) = \mathfrak{f}_\mathrm{n}(v).$$

Remark 3.32 (Departure from admissible set). The space semidiscrete solution u_h can depart from the set of admissible states $\mathcal{U}$. This is especially the case when approximating a rough solution by high-degree polynomials, whereby spurious oscillations often appear at the discrete level. These oscillations can be tamed, but often not fully eliminated, using limiters; cf. Sect. 3.2.4 below. In practice, it can be necessary to extend the domain of the flux function $\mathfrak{f}$ and of the numerical flux Φ beyond the set $\mathcal{U}$.

Localizing the test function v_h in (3.66) to a single mesh element, we obtain a local formulation similar to that derived in Sects. 2.2.3 and 2.3.4 for the advection-reaction equation. Specifically, let $T \in \mathcal{T}_h$ and let $\xi \in \mathbb{P}_d^k(T)$. Using $v_h := \xi \chi_T$ in (3.66) (where χ_T denotes the characteristic function of T) yields

$$d_t \int_T u_h \xi - \int_T \mathfrak{f}(u_h) \cdot \nabla \xi + \sum_{F \in \mathcal{F}_T} \epsilon_{T,F} \int_F \Phi(\mathrm{n}_F, u_F^-, u_F^+) \xi = 0, \qquad (3.67)$$

where, as usual, $\epsilon_{T,F} := \mathrm{n}_T \cdot \mathrm{n}_F$. In particular, taking $v_h \equiv 1$ in (3.67), we infer

$$d_t \int_T u_h + \sum_{F \in \mathcal{F}_T} \epsilon_{T,F} \int_F \Phi(\mathrm{n}_F, u_F^-, u_F^+) = 0, \qquad (3.68)$$

so that the time variation of the mean value of u_h inside each mesh element $T \in \mathcal{T}_h$ is balanced by the fluxes through the boundary ∂T. Conservativity is important to assert that, whenever the sequence of discrete solutions converges to a function u, this limit is a weak solution to (3.64) (without boundary conditions); see Hou and LeFloch [196].

3.2.2.1 Stability

In the context of nonlinear conservation laws, bounding the entropy of the space semidiscrete solution is crucial. For simplicity, we focus on the quadratic entropy $U(v) := \frac{1}{2} v^2$ for which a bound can be derived (at least formally) by testing the nonlinear conservation law with the exact solution itself. We assume for simplicity that $g \equiv 0$ on $\partial \Omega$. Introducing the $\mathbb{R}^d$-valued primitive $\gamma(v) := \int_0^v \mathfrak{f}(w) \, \mathrm{d}w$

and observing that $\mathfrak{f}(v)\cdot\nabla v = \nabla\cdot\gamma(v)$ for all $v \in \mathcal{U}$, we obtain after integrating by parts,

$$d_t \int_\Omega U(u) = \int_\Omega u\partial_t u = \int_\Omega \mathfrak{f}(u)\cdot\nabla u - \int_{\partial\Omega} \mathfrak{f}_n(u)u$$
$$= \int_\Omega \nabla\cdot\gamma(u) - \int_{\partial\Omega} \mathfrak{f}_n(u)u$$
$$= \int_{\partial\Omega} \big\{\gamma_n(u) - \mathfrak{f}_n(u)u\big\},$$

with $\gamma_n(v) := \int_0^v \mathfrak{f}_n(w)\,\mathrm{d}w$. Let $\partial\Omega_0 := \{x \in \partial\Omega \mid u = 0\}$ be the part of the boundary where the boundary condition is enforced. It is clear that $\gamma_n(u) - \mathfrak{f}_n(u)u = 0 - 0 = 0$ on $\partial\Omega_0$. On the remaining part of the boundary $\partial\Omega \setminus \partial\Omega_0$, we use the boundary condition (3.65) to infer $\gamma_n(u) - \mathfrak{f}_n(u)u \le 0$. For instance, if $u > 0$, we obtain $\gamma_n(u) = \int_0^u \mathfrak{f}_n(w)\,\mathrm{d}w \le \mathfrak{f}_n(u)u$ since, for all $w \in [0, u]$, $\mathfrak{f}_n(w) \le \mathfrak{f}_n(u)$. As a result,

$$d_t \int_\Omega U(u) = \int_{\partial\Omega\setminus\partial\Omega_0} \big\{\gamma_n(u) - \mathfrak{f}_n(u)u\big\} \le 0.$$

Recalling the expression of U, we infer, for all $t \in (0, t_\mathrm{F})$,

$$\|u(t)\|_{L^2(\Omega)} \le \|u_0\|_{L^2(\Omega)}. \tag{3.69}$$

Remark 3.33 (Linear case). In the linear case where $\mathfrak{f}_n(v) = (\beta\cdot n)v$, we obtain $\int_{\partial\Omega\setminus\partial\Omega_0}\{\gamma_n(u) - \mathfrak{f}_n(u)u\} = -\int_{\{\beta\cdot n \ge 0\}} \frac{1}{2}(\beta\cdot n)u^2 \le 0$.

Remark 3.34 (Kruzhkov entropies). The analysis of nonlinear conservation laws with boundary conditions considers the Kruzhkov-type entropies $(u - \kappa)^\oplus$ and $(u-\kappa)^\ominus$ for any real number κ (see, e.g., Vovelle [303]). The difficulty with such entropies at the discrete level is that the derivation of entropy bounds requires testing the space semidiscrete problem with functions that are not in the discrete space V_h (except for polynomial order $k = 0$). This difficulty is avoided with the quadratic entropy where the discrete solution can be used as test function.

Our aim is now to reproduce the L^2-stability property (3.69) for the space semidiscrete solution. Achieving this result depends on the properties of the selected numerical flux Φ. Following Osher [253], we introduce the concept of E-flux.

Definition 3.35 (E-flux). *We say that the numerical flux $\Phi : \mathbb{R}^d \times \mathcal{U} \times \mathcal{U} \to \mathbb{R}$ is an E-flux if for all $\mathrm{n} \in \mathbb{R}^d$, all $u^-, u^+ \in \mathcal{U}$, and all $v \in \lfloor u^-, u^+\rceil$ with*

$$\lfloor u^-, u^+\rceil := \{\theta u^- + (1-\theta)u^+\}_{\theta\in[0,1]},$$

there holds

$$\big(\Phi(\mathrm{n}, u^-, u^+) - \mathfrak{f}_n(v)\big)\big(u^- - u^+\big) \ge 0. \tag{3.70}$$

3.2. Nonlinear Conservation Laws

Assuming that the numerical flux Φ is Lipschitz continuous with respect to its second and third arguments, the E-flux property implies consistency. Moreover, the E-flux property is instrumental in deriving L^2-stability for the space semidiscrete solution u_h, as shown by Jiang and Shu [202] for dG methods. We now state and prove this result.

Lemma 3.36 (Energy estimate). *Assume that $g \equiv 0 \in \mathcal{U}$ and that Φ is an E-flux. Then, for all $t \in (0, t_F)$,*
$$\|u_h(t)\|_{L^2(\Omega)} \leq \|\pi_h u_0\|_{L^2(\Omega)}.$$

Proof. Using u_h as test function in (3.66), we infer
$$\frac{1}{2} d_t \|u_h\|_{L^2(\Omega)}^2 = \sum_{T \in \mathcal{T}_h} \int_T \mathfrak{f}(u_h) \cdot \nabla u_h - \sum_{F \in \mathcal{F}_h} \int_F \Phi(\mathbf{n}_F, u_F^-, u_F^+) [\![u_h]\!].$$

The first term on the right-hand side can be reformulated as a boundary term using the $\mathbb{R}^d$-valued primitive $\gamma(v)$ defined above. This yields
$$\sum_{T \in \mathcal{T}_h} \int_T \mathfrak{f}(u_h) \cdot \nabla u_h = \sum_{T \in \mathcal{T}_h} \int_{\partial T} \gamma(u_h|_T) \cdot \mathbf{n}_T = \sum_{F \in \mathcal{F}_h} \int_F [\![\gamma(u_h)]\!] \cdot \mathbf{n}_F.$$

As a result,
$$\frac{1}{2} d_t \|u_h\|_{L^2(\Omega)}^2 = \sum_{F \in \mathcal{F}_h} \int_F \left\{ [\![\gamma(u_h)]\!] \cdot \mathbf{n}_F - \Phi(\mathbf{n}_F, u_F^-, u_F^+) [\![u_h]\!] \right\}.$$

Since $\gamma' = \mathfrak{f}$, the mean-value theorem implies that, for all $F \in \mathcal{F}_h$ and a.e. $x \in F$, there exists $v_F \in \lfloor u_F^-, u_F^+ \rceil$ such that
$$[\![\gamma(u_h)]\!]_F \cdot \mathbf{n}_F = \mathfrak{f}(v_F) \cdot \mathbf{n}_F [\![u_h]\!]_F. \tag{3.71}$$

Using the fact that Φ is an E-flux, cf. (3.70), we then infer
$$\frac{1}{2} d_t \|u_h\|_{L^2(\Omega)}^2 = \sum_{F \in \mathcal{F}_h} \int_F \left\{ \mathfrak{f}_{\mathbf{n}_F}(v_F) - \Phi(\mathbf{n}_F, u_F^-, u_F^+) \right\} [\![u_h]\!] \leq 0.$$

Integrating in time concludes the proof. □

A more stringent requirement on the numerical flux than being an E-flux is monotonicity (see, e.g., LeVeque [230, p. 169]). This concept is motivated by the requirement that the dG scheme yield a monotone finite volume scheme when $V_h = \mathbb{P}_d^0(\mathcal{T}_h)$ is spanned by piecewise constant functions.

Definition 3.37 (Monotone numerical flux). *We say that the numerical flux $\Phi : \mathbb{R}^d \times \mathcal{U} \times \mathcal{U} \to \mathbb{R}$ is monotone if it is nondecreasing in its second argument and nonincreasing in its third argument, that is, for all $\mathbf{n} \in \mathbb{R}^d$,*
$$\partial_{u^-} \Phi(\mathbf{n}, u^-, u^+) \geq 0, \qquad \partial_{u^+} \Phi(\mathbf{n}, u^-, u^+) \leq 0.$$

It is easily verified that a (consistent) monotone numerical flux is an E-flux. For instance, in the case $u^- \le u^+$, using consistency yields, for all $v \in [u^-, u^+]$,

$$\Phi(\mathrm{n}, u^-, u^+) - \mathfrak{f}_\mathrm{n}(v) = \left\{\Phi(\mathrm{n}, u^-, u^+) - \Phi(\mathrm{n}, v, u^+)\right\}$$
$$+ \left\{\Phi(\mathrm{n}, v, u^+) - \Phi(\mathrm{n}, v, v)\right\},$$

and the two addends on the right-hand side are ≤ 0 owing to monotonicity. Since $u^- - u^+ \le 0$, this yields (3.70). The case $u^- \ge u^+$ is treated similarly.

3.2.2.2 Examples

We now present various classical examples of numerical fluxes. We focus on the design of such fluxes at interfaces.

Centered and Upwind Fluxes for Linear Problems In the linear case, there holds $\mathfrak{f}(u) = \beta u$ for some $\mathbb{R}^d$-valued advective field β. As in Sects. 2.2.3 and 2.3.4, a possible choice for the numerical flux in the linear case is, for all $F \in \mathcal{F}_h$,

$$\Phi(\mathrm{n}_F, u^-, u^+) = (\beta \cdot \mathrm{n}_F)\frac{u^- + u^+}{2} + \eta\frac{|\beta \cdot \mathrm{n}_F|}{2}(u^- - u^+), \qquad (3.72)$$

where $\eta \ge 0$ is a user-defined parameter (for the states, the face index is omitted to alleviate the notation). The choice $\eta = 0$ corresponds to the centered flux discussed in Sect. 2.2.3, whereas taking $\eta = 1$ yields the upwind flux of Sect. 2.3.4. While the centered flux is not an E-flux, the upwind flux is a monotone flux since

$$\Phi(\mathrm{n}_F, u^-, u^+) = (\beta \cdot \mathrm{n}_F)^\oplus u^- - (\beta \cdot \mathrm{n}_F)^\ominus u^+,$$

where $(\beta \cdot \mathrm{n}_F)^\oplus$ and $(\beta \cdot \mathrm{n}_F)^\ominus$ denote the positive and negative parts of $\beta \cdot \mathrm{n}_F$; cf. (2.12). However, an energy estimate can also be derived for the centered flux. Indeed, following the proof of Lemma 3.36 for the numerical flux Φ defined by (3.72) with $\eta \ge 0$, we deduce $\gamma(v) = \frac{1}{2}\beta v^2$, so that the mean-value theorem yields, for all $F \in \mathcal{F}_h$, $v_F = \frac{1}{2}(u_F^- + u_F^+)$ in (3.71).

Proposition 3.38. *For the numerical flux (3.72) with $\eta \ge 0$, there holds, for all $t \in (0, t_\mathrm{F})$,*

$$\|u_h(t)\|_{L^2(\Omega)}^2 = \|\pi_h u_0\|_{L^2(\Omega)}^2 - \frac{\eta}{2}\sum_{F \in \mathcal{F}_h}\int_0^t\int_F |\beta \cdot \mathrm{n}_F|\,|[\![u_h(s)]\!]|^2\,\mathrm{d}s, \qquad (3.73)$$

yielding, for centered fluxes ($\eta = 0$), exact energy conservation in the form

$$\|u_h(t)\|_{L^2(\Omega)} = \|\pi_h u_0\|_{L^2(\Omega)}.$$

Equation (3.73) shows that the rate of energy decrease is proportional to the parameter η in the numerical flux. Thus, this parameter controls the amount of numerical dissipation; the larger η, the more dissipative the scheme.

3.2. Nonlinear Conservation Laws

Godunov Numerical Flux for Nonlinear Problems In the nonlinear case, a monotone flux can be designed by considering the Godunov numerical flux such that, for all $F \in \mathcal{F}_h$,

$$\Phi(\mathrm{n}_F, u^-, u^+) := \begin{cases} \min_{w \in [u^-, u^+]} \mathfrak{f}_{\mathrm{n}_F}(w) & \text{if } u^- \leq u^+, \\ \max_{w \in [u^+, u^-]} \mathfrak{f}_{\mathrm{n}_F}(w) & \text{otherwise.} \end{cases} \qquad (3.74)$$

Monotonicity results from the dependency of the minimizing and maximizing intervals on u^- and u^+. In the linear case, the Godunov numerical flux (3.74) coincides with the upwind flux.

The Godunov numerical flux can be associated with the solution of a one-dimensional *Riemann problem* in the direction normal to the face F. Such a problem amounts to finding the weak entropy solution for the one-dimensional evolution problem $\partial_t r + \partial_x \mathfrak{f}_{\mathrm{n}_F}(r) = 0$ with initial data consisting of the two constant states u^- and u^+. Such a solution is selfsimilar and, assuming there is no stationary shock at the interface, takes a constant value at the interface for any positive time, which we denote by $\mathfrak{R}(u^-, u^+)$. Then, a classical result [230, p. 145] states that

$$\Phi(\mathrm{n}_F, u^-, u^+) = \mathfrak{f}_{\mathrm{n}_F}(\mathfrak{R}(u^-, u^+)). \qquad (3.75)$$

Rusanov, Lax–Friedrichs, and Roe Numerical Fluxes In practice, the numerical flux must be evaluated at quadrature nodes of each mesh face to compute integrals of the form $\int_F \Phi(\mathrm{n}_F, u_F^-, u_F^+)[\![v_h]\!]$; cf. (3.66). Using the Godunov numerical flux then requires solving a scalar nonlinear optimization problem at each quadrature node of each mesh face, and this can be computationally demanding. Alternative approaches have been devised based on approximate Riemann solvers (see, among others, the textbooks by LeVeque [230, p. 146] and Toro [295, p. 293]). A possible work-around to having to solve a Riemann problem that, in some situations, produces satisfactory results consists in using the *Rusanov (or generalized Lax–Friedrichs) numerical flux*

$$\Phi(\mathrm{n}_F, u^-, u^+) := \frac{\mathfrak{f}_{\mathrm{n}_F}(u^-) + \mathfrak{f}_{\mathrm{n}_F}(u^+)}{2} + \frac{\eta_{\mathfrak{f}}}{2}(u^- - u^+), \qquad (3.76)$$

with stabilization parameter $\eta_{\mathfrak{f}} > 0$ large enough. In the linear case, the Rusanov numerical flux (3.76) coincides with the upwind flux (3.72) for the choice $\eta_{\mathfrak{f}} = \eta|\beta \cdot \mathrm{n}_F|$. In the nonlinear case, setting $\eta_{\mathfrak{f}}$ to

$$\eta_{\mathfrak{f}} = \sup_{v \in \mathcal{U}} |\mathfrak{f}'_{\mathrm{n}_F}(v)| \qquad (3.77)$$

yields a monotone numerical flux. Indeed, for all $u^-, u^+ \in \mathcal{U}$,

$$\partial_{u^-} \Phi(\mathrm{n}_F, u^-, u^+) = \frac{1}{2}\left(\mathfrak{f}'_{\mathrm{n}_F}(u^-) + \sup_{v \in \mathcal{U}}|\mathfrak{f}'_{\mathrm{n}_F}(v)|\right) \geq 0,$$

$$\partial_{u^+} \Phi(\mathrm{n}_F, u^-, u^+) = \frac{1}{2}\left(\mathfrak{f}'_{\mathrm{n}_F}(u^+) - \sup_{v \in \mathcal{U}}|\mathfrak{f}'_{\mathrm{n}_F}(v)|\right) \leq 0.$$

However, choosing the stabilization parameter as in (3.77) generally introduces a large amount of numerical diffusion, which in turn spoils the quality of the approximate solution. Indeed, while stability pleads for large values of $\eta_\mathfrak{f}$, this parameter can have a sizable impact on the dissipative properties of the scheme, and excessive dissipation ruins accuracy. Therefore, tuning the parameter $\eta_\mathfrak{f}$ in the Rusanov numerical flux is often a relevant issue.

A popular choice to reduce numerical dissipation is provided by the *local Lax–Friedrichs numerical flux* for which

$$\eta_\mathfrak{f} = \sup_{v \in \lfloor u^-, u^+ \rceil} |\mathfrak{f}'_{\mathrm{n}_F}(v)|. \tag{3.78}$$

The supremum is taken here over a smaller interval, which suffices to ensure monotonicity. In practice, the supremum in (3.78) is often replaced by

$$\eta_\mathfrak{f} = \begin{cases} \left| \dfrac{\mathfrak{f}_{\mathrm{n}_F}(u^-) - \mathfrak{f}_{\mathrm{n}_F}(u^+)}{u^- - u^+} \right| & \text{if } u^- \neq u^+, \\ \mathfrak{f}'_{\mathrm{n}_F}(u^+) & \text{otherwise,} \end{cases} \tag{3.79}$$

yielding the so-called *Roe numerical flux*.

Remark 3.39 (Roe numerical flux as approximate Riemann solver). The Roe numerical flux can be derived from the solution of a Riemann problem associated with the linearized equation

$$\partial_t w + V(\mathrm{n}_F, u^-, u^+) \partial_x w = 0, \qquad V(\mathrm{n}_F, u^-, u^+) := \frac{\mathfrak{f}_{\mathrm{n}_F}(u^-) - \mathfrak{f}_{\mathrm{n}_F}(u^+)}{u^- - u^+},$$

with initial data $u^- \neq u^+$. The unique entropy solution of the above linearized problem features a shock of speed $V(\mathrm{n}_F, u^-, u^+)$. Plugging this solution into the expression of $\mathfrak{f}_{\mathrm{n}_F}$ as in (3.74) yields the Rusanov numerical flux (3.76) with stabilization parameter $\eta_\mathfrak{f}$ given by (3.79). In the case of concave (resp., convex) flux functions, the solution of the linearized problem is exact provided $u^- < u^+$ (resp. $u^- > u^+$).

3.2.3 Time Discretization

As in the linear case, we formulate the space semidiscrete problem (3.66) in operator form. Specifically, we define the (nonlinear) operator A_h such that, for all $v_h, w_h \in V_h$,

$$(A_h(v_h), w_h)_{L^2(\Omega)} = -\int_\Omega \mathfrak{f}(v_h) \cdot \nabla_h w_h + \sum_{F \in \mathcal{F}_h} \int_F \Phi(\mathrm{n}_F, v_F^-, v_F^+) \llbracket w_h \rrbracket.$$

Problem (3.66) is then equivalent to

$$d_t u_h + A_h(u_h) = 0. \tag{3.80}$$

In component form, (3.80) amounts to a system of coupled nonlinear ODEs with the mass matrix in front of the time derivative. As the mass matrix for dG

3.2. Nonlinear Conservation Laws

methods is easily invertible (cf. Sect. A.1.2), problem (3.80) can be effectively solved by explicit time integration methods as the ones discussed in Sect. 3.1.3.

Let δt be the time step, taken to be constant for simplicity, and such that $t_F = N\delta t$ where N is an integer. For $n \in \{0, \ldots, N\}$, we define the discrete times $t^n = n\delta t$. Following Shu and Osher [279] and Gottlieb, Shu, and Tadmor [175, 176], we consider s-stage explicit RK schemes in the form

$$u_h^{n,0} = u_h^n,$$
$$u_h^{n,i} = \sum_{j=0}^{i-1} \left(\alpha_{ij} u_h^{n,j} - \delta t \beta_{ij} A_h(u_h^{n,j}) \right) \quad \forall i \in \{1, \ldots, s\},$$
$$u_h^{n+1} = u_h^{n,s},$$

and we recall that the scheme is specified through the lower triangular matrices $\alpha := (\alpha_{ij})_{1 \le i \le s, 0 \le j \le i-1}$ and $\beta := (\beta_{ij})_{1 \le i \le s, 0 \le j \le i-1}$. We are interested in RK schemes where the real numbers α_{ij} and β_{ij} are nonnegative, and such that α_{ij} is nonzero whenever β_{ij} is nonzero. A basic consistency requirement is

$$\sum_{j=0}^{i-1} \alpha_{ij} = 1 \quad \forall i \in \{1, \ldots, s\}.$$

A consequence of these properties is that the intermediate stages $u_h^{n,i}$ amount to convex combinations of forward Euler substeps with time step δt replaced by $\frac{\beta_{ij}}{\alpha_{ij}} \delta t$. This leads to *Strong Stability-Preserving (SSP) RK* schemes. In particular, if the forward Euler scheme is L^2-stable under a CFL condition of the form

$$\delta t \le \sigma^{-1} h$$

with reference velocity σ, that is, if $\|u_h^n - \delta t A_h(u_h^n)\|_{L^2(\Omega)} \le \|u_h^n\|_{L^2(\Omega)}$ under the above CFL condition, then the s-stage RK scheme is such that $\|u_h^{n+1}\|_{L^2(\Omega)} \le \|u_h^n\|_{L^2(\Omega)}$ under the CFL condition

$$\delta t \le \min_{i,j} \left(\frac{\alpha_{ij}}{\beta_{ij}} \right) \sigma^{-1} h.$$

Optimal (in the sense of CFL condition and computational costs) SSP-RK schemes are investigated by Gottlieb and Shu [175]. An optimal two-stage SSP-RK scheme is the scheme (3.17), that is,

$$\alpha = \begin{bmatrix} 1 & \\ 1/2 & 1/2 \end{bmatrix}, \quad \beta = \begin{bmatrix} 1 & \\ 0 & 1/2 \end{bmatrix},$$

which, applied to (3.80), yields

$$u_h^{n,1} = u_h^n - \delta t A_h(u_h^n),$$
$$u_h^{n+1} = \tfrac{1}{2}(u_h^n + u_h^{n,1}) - \tfrac{1}{2} \delta t A_h(u_h^{n,1}).$$

An optimal three-stage SSP-RK scheme results from the choice

$$\alpha = \begin{bmatrix} 1 & & \\ 3/4 & 1/4 & \\ 1/3 & 0 & 2/3 \end{bmatrix}, \qquad \beta = \begin{bmatrix} 1 & & \\ 0 & 1/4 & \\ 0 & 0 & 2/3 \end{bmatrix},$$

which, applied to (3.80), yields

$$u_h^{n,1} = u_h^n - \delta t A_h(u_h^n),$$
$$u_h^{n,2} = \tfrac{1}{4}(3u_h^n + u_h^{n,1}) - \tfrac{1}{4}\delta t A_h(u_h^{n,1}),$$
$$u_h^{n+1} = \tfrac{1}{3}(u_h^n + 2u_h^{n,2}) - \tfrac{2}{3}\delta t A_h(u_h^{n,2}).$$

The convergence analysis of dG space semidiscretization combined with SSP-RK time schemes is far from being complete, especially for rough exact solutions. For smooth exact solutions, convergence rates for the L^2-error (identical to those achieved in the linear case) are derived by Zhang and Shu for the SSP-RK2 and RK3 schemes [310, 311]. For the SSP-RK2 scheme (cf. Theorem 3.10), a 4/3-CFL condition is invoked for piecewise polynomials with degree $k \geq 2$, while the usual CFL condition is sufficient for $k = 1$. Instead, for the SSP-RK3 scheme (cf. Theorem 3.13), the usual CFL condition is sufficient for all $k \geq 1$.

3.2.4 Limiters

Using high-order discretizations to approximate rough solutions triggers the Gibbs phenomenon (see, e.g., Gottlieb and Shu [174]), whereby the numerical solution is polluted by spurious oscillations. Oscillations can cause the departure from the set of admissible values $\mathcal{U}$, thereby violating the physical principle of the conservation law (requiring, for instance, nonnegative densities). Moreover, they can lead, e.g., to convergence towards non-physical shocks. We first address the one-dimensional case and focus for simplicity on a periodic setting. We examine the stability of fully discrete schemes in terms of the total variation in space of the discrete solution. This concept can be used as a measure of the spurious oscillations in the discrete solution, the exact solution being total variation-diminishing in time. Unfortunately, this property does not transfer to the discrete level, so that we content ourselves with the more moderate requirement that the total-variation of the elementwise averaged discrete solution be diminishing in time, leading to so-called TVDM schemes. A sufficient means to achieve the TVDM property is the use of minmod limiters. Finally, we briefly address a heuristic extension to the multi-dimensional case (the extension of the one-dimensional analysis to multiple space dimensions is still an open problem).

3.2.4.1 TVDM Methods

We consider the one-dimensional case. Let $\Omega = (a, b)$ denote the space domain. In what follows, we replace for simplicity the boundary condition (3.65) with the periodic boundary condition

$$u(a, t) = u(b, t) \text{ for all } t \in (0, t_F).$$

3.2. Nonlinear Conservation Laws

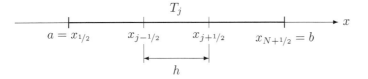

Fig. 3.2: Notation for the one-dimensional case

The mesh $\mathcal{T}_h$ of Ω is composed of elements $\{T_j\}_{1\leq j\leq N}$ of the form (cf. Fig. 3.2)

$$T_j = (x_{j-1/2}, x_{j+1/2}) \quad \forall j \in \{1,\ldots,N\},$$

resulting from a set of equally spaced points $\{x_{j+1/2}\}_{0\leq j\leq N}$ such that

$$a = x_{1/2} < x_{3/2} < \ldots < x_{N-1/2} < x_{N+1/2} = b.$$

Thus, the meshsize is $h = |b-a|/N$. At each point $x_{j+1/2}$, $j \in \{0,\ldots,N\}$, we define the left and right values as

$$u_{j+1/2}^- := \begin{cases} u_h|_{T_j}(x_{j+1/2}) & \text{if } j > 0, \\ u_h|_{T_N}(x_{N+1/2}) & \text{if } j = 0, \end{cases} \quad u_{j+1/2}^+ := \begin{cases} u_h|_{T_{j+1}}(x_{j+1/2}) & \text{if } j < N, \\ u_h|_{T_1}(x_{1/2}) & \text{if } j = N, \end{cases}$$

where the extensions at the endpoints of Ω reflect the periodic setting. For $j \in \{0,\ldots,N\}$, the jump across $x_{j+1/2}$ is defined as

$$[\![u_h]\!]_{j+1/2} := u_{j+1/2}^- - u_{j+1/2}^+,$$

which corresponds to taking the positive normal $\mathrm{n}_{j+1/2} = 1$ at each point $x_{j+1/2}$. We introduce the concept of total variation for the elementwise averaged solution. The discrete solution u_h can be decomposed, inside each mesh element T_j, as the sum of its average value $\overline{u}_j$ and an oscillatory component $\widetilde{u}_j(x)$ with zero mean-value on T_j, so that

$$u_h|_{T_j}(x) = \overline{u}_j + \widetilde{u}_j(x) \quad \forall j \in \{1,\ldots,N\}, \forall x \in T_j. \tag{3.81}$$

Recalling that periodic boundary conditions are enforced, we also define

$$\overline{u}_0 := \overline{u}_N, \quad \overline{u}_{N+1} := \overline{u}_1.$$

Definition 3.40 (TVDM scheme). We say that a numerical scheme is *total variation diminishing in the means* (in short, TVDM) if there holds

$$\mathrm{TVM}(u_h^{n+1}) := \sum_{1\leq j\leq N} |\overline{u}_{j+1}^{n+1} - \overline{u}_j^{n+1}| \leq \mathrm{TVM}(u_h^n).$$

We discretize in time problem (3.66) using the forward Euler method, which constitutes the basic building block for the explicit Runge–Kutta methods (cf. Sect. 3.2.3). For brevity of notation, the index n for the solution at time t^n is omitted. The solution $u_h^{n+1} \in V_h$ at time t^{n+1} is computed by solving, for all $v_h \in V_h$,

$$\int_\Omega \left\{ (u_h^{n+1} - u_h)\, v_h - \lambda \mathfrak{f}(u_h) d_x v_h \right\} + \lambda \sum_{0 \le j \le N} \Phi(u_{j+1/2}^-, u_{j+1/2}^+) [\![v_h]\!]_{j+1/2} = 0, \tag{3.82}$$

where

$$\lambda := \frac{\delta t}{h}.$$

When dealing with the one-dimensional case, the dependence of the numerical flux Φ on the normal is irrelevant since the flux function $\mathfrak{f}$ is scalar-valued. Taking $v_h = \chi_{T_j}$ for $j \in \{1, \ldots, N\}$, we infer

$$\overline{u}_j^{n+1} - \overline{u}_j + \lambda \left[\Phi(u_{j+1/2}^-, u_{j+1/2}^+) - \Phi(u_{j-1/2}^-, u_{j-1/2}^+) \right] = 0, \tag{3.83}$$

which is the forward Euler time discretization of (3.68) in the present one-dimensional setting.

In what follows, we investigate a sufficient condition for the TVDM property, and we propose a modification of the scheme (3.82) to achieve it. Using the TVDM property, minimal convergence results can be proven [94, Theorem 2.13], e.g., the convergence in $L^\infty(0, t_F; L^1(\Omega))$ as the meshsize $h \in \mathcal{H}$ goes to zero with fixed parameter λ, up to a subsequence, of the sequence $(\overline{u}_h)_{h \in \mathcal{H}}$ to a weak solution u of the continuous problem (which, however, may not be the weak entropy solution). From a practical viewpoint, the TVDM property enhances the possibility that oscillations are sufficiently well controlled to retain the convergence of the scheme towards the entropy solution.

The following result is the basic tool to identify a set of sufficient conditions for the TVDM property to hold (see Abgrall [1, Chap. 5] or Godlewski and Raviart [173, Chap. 3]). Given N real numbers $\{v_j\}_{1 \le j \le N}$ with $v_{N+1} := v_1$ by periodicity, we set $\mathrm{TV}(v) := \sum_{1 \le j \le N} |v_{j+1} - v_j|$.

Lemma 3.41 (Harten's Lemma). *Let $\{v_j\}_{1 \le j \le N}$ with $v_0 := v_N$ and $v_{N+1} := v_1$ by periodicity, and let $\lambda > 0$. Let $\{w_j\}_{1 \le j \le N}$ be defined by*

$$w_j = v_j + \lambda \Big\{ C_{j+1/2}(v_{j+1} - v_j) - D_{j-1/2}(v_j - v_{j-1}) \Big\}, \tag{3.84}$$

with given real numbers $\{C_{j+1/2}\}_{1 \le j \le N}$ and $\{D_{j-1/2}\}_{1 \le j \le N}$. Assume that, for all $j \in \{1, \ldots, N\}$,

$$C_{j+1/2} \ge 0, \qquad D_{j-1/2} \ge 0, \qquad \lambda \left(C_{j+1/2} + D_{j+1/2} \right) \le 1, \tag{3.85}$$

with $D_{N+1/2} := D_{1/2}$ by periodicity. Then,

$$\mathrm{TV}(w) \le \mathrm{TV}(v).$$

3.2. Nonlinear Conservation Laws

Proof. Using (3.84), we infer, for all $j \in \{1, \ldots, N\}$,

$$w_{j+1} - w_j = \Big\{1 - \lambda(C_{j+1/2} + D_{j+1/2})\Big\}(v_{j+1} - v_j)$$
$$+ \lambda C_{j+3/2}(v_{j+2} - v_{j+1}) + \lambda D_{j-1/2}(v_j - v_{j-1}),$$

with $w_{N+1} := w_1$, $v_{N+2} := v_2$, and $C_{N+3/2} := C_{3/2}$ by periodicity. Taking the absolute value, using assumption (3.85) so that $C_{j+3/2} \geq 0$, $D_{j-1/2} \geq 0$, and $\lambda(C_{j+1/2} + D_{j+1/2}) \leq 1$, and summing over j yields

$$\text{TV}(w) \leq \sum_{1 \leq j \leq N} \Big\{1 - \lambda(C_{j+1/2} + D_{j+1/2})\Big\} |v_{j+1} - v_j|$$
$$+ \lambda \sum_{1 \leq j \leq N} \Big\{C_{j+3/2}|v_{j+2} - v_{j+1}| + D_{j-1/2}|v_j - v_{j-1}|\Big\}$$
$$= \text{TV}(v),$$

where we have used periodicity to shift indices in the last two summations. □

Theorem 3.42 (TVDM scheme). *Let Φ be a monotone, Lipschitz continuous, numerical flux. Assume that there is $\theta > 0$ such that, for all $j \in \{1, \ldots, N\}$,*

$$-1 \leq \frac{\widetilde{u}^+_{j+1/2} - \widetilde{u}^+_{j-1/2}}{\overline{u}_{j+1} - \overline{u}_j} \leq \theta, \qquad -1 \leq \frac{\widetilde{u}^-_{j+1/2} - \widetilde{u}^-_{j-1/2}}{\overline{u}_j - \overline{u}_{j-1}} \leq \theta. \tag{3.86}$$

Assume the CFL condition

$$\delta t \leq \frac{1}{1+\theta} \frac{1}{L_1 + L_2} h, \tag{3.87}$$

where L_1 and L_2 denote the Lipschitz modules of Φ with respect to its first and second argument, respectively. Then, the scheme (3.83) is TVDM.

Proof. We check the assumptions of Harten's Lemma. Provided the denominators in (3.86) are nonzero (which yields a trivial case), (3.83) can be rewritten as follows: For all $j \in \{1, \ldots, N\}$,

$$\overline{u}_j^{n+1} = \overline{u}_j - \lambda\Big\{\Phi(u^-_{j+1/2}, u^+_{j+1/2}) - \Phi(u^-_{j+1/2}, u^+_{j-1/2})\Big\}$$
$$- \lambda\Big\{\Phi(u^-_{j+1/2}, u^+_{j-1/2}) - \Phi(u^-_{j-1/2}, u^+_{j-1/2})\Big\}$$
$$= \overline{u}_j + \lambda\Big\{C_{j+1/2}(\overline{u}_{j+1} - \overline{u}_j) - D_{j-1/2}(\overline{u}_j - \overline{u}_{j-1})\Big\},$$

with

$$C_{j+1/2} = \frac{\Phi(u^-_{j+1/2}, u^+_{j+1/2}) - \Phi(u^-_{j+1/2}, u^+_{j-1/2})}{u^+_{j-1/2} - u^+_{j+1/2}} \left(1 + \frac{\widetilde{u}^+_{j+1/2} - \widetilde{u}^+_{j-1/2}}{\overline{u}_{j+1} - \overline{u}_j}\right),$$

$$D_{j-1/2} = \frac{\Phi(u^-_{j+1/2}, u^+_{j-1/2}) - \Phi(u^-_{j-1/2}, u^+_{j-1/2})}{u^-_{j+1/2} - u^-_{j-1/2}} \left(1 + \frac{\widetilde{u}^-_{j+1/2} - \widetilde{u}^-_{j-1/2}}{\overline{u}_j - \overline{u}_{j-1}}\right),$$

since $\overline{u}_j + \widetilde{u}^+_{j-1/2} = u^+_{j-1/2}$, and so on. Both $C_{j+1/2}$ and $D_{j-1/2}$ are nonegative owing to the monotonicity of Φ together with (3.86). To check the remaining condition of Harten's Lemma, we observe that, for all $j \in \{1,\ldots,N\}$,

$$C_{j+1/2} \leq L_2(1+\theta), \qquad D_{j+1/2} \leq L_1(1+\theta).$$

Therefore,

$$\lambda \left(C_{j+1/2} + D_{j+1/2}\right) \leq \delta t h^{-1}(L_1 + L_2)(1+\theta) \leq 1,$$

where we have used the CFL condition (3.87). □

3.2.4.2 Enforcing the TVDM Property

Condition (3.86) has to be enforced since the scheme (3.82) does not provide it, and this is where limiters come into play. We follow here the presentation of Cockburn [94, Sect. 2.4]. For real numbers $v_1,\ldots,v_k$, we define the generalized minmod function

$$\operatorname{minmod}(v_1,\ldots,v_k) = \begin{cases} \alpha \min(|v_1|,\ldots,|v_k|) & \text{if } \alpha = \operatorname{sgn}(v_1) = \ldots = \operatorname{sgn}(v_k), \\ 0 & \text{otherwise.} \end{cases}$$

We consider first the forward Euler scheme. Fix a time step $n \in \{0,\ldots,N\}$ and let u_h denote the discrete solution at time t^n. We replace u_h by a limited function $v_h := \Lambda u_h$ (for $n = 0$, this means that the initial condition becomes $\Lambda \pi_h u_0$). The goal is to ensure that the limited function v_h is such that, for all $j \in \{1,\ldots,N\}$,

$$v^-_{j+1/2} = \overline{v}_j + \operatorname{minmod}(v^-_{j+1/2} - \overline{v}_j, \overline{v}_{j+1} - \overline{v}_j, \overline{v}_j - \overline{v}_{j-1}), \tag{3.88a}$$

$$v^+_{j-1/2} = \overline{v}_j - \operatorname{minmod}(\overline{v}_j - v^+_{j-1/2}, \overline{v}_{j+1} - \overline{v}_j, \overline{v}_j - \overline{v}_{j-1}). \tag{3.88b}$$

For instance, owing to the minmod function in (3.88a), the higher-order contribution at the interface, $(v^-_{j+1/2} - \overline{v}_j)$, is only taken into account when (1) It has the same sign as the variations of the mean values in adjoining cells, $\overline{v}_{j+1} - \overline{v}_j$ and $\overline{v}_j - \overline{v}_{j-1}$, and (2) It is smaller in magnitude. Conditions (3.88) imply

$$-1 \leq \frac{\widetilde{v}^+_{j+1/2} - \widetilde{v}^+_{j-1/2}}{\overline{v}_{j+1} - \overline{v}_j} \leq 1, \qquad -1 \leq \frac{\widetilde{v}^-_{j+1/2} - \widetilde{v}^-_{j-1/2}}{\overline{v}_j - \overline{v}_{j-1}} \leq 1,$$

so that the condition (3.86) in Harten's Lemma holds true with $\theta = 1$. Within higher-order RK methods, the limiter is applied at each forward Euler substep. The general TVDM-RK method (cf. Sect. 3.2.3) then reads

$$u_h^{n,0} = u_h^n,$$

$$u_h^{n,i} = \Lambda \left(\sum_{j=0}^{i-1} \left(\alpha_{ij} u_h^{n,j} - \delta t \beta_{ij} A_h(u_h^{n,j}) \right) \right) \qquad \forall i \in \{1,\ldots,s\},$$

$$u_h^{n+1} = u_h^{n,s}.$$

3.2. Nonlinear Conservation Laws

We now discuss how to construct the limited function $v_h = \Lambda u_h$. In the piecewise affine case, one possibility consists in setting, for all $j \in \{1, \ldots, N\}$,

$$v_h|_{T_j}(x) = \overline{u}_j + (x - x_j) \operatorname{minmod}(d_x \tilde{u}_j, 2h^{-1}(\overline{u}_{j+1} - \overline{u}_j), 2h^{-1}(\overline{u}_j - \overline{u}_{j-1})).$$

Besides ensuring (3.88) (as can be checked directly), this limiter preserves the mean value element by element, that is, for all $j \in \{1, \ldots, N\}$,

$$\overline{v}_j = \overline{u}_j.$$

This is a sensible property in the context of conservation laws. Moreover, if u_h is globally linear on the macro-element $(x_{j-3/2}, x_{j+3/2})$, then u_h is not modified in T_j. This property can be interpreted as a minimal accuracy requirement on the limiter. Additionally, the slope of v_h in each mesh element is smaller than that of u_h (and given by $d_x \tilde{u}_j$).

When working with high-degree polynomials ($k \geq 2$), one heuristic approach is to assume that spurious oscillations are present in u_h only if they are present in $\pi_h^1 u_h$, where $\pi_h^1 u_h$ is the L^2-orthogonal projection of u_h onto piecewise affine functions. Thus, if spurious oscillations are not present in $\pi_h^1 u_h$, i.e., if $\Lambda \pi_h^1 u_h = \pi_h^1 u_h$, we set $\Lambda u_h = u_h$, whereas, if $\Lambda \pi_h^1 u_h$ and $\pi_h^1 u_h$ differ, we discard the higher-degree part of u_h and limit its affine part, that is, we set $\Lambda u_h = \Lambda \pi_h^1 u_h$.

The main drawback of using limiters is reducing the accuracy of the approximation in smooth regions. A possible remedy consists in estimating the regularity of the solution starting from the highest polynomial degree and progressively removing oscillating components, as discussed by Biswas, Devine, and Flaherty [43]. With this technique, the polynomial degree may be lowered to some positive integer, possibly not too low, thereby limiting the precision loss. Another improvement, proposed by Shu [278], consists in modifying the minmod function so as to avoid degradation at smooth local extrema. The basic idea here is to desactivate the limiter when space derivatives are of order h^2 by defining

$$\operatorname{minmod}_{\text{TVB}}(v_1, \ldots, v_k) := \begin{cases} v_1 & \text{if } |v_1| \leq Mh^2, \\ \operatorname{minmod}(v_1, \ldots, v_k) & \text{otherwise,} \end{cases} \quad (3.89)$$

where $M > 0$ is a tunable parameter which can be evaluated from the curvature of the initial datum at its extrema by setting

$$M := \sup_{y \in \Omega,\, u_0'(y)=0} |u_0''(y)|.$$

Such limiters do not enforce the TVDM property, but they yield solutions that are total variation bounded in the means (TVBM).

3.2.4.3 Limiters in Multiple Space Dimensions

Limiters in multiple space dimensions are mainly based on heuristics. To convey the main ideas, we present a limiter for matching triangular meshes based on a barycentric reconstruction (see Cockburn [94, Sect. 3.3.10]). For more elaborate

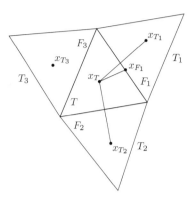

Fig. 3.3: Notation for the limiter in two space dimensions

constructions of multidimensional limiters, we refer the reader, e.g., to Cockburn, Hou, and Shu [99], Hoteit, Ackerer, Mosé, Erhel, and Philippe [195], Krivodonova [215], Krivodonova, Xin, Remacle, Chevaugeon, and Flaherty [216], Luo, Baum, and Löhner [234], and Kuzmin [217].

The limiting procedure is performed independently on each mesh element. We consider the situation of Fig. 3.3: let T be a triangular mesh element with faces F_1, F_2, F_3. Let x_L, $L \in \{T, T_1, T_2, T_3, F_1, F_2, F_3\}$, denote the center of mass of the corresponding geometric item, defined as $x_L := \int_L x / \int_L 1$. Let Ω_T collect the points in (the closure of) T and its three neighbors. We assume that there are two real numbers λ_1 and λ_2 such that

$$x_{F_1} - x_T = \lambda_1(x_{T_1} - x_T) + \lambda_2(x_{T_2} - x_T).$$

This construction is possible if we assume that the points $\{x_T, x_{T_1}, x_{T_2}\}$ are not aligned; it is also reasonable to assume that the resulting triangle, say T_{12}, is not too flat, i.e., that the largest inscribed ball in T_{12} has diameter comparable to the diameter of T_{12}. In the situation depicted in Fig. 3.3, the point x_{F_1} lies in the interior of T_{12}, so that the real numbers λ_1 and λ_2 are both nonnegative.

If we consider now a linear function in Ω_T, say $\xi \in \mathbb{P}_2^1(\Omega_T)$, letting

$$\delta_1\xi := \lambda_1\big\{\xi(x_{T_1}) - \xi(x_T)\big\} + \lambda_2\big\{\xi(x_{T_2}) - \xi(x_T)\big\}, \tag{3.90}$$

and since $\overline{\xi}_L = \xi(x_L)$, $L \in \{T, T_1, T_2\}$, we infer

$$\delta_1\xi = \xi(x_{F_1}) - \overline{\xi}_T = \widetilde{\xi}(x_{F_1}).$$

In other terms, $\delta_1\xi$ represents the departure of the value of ξ at x_{F_1} from its average on the element T. Using a similar construction, we can compute the values $\delta_2\xi$ and $\delta_3\xi$ for the other two faces of T. Let now $\{\varphi_i\}_{1 \leq i \leq 3}$ denote the Lagrangian basis in T associated with the set of nodes $\{x_{F_i}\}_{1 \leq i \leq 3}$. The linear

3.2. Nonlinear Conservation Laws

function $\xi|_T$ can be expressed in terms of the functions $\{\varphi_i\}_{1\leq i\leq 3}$ as

$$\xi|_T(x) = \overline{\xi}_T + \sum_{1\leq i\leq 3} \delta_i \xi \varphi_i(x) = \overline{\xi}_T + \sum_{1\leq i\leq 3} \widetilde{\xi}(x_{F_i})\varphi_i(x).$$

Let now $u_h \in \mathbb{P}_2^k(\mathcal{T}_h)$. As in the one-dimensional case, the limiter acts elementwise and only on the piecewise affine projection $\pi_h^1 u_h$, and, therefore, it suffices to define $v_h = \Lambda u_h$ in the case where $k = 1$. Let $T \in \mathcal{T}_h$. As in (3.81), we consider the decomposition

$$u_h|_T(x) = \overline{u}_T + \widetilde{u}_T(x) \qquad \forall x \in T.$$

Similarly to (3.90), we set

$$\delta_1 u_h := \lambda_1(\overline{u}_{T_1} - \overline{u}_T) + \lambda_2(\overline{u}_{T_2} - \overline{u}_T),$$

along with similar definitions for $\delta_2 u_h$ and $\delta_3 u_h$. Since u_h is, in general, not globally linear in Ω_T, the value of $\delta_i u_h$, $i \in \{1,2,3\}$, differs from that of $\widetilde{u}_T(x_{F_i})$. The idea is then to capture oscillations by controlling the departure of the values $\widetilde{u}_T(x_{F_i})$ from the values $\delta_i u_h$. For all $i \in \{1,2,3\}$, we compute the quantities

$$\Delta_i := \text{minmod}_{\text{TVB}}\left(\widetilde{u}_T(x_{F_i}), \theta \delta_i u_h\right),$$

$$\widehat{\Delta}_i := \min\left(1, \frac{\sum_{1\leq j\leq 3} \Delta_j^{\ominus}}{\sum_{1\leq j\leq 3} \Delta_j^{\oplus}}\right)\Delta_i^{\oplus} - \min\left(1, \frac{\sum_{1\leq j\leq 3} \Delta_j^{\oplus}}{\sum_{1\leq j\leq 3} \Delta_j^{\ominus}}\right)\Delta_i^{\ominus},$$

where $\theta \geq 1$ is a user-defined parameter and the TVB minmod function is used, as in the one-dimensional case, to improve the limiter near smooth local extrema. Moreover, the superscripts denote positive and negative parts; cf. (2.12). Finally, the limited function $v_h := \Lambda u_h$ is defined as

$$v_h|_T = \begin{cases} \overline{u}_T + \sum_{1\leq i\leq 3} \Delta_i \varphi_i & \text{if } \sum_{1\leq i\leq 3} \Delta_i = 0, \\ \overline{u}_T + \sum_{1\leq i\leq 3} \widehat{\Delta}_i \varphi_i & \text{otherwise.} \end{cases}$$

By its construction, the limiter is mass-preserving since, for all $T \in \mathcal{T}_h$, $\overline{v}_T = \overline{u}_T$. Indeed, if $\sum_{1\leq i\leq 3} \Delta_i = 0$, we obtain

$$\overline{v}_T = \overline{u}_T + \frac{1}{3} \sum_{1\leq i\leq 3} \Delta_i = \overline{u}_T,$$

while, otherwise, the same result is obtained owing to the fact that $\sum_{1\leq i\leq 3} \widehat{\Delta}_i = 0$. Moreover, the limiter preserves functions that are globally affine on Ω_T. Finally, for all $i \in \{1,2,3\}$, we observe that

$$|v_h(x_{F_i}) - \overline{v}_T| \leq \max(|\widehat{\Delta}_i|, |\Delta_i|) \leq |\Delta_i| \leq |\widetilde{u}_T(x_{F_i})|,$$

meaning that, in each mesh element, the gradient of the limited function v_h is not "larger" than that of the original function u_h.

Part II

Scalar Second-Order PDEs

Chapter 4
PDEs with Diffusion

The aim of this chapter is to investigate the approximation of scalar PDEs with diffusion. As a first step, we consider the *Poisson problem* with homogeneous Dirichlet boundary condition

$$-\triangle u = f \quad \text{in } \Omega, \tag{4.1a}$$
$$u = 0 \quad \text{on } \partial\Omega, \tag{4.1b}$$

and source term $f \in L^2(\Omega)$. The scalar-valued function u is termed the potential and the vector-valued function $-\nabla u$ the diffusive flux.

In Sect. 4.1, we briefly describe the continuous setting for the model problem (4.1). Then, in the following sections, we discuss three possible approaches to design a dG approximation for this problem. In Sect. 4.2, we present a heuristic derivation of a suitable discrete bilinear form loosely following the same path of ideas as in Chap. 2 hinging on consistency and discrete coercivity. There are, however, substantial differences: a specific term needs to be added to recover consistency, interface jumps as well as boundary values are penalized, and the penalty term scales as the reciprocal of the local meshsize so that discrete coercivity is expressed using a mesh-dependent norm. A further important difference is that we require to work at least with piecewise affine polynomials, thereby excluding, for the time being, methods of finite volume-type. This derivation yields the so-called *Symmetric Interior Penalty (SIP)* method of Arnold [14]. The error analysis follows fairly standard arguments and leads to optimal error estimates for smooth exact solutions. We also present a more recent analysis by the authors [132] in the case of low-regularity exact solutions. Then, using liftings of the interface jumps and boundary values, we introduce in Sect. 4.3 the important concept of discrete gradient. Applications include (1) a reformulation of the SIP bilinear form that plays a central role in Sect. 5.2 to analyze the convergence to minimal regularity solutions (as shown recently by the authors in [131]), and (2) an elementwise formulation of the discrete problem leading to a local conservation property in terms of numerical fluxes. Finally, the third approach is pursued in Sect. 4.4 where we consider mixed dG methods that approximate both the potential and the diffusive flux. In such methods, local problems for the discrete potential and diffusive flux are formulated using numerical fluxes for both quantities, following the pioneering works of Bassi, Rebay, and coworkers [34, 35] and Cockburn and Shu [112]. This viewpoint has been adopted by Arnold, Brezzi, Cockburn, and Marini in [16] for a unified presentation of dG methods for the Poisson problem. For simplicity, we focus here on the SIP method and so-called *Local Discontinuous Galerkin (LDG)* methods [112]. In both methods, the discrete diffusive flux can be eliminated locally. We postpone the study of two-field dG methods to Sect. 7.3 in the more general context of Friedrichs' systems. Finally, we discuss hybrid mixed dG methods where additional degrees of freedom are introduced at interfaces, thereby allowing one to eliminate locally both the discrete potential and the discrete diffusive flux. This leads, in particular, to the so-called *Hybridized Discontinuous Galerkin (HDG)* methods introduced by Cockburn, Gopalakrishnan, and Lazarov [97]; see also Causin and Sacco [83] for a different approach based on a discontinuous Petrov–Galerkin formulation and Droniou and Eymard [135] for similar ideas in the context of hybrid mixed finite volume schemes.

The rest of this chapter is devoted to the study of diffusive PDEs that comprise additional terms with respect to the model problem (4.1). In Sect. 4.5, we extend the SIP method analyzed in Sect. 4.2 to heterogeneous (anisotropic) diffusion problems. The main ingredients are diffusion-dependent weighted averages to formulate the consistency and symmetry terms in the discrete bilinear form together with diffusion-dependent penalty parameters using the harmonic mean of the diffusion coefficient at each interface. In Sect. 4.6, we analyze heterogeneous diffusion-advection-reaction problems. We combine the ideas of Sect. 4.5 to handle the diffusion term with those developed in Sect. 2.3 for the upwind dG method to handle the advection-reaction term. The goal is a convergence analysis that covers both diffusion-dominated and advection-dominated regimes. The present analysis includes the case where the diffusion vanishes locally in some parts of the domain. Finally, in Sect. 4.7, we consider the heat equation as a prototype for time-dependent scalar PDEs with diffusion (that is, parabolic PDEs). The approximation is based on the SIP method for space discretization and an A-stable finite difference scheme in time; for simplicity, we focus on backward (or implicit) Euler and BDF2 schemes.

4.1 Pure Diffusion: The Continuous Setting

In this section, we present some basic facts concerning the model problem (4.1).

4.1.1 Weak Formulation and Well-Posedness

The weak formulation of (4.1) is classical:

$$\text{Find } u \in V \text{ s.t. } a(u,v) = \int_\Omega fv \text{ for all } v \in V, \qquad (4.2)$$

4.1. Pure Diffusion: The Continuous Setting

with energy space $V = H_0^1(\Omega) := \{v \in H^1(\Omega) \mid v|_{\partial\Omega} = 0\}$ and bilinear form

$$a(u,v) := \int_\Omega \nabla u \cdot \nabla v. \qquad (4.3)$$

Recalling the *Poincaré inequality* (see, e.g., Evans [153, p. 265] or Brézis [55, p. 174]) stating that there is C_Ω such that, for all $v \in H_0^1(\Omega)$,

$$\|v\|_{L^2(\Omega)} \le C_\Omega \|\nabla v\|_{[L^2(\Omega)]^d}, \qquad (4.4)$$

we infer that the bilinear form a is coercive on V. Therefore, owing to the Lax–Milgram Lemma, the weak problem (4.2) is well-posed.

4.1.2 Potential and Diffusive Flux

The PDE (4.1a) can be rewritten in *mixed form* as a system of first-order PDEs:

$$\sigma + \nabla u = 0 \quad \text{in } \Omega, \qquad (4.5a)$$
$$\nabla \cdot \sigma = f \quad \text{in } \Omega. \qquad (4.5b)$$

Definition 4.1 (Potential and diffusive flux). In the context of the mixed formulation (4.5), the scalar-valued function u is termed the *potential* and the vector-valued function $\sigma := -\nabla u$ is termed the *diffusive flux*.

The derivation of dG methods to approximate the model problems (4.1) on a given mesh $\mathcal{T}_h$ hinges on the fact that the jumps of the potential and of the normal component of the diffusive flux vanish across interfaces. To allow for a more compact notation, we define boundary averages and jumps (cf. Definition 1.17 for interface averages and jumps).

Definition 4.2 (Boundary averages and jumps). For a smooth function v, for all $F \in \mathcal{F}_h^b$, and for a.e. $x \in F$, we define the *average and jump* of v as

$$\{\!\{v\}\!\}_F(x) = [\![v]\!]_F(x) := v(x).$$

The subscript as well as the dependence on x are omitted unless necessary.

Since the potential u is in the energy space V, we infer that, for all $T \in \mathcal{T}_h$ and all $F \in \mathcal{F}_T$, letting $u_T := u|_T$, the trace $u_T|_F$ is in $L^2(F)$. Furthermore, the diffusive flux σ is in the space $H(\text{div}; \Omega)$ defined by (1.23). Traces on mesh faces of the normal component of functions in $H(\text{div}; \Omega)$ are discussed in Sect. 1.2.6. In particular, under the regularity assumption $u \in W^{2,1}(\Omega)$, there holds $\sigma \in [W^{1,1}(\Omega)]^d$, so that, for all $T \in \mathcal{T}_h$ and all $F \in \mathcal{F}_T$, letting $\sigma_T := \sigma|_T$ and $\sigma_{\partial T} := \sigma_T \cdot n_T$ on ∂T, the trace $\sigma_{\partial T}|_F$ is in $L^1(F)$. This trace is in $L^2(F)$ under the stronger regularity assumption $u \in H^2(\Omega)$ (the assumption $u \in H^{3/2+\epsilon}(\Omega)$, $\epsilon > 0$, is actually sufficient).

We can now examine the jumps of the potential and of the normal component of the diffusive flux.

Lemma 4.3 (Jumps of potential and diffusive flux). *Assume $u \in V \cap W^{2,1}(\Omega)$. Then, there holds*

$$[\![u]\!] = 0 \quad \forall F \in \mathcal{F}_h, \tag{4.6a}$$

$$[\![\sigma]\!]\cdot n_F = 0 \quad \forall F \in \mathcal{F}_h^i. \tag{4.6b}$$

Proof. Assertion (4.6a) results from Lemma 1.23 for interfaces and from Definition 4.2 for boundary faces. Assertion (4.6b) results from Lemma 1.24. □

4.2 Symmetric Interior Penalty

Our goal is to approximate the solution of the model problem (4.2) using dG methods in the broken polynomial space $\mathbb{P}_d^k(\mathcal{T}_h)$ defined by (1.15). We set

$$V_h := \mathbb{P}_d^k(\mathcal{T}_h),$$

with polynomial degree $k \geq 1$ and $\mathcal{T}_h$ belonging to an admissible mesh sequence. The focus of this section is on a specific dG method, the Symmetric Interior Penalty (SIP) method introduced by Arnold [14].

For simplicity, we enforce a somewhat strong regularity assumption on the exact solution. A weaker regularity assumption is made in Sect. 4.2.5.

Assumption 4.4 (Regularity of exact solution and space V_*). *We assume that the exact solution u is such that*

$$u \in V_* := V \cap H^2(\Omega).$$

In the spirit of Sect. 1.3, *we set* $V_{*h} := V_* + V_h$.

Without further knowledge on the exact solution u apart from the domain Ω and the datum $f \in L^2(\Omega)$, Assumption 4.4 can be asserted for instance if the domain Ω is convex; see Grisvard [177]. Assumption 4.4 differs from the concept of elliptic regularity (cf. Definition 4.24 below) since Assumption 4.4 only concerns the exact solution u.

4.2.1 Heuristic Derivation

To derive a suitable discrete bilinear form, we loosely follow the same path of ideas as in Chap. 2 aiming at a discrete bilinear form that satisfies the consistency requirement (1.32) and enjoys discrete coercivity. Moreover, we add a (consistent) term to recover, at the discrete level, the symmetry of the continuous problem.

4.2.1.1 Consistency

We begin localizing gradients to mesh elements in the exact bilinear form a, that is, we set, for all $(v, w_h) \in V_{*h} \times V_h$,

$$a_h^{(0)}(v, w_h) := \int_\Omega \nabla_h v \cdot \nabla_h w_h = \sum_{T \in \mathcal{T}_h} \int_T \nabla v \cdot \nabla w_h.$$

4.2. Symmetric Interior Penalty

To examine the consistency requirement (1.32), we integrate by parts on each mesh element. This leads to

$$a_h^{(0)}(v,w_h) = -\sum_{T\in\mathcal{T}_h}\int_T (\Delta v)w_h + \sum_{T\in\mathcal{T}_h}\int_{\partial T}(\nabla v\cdot\mathbf{n}_T)w_h.$$

The second term on the right-hand side can be reformulated as a sum over mesh faces in the form

$$\sum_{T\in\mathcal{T}_h}\int_{\partial T}(\nabla v\cdot\mathbf{n}_T)w_h = \sum_{F\in\mathcal{F}_h^i}\int_F [\![(\nabla_h v)w_h]\!]\cdot\mathbf{n}_F + \sum_{F\in\mathcal{F}_h^b}\int_F (\nabla v\cdot\mathbf{n}_F)w_h,$$

since for all $F\in\mathcal{F}_h^i$ with $F=\partial T_1\cap\partial T_2$, $\mathbf{n}_F = \mathbf{n}_{T_1} = -\mathbf{n}_{T_2}$. Moreover,

$$[\![(\nabla_h v)w_h]\!] = \{\!\{\nabla_h v\}\!\}[\![w_h]\!] + [\![\nabla_h v]\!]\{\!\{w_h\}\!\},$$

since letting $a_i = (\nabla v)|_{T_i}$, $b_i = w_h|_{T_i}$, $i\in\{1,2\}$, yields

$$[\![(\nabla_h v)w_h]\!] = a_1 b_1 - a_2 b_2$$
$$= \tfrac{1}{2}(a_1+a_2)(b_1-b_2) + (a_1-a_2)\tfrac{1}{2}(b_1+b_2)$$
$$= \{\!\{\nabla_h v\}\!\}[\![w_h]\!] + [\![\nabla_h v]\!]\{\!\{w_h\}\!\}.$$

As a result, and accounting for boundary faces using Definition 4.2, yields

$$\sum_{T\in\mathcal{T}_h}\int_{\partial T}(\nabla v\cdot\mathbf{n}_T)w_h = \sum_{F\in\mathcal{F}_h}\int_F \{\!\{\nabla_h v\}\!\}\cdot\mathbf{n}_F[\![w_h]\!] + \sum_{F\in\mathcal{F}_h^i}\int_F [\![\nabla_h v]\!]\cdot\mathbf{n}_F\{\!\{w_h\}\!\}.$$

Hence,

$$a_h^{(0)}(v,w_h) = -\sum_{T\in\mathcal{T}_h}\int_T (\Delta v)w_h + \sum_{F\in\mathcal{F}_h}\int_F \{\!\{\nabla_h v\}\!\}\cdot\mathbf{n}_F[\![w_h]\!]$$
$$+ \sum_{F\in\mathcal{F}_h^i}\int_F [\![\nabla_h v]\!]\cdot\mathbf{n}_F\{\!\{w_h\}\!\}. \qquad (4.7)$$

To check consistency, we set $v=u$ in (4.7). A consequence of (4.6b) is that, for all $w_h\in V_h$,

$$a_h^{(0)}(u,w_h) = \int_\Omega fw_h + \sum_{F\in\mathcal{F}_h}\int_F (\nabla u\cdot\mathbf{n}_F)[\![w_h]\!].$$

In order to match the consistency requirement (1.32), we are prompted to modify $a_h^{(0)}$ as follows: For all $(v,w_h)\in V_{*h}\times V_h$,

$$a_h^{(1)}(v,w_h) := \int_\Omega \nabla_h v\cdot\nabla_h w_h - \sum_{F\in\mathcal{F}_h}\int_F \{\!\{\nabla_h v\}\!\}\cdot\mathbf{n}_F[\![w_h]\!].$$

It is clear that $a_h^{(1)}$ is consistent in the sense of (1.32), i.e., for all $w_h\in V_h$,

$$a_h^{(1)}(u,w_h) = \int_\Omega fw_h.$$

4.2.1.2 Symmetry

A desirable property of the discrete bilinear form is to preserve the original symmetry of the exact bilinear form. Indeed, symmetry can simplify the solution of the resulting linear system and furthermore, it is a natural ingredient to derive optimal L^2-norm error estimates (cf. Sect. 4.2.4); nonsymmetric variants are discussed in Sect. 5.3. In view of this remark, we set, for all $(v, w_h) \in V_{*h} \times V_h$,

$$a_h^{\mathrm{cs}}(v, w_h) := \int_\Omega \nabla_h v \cdot \nabla_h w_h - \sum_{F \in \mathcal{F}_h} \int_F \left(\{\!\{\nabla_h v\}\!\} \cdot \mathbf{n}_F [\![w_h]\!] + [\![v]\!] \{\!\{\nabla_h w_h\}\!\} \cdot \mathbf{n}_F \right), \quad (4.8)$$

so that a_h^{cs} is symmetric on $V_h \times V_h$. The bilinear form a_h^{cs} remains consistent owing to (4.6a). The superscript in a_h^{cs} indicates the consistency and symmetry achieved so far. For future use, we record the following equivalent expression of a_h^{cs} resulting from (4.7),

$$a_h^{\mathrm{cs}}(v, w_h) = - \sum_{T \in \mathcal{T}_h} \int_T (\Delta v) w_h + \sum_{F \in \mathcal{F}_h^i} \int_F [\![\nabla_h v]\!] \cdot \mathbf{n}_F \{\!\{w_h\}\!\}$$

$$- \sum_{F \in \mathcal{F}_h} \int_F [\![v]\!] \{\!\{\nabla_h w_h\}\!\} \cdot \mathbf{n}_F. \quad (4.9)$$

4.2.1.3 Penalties on Interface Jumps and Boundary Values

The last requirement to match is discrete coercivity on the broken polynomial space V_h with respect to a suitable norm. The difficulty with the discrete bilinear form a_h^{cs} defined by (4.8) is that, for all $v_h \in V_h$,

$$a_h^{\mathrm{cs}}(v_h, v_h) = \|\nabla_h v_h\|_{[L^2(\Omega)]^d}^2 - 2 \sum_{F \in \mathcal{F}_h} \int_F \{\!\{\nabla_h v_h\}\!\} \cdot \mathbf{n}_F [\![v_h]\!],$$

and the second term on the right-hand side has no a priori sign so that, without adding a further term, there is no hope for discrete coercivity (in some situations, discrete inf-sup stability can be achieved without penalty; cf. Remark 4.14). To achieve discrete coercivity, we add to a_h^{cs} a term penalizing interface and boundary jumps, namely we set, for all $(v, w_h) \in V_{*h} \times V_h$,

$$a_h^{\mathrm{sip}}(v, w_h) := a_h^{\mathrm{cs}}(v, w_h) + s_h(v, w_h), \quad (4.10)$$

with the stabilization bilinear form

$$s_h(v, w_h) := \sum_{F \in \mathcal{F}_h} \frac{\eta}{h_F} \int_F [\![v]\!] [\![w_h]\!], \quad (4.11)$$

where $\eta > 0$ is a user-dependent parameter and h_F a local length scale associated with the mesh face $F \in \mathcal{F}_h$. We observe that, owing to (4.6a), adding the bilinear form s_h to a_h^{cs} does not alter the consistency and symmetry achieved so far.

4.2. Symmetric Interior Penalty

Moreover, Lemma 4.12 below shows that, provided the penalty parameter η is large enough, the discrete bilinear form a_h^{sip} enjoys discrete coercivity on V_h.

We now present a simple choice for the local length scale h_F.

Definition 4.5 (Local length scale h_F). For all $F \in \mathcal{F}_h$, in dimension $d \geq 2$, we set h_F to be equal to the diameter of the face F, while, in dimension 1, we set $h_F := \min(h_{T_1}, h_{T_2})$ if $F \in \mathcal{F}_h^i$ with $F = \partial T_1 \cap \partial T_2$ and $h_F := h_T$ if $F \in \mathcal{F}_h^b$ with $F = \partial T \cap \partial \Omega$. In all cases, for a mesh element $T \in \mathcal{T}_h$, h_T denotes its diameter (cf. Definition 1.13).

Remark 4.6 (Local length scale h_F). Other choices are possible for the local length scale h_F weighting the face penalties in the stabilization bilinear form s_h, e.g., the choice $h_F = \{\!\{h\}\!\} := \frac{1}{2}(h_{T_1} + h_{T_2})$ for all $F \in \mathcal{F}_h^i$, or the choice $h_F = \frac{\{\!\{|T|_d\}\!\}}{|F|_{d-1}}$ (that is, the mean value of the d-dimensional Hausdorff measures of the neighboring elements divided by the $(d-1)$-dimensional Hausdorff measure of the face, recalling that for $d = 1$, $|F|_0 = 1$). Incidentally, we observe that modifying the choice for the local length scale impacts the value of the minimal threshold on the penalty parameter η for which discrete coercivity is achieved.

Combining (4.10) with (4.11) yields, for all $(v, w_h) \in V_{*h} \times V_h$,

$$a_h^{\text{sip}}(v, w_h) = \int_\Omega \nabla_h v \cdot \nabla_h w_h - \sum_{F \in \mathcal{F}_h} \int_F (\{\!\{\nabla_h v\}\!\} \cdot \mathbf{n}_F [\![w_h]\!] + [\![v]\!] \{\!\{\nabla_h w_h\}\!\} \cdot \mathbf{n}_F)$$
$$+ \sum_{F \in \mathcal{F}_h} \frac{\eta}{h_F} \int_F [\![v]\!] [\![w_h]\!], \quad (4.12)$$

or, equivalently using (4.9),

$$a_h^{\text{sip}}(v, w_h) = -\sum_{T \in \mathcal{T}_h} \int_T (\triangle v) w_h + \sum_{F \in \mathcal{F}_h^i} \int_F [\![\nabla_h v]\!] \cdot \mathbf{n}_F \{\!\{w_h\}\!\}$$
$$- \sum_{F \in \mathcal{F}_h} \int_F [\![v]\!] \{\!\{\nabla_h w_h\}\!\} \cdot \mathbf{n}_F + \sum_{F \in \mathcal{F}_h} \frac{\eta}{h_F} \int_F [\![v]\!] [\![w_h]\!]. \quad (4.13)$$

The idea of weakly enforcing boundary and jump conditions on the discrete solution using penalties can be traced back to the seventies, in particular the work of Nitsche [248, 249], Babuška [20], Babuška and Zlámal [24], Douglas and Dupont [134], Baker [25], and Wheeler [306]. The discrete bilinear form a_h^{sip} defined by (4.12) corresponds to the Symmetric Interior Penalty (SIP) method introduced by Arnold [14]; henceforth, a_h^{sip} is called the SIP bilinear form. In the present context, interior penalty means interior as well as boundary penalties.

Definition 4.7 (Consistency, symmetry, and penalty terms). The second, third, and fourth terms on the right-hand side of (4.12) are respectively called *consistency*, *symmetry*, and *penalty* terms.

4.2.1.4 The Discrete Problem

The discrete problem is

$$\text{Find } u_h \in V_h \text{ s.t. } a_h^{\text{sip}}(u_h, v_h) = \int_\Omega f v_h \text{ for all } v_h \in V_h. \tag{4.14}$$

Lemma 4.12 below states that provided the penalty parameter η is large enough, the SIP bilinear form is coercive on V_h. Thus, owing to the Lax–Milgram Lemma, the discrete problem (4.14) is well-posed. Moreover, a straightforward consequence of the above derivation is consistency.

Lemma 4.8 (Consistency). *Assume $u \in V_*$. Then, for all $v_h \in V_h$,*

$$a_h^{\text{sip}}(u, v_h) = \int_\Omega f v_h.$$

Remark 4.9 (Rough right-hand side). At the continuous level, the Poisson problem can be posed for a right-hand side $f \in V' = H^{-1}(\Omega)$, the dual space of the energy space $V = H_0^1(\Omega)$, leading to the weak formulation

$$a(u, v) = \langle f, v \rangle_{V', V} \qquad \forall v \in V.$$

Since the discrete space V_h is nonconforming in V, it is not possible, at the discrete level, to take $\langle f, v_h \rangle_{V', V}$ as right-hand side in (4.14). One possibility is to use a smoothing operator $\mathcal{I}_h : V_h \to V_h \cap H_0^1(\Omega)$ and to consider the discrete problem

$$\text{Find } u_h \in V_h \text{ s.t. } a_h^{\text{sip}}(u_h, v_h) = \langle f, \mathcal{I}_h v_h \rangle_{V', V} \text{ for all } v_h \in V_h. \tag{4.15}$$

One example of smoothing operator is the averaging operator considered in Sect. 5.5.2. An important observation is that (4.15) is no longer consistent.

Remark 4.10 (Stencil). With an eye toward implementation, we identify the elementary stencil (cf. Definition 2.26) associated with the SIP bilinear form. For all $T \in \mathcal{T}_h$, the stencil of the volume contribution is just the element T, while the stencil associated with the consistency, symmetry, and penalty terms consists of T and its neighbors in the sense of faces. Figure 4.1 illustrates the stencil for a matching triangular mesh; cf. Sect. A.1.3 for further insight.

4.2.2 Other Boundary Conditions

The discrete problem (4.14), which was derived in the context of homogeneous Dirichlet boundary conditions, needs to be slightly modified when dealing with other boundary conditions. The modifications are designed so as to maintain consistency when the exact solution satisfies other boundary conditions. For instance, when (weakly) enforcing the nonhomogeneous Dirichlet boundary condition $u = g$ on $\partial \Omega$ with $g \in H^{1/2}(\partial \Omega)$, the discrete problem becomes

$$\text{Find } u_h \in V_h \text{ s.t. } a_h^{\text{sip}}(u_h, v_h) = l_h^D(g; v_h) \text{ for all } v_h \in V_h,$$

4.2. Symmetric Interior Penalty

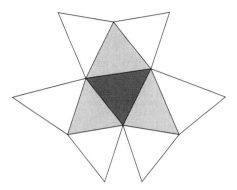

Fig. 4.1: Example of stencil of an element $T \in \mathcal{T}_h$ when $\mathcal{T}_h$ is a matching triangular mesh; the mesh element T is highlighted in *dark grey*, and its three neighbors, which all belong to the stencil, are highlighted in *light grey*; the other triangles do not belong to the stencil

with a_h^{sip} still defined by (4.12) and the new right-hand side

$$l_h^D(g; v_h) := \int_\Omega f v_h - \int_{\partial\Omega} g \nabla_h v_h \cdot \mathrm{n} + \sum_{F \in \mathcal{F}_h^b} \frac{\eta}{h_F} \int_F g v_h.$$

As a result, for the exact solution $u \in V_*$, $a_h^{\text{sip}}(u, v_h) = l_h^D(g; v_h)$, for all $v_h \in V_h$. Furthermore, when (weakly) enforcing the Robin boundary condition $\gamma u + \nabla u \cdot \mathrm{n} = g$ on $\partial\Omega$ with $g \in L^2(\partial\Omega)$ and $\gamma \in L^\infty(\partial\Omega)$ such that γ is nonnegative a.e. on $\partial\Omega$, the discrete problem becomes

Find $u_h \in V_h$ s.t. $a_h^R(u_h, v_h) = l_h^R(g; v_h)$ for all $v_h \in V_h$,

where, for all $(v, w_h) \in V_{*h} \times V_h$,

$$a_h^R(v, w_h) := \int_\Omega \nabla_h v \cdot \nabla_h w_h - \sum_{F \in \mathcal{F}_h^i} \int_F (\{\!\{\nabla_h v\}\!\} \cdot \mathrm{n}_F [\![w_h]\!] + [\![v]\!] \{\!\{\nabla_h w_h\}\!\} \cdot \mathrm{n}_F)$$
$$+ \sum_{F \in \mathcal{F}_h^i} \frac{\eta}{h_F} \int_F [\![v]\!][\![w_h]\!] + \sum_{F \in \mathcal{F}_h^b} \int_F \gamma v_h w_h, \qquad (4.16)$$

and

$$l_h^R(g; w_h) := \int_\Omega f w_h + \int_{\partial\Omega} g w_h.$$

As a result, for the exact solution $u \in V_*$, $a_h^R(u, v_h) = l_h^R(g; v_h)$, for all $v_h \in V_h$. Moreover, we observe that, unlike in the Dirichlet case, the summations in the consistency and symmetry terms are restricted to interfaces. Finally, the case $\gamma \equiv 0$ corresponds to the Neumann problem. For this problem, the data must comply with the compatibility condition $\int_\Omega f = -\int_{\partial\Omega} g$, and the solution is determined up to an additive constant. One possibility is to additionally enforce $\int_\Omega u_h = 0$. In practice, the discrete problem can still be formulated using V_h as trial and test space, and the additional constraint can be enforced by postprocessing. Observing that the rank of the problem matrix is one unit less than its size, the discrete solution can be obtained (1) using a direct solver with full pivoting, so that the zero pivot is encountered when processing the last line of the matrix and a solution can be obtained by fixing an arbitrary value for the degree of freedom left or (2) using an iterative solver which only requires matrix-vector products. The most common linear algebra libraries (e.g., the PETSc library [26]) offer specific functionalities to handle this case efficiently.

4.2.3 Basic Energy-Error Estimate

Let u solve the weak problem (4.2) and let u_h solve the discrete problem (4.14). The aim of this section is to estimate the approximation error $(u - u_h)$. The convergence analysis is performed in the spirit of Theorem 1.35. We recall that the space V_* is specified in Assumption 4.4 and that $V_{*h} = V_* + V_h$.

4.2.3.1 Discrete Coercivity

We aim at asserting this property using the following norm: For all $v \in V_{*h}$,

$$\|v\|_{\text{sip}} := \left(\|\nabla_h v\|^2_{[L^2(\Omega)]^d} + |v|^2_{\text{J}} \right)^{1/2}, \tag{4.17}$$

with the *jump seminorm*

$$|v|_{\text{J}} := (\eta^{-1} s_h(v,v))^{1/2} = \left(\sum_{F \in \mathcal{F}_h} \frac{1}{h_F} \|[\![v]\!]\|^2_{L^2(F)} \right)^{1/2}. \tag{4.18}$$

We observe that $\|\cdot\|_{\text{sip}}$ is indeed a norm on V_{*h}, and even on the broken Sobolev space $H^1(\mathcal{T}_h)$. The only nontrivial property to check is whether, for all $v \in H^1(\mathcal{T}_h)$, $\|v\|_{\text{sip}} = 0$ implies $v = 0$. Clearly, $\|v\|_{\text{sip}} = 0$ implies $\|\nabla_h v\|_{[L^2(\Omega)]^d} = 0$ and $|v|_{\text{J}} = 0$. The first property yields $\nabla_h v = 0$ so that v is piecewise constant. The second property implies that the interface and boundary jumps of v vanish. Hence, $v = 0$.

Our first step toward establishing discrete coercivity for the SIP bilinear form is a bound on the consistency term using the jump seminorm $|\cdot|_{\text{J}}$.

Lemma 4.11 (Bound on consistency term). *For all* $(v, w_h) \in V_{*h} \times V_h$,

$$\left| \sum_{F \in \mathcal{F}_h} \int_F \{\!\!\{\nabla_h v\}\!\!\} \cdot n_F [\![w_h]\!] \right| \leq \left(\sum_{T \in \mathcal{T}_h} \sum_{F \in \mathcal{F}_T} h_F \|\nabla v|_T \cdot n_F\|^2_{L^2(F)} \right)^{1/2} |w_h|_{\text{J}}. \tag{4.19}$$

4.2. Symmetric Interior Penalty

Proof. For all $F \in \mathcal{F}_h^i$ with $F = \partial T_1 \cap \partial T_2$, and $a_i = (\nabla v)|_{T_i} \cdot \mathbf{n}_F$, $i \in \{1,2\}$, the Cauchy–Schwarz inequality yields

$$\int_F \{\!\!\{\nabla_h v\}\!\!\} \cdot \mathbf{n}_F [\![w_h]\!] = \int_F \frac{1}{2}(a_1 + a_2)[\![w_h]\!]$$

$$\leq \left(\frac{1}{2} h_F (\|a_1\|_{L^2(F)}^2 + \|a_2\|_{L^2(F)}^2)\right)^{1/2} h_F^{-1/2} \|[\![w_h]\!]\|_{L^2(F)}.$$

Moreover, for all $F \in \mathcal{F}_h^b$ with $F = \partial T \cap \partial \Omega$,

$$\int_F \{\!\!\{\nabla_h v\}\!\!\} \cdot \mathbf{n}_F [\![w_h]\!] \leq h_F^{1/2} \|\nabla v|_T \cdot \mathbf{n}_F\|_{L^2(F)} \times h_F^{-1/2} \|[\![w_h]\!]\|_{L^2(F)}.$$

Summing over mesh faces, using the Cauchy–Schwarz inequality, and regrouping the face contributions for each mesh element yields the assertion. □

We can now turn to the discrete coercivity of the SIP bilinear form. We recall that N_∂, defined by (1.12), denotes the maximum number of mesh faces composing the boundary of a generic mesh element and that this quantity is bounded uniformly in h; cf. Lemma 1.41.

Lemma 4.12 (Discrete coercivity). *For all $\eta > \underline{\eta} := C_{\text{tr}}^2 N_\partial$ where C_{tr} results from the discrete trace inequality (1.37) and the parameter N_∂ is defined by (1.12), the SIP bilinear form defined by (4.12) is coercive on V_h with respect to the $\|\cdot\|_{\text{sip}}$-norm, i.e.,*

$$\forall v_h \in V_h, \qquad a_h^{\text{sip}}(v_h, v_h) \geq C_\eta \|v_h\|_{\text{sip}}^2,$$

with $C_\eta := (\eta - C_{\text{tr}}^2 N_\partial)(1+\eta)^{-1}$.

Proof. Let $v_h \in V_h$. Since, for all $T \in \mathcal{T}_h$ and all $F \in \mathcal{F}_T$, $h_F \leq h_T$ (cf. Definition 4.5), we obtain using the discrete trace inequality (1.40),

$$\sum_{T \in \mathcal{T}_h} \sum_{F \in \mathcal{F}_T} h_F \|\nabla v_h|_T \cdot \mathbf{n}_F\|_{L^2(F)}^2 \leq \sum_{T \in \mathcal{T}_h} h_T \|\nabla v_h|_T \cdot \mathbf{n}_F\|_{L^2(\partial T)}^2$$

$$\leq C_{\text{tr}}^2 N_\partial \|\nabla_h v_h\|_{[L^2(\Omega)]^d}^2,$$

whence we infer from (4.19) that

$$\left|\sum_{F \in \mathcal{F}_h} \int_F \{\!\!\{\nabla_h v_h\}\!\!\} \cdot \mathbf{n}_F [\![v_h]\!]\right| \leq C_{\text{tr}} N_\partial^{1/2} \|\nabla_h v_h\|_{[L^2(\Omega)]^d} |v_h|_{\text{J}}.$$

As a result,

$$a_h^{\text{sip}}(v_h, v_h) \geq \|\nabla_h v_h\|_{[L^2(\Omega)]^d}^2 - 2C_{\text{tr}} N_\partial^{1/2} \|\nabla_h v_h\|_{[L^2(\Omega)]^d} |v_h|_{\text{J}} + \eta |v_h|_{\text{J}}^2.$$

We now use the following inequality: Let β be a positive real number, let $\eta > \beta^2$; then, for all $x, y \in \mathbb{R}$,

$$x^2 - 2\beta xy + \eta y^2 \geq \frac{\eta - \beta^2}{1 + \eta}(x^2 + y^2).$$

Applying this inequality with $\beta = C_{\mathrm{tr}} N_\partial^{1/2}$, $x = \|\nabla_h v_h\|_{[L^2(\Omega)]^d}$, and $y = |v_h|_{\mathrm{J}}$ yields the assertion. □

Remark 4.13 (Modifying the local length scale). Recalling Remark 4.6, other choices for the local length scale h_F can be made when defining the stabilization bilinear form s_h. The jump seminorm $|\cdot|_{\mathrm{J}}$ is still defined by $|v|_{\mathrm{J}} := (\eta^{-1} s_h(v,v))^{-1/2}$, and the proof of Lemma 4.12 can be deployed as above as long as the chosen length scale is a lower bound for the diameter of both neighboring elements; otherwise, an additional factor appears in the definition of the minimal threshold η.

Remark 4.14 (Discrete stability without penalty). In one space dimension, the discrete bilinear form a_h^{cs} enjoys discrete inf-sup stability without adding the stabilization bilinear form s_h for polynomial degrees $k \geq 2$; see Burman, Ern, Mozolevski, and Stamm [67]. Furthermore, in two and three space dimensions and using piecewise affine discrete functions supplemented by suitable element bubble functions, discrete inf-sup stability can be proven for the discrete bilinear form a_h^{cs} again without adding the stabilization bilinear form s_h; see Burman and Stamm [70]. Finally, it is also possible to devise penalty strategies acting only on the low-degree part of the jumps; see, e.g., Hansbo and Larson [184] and Burman and Stamm [71].

Remark 4.15 (Poincaré inequality using the $\|\cdot\|_{\mathrm{sip}}$-norm). It can be proven (cf. Corollary 5.4) that there exists σ_2, independent of h, such that

$$\forall v_h \in V_h, \qquad \|v_h\|_{L^2(\Omega)} \leq \sigma_2 \|v_h\|_{\mathrm{sip}}. \tag{4.20}$$

More generally, on the broken Sobolev space $H^1(\mathcal{T}_h)$, it is proven by Brenner [51] (see also Arnold [14]) that, for $d \in \{2,3\}$, there is σ_2', independent of h, such that

$$\forall v \in H^1(\mathcal{T}_h), \qquad \|v\|_{L^2(\Omega)} \leq \sigma_2' \|v\|_{\mathrm{sip}}. \tag{4.21}$$

4.2.3.2 Boundedness

We define on V_{*h} the norm

$$\|v\|_{\mathrm{sip},*} := \left(\|v\|_{\mathrm{sip}}^2 + \sum_{T \in \mathcal{T}_h} h_T \|\nabla v|_T \cdot \mathbf{n}_T\|_{L^2(\partial T)}^2 \right)^{1/2}. \tag{4.22}$$

Lemma 4.16 (Boundedness). *There is C_{bnd}, independent of h, such that*

$$\forall (v, w_h) \in V_{*h} \times V_h, \qquad a_h^{\mathrm{sip}}(v, w_h) \leq C_{\mathrm{bnd}} \|v\|_{\mathrm{sip},*} \|w_h\|_{\mathrm{sip}}. \tag{4.23}$$

Proof. Let $(v, w_h) \in V_{*h} \times V_h$. We observe that

$$a_h^{\mathrm{sip}}(v, w_h) = \int_\Omega \nabla_h v \cdot \nabla_h w_h - \sum_{F \in \mathcal{F}_h} \int_F \{\!\!\{\nabla_h v\}\!\!\} \cdot \mathbf{n}_F [\![w_h]\!] - \sum_{F \in \mathcal{F}_h} \int_F [\![v]\!] \{\!\!\{\nabla_h w_h\}\!\!\} \cdot \mathbf{n}_F$$

$$+ \sum_{F \in \mathcal{F}_h} \frac{\eta}{h_F} \int_F [\![v]\!] [\![w_h]\!] := \mathfrak{T}_1 + \mathfrak{T}_2 + \mathfrak{T}_3 + \mathfrak{T}_4. \tag{4.24}$$

4.2. Symmetric Interior Penalty

Using the Cauchy–Schwarz inequality yields $|\mathfrak{T}_1| \leq \|\nabla_h v\|_{[L^2(\Omega)]^d} \|\nabla_h w_h\|_{[L^2(\Omega)]^d}$ and $|\mathfrak{T}_4| \leq \eta |v|_{\text{J}} |w_h|_{\text{J}}$. Moreover, owing to the bound (4.19) and since $h_F \leq h_T$,

$$|\mathfrak{T}_2| \leq \left(\sum_{T \in \mathcal{T}_h} h_T \| \nabla v|_T \cdot \mathbf{n}_T \|^2_{L^2(\partial T)} \right)^{1/2} |w_h|_{\text{J}} \leq \|v\|_{\text{sip},*} |w_h|_{\text{J}} \leq \|v\|_{\text{sip},*} \|w_h\|_{\text{sip}},$$

by definition of the $\|\cdot\|_{\text{sip},*}$-norm. Finally, still owing to the bound (4.19) and proceeding as in the proof of Lemma 4.12 leads to

$$|\mathfrak{T}_3| \leq C_{\text{tr}} N_\partial^{1/2} |v|_{\text{J}} \|\nabla_h w_h\|_{[L^2(\Omega)]^d} \leq C_{\text{tr}} N_\partial^{1/2} \|v\|_{\text{sip}} \|w_h\|_{\text{sip}}.$$

Collecting the above bounds yields (4.23) with $C_{\text{bnd}} = 2 + \eta + C_{\text{tr}} N_\partial^{1/2}$. □

4.2.3.3 $\|\cdot\|_{\text{sip}}$-Norm Error Estimate and Convergence

A straightforward consequence of the above results together with Theorem 1.35 is the following error estimate.

Theorem 4.17 ($\|\cdot\|_{\text{sip}}$-norm error estimate). *Let $u \in V_*$ solve (4.2). Let u_h solve (4.14) with a_h^{sip} defined by (4.12) and penalty parameter as in Lemma 4.12. Then, there is C, independent of h, such that*

$$\|u - u_h\|_{\text{sip}} \leq C \inf_{v_h \in V_h} \|u - v_h\|_{\text{sip},*}. \tag{4.25}$$

To infer a convergence result from (4.25), we assume that the exact solution is smooth enough and use Lemmata 1.58 and 1.59. The resulting estimate is optimal both for the broken gradient and the jump seminorm.

Corollary 4.18 (Convergence rate in $\|\cdot\|_{\text{sip}}$-norm). *Besides the hypotheses of Theorem 4.17, assume $u \in H^{k+1}(\Omega)$. Then, there holds*

$$\|u - u_h\|_{\text{sip}} \leq C_u h^k, \tag{4.26}$$

with $C_u = C\|u\|_{H^{k+1}(\Omega)}$ and C independent of h.

Remark 4.19 (Bound on the jumps). The contribution of the jump seminorm to the error $\|u - u_h\|_{\text{sip}}$ can be controlled by the contribution of the broken gradient under some assumptions. For instance, Lemma 5.30 shows that, on matching simplicial meshes and for a large enough penalty parameter, there holds, up to a positive factor independent of h,

$$|u - u_h|_{\text{J}} = |u_h|_{\text{J}} \lesssim \|\nabla_h (u - u_h)\|_{[L^2(\Omega)]^d} + \mathcal{R}_{\text{osc},\Omega},$$

where the data oscillation term $\mathcal{R}_{\text{osc},\Omega}$, defined by (5.34), converges to zero at order h^{k+1} if $f \in H^k(\Omega)$ and at order h^{k+2} if $f \in H^{k+1}(\Omega)$. We also refer the reader to Bonito and Nochetto [46] for a similar bound on the jumps on general meshes, and to Ainsworth and Rankin [7,9] for a sharper condition on the penalty parameter on triangular meshes with hanging nodes.

4.2.3.4 Analysis Using Only the $\|\cdot\|_{\text{sip},*}$-Norm

The convergence analysis of elliptic problems is often performed using a single norm. Such an approach is possible here by working only with the $\|\cdot\|_{\text{sip}}$-norm which turns out to be uniformly equivalent to the $\|\cdot\|_{\text{sip},*}$-norm on V_h.

Lemma 4.20 (Uniform equivalence of $\|\cdot\|_{\text{sip}}$- and $\|\cdot\|_{\text{sip},*}$-norms on V_h). *The $\|\cdot\|_{\text{sip}}$- and $\|\cdot\|_{\text{sip},*}$-norms are uniformly equivalent on V_h. Specifically,*

$$C_{\text{sip}} \|v_h\|_{\text{sip},*} \leq \|v_h\|_{\text{sip}} \leq \|v_h\|_{\text{sip},*} \qquad \forall v_h \in V_h,$$

with C_{sip} independent of h.

Proof. The upper bound is immediate, while the lower bound results from the discrete trace inequality (1.40) and the uniform bound on N_∂. □

A consequence of Lemma 4.20 is discrete coercivity on V_h in the form

$$\forall v_h \in V_h, \qquad a_h^{\text{sip}}(v_h, v_h) \geq C'_\eta \|v_h\|_{\text{sip},*}^2,$$

with C'_η independent of h. Moreover, an inspection at the proof of Lemma 4.16 leads to boundedness on $V_{*h} \times V_h$ in the form

$$\forall (v, w_h) \in V_{*h} \times V_h, \qquad a_h^{\text{sip}}(v, w_h) \leq C'_{\text{bnd}} \|v\|_{\text{sip},*} \|w_h\|_{\text{sip},*},$$

with C'_{bnd} independent of h. Using the above results leads to the following convergence results in the $\|\cdot\|_{\text{sip},*}$-norm.

Theorem 4.21 ($\|\cdot\|_{\text{sip},*}$-norm error estimate). *Under the hypotheses of Theorem 4.17, there is C, independent of h, such that*

$$\|u - u_h\|_{\text{sip},*} \leq C \inf_{v_h \in V_h} \|u - v_h\|_{\text{sip},*}. \tag{4.27}$$

Corollary 4.22 (Convergence rate in $\|\cdot\|_{\text{sip},*}$-norm). *Besides the hypotheses of Theorem 4.17, assume $u \in H^{k+1}(\Omega)$. Then, there holds*

$$\|u - u_h\|_{\text{sip},*} \leq C_u h^k, \tag{4.28}$$

with $C_u = C\|u\|_{H^{k+1}(\Omega)}$ and C independent of h.

Remark 4.23 (Comparison with $\|\cdot\|_{\text{sip}}$-norm estimates). The discrete coercivity of a_h^{sip} is naturally expressed using the $\|\cdot\|_{\text{sip}}$-norm, whereas using the $\|\cdot\|_{\text{sip},*}$-norm leads to the inclusion in the error estimate of the additional factor C_{sip} related to norm equivalence (cf. Lemma 4.20). Therefore, estimates (4.25) and (4.26) deliver a sharper bound on the broken gradient and the jump seminorm. However, estimates (4.27) and (4.28) convey additional information regarding the convergence of the normal gradient at mesh element boundaries.

4.2.4 L^2-Norm Error Estimate

Using the broken Poincaré inequality (4.21), the $\|\cdot\|_{\mathrm{sip}}$-norm estimate (4.26) yields the L^2-norm estimate

$$\|u - u_h\|_{L^2(\Omega)} \leq \sigma_2' C_u h^k.$$

This estimate is suboptimal by one power in h. To remedy this drawback and recover optimality, it is possible to resort to a duality argument (the so-called Aubin–Nitsche argument [17]) under the following assumption.

Definition 4.24 (Elliptic regularity). We say that *elliptic regularity* holds true for the model problem (4.2) if there is C_{ell}, only depending on Ω, such that, for all $\psi \in L^2(\Omega)$, the solution to the problem:

$$\text{Find } \zeta \in H_0^1(\Omega) \text{ s.t. } a(\zeta, v) = \int_\Omega \psi v \text{ for all } v \in H_0^1(\Omega),$$

is in V_* and satisfies

$$\|\zeta\|_{H^2(\Omega)} \leq C_{\mathrm{ell}} \|\psi\|_{L^2(\Omega)}.$$

Elliptic regularity can be asserted if, for instance, the polygonal domain Ω is convex; see Grisvard [177]. To derive an L^2-norm error estimate, we extend the SIP bilinear form to $V_{*h} \times V_{*h}$, so that both arguments of a_h^{sip} can belong to V_*.

Theorem 4.25 (L^2-norm error estimate). *Let $u \in V_*$ solve (4.2). Let u_h solve (4.14) with a_h^{sip} defined by (4.12). Assume elliptic regularity. Then, there is C, independent of h, such that*

$$\|u - u_h\|_{L^2(\Omega)} \leq Ch \|u - u_h\|_{\mathrm{sip},*}. \tag{4.29}$$

Proof. We consider the auxiliary problem

$$\text{Find } \zeta \in H_0^1(\Omega) \text{ s.t. } a(\zeta, v) = \int_\Omega (u - u_h) v \text{ for all } v \in H_0^1(\Omega),$$

and use elliptic regularity to infer $\|\zeta\|_{H^2(\Omega)} \leq C_{\mathrm{ell}} \|u - u_h\|_{L^2(\Omega)}$. Since $\zeta \in V_*$, $[\![\nabla \zeta]\!] \cdot n_F = 0$ on all $F \in \mathcal{F}_h^i$ and $[\![\zeta]\!] = 0$ on all $F \in \mathcal{F}_h$. Hence, (4.13) implies

$$a_h^{\mathrm{sip}}(\zeta, u - u_h) = \int_\Omega (-\triangle \zeta)(u - u_h).$$

Exploiting the symmetry of a_h^{sip} and since $-\triangle \zeta = u - u_h$, we obtain

$$a_h^{\mathrm{sip}}(u - u_h, \zeta) = \|u - u_h\|_{L^2(\Omega)}^2.$$

Furthermore, since consistency implies Galerkin orthogonality (cf. Remark 1.32) and letting π_h^1 be the L^2-orthogonal projection onto $\mathbb{P}_d^1(\mathcal{T}_h) \subset V_h$ (since $k \geq 1$), we infer

$$a_h^{\mathrm{sip}}(u - u_h, \pi_h^1 \zeta) = 0.$$

Hence, using the boundedness of $a_h^{\rm sip}$ on $V_{*h} \times V_{*h}$ which results from the fact that $a_h^{\rm sip}(v,w) \lesssim \|v\|_{{\rm sip},*}\|w\|_{{\rm sip},*}$ for all $v,w \in V_{*h}$, the approximation properties of π_h^1 in the $\|\cdot\|_{{\rm sip},*}$-norm, and the regularity of ζ, we obtain, up to multiplicative factors independent of h,

$$\begin{aligned}\|u-u_h\|_{L^2(\Omega)}^2 &= a_h^{\rm sip}(u-u_h, \zeta - \pi_h^1\zeta) \\ &\lesssim \|u-u_h\|_{{\rm sip},*}\|\zeta - \pi_h^1\zeta\|_{{\rm sip},*} \\ &\lesssim \|u-u_h\|_{{\rm sip},*} h \|\zeta\|_{H^2(\mathcal{T}_h)} \\ &\lesssim \|u-u_h\|_{{\rm sip},*} h \|u-u_h\|_{L^2(\Omega)}.\end{aligned}$$

Simplifying by $\|u-u_h\|_{L^2(\Omega)}$ yields (4.29). □

A straightforward consequence of (4.28) and (4.29) is the following convergence result for smooth solutions.

Corollary 4.26 (Convergence rate in L^2-norm). *Besides the hypotheses of Theorem 4.17, assume elliptic regularity and $u \in H^{k+1}(\Omega)$. Then, there holds*

$$\|u-u_h\|_{L^2(\Omega)} \leq C_u h^{k+1}, \qquad (4.30)$$

with $C_u = C\|u\|_{H^{k+1}(\Omega)}$ and C independent of h.

Estimate (4.30) is optimal. We emphasize that the symmetry of $a_h^{\rm sip}$ has been used in the proof of Theorem 4.25.

Remark 4.27 (Adjoint-consistency). Following Arnold, Brezzi, Cockburn, and Marini [16], the property $a_h^{\rm sip}(u-u_h, \zeta) = \int_\Omega (-\triangle\zeta)(u-u_h)$, which results from symmetry and consistency, can be termed *adjoint consistency*.

Remark 4.28 (Error estimates in other norms). We refer the reader, e.g., to Chen and Chen [89] and to Guzmán [181] for pointwise error estimates on the discrete solution and its broken gradient using weighted broken Sobolev norms.

4.2.5 Analysis for Low-Regularity Solutions

This section is devoted to the analysis of the SIP method under a regularity assumption on the exact solution that is weaker than Assumption 4.4.

Assumption 4.29 (Regularity of exact solution and space V_*). *We assume that $d \geq 2$ and that there is $p \in (\frac{2d}{d+2}, 2]$ such that, for the exact solution u,*

$$u \in V_* := V \cap W^{2,p}(\Omega).$$

In the spirit of Sect. 1.3, *we set* $V_{*h} := V_* + V_h$.

Assumption 4.29 requires $p > 1$ for $d = 2$ and $p > \frac{6}{5}$ for $d = 3$. In particular, we observe that, in two space dimensions, $u \in W^{2,p}(\Omega)$ with $p > 1$ holds true in

4.2. Symmetric Interior Penalty

polygonal domains; see, e.g., Dauge [119]. Moreover, using Sobolev embeddings (see Evans [153, Sect. 5.6] or Brézis [55, Sect. IX.3]), Assumption 4.29 implies

$$u \in H^{1+\alpha_p}(\Omega), \qquad \alpha_p = \frac{d+2}{2} - \frac{d}{p} > 0. \qquad (4.31)$$

We still consider the discrete problem (4.14) with the discrete bilinear form a_h^{sip} defined by (4.12). The convergence analysis under the regularity assumption 4.29 has been performed recently by Wihler and Rivière [308] in two space dimensions and, using slightly different techniques, by the authors [132] in the context of heterogeneous diffusion in any space dimension; cf. Sect. 4.5. We follow here the approach of [132], building up on the analysis presented in Sect. 4.2.3 for smooth solutions. In the present context of an exact solution with low-regularity, we assume for simplicity $k = 1$. We also assume $p < 2$ since in the case $p = 2$, Assumption 4.29 amounts to Assumption 4.4.

We already know that discrete coercivity holds true provided the penalty parameter is chosen as in Lemma 4.12. Moreover, owing to Lemma 4.3, the discrete bilinear form a_h^{sip} can be extended to $V_{*h} \times V_h$, and consistency can be asserted as in Lemma 4.8. Thus, it only remains to prove boundedness. To this purpose, we need to redefine the $\|\cdot\|_{\text{sip},*}$-norm since functions in V_* are such that, for all $T \in \mathcal{T}_h$, $\nabla v|_T \cdot n_T$ is in $L^p(\partial T)$, but not necessarily in $L^2(\partial T)$. Thus, we now define on V_{*h} the norm

$$\|v\|_{\text{sip},*} := \left(\|v\|_{\text{sip}}^p + \sum_{T \in \mathcal{T}_h} h_T^{1+\gamma_p} \|\nabla v|_T \cdot n_T\|_{L^p(\partial T)}^p \right)^{1/p}, \qquad (4.32)$$

where $\gamma_p := \frac{1}{2}d(p-2)$. We observe that, for $p = 2$, we recover the previous definition (4.22) of the $\|\cdot\|_{\text{sip},*}$-norm. The value for γ_p is motivated by the following boundedness result.

Lemma 4.30 (Boundedness). *There is C_{bnd}, independent of h, such that*

$$\forall (v, w_h) \in V_{*h} \times V_h, \qquad a_h^{\text{sip}}(v, w_h) \leq C_{\text{bnd}} \|v\|_{\text{sip},*} \|w_h\|_{\text{sip}}. \qquad (4.33)$$

Proof. Let $(v, w_h) \in V_{*h} \times V_h$. We need to bound the four terms $\mathfrak{T}_1, \ldots, \mathfrak{T}_4$ in (4.24). Proceeding as in the proof of Lemma 4.16, we obtain

$$|\mathfrak{T}_1 + \mathfrak{T}_3 + \mathfrak{T}_4| \leq C \|v\|_{\text{sip}} \|w_h\|_{\text{sip}},$$

with C independent of h, so that it only remains to bound the consistency term $\mathfrak{T}_2$. To this purpose, we proceed similarly to the proof of (4.19), but use Hölder's inequality instead of the Cauchy–Schwarz inequality. For all $F \in \mathcal{F}_h^i$ with $F = \partial T_1 \cap \partial T_2$, and $a_i = (\nabla v)|_{T_i} \cdot n_F$, $i \in \{1, 2\}$, Hölder's inequality yields

$$\int_F \{\!\!\{\nabla_h v\}\!\!\} \cdot n_F [\![w_h]\!] = \int_F \frac{1}{2}(a_1 + a_2)[\![w_h]\!]$$

$$\leq \left(\frac{1}{2} h_F^{1+\gamma_p}(\|a_1\|_{L^p(F)}^p + \|a_2\|_{L^p(F)}^p) \right)^{1/p} h_F^{-\beta_p} \|[\![w_h]\!]\|_{L^q(F)},$$

with $\beta_p = \frac{1+\gamma_p}{p}$ and $q = \frac{p}{p-1}$. Moreover, for all $F \in \mathcal{F}_h^b$ with $F = \partial T \cap \partial \Omega$,

$$\int_F \{\nabla_h v\} \cdot n_F [\![w_h]\!] \leq \left(h_F^{1+\gamma_p} \|\nabla v|_T \cdot n_F\|_{L^p(F)}^p \right)^{1/p} h_F^{-\beta_p} \|[\![w_h]\!]\|_{L^q(F)}.$$

Owing to the inverse inequality (1.43) and since $\beta_p - \frac{1}{2} = (d-1)(\frac{1}{q} - \frac{1}{2})$, we infer

$$h_F^{-\beta_p} \|[\![w_h]\!]\|_{L^q(F)} \leq C_{\mathrm{inv},q,2} h_F^{-1/2} \|[\![w_h]\!]\|_{L^2(F)},$$

where $C_{\mathrm{inv},q,2}$ is independent of h and can be bounded uniformly in q (cf. Remark 1.51). Combining the above bounds, summing over mesh faces, and using Hölder's inequality yields

$$\left| \sum_{F \in \mathcal{F}_h} \int_F \{\nabla_h v\} \cdot n_F [\![w_h]\!] \right| \leq \left(\sum_{T \in \mathcal{T}_h} h_T^{1+\gamma_p} \|\nabla v|_T \cdot n_T\|_{L^p(\partial T)}^p \right)^{1/p}$$

$$\times C_{\mathrm{inv},q,2} \left(\sum_{F \in \mathcal{F}_h} \left(h_F^{-1/2} \|[\![w_h]\!]\|_{L^2(F)} \right)^q \right)^{1/q}.$$

Since $q \geq 2$, we obtain

$$\left(\sum_{F \in \mathcal{F}_h} \left(h_F^{-1/2} \|[\![w_h]\!]\|_{L^2(F)} \right)^q \right)^{1/q} \leq \left(\sum_{F \in \mathcal{F}_h} \left(h_F^{-1/2} \|[\![w_h]\!]\|_{L^2(F)} \right)^2 \right)^{1/2} = |w_h|_\mathrm{J}.$$

Hence,

$$\left| \sum_{F \in \mathcal{F}_h} \int_F \{\nabla_h v\} \cdot n_F [\![w_h]\!] \right| \leq \left(\sum_{T \in \mathcal{T}_h} h_T^{1+\gamma_p} \|\nabla v|_T \cdot n_T\|_{L^p(\partial T)}^p \right)^{1/p} C_{\mathrm{inv},q,2} |w_h|_\mathrm{J},$$

whence we infer (4.33). □

To state a convergence result, we need optimal polynomial approximation for functions in $V_* = W^{2,p}(\Omega)$. For simplicity, we restricted the presentation of Sect. 1.4.4 to the Hilbertian setting. In the present non-Hilbertian setting, we make the following assumption.

Assumption 4.31 (Optimal polynomial approximation in $W^{2,p}(T)$). *The mesh sequence $(\mathcal{T}_h)_{h \in \mathcal{H}}$ is such that, for all $h \in \mathcal{H}$, all $T \in \mathcal{T}_h$, and all $v \in W^{2,p}(T)$, there holds*

$$|v - \pi_h v|_{W^{m,p}(T)} \leq C_{\mathrm{app}} h_T^{2-m} |v|_{W^{2,p}(T)} \qquad m \in \{0,1,2\}, \qquad (4.34\mathrm{a})$$

$$|v - \pi_h v|_{H^m(T)} \leq C_{\mathrm{app}} h_T^{1+\alpha_p - m} |v|_{W^{2,p}(T)} \qquad m \in \{0,1\}, \qquad (4.34\mathrm{b})$$

with C_{app} independent of both T and h, while α_p is defined by (4.31).

Assumption 4.31 can be asserted for mesh sequences with star-shaped property or finitely-shaped property; cf. Lemmata 1.61 and 1.62. We can now turn to our main convergence result.

Theorem 4.32 ($\|\cdot\|_{\text{sip}}$-norm error estimate and convergence rate). *Let $u \in V_*$ solve (4.2). Let u_h solve (4.14) with a_h^{sip} defined by (4.12) and penalty parameter as in Lemma 4.12. Then, there is C, independent of h, such that*

$$\|u - u_h\|_{\text{sip}} \leq C \inf_{v_h \in V_h} \|u - v_h\|_{\text{sip},*}, \qquad (4.35)$$

where the $\|\cdot\|_{\text{sip},*}$-norm is defined by (4.32). Moreover, under Assumption 4.31, there holds

$$\|u - u_h\|_{\text{sip}} \leq C_u h^{\alpha_p}, \qquad (4.36)$$

with $C_u = C|u|_{W^{2,p}(\Omega)}$ and C independent of h.

Proof. Estimate (4.35) is a direct consequence of Theorem 1.35 since we established discrete coercivity, consistency, and boundedness. We now take $v_h = \pi_h u$ in (4.35). We first observe that, for all $T \in \mathcal{T}_h$, using (4.34a) together with the continuous trace inequality (1.18) yields

$$\|\nabla(u - \pi_h u)|_T \cdot \mathbf{n}_T\|_{L^p(\partial T)} \lesssim h_T^{1-1/p} |u|_{W^{2,p}(T)},$$

where $a \lesssim b$ means the inequality $a \leq Cb$ with generic positive C independent of h and T. Since $\frac{1+\gamma_p}{p} + 1 - \frac{1}{p} = \alpha_p$, we infer

$$\left(\sum_{T \in \mathcal{T}_h} h_T^{1+\gamma_p} \|\nabla(u - \pi_h u)|_T \cdot \mathbf{n}_T\|_{L^p(\partial T)}^p \right)^{1/p} \lesssim h^{\alpha_p} |u|_{W^{2,p}(\Omega)}.$$

Moreover, using (4.34b) together with the continuous trace inequality (1.19) yields

$$\|u - \pi_h u\|_{\text{sip}} \lesssim h^{\alpha_p} |u|_{W^{2,p}(\Omega)}.$$

Combining the two above bounds leads to (4.36). □

The convergence rate in the error estimate (4.36) is optimal both for the broken gradient and the jump seminorm.

4.3 Liftings and Discrete Gradients

Liftings are operators that map scalar-valued functions defined on mesh faces to vector-valued functions defined on mesh elements. In the context of dG methods, liftings act on interface and boundary jumps. They were introduced by Bassi, Rebay, and coworkers [34, 35] in the context of compressible flows and analyzed by Brezzi, Manzini, Marini, Pietra, and Russo [58, 59] in the context of the Poisson problem (see also Perugia and Schötzau [257] for the hp-analysis). Liftings have many useful applications. They can be combined with the broken gradient to define discrete gradients. Discrete gradients play an important role in the design and analysis of dG methods. Indeed, they can be used to formulate the discrete problem locally on each mesh element using numerical fluxes. Moreover, as detailed in Sect. 5.1, they are instrumental in the derivation of discrete functional analysis results, that, in turn, play a central role in the convergence analysis to minimal regularity solutions (cf. Sect. 5.2). Liftings can also be employed to define the stabilization bilinear form [35], yielding a more convenient lower bound for the penalty parameter η; cf. Sect. 5.3.2.

4.3.1 Liftings: Definition and Stability

As before, we assume that the mesh $\mathcal{T}_h$ belongs to an admissible mesh sequence. For any mesh face $F \in \mathcal{F}_h$ and for any integer $l \geq 0$, we define the (local) lifting operator
$$\mathrm{r}_F^l : L^2(F) \longrightarrow [\mathbb{P}_d^l(\mathcal{T}_h)]^d$$
as follows: For all $\varphi \in L^2(F)$,
$$\int_\Omega \mathrm{r}_F^l(\varphi)\cdot\tau_h = \int_F \{\!\!\{\tau_h\}\!\!\}\cdot\mathrm{n}_F\varphi \qquad \forall \tau_h \in [\mathbb{P}_d^l(\mathcal{T}_h)]^d. \tag{4.37}$$

We observe that the support of $\mathrm{r}_F^l(\varphi)$ consists of the one or two mesh elements of which F is part of the boundary; using the set $\mathcal{T}_F$ defined by (1.13) yields
$$\mathrm{supp}(\mathrm{r}_F^l) = \bigcup_{T\in\mathcal{T}_F} \overline{T}. \tag{4.38}$$

Moreover, whenever the mesh face F is a portion of a hyperplane (this happens, for instance, when working with simplicial meshes or with general meshes consisting of convex elements), $\mathrm{r}_F^l(\varphi)$ is colinear to the normal vector n_F.

Lemma 4.33 (Bound on local lifting). *Let $F \in \mathcal{F}_h$ and let $l \geq 0$. For all $\varphi \in L^2(F)$, there holds*
$$\|\mathrm{r}_F^l(\varphi)\|_{[L^2(\Omega)]^d} \leq C_{\mathrm{tr}} h_F^{-1/2} \|\varphi\|_{L^2(F)}. \tag{4.39}$$

Proof. Let $\varphi \in L^2(F)$. Equation (4.37), the fact that $h_F \leq h_T$ for all $T \in \mathcal{T}_F$, and the discrete trace inequality (1.37) yield
$$\|\mathrm{r}_F^l(\varphi)\|_{[L^2(\Omega)]^d}^2 = \int_\Omega \mathrm{r}_F^l(\varphi)\cdot\mathrm{r}_F^l(\varphi) = \int_F \{\!\!\{\mathrm{r}_F^l(\varphi)\}\!\!\}\cdot\mathrm{n}_F\varphi$$
$$\leq \left(\frac{1}{h_F}\int_F |\varphi|^2\right)^{1/2} \times \left(h_F\int_F |\{\!\!\{\mathrm{r}_F^l(\varphi)\}\!\!\}|^2\right)^{1/2}$$
$$\leq h_F^{-1/2}\|\varphi\|_{L^2(F)} \times C_{\mathrm{tr}}\left(\mathrm{card}(\mathcal{T}_F)^{-1}\sum_{T\in\mathcal{T}_F}\int_T |\mathrm{r}_F^l(\varphi)|^2\right)^{1/2},$$
whence (4.39) follows since $\mathrm{card}(\mathcal{T}_F)^{-1} \leq 1$ and since $\sum_{T\in\mathcal{T}_F}\int_T |\mathrm{r}_F^l(\varphi)|^2 = \|\mathrm{r}_F^l(\varphi)\|_{[L^2(\Omega)]^d}^2$ owing to (4.38). □

4.3. Liftings and Discrete Gradients

For any integer $l \geq 0$ and for any function $v \in H^1(\mathcal{T}_h)$, we define the (global) lifting of its interface and boundary jumps as

$$\mathrm{R}_h^l(\llbracket v \rrbracket) := \sum_{F \in \mathcal{F}_h} \mathrm{r}_F^l(\llbracket v \rrbracket) \in [\mathbb{P}_d^l(\mathcal{T}_h)]^d, \qquad (4.40)$$

being implicitly understood that r_F^l acts on the function $\llbracket v \rrbracket_F$ (which is in $L^2(F)$ since $v \in H^1(\mathcal{T}_h)$).

Lemma 4.34 (Bound on global lifting). *Let $l \geq 0$. For all $v \in H^1(\mathcal{T}_h)$, there holds*

$$\| \mathrm{R}_h^l(\llbracket v \rrbracket) \|_{[L^2(\Omega)]^d} \leq N_\partial^{1/2} \left(\sum_{F \in \mathcal{F}_h} \| \mathrm{r}_F^l(\llbracket v \rrbracket) \|_{[L^2(\Omega)]^d}^2 \right)^{1/2}, \qquad (4.41)$$

so that

$$\| \mathrm{R}_h^l(\llbracket v \rrbracket) \|_{[L^2(\Omega)]^d} \leq C_{\mathrm{tr}} N_\partial^{1/2} |v|_\mathrm{J}. \qquad (4.42)$$

Proof. Let $v \in H^1(\mathcal{T}_h)$. Owing to (4.38), we infer $(\mathrm{R}_h^l(\llbracket v \rrbracket))|_T = \sum_{F \in \mathcal{F}_T} (\mathrm{r}_F^l(\llbracket v \rrbracket))|_T$, so that using the Cauchy–Schwarz inequality, we obtain

$$\| \mathrm{R}_h^l(\llbracket v \rrbracket) \|_{[L^2(\Omega)]^d}^2 = \sum_{T \in \mathcal{T}_h} \int_T \left| \sum_{F \in \mathcal{F}_T} \mathrm{r}_F^l(\llbracket v \rrbracket) \right|^2$$

$$\leq \sum_{T \in \mathcal{T}_h} \mathrm{card}(\mathcal{F}_T) \sum_{F \in \mathcal{F}_T} \int_T |\mathrm{r}_F^l(\llbracket v_h \rrbracket)|^2$$

$$\leq \max_{T \in \mathcal{T}_h} \mathrm{card}(\mathcal{F}_T) \sum_{T \in \mathcal{T}_h} \sum_{F \in \mathcal{F}_T} \int_T |\mathrm{r}_F^l(\llbracket v_h \rrbracket)|^2$$

$$= \max_{T \in \mathcal{T}_h} \mathrm{card}(\mathcal{F}_T) \sum_{F \in \mathcal{F}_h} \| \mathrm{r}_F^l(\llbracket v \rrbracket) \|_{[L^2(\Omega)]^d}^2,$$

and the bound (4.41) follows using the definition (1.12) of N_∂. Finally, (4.42) results from (4.41) and the fact that

$$\left(\sum_{F \in \mathcal{F}_h} \| \mathrm{r}_F^l(\llbracket v \rrbracket) \|_{[L^2(\Omega)]^d}^2 \right)^{1/2} \leq C_{\mathrm{tr}} |v|_\mathrm{J},$$

owing to Lemma 4.33. $\square$

To illustrate in the case $l = 0$ (piecewise constant liftings), we obtain, for all $v \in H^1(\mathcal{T}_h)$ and all $T \in \mathcal{T}_h$,

$$\mathrm{R}_h^0(\llbracket v \rrbracket)|_T = \sum_{F \in \mathcal{F}_T} \frac{|F|_{d-1}}{|T|_d} (v_F - v_T) \mathrm{n}_{T,F}, \qquad (4.43)$$

where $\mathrm{n}_{T,F}$ is the outward normal to T on F, $v_T := v|_T$, and $v_F := \frac{1}{2}(v_T + v|_{T'})$ whenever $F = \partial T \cap \partial T'$, $T \neq T'$, while $v_F := 0$ if $F \in \mathcal{F}_h^b$. The (opposite of the) above expression has been used as a gradient reconstruction in the context of finite volume methods replacing v_F by a consistent trace reconstruction (see Eymard, Gallouët, and Herbin [158]); cf. also formula (5.28).

4.3.2 Discrete Gradients: Definition and Stability

For any integer $l \geq 0$, we define the discrete gradient operator

$$G_h^l : H^1(\mathcal{T}_h) \longrightarrow [L^2(\Omega)]^d,$$

as follows: For all $v \in H^1(\mathcal{T}_h)$,

$$G_h^l(v) := \nabla_h v - \mathrm{R}_h^l(\llbracket v \rrbracket). \qquad (4.44)$$

Proposition 4.35 (Bound on discrete gradient). *Let $l \geq 0$. For all $v \in H^1(\mathcal{T}_h)$, there holds*

$$\|G_h^l(v)\|_{[L^2(\Omega)]^d} \leq (1 + C_{\mathrm{tr}}^2 N_\partial)^{1/2} \|v\|_{\mathrm{sip}},$$

where the $\|\cdot\|_{\mathrm{sip}}$-norm is defined by (4.17).

Proof. Let $v \in H^1(\mathcal{T}_h)$. Using the triangle inequality together with (4.42) yields

$$\|G_h^l(v)\|_{[L^2(\Omega)]^d} \leq \|\nabla_h v\|_{[L^2(\Omega)]^d} + \|\mathrm{R}_h^l(\llbracket v \rrbracket)\|_{[L^2(\Omega)]^d}$$
$$\leq \|\nabla_h v\|_{[L^2(\Omega)]^d} + C_{\mathrm{tr}} N_\partial^{1/2} |v|_\mathrm{J},$$

whence the assertion. $\square$

4.3.3 Reformulation of the SIP Bilinear Form

Let $l \in \{k-1, k\}$ and set, as in Sect. 4.2, $V_h = \mathbb{P}_d^k(\mathcal{T}_h)$ with $k \geq 1$ and $\mathcal{T}_h$ belonging to an admissible mesh sequence. Following Brezzi, Manzini, Marini, Pietra, and Russo [58], it is interesting to observe that the bilinear form a_h^{cs} defined by (4.8) can be equivalently written as follows: For all $v_h, w_h \in V_h$,

$$a_h^{\mathrm{cs}}(v_h, w_h) = \int_\Omega \nabla_h v_h \cdot \nabla_h w_h - \int_\Omega \nabla_h v_h \cdot \mathrm{R}_h^l(\llbracket w_h \rrbracket) - \int_\Omega \nabla_h w_h \cdot \mathrm{R}_h^l(\llbracket v_h \rrbracket). \qquad (4.45)$$

This results from definitions (4.37) and (4.40) and the fact that $\nabla_h v_h$ and $\nabla_h w_h$ are in $[\mathbb{P}_d^l(\mathcal{T}_h)]^d$ since $l \geq k-1$, so that, for all $F \in \mathcal{F}_h$,

$$\int_F \{\nabla_h v_h\} \cdot \mathbf{n}_F \llbracket w_h \rrbracket = \int_\Omega \nabla_h v_h \cdot \mathrm{r}_F^l(\llbracket w_h \rrbracket).$$

Starting from (4.45) and using the definition (4.44) of the discrete gradient, we infer, for all $v_h, w_h \in V_h$,

$$a_h^{\mathrm{cs}}(v_h, w_h) = \int_\Omega G_h^l(v_h) \cdot G_h^l(w_h) - \int_\Omega \mathrm{R}_h^l(\llbracket v_h \rrbracket) \cdot \mathrm{R}_h^l(\llbracket w_h \rrbracket).$$

As a result, recalling that the SIP bilinear form considered in Sect. 4.2 is such that $a_h^{\mathrm{sip}} = a_h^{\mathrm{cs}} + s_h$ with s_h defined by (4.11), we obtain, for all $v_h, w_h \in V_h$,

$$a_h^{\mathrm{sip}}(v_h, w_h) = \int_\Omega G_h^l(v_h) \cdot G_h^l(w_h) + \hat{s}_h^{\mathrm{sip}}(v_h, w_h), \qquad (4.46)$$

4.3. Liftings and Discrete Gradients

with

$$\hat{s}_h^{\text{sip}}(v_h, w_h) := \sum_{F \in \mathcal{F}_h} \frac{\eta}{h_F} \int_F [\![v_h]\!] [\![w_h]\!] - \int_\Omega \mathrm{R}_h^l([\![v_h]\!]) \cdot \mathrm{R}_h^l([\![w_h]\!]). \qquad (4.47)$$

The most natural choice for l appears to be $l = k-1$ since the broken gradient is in $[\mathbb{P}_d^{k-1}(\mathcal{T}_h)]^d$. The choice $l = k$ can facilitate the implementation of the method in that it allows one to use the same polynomial basis for computing the liftings and assembling the matrix.

The interest in using discrete gradients to formulate dG methods has been recognized recently in various contexts, e.g., by Lew, Neff, Sulsky, and Ortiz [232] and Ten Eyck and Lew [293] for linear and nonlinear elasticity, Buffa and Ortner [61] and Burman and Ern [65] for nonlinear variational problems, and the authors [131] for the Navier–Stokes equations; see also Agélas, Di Pietro, Eymard, and Masson [6]. The expression (4.46) of the SIP bilinear form plays a central role in Sect. 5.2 when analyzing the convergence to minimal regularity solutions. This expression is also useful in Sect. 4.4 in the context of a mixed dG approximation.

It is interesting to notice the following straightforward consequence of the bound (4.42).

Proposition 4.36 (Discrete coercivity). *For all $v_h \in V_h$,*

$$a_h^{\text{sip}}(v_h, v_h) \geq \|G_h^l(v_h)\|_{[L^2(\Omega)]^d}^2 + (\eta - C_{\text{tr}}^2 N_\partial)|v_h|_J^2.$$

In view of this result, the expression (4.46) for a_h^{sip} consists of two terms, both yielding a nonnegative contribution whenever $w_h = v_h$ and, as in Lemma 4.12, $\eta > C_{\text{tr}}^2 N_\partial$. The first term can be seen as the discrete counterpart of the exact bilinear form a (such that $a(v,w) = \int_\Omega \nabla v \cdot \nabla w$) and provides a control on the discrete gradient in $[L^2(\Omega)]^d$. The role of the second term is to strengthen the discrete stability of the method.

Remark 4.37 (Extension to broken Sobolev spaces). We emphasize that the definition (4.46) of a_h^{sip} is equivalent to (4.12) only at the discrete level. Differences occur when extending the definitions (4.12) and (4.46) to larger spaces, e.g., broken Sobolev spaces. As discussed in Sect. 4.2.1, the SIP bilinear form defined by (4.12) cannot be extended to the minimum regularity space $H^1(\Omega)$ because traces of gradients on mesh faces are used. Instead, the bilinear form defined by (4.46) can be extended to the broken Sobolev space $H^1(\mathcal{T}_h)$. We denote this extension by $\tilde{a}_h^{\text{sip}}$. Incidentally, $\tilde{a}_h^{\text{sip}}$ is no longer consistent. For convergence analysis to smooth solutions, Strang's First Lemma (see [285] or, e.g., Braess [49, p. 106]) dedicated to nonconsistent finite element methods can be used, whereby the consistency error is estimated for $u \in H^{k+1}(\Omega)$ as follows: For all $v_h \in V_h$,

$$\tilde{a}_h^{\text{sip}}(u - u_h, v_h) = \sum_{F \in \mathcal{F}_h} \int_F \{\!\!\{\nabla u - \pi_h(\nabla u)\}\!\!\} \cdot \mathbf{n}_F [\![v_h]\!] \leq C_u h^k |v_h|_J,$$

where π_h denotes the L^2-orthogonal projection onto V_h. As a result, the consistency error tends optimally to zero as the meshsize goes to zero.

4.3.4 Numerical Fluxes

DG methods can be viewed as high-order finite volume methods. The aim of this section is to identify the local conservation properties associated with dG methods. Such properties are important when the diffusive flux is to be used as an advective velocity in a transport problem, as discussed, e.g., by Dawson, Sun, and Wheeler [121] in the context of coupled porous media flow and contaminant transport. While most discretization methods possess local conservation properties, the specificity of dG methods, together with finite volume and mixed finite element methods, is to achieve local conservation at the element level as opposed to vertex-centered or face-centered macro-elements; see, e.g., Eymard, Hilhorst, and Vohralík [161].

Let $T \in \mathcal{T}_h$ and let $\xi \in \mathbb{P}_d^k(T)$. Integration by parts shows that, for the exact solution u,

$$\int_T f\xi = -\int_T (\triangle u)\xi = \int_T \nabla u \cdot \nabla \xi - \int_{\partial T} (\nabla u \cdot \mathrm{n}_T)\xi.$$

Therefore, defining on each mesh face $F \in \mathcal{F}_h$ the exact flux as

$$\Phi_F(u) := -\nabla u \cdot \mathrm{n}_F, \qquad (4.48)$$

and recalling the notation $\epsilon_{T,F} = \mathrm{n}_T \cdot \mathrm{n}_F$ introduced in Sect. 2.2.3, we infer

$$\int_T \nabla u \cdot \nabla \xi + \sum_{F \in \mathcal{F}_T} \epsilon_{T,F} \int_F \Phi_F(u)\xi = \int_T f\xi.$$

This is a local conservation property satisfied by the exact solution. Our goal is to identify a similar relation satisfied by the discrete solution u_h solving (4.14). Using $v_h = \xi\chi_T$ as test function in (4.14) (where χ_T denotes the characteristic function of T), observing that $\nabla_h(\xi\chi_T) = (\nabla\xi)\chi_T$, and recalling the definition (4.12) of a_h^{sip}, we obtain

$$\int_T f\xi = a_h^{\mathrm{sip}}(u_h, \xi\chi_T) = \int_T \nabla u_h \cdot \nabla\xi - \sum_{F \in \mathcal{F}_T} \int_F \{\!\{\nabla_h u_h\}\!\} \cdot \mathrm{n}_F [\![\xi\chi_T]\!]$$

$$- \sum_{F \in \mathcal{F}_T} \int_F \{\!\{(\nabla\xi)\chi_T\}\!\} \cdot \mathrm{n}_F [\![u_h]\!] + \sum_{F \in \mathcal{F}_T} \frac{\eta}{h_F} \int_F [\![u_h]\!][\![\xi\chi_T]\!].$$

Let $l \in \{k-1, k\}$. The first and third terms on the right-hand side sum up to $\int_T G_h^{k-1}(u_h) \cdot \nabla\xi$ since $\nabla\xi \in [\mathbb{P}_d^{k-1}(T)]^d$ and $l \geq k-1$, while in the second and fourth terms, we observe that $[\![\xi\chi_T]\!] = \epsilon_{T,F}\xi$. As a result, for all $T \in \mathcal{T}_h$ and all $\xi \in \mathbb{P}_d^k(T)$,

$$\int_T G_h^l(u_h) \cdot \nabla\xi + \sum_{F \in \mathcal{F}_T} \epsilon_{T,F} \int_F \phi_F(u_h)\xi = \int_T f\xi, \qquad (4.49)$$

with the numerical flux $\phi_F(u_h)$ defined as

$$\phi_F(u_h) := -\{\!\{\nabla_h u_h\}\!\} \cdot \mathrm{n}_F + \frac{\eta}{h_F}[\![u_h]\!]. \qquad (4.50)$$

4.4. Mixed dG Methods

We notice that the two contributions to $\phi_F(u_h)$ in (4.50) respectively stem from the consistency term and the penalty term (cf. Definition 4.7). Equation (4.49) is the local conservation property satisfied by the dG approximation. Interestingly, the expression (4.50) is consistent with (4.48) since, for the exact solution u, $\phi_F(u) = \Phi_F(u)$. We also observe that the local conservation property (4.49) is richer than that encountered in finite volume methods, which can be recovered by just taking $\xi \equiv 1$, i.e.,

$$\sum_{F \in \mathcal{F}_T} \epsilon_{T,F} \int_F \phi_F(u_h) = \int_T f. \tag{4.51}$$

4.4 Mixed dG Methods

In this section, we discuss mixed dG methods, that is, dG approximations to the mixed formulation (4.5) with the homogeneous Dirichlet boundary condition (4.1b). Other boundary conditions can be considered. Such methods produce an approximation u_h for the potential u and an approximation σ_h for the diffusive flux σ.

Definition 4.38 (Discrete potential and discrete diffusive flux). Consistently with Definition 4.1, the scalar-valued function u_h is termed the *discrete potential* and the vector-valued function σ_h the *discrete diffusive flux*.

First, we reformulate the SIP method of Sect. 4.2 as a mixed dG method and show how the discrete diffusive flux can be eliminated locally. Then, we formulate more general mixed dG methods in terms of local problems using numerical fluxes for the discrete potential and the discrete diffusive flux following Bassi, Rebay, and coworkers [34,35]. This leads, in particular, to the LDG methods introduced by Cockburn and Shu [112]. In these methods, the discrete diffusive flux can also be eliminated locally. Finally, we discuss hybrid mixed dG methods where additional degrees of freedom are introduced at interfaces, thereby allowing one to eliminate locally both the discrete potential and the discrete diffusive flux.

4.4.1 The SIP Method As a Mixed dG Method

One possible weak formulation of the mixed formulation (4.5) with the homogeneous Dirichlet boundary condition (4.1b) consists in finding $(\sigma, u) \in X := [L^2(\Omega)]^d \times H_0^1(\Omega)$ such that

$$\begin{cases} m(\sigma, \tau) + b(\tau, u) = 0 & \forall \tau \in [L^2(\Omega)]^d, \\ -b(\sigma, v) = \int_\Omega fv & \forall v \in H_0^1(\Omega), \end{cases} \tag{4.52}$$

where, for all $\sigma, \tau \in [L^2(\Omega)]^d$ and all $v \in H_0^1(\Omega)$, we have introduced the bilinear forms

$$m(\sigma, \tau) := \int_\Omega \sigma \cdot \tau, \qquad b(\tau, v) := \int_\Omega \tau \cdot \nabla v.$$

It is easily seen that $(\sigma, u) \in X$ solves (4.52) if and only if $\sigma = -\nabla u$ and u solves the weak problem (4.2).

At the discrete level, a mixed dG approximation can be designed as follows. We consider a polynomial degree $k \geq 1$ for the approximation of the potential and choose the polynomial degree l for the approximation of the diffusive flux such that $l \in \{k-1, k\}$. The relevant discrete spaces are

$$\Sigma_h := [\mathbb{P}_d^l(\mathcal{T}_h)]^d, \qquad U_h := \mathbb{P}_d^k(\mathcal{T}_h), \qquad X_h := \Sigma_h \times U_h.$$

The discrete problem consists in finding $(\sigma_h, u_h) \in X_h$ such that

$$\begin{cases} m(\sigma_h, \tau_h) + b_h(\tau_h, u_h) = 0 & \forall \tau_h \in \Sigma_h, \\ -b_h(\sigma_h, v_h) + \hat{s}_h^{\mathrm{sip}}(u_h, v_h) = \int_\Omega f v_h & \forall v_h \in U_h, \end{cases} \qquad (4.53)$$

with discrete bilinear form

$$b_h(\tau_h, v_h) := \int_\Omega \tau_h \cdot G_h^l(v_h),$$

where the discrete gradient operator G_h^l is defined by (4.44) and the stabilization bilinear form $\hat{s}_h^{\mathrm{sip}}$ by (4.47).

Proposition 4.39 (Elimination of discrete diffusive flux). *The pair $(\sigma_h, u_h) \in X_h$ solves (4.53) if and only if*

$$\sigma_h = -G_h^l(u_h), \qquad (4.54)$$

and $u_h \in U_h$ is such that

$$\int_\Omega G_h^l(u_h) \cdot G_h^l(v_h) + \hat{s}_h^{\mathrm{sip}}(u_h, v_h) = \int_\Omega f v_h \qquad \forall v_h \in U_h. \qquad (4.55)$$

Proof. The first equation in (4.53) yields

$$\int_\Omega (\sigma_h + G_h^l(u_h)) \cdot \tau_h = 0 \qquad \forall \tau_h \in \Sigma_h.$$

Recalling that $G_h^l(u_h) = \nabla_h u_h - R_h^l(\llbracket u_h \rrbracket)$ and since $l \geq k-1$, we infer that $G_h^l(u_h) \in \Sigma_h$; therefore, (4.54) is satisfied. Substituting this relation into the second equation of (4.53) yields (4.55). The converse is straightforward. □

Proposition 4.39 shows that the mixed dG method (4.53) is in fact equivalent to a problem in the sole unknown u_h. In particular, the above choice for b_h and $\hat{s}_h^{\mathrm{sip}}$ yields the SIP method of Sect. 4.2; cf. (4.46).

Remark 4.40 ($H(\mathrm{div}; \Omega)$-conformity of discrete diffusive flux). One drawback of mixed dG approximations, and in particular (4.53), is that the discrete diffusive flux $\sigma_h = -G_h^l(u_h)$ is not in $H(\mathrm{div}; \Omega)$ because its normal component is, in general, discontinuous across interfaces. This point is further examined in Sect. 5.5 where we discuss a cost-effective, locally conservative diffusive flux reconstruction obtained by postprocessing the discrete potential.

4.4.2 Numerical Fluxes

In what follows, we focus for simplicity on equal-order approximations for the potential and the diffusive flux, that is, we set $l = k$ so that $\Sigma_h := [\mathbb{P}_d^k(\mathcal{T}_h)]^d$, while, as before, $U_h := \mathbb{P}_d^k(\mathcal{T}_h)$. Similarly to Sect. 4.3.4, we can derive a local formulation by localizing test functions to a single mesh element. Let $T \in \mathcal{T}_h$, let $\zeta \in [\mathbb{P}_d^k(T)]^d$, and let $\xi \in \mathbb{P}_d^k(T)$. Integrating by parts in T, splitting the boundary integral on ∂T as a sum over the mesh faces $F \in \mathcal{F}_T$, and setting $\epsilon_{T,F} = n_T \cdot n_F$, we infer for the exact solution that

$$\int_T \sigma \cdot \zeta - \int_T u \nabla \cdot \zeta + \sum_{F \in \mathcal{F}_T} \epsilon_{T,F} \int_F u_F (\zeta \cdot n_F) = 0,$$

$$-\int_T \sigma \cdot \nabla \xi + \sum_{F \in \mathcal{F}_T} \epsilon_{T,F} \int_F (\sigma_F \cdot n_F) \xi = \int_T f \xi,$$

since $\sigma = -\nabla u$ and $\nabla \cdot \sigma = f$. The traces u_F and $\sigma_F \cdot n_F$ are single-valued on each interface; cf. Lemma 4.3.

At the discrete level, the general form of the mixed dG approximation is derived by introducing numerical fluxes for the discrete potential and for the discrete diffusive flux. These two numerical fluxes, which are denoted by $\hat{u}_F$ and $\hat{\sigma}_F$ for all $F \in \mathcal{F}_h$, are single-valued on each $F \in \mathcal{F}_h$. The numerical flux $\hat{u}_F$ is scalar-valued and the numerical flux $\hat{\sigma}_F$ is vector-valued. We obtain, for all $T \in \mathcal{T}_h$, all $\zeta \in [\mathbb{P}_d^k(T)]^d$, and all $\xi \in \mathbb{P}_d^k(T)$,

$$\int_T \sigma_h \cdot \zeta - \int_T u_h \nabla \cdot \zeta + \sum_{F \in \mathcal{F}_T} \epsilon_{T,F} \int_F \hat{u}_F (\zeta \cdot n_F) = 0, \qquad (4.56\text{a})$$

$$-\int_T \sigma_h \cdot \nabla \xi + \sum_{F \in \mathcal{F}_T} \epsilon_{T,F} \int_F (\hat{\sigma}_F \cdot n_F) \xi = \int_T f \xi. \qquad (4.56\text{b})$$

Lemma 4.41 (Numerical fluxes for SIP)**.** *For the SIP method, the numerical fluxes are given by*

$$\hat{u}_F = \begin{cases} \{\!\{u_h\}\!\} & \forall F \in \mathcal{F}_h^i, \\ 0 & \forall F \in \mathcal{F}_h^b, \end{cases} \qquad (4.57\text{a})$$

$$\hat{\sigma}_F = -\{\!\{\nabla_h u_h\}\!\} + \eta h_F^{-1} [\![u_h]\!] n_F \quad \forall F \in \mathcal{F}_h. \qquad (4.57\text{b})$$

Proof. The assertion is obtained by testing the first equation in (4.53) with $\tau_h = \zeta \chi_T$, where χ_T denotes the characteristic function of T, and testing the second equation with $v_h = \xi \chi_T$. □

A first possible variant of the SIP method consists in keeping the definition (4.57a) for the numerical flux $\hat{u}_F$ and defining the numerical flux $\hat{\sigma}_F$ as

$$\hat{\sigma}_F = \{\!\{\sigma_h\}\!\} + \eta h_F^{-1} [\![u_h]\!] n_F.$$

The resulting dG method belongs to the class of LDG methods. The discrete diffusive flux σ_h can still be eliminated locally (since the numerical flux $\hat{u}_F$ only depends on u_h), and the discrete potential $u_h \in U_h$ is such that

$$a_h^{\text{ldg}}(u_h, v_h) = \int_\Omega f v_h \qquad \forall v_h \in U_h,$$

with the discrete bilinear form

$$a_h^{\text{ldg}}(u_h, v_h) = \int_\Omega \nabla_h u_h \cdot \nabla_h v_h - \sum_{F \in \mathcal{F}_h} \int_F (\{\!\{\nabla_h u_h\}\!\} \cdot \mathbf{n}_F [\![v_h]\!] + \{\!\{\nabla_h v_h\}\!\} \cdot \mathbf{n}_F [\![u_h]\!])$$
$$+ \int_\Omega \mathrm{R}_h^k([\![u_h]\!]) \cdot \mathrm{R}_h^k([\![v_h]\!]) + \sum_{F \in \mathcal{F}_h} \frac{\eta}{h_F} \int_F [\![u_h]\!][\![v_h]\!]$$
$$= \int_\Omega G_h^k(u_h) \cdot G_h^k(v_h) + \sum_{F \in \mathcal{F}_h} \frac{\eta}{h_F} \int_F [\![u_h]\!][\![v_h]\!].$$

A nice feature of the discrete bilinear form a_h^{ldg} is that discrete coercivity holds on U_h with respect to the $\|\cdot\|_{\text{sip}}$-norm for any $\eta > 0$ (a simple choice is $\eta = 1$). The drawback is that the elementary *stencil* associated with the term $\int_\Omega \mathrm{R}_h^k([\![u_h]\!]) \cdot \mathrm{R}_h^k([\![v_h]\!])$ consists of a given mesh element, its neighbors, and the neighbors of its neighbors in the sense of faces; cf. Fig. 4.2. Such a stencil is considerably larger than that associated with the SIP method; cf. Fig. 4.1.

More general forms of the LDG method can be designed with the numerical fluxes

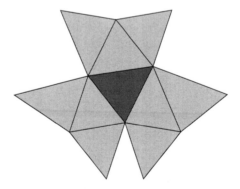

Fig. 4.2: Example of LDG stencil of an element $T \in \mathcal{T}_h$ when $\mathcal{T}_h$ is a matching triangular mesh; the mesh element is highlighted in *dark grey*, and all the nine other elements, highlighted in *light grey*, also belong to the stencil (compare with Fig. 4.1)

4.4. Mixed dG Methods

$$\hat{u}_F = \begin{cases} \{\!\{u_h\}\!\} + \Upsilon \cdot \mathrm{n}_F [\![u_h]\!] & \forall F \in \mathcal{F}_h^i, \\ 0 & \forall F \in \mathcal{F}_h^b, \end{cases}$$

$$\hat{\sigma}_F = \begin{cases} \{\!\{\sigma_h\}\!\} - \Upsilon[\![\sigma_h]\!] \cdot \mathrm{n}_F + \eta h_F^{-1} [\![u_h]\!] \mathrm{n}_F & \forall F \in \mathcal{F}_h^i, \\ \sigma_h + \eta h_F^{-1} u_h \mathrm{n} & \forall F \in \mathcal{F}_h^b, \end{cases}$$

where Υ is vector-valued and $\eta > 0$ is scalar-valued (in LDG methods, ηh_F^{-1} is often denoted by C_{11} and Υ by C_{12}). Since the numerical flux $\hat{u}_F$ only depends on u_h, the discrete diffusive flux σ_h can be eliminated locally. The above form of the diffusive fluxes ensures symmetry and discrete stability for the resulting dG method. A simple choice for the penalty parameter is again $\eta = 1$, while the auxiliary vector-parameter Υ can be freely chosen. LDG methods for the Poisson problem have been extensively analyzed by Castillo, Cockburn, Perugia, and Schötzau [80] (see also Castillo, Cockburn, Schötzau, and Schwab [81] for the hp-analysis of LDG methods applied to diffusion-advection problems). In [80], various choices of the penalty parameter C_{11} are discussed; the above choice $C_{11} = \eta h_F^{-1}$ leads to the same energy-norm error estimates as for the SIP method. A particular choice for the vector-parameter Υ leading to superconvergence on Cartesian grids has been studied by Cockburn, Kanschat, Perugia, and Schötzau [100]. Variants of the LDG method aiming at reducing the stencil have been discussed by Sherwin, Kirby, Peiró, Taylor, and Zienkiewicz [277], Peraire and Persson [256], and Castillo [79].

A further variant of the SIP and LDG methods consists in considering the numerical fluxes

$$\hat{u}_F = \begin{cases} \{\!\{u_h\}\!\} + \eta_\sigma [\![\sigma_h]\!] \cdot \mathrm{n}_F & \forall F \in \mathcal{F}_h^i, \\ 0 & \forall F \in \mathcal{F}_h^b, \end{cases}$$

$$\hat{\sigma}_F = \{\!\{\sigma_h\}\!\} + \eta_u [\![u_h]\!] \mathrm{n}_F \quad \forall F \in \mathcal{F}_h.$$

Here, the penalty parameters η_u and η_σ are positive user-dependent real numbers, and a simple choice is to set $\eta_u = \eta_\sigma = 1$. This method is analyzed in Sect. 7.3 in the more general context of Friedrichs' systems. Because the numerical flux $\hat{u}_F$ depends on σ_h, (4.56a) can no longer be used to express locally the discrete diffusive flux σ_h in terms of the discrete potential u_h. This precludes the local elimination of σ_h and, therefore, enhances the computational cost of the approximation method. The approach presents, however, some advantages since it can be used with polynomial degree $k = 0$ and there is no minimal threshold on the penalty parameters (apart from being positive). Moreover, the approximation on the diffusive flux is more accurate yielding convergence rates in the L^2-norm of order $h^{k+1/2}$ for smooth solutions, as opposed to the convergence rates of order h^k delivered by the SIP method (cf. (4.26)).

Finally, we mention that an even more general presentation can allow for two-valued numerical fluxes at interfaces; see Arnold, Brezzi, Cockburn, and Marini [16] for a unified analysis of dG methods. Two-valued numerical fluxes are obtained, for instance, when rewriting the nonsymmetric dG methods discussed in Sect. 5.3.1 as mixed dG methods.

4.4.3 Hybrid Mixed dG Methods

The key idea in hybrid mixed dG methods is to introduce additional degrees of freedom at interfaces, thereby allowing one to eliminate locally both the discrete potential and the discrete diffusive flux. Herein, we focus on the HDG methods introduced by Cockburn, Gopalakrishnan, and Lazarov [97]; see also Causin and Sacco [83] for a different approach based on a discontinuous Petrov–Galerkin formulation, Droniou and Eymard [135] for similar ideas in the context of hybrid mixed finite volume schemes, and Ewing, Wang, and Yang for hybrid primal dG methods [154].

In the HDG method, the additional degrees of freedom are used to enforce the continuity of the normal component of the discrete diffusive flux. These additional degrees of freedom act as Lagrange multipliers in the discrete problem and can be interpreted as single-valued traces of the discrete potential on interfaces. We introduce the discrete space

$$\Lambda_h := \bigoplus_{F \in \mathcal{F}_h^i} \mathbb{P}_{d-1}^k(F).$$

A function $\mu_h \in \Lambda_h$ is such that, for all $F \in \mathcal{F}_h^i$, $\mu_h|_F \in \mathbb{P}_{d-1}^k(F)$. The discrete unknowns $(\sigma_h, u_h, \lambda_h) \in \Sigma_h \times U_h \times \Lambda_h$ satisfy the following local problems: For all $T \in \mathcal{T}_h$, all $\zeta \in [\mathbb{P}_d^k(T)]^d$, and all $\xi \in \mathbb{P}_d^k(T)$,

$$\int_T \sigma_h \cdot \zeta - \int_T u_h \nabla \cdot \zeta + \sum_{F \in \mathcal{F}_T} \epsilon_{T,F} \int_F \hat{u}_F(\zeta \cdot n_F) = 0, \qquad (4.58a)$$

$$-\int_T \sigma_h \cdot \nabla \xi + \sum_{F \in \mathcal{F}_T} \epsilon_{T,F} \int_F (\hat{\sigma}_{T,F} \cdot n_F) \xi = \int_T f\xi, \qquad (4.58b)$$

while normal diffusive flux continuity is enforced by setting, for all $F \in \mathcal{F}_T \cap \mathcal{F}_h^i$ and all $\mu \in \mathbb{P}_{d-1}^k(F)$,

$$\int_F [\![\hat{\sigma}_{T,F}]\!] \cdot n_F \mu = 0. \qquad (4.59)$$

Here, the numerical fluxes are such that

$$\hat{u}_F = \begin{cases} \lambda_h & \forall F \in \mathcal{F}_h^i, \\ 0 & \forall F \in \mathcal{F}_h^b, \end{cases} \qquad (4.60a)$$

$$\hat{\sigma}_{T,F} = \sigma_h|_T + \tau_T(u_h|_T - \hat{u}_F)n_T \quad \forall F \in \mathcal{F}_h, \qquad (4.60b)$$

with penalty parameter τ_T defined elementwise. We observe that (4.59) indeed enforces $[\![\hat{\sigma}_{T,F}]\!] \cdot n_F = 0$ for all $F \in \mathcal{F}_h^i$ since $[\![\hat{\sigma}_{T,F}]\!] \cdot n_F \in \mathbb{P}_{d-1}^k(F)$. As a result, the quantity $(\hat{\sigma}_{T,F} \cdot n_F)$ in (4.58b) is actually single-valued.

4.4. Mixed dG Methods

Lemma 4.42 (HDG as mixed dG method). *Let $(\sigma_h, u_h, \lambda_h) \in \Sigma_h \times U_h \times \Lambda_h$ solve (4.58) and (4.59). Then, the pair $(\sigma_h, u_h) \in \Sigma_h \times U_h$ solves the local problems of the mixed dG formulation (4.56) with numerical fluxes such that, for all $F \in \mathcal{F}_h^i$ with $F = \partial T_1 \cap \partial T_2$,*

$$\hat{u}_F = \{u_h\} + C_{12} \cdot [\![u_h]\!] \mathrm{n}_F + C_{22} [\![\sigma_h]\!] \cdot \mathrm{n}_F, \qquad (4.61a)$$
$$\hat{\sigma}_F = \{\sigma_h\} + C_{11} [\![u_h]\!] \mathrm{n}_F - C_{12} [\![\sigma_h]\!] \cdot \mathrm{n}_F, \qquad (4.61b)$$

with the parameters

$$C_{11} = \frac{\tau_1 \tau_2}{\tau_1 + \tau_2}, \qquad C_{12} = \frac{\tau_1 - \tau_2}{2(\tau_1 + \tau_2)} \mathrm{n}_F, \qquad C_{22} = \frac{1}{\tau_1 + \tau_2},$$

where $\tau_i := \tau_{T_i}$, $i \in \{1, 2\}$. Moreover, for all $F \in \mathcal{F}_h^b$ with $F = \partial T \cap \partial \Omega$, $\hat{u}_F = 0$ and $\hat{\sigma}_F = \sigma_h + \tau_T u_h$.

Proof. Since $[\![\hat{\sigma}_{T,F}]\!] \cdot \mathrm{n}_F = 0$, we infer from (4.60b) that

$$[\![\sigma_h]\!] \cdot \mathrm{n}_F + 2\{\tau u_h\} - 2\{\tau\} \hat{u}_F = 0.$$

Observing that $\{\tau u_h\} = \{\tau\}\{u_h\} + \frac{1}{4}[\![\tau]\!][\![u_h]\!]$, we obtain

$$\hat{u}_F = \{u_h\} + \frac{1}{4} \frac{[\![\tau]\!]}{\{\tau\}} [\![u_h]\!] + \frac{1}{2\{\tau\}} [\![\sigma_h]\!] \cdot \mathrm{n}_F,$$

which yields (4.61a). Moreover, since the normal component of $\hat{\sigma}_{T,F}$ is single-valued, we infer

$$\hat{\sigma}_{T,F} \cdot \mathrm{n}_F = \{\sigma_h\} \cdot \mathrm{n}_F + \frac{1}{2}[\![\tau u_h]\!] - \frac{1}{2}[\![\tau]\!] \hat{u}_F.$$

Observing that $[\![\tau u_h]\!] = [\![\tau]\!]\{u_h\} + \{\tau\}[\![u_h]\!]$, we obtain

$$\hat{\sigma}_{T,F} \cdot \mathrm{n}_F = \{\sigma_h\} \cdot \mathrm{n}_F + \frac{1}{2}[\![\tau]\!](\{u_h\} - \hat{u}_F) + \frac{1}{2}\{\tau\}[\![u_h]\!].$$

Using (4.61a) to evaluate $\hat{u}_F$ in this expression and rearranging terms leads to

$$\hat{\sigma}_{T,F} \cdot \mathrm{n}_F = \{\sigma_h\} \cdot \mathrm{n}_F + \frac{\tau_1 \tau_2}{\tau_1 + \tau_2} [\![u_h]\!] - \frac{\tau_1 - \tau_2}{2(\tau_1 + \tau_2)} [\![\sigma_h]\!] \cdot \mathrm{n}_F.$$

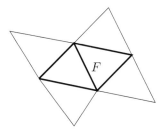

Fig. 4.3: Stencil $\mathcal{S}(F)$ for HDG methods

An inspection of (4.61b) shows that $\hat{\sigma}_{T,F}\cdot n_F = \hat{\sigma}_F\cdot n_F$, and this concludes the proof. □

We observe that the numerical flux $\hat{u}_F$ in (4.61a) depends on σ_h since $C_{22} \neq 0$. As a result, the discrete diffusive flux cannot be eliminated locally to derive a discrete problem for the sole discrete potential. Instead, a computationally efficient implementation of HDG methods consists in using (4.58) to eliminate locally both the discrete potential and the discrete diffusive flux by static condensation (similarly to mixed finite element methods; see, e.g., Arnold and Brezzi [15]), so as to obtain, using (4.59), a discrete problem where the sole unknown is $\lambda_h \in \Lambda_h$. For a given interface $F \in \mathcal{F}_h^i$ with $F = \partial T_1 \cap \partial T_2$, the stencil associated with this interface is (cf. Fig. 4.3)

$$\mathcal{S}(F) = \{F' \in \mathcal{F}_h^i \mid F' \in \mathcal{F}_{T_1} \cup \mathcal{F}_{T_2}\}.$$

For matching simplicial meshes, the set $\mathcal{S}(F)$ contains five interfaces for $d = 2$ and seven interfaces for $d = 3$.

HDG methods for elliptic problems have been analyzed by Cockburn, Dong, and Guzmán [95] and Cockburn, Guzmán, and Wang [98] where error estimates in various norms are derived for various choices of the penalty parameter τ. In particular, L^2-norm error estimates of order h^{k+1} can be derived both for the potential and the diffusive flux for smooth solutions and polynomial order $k \geq 0$. Moreover, for $k \geq 1$, a postprocessed potential converging at order h^{k+2} can be derived, similarly to classical mixed finite element methods. The extension of HDG methods to diffusion-advection methods is investigated by Cockburn, Dong, Guzmán, Restelli, and Sacco [96] and Nguyen, Peraire, and Cockburn [245].

4.5 Heterogeneous Diffusion

In this section, we consider a model problem with heterogeneous diffusion. To approximate this problem using dG methods, we revisit the design and analysis of the SIP method considered in Sect. 4.2 for the Poisson problem. Following Dryja [136], we use diffusion-dependent weights to formulate the consistency and symmetry terms in the discrete bilinear form and we penalize interface and boundary jumps using a diffusion-dependent parameter scaling as the harmonic mean of the diffusion coefficient. Such a penalty strategy is particularly important in heterogeneous diffusion-advection-reaction equations (cf. Sect. 4.6) where the diffusion coefficient takes locally small values leading to so-called advection-dominated regimes. In this context, the exact solution exhibits sharp inner layers which, in practice, are not resolved by the underlying meshes, so that excessive penalty at such layers triggers spurious oscillations. Using the harmonic mean of the diffusion coefficient to penalize jumps turns out to tune automatically the amount of penalty and thereby avoid such oscillations. Incidentally, we also observe that, in finite volume and mixed finite element schemes, the harmonic mean of the diffusion coefficient is often considered at interfaces.

4.5.1 The Continuous Setting

Let $\kappa \in L^\infty(\Omega)$ be the diffusion coefficient and assume that κ is uniformly bounded from below in Ω by a positive real number. The anisotropic case, where κ is actually $\mathbb{R}^{d,d}$-valued, is examined in Sect. 4.5.6. We are interested in the problem

$$-\nabla\cdot(\kappa \nabla u) = f \quad \text{in } \Omega,$$
$$u = 0 \quad \text{on } \partial\Omega,$$

with source term $f \in L^2(\Omega)$. The weak form of this problem is

$$\text{Find } u \in V \text{ s.t. } a(u,v) = \int_\Omega fv \text{ for all } v \in V, \quad (4.62)$$

with energy space $V = H_0^1(\Omega)$ and bilinear form

$$a(u,v) := \int_\Omega \kappa \nabla u \cdot \nabla v.$$

Owing to the above assumptions on the diffusion coefficient κ, the Lax–Milgram Lemma implies that (4.62) is well-posed. The case where κ is constant in Ω yields, up to rescaling, the Poisson problem; the latter can thus be viewed as a prototype for homogeneous diffusion problems.

Adopting the terminology used for the Poisson problem (cf. Definition 4.1), the $\mathbb{R}^d$-valued function

$$\sigma := -\kappa \nabla u$$

is termed the *diffusive flux*. By construction, $\sigma \in H(\text{div}; \Omega)$.

In practice, the diffusion coefficient has more regularity than just belonging to $L^\infty(\Omega)$. Henceforth, we make the following assumption.

Assumption 4.43 (Partition of Ω). *There is a partition $P_\Omega := \{\Omega_i\}_{1 \leq i \leq N_\Omega}$ of Ω such that*

(i) *Each Ω_i, $1 \leq i \leq N_\Omega$, is a polyhedron;*

(ii) *The restriction of κ to each Ω_i, $1 \leq i \leq N_\Omega$, is constant.*

Remark 4.44 (Motivation for assumption 4.43). In groundwater flow applications, the partition P_Ω results for instance from the partitioning of the porous medium into various geological layers.

From a physical viewpoint, the normal component of the diffusive flux σ is continuous across any interface $\partial\Omega_i \cap \partial\Omega_j$ with positive $(d-1)$-dimensional Hausdorff measure. Assuming $\kappa|_{\Omega_i} \neq \kappa|_{\Omega_j}$, this implies that the normal component of ∇u cannot be continuous across this interface. This fact modifies the regularity that can be expected for the exact solution in heterogeneous diffusion problems with respect to the Poisson problem.

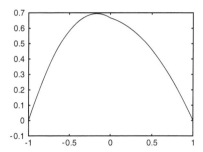 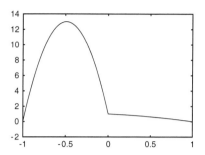

Fig. 4.4: Exact solution with diffusion heterogeneity parameter $\alpha = 0.5$ (*left*) and $\alpha = 0.01$ (*right*); the two panels use different vertical scales

4.5.1.1 One-Dimensional Example

Let $\Omega = (-1, 1)$ be partitioned into two subdomains $\Omega_1 = (-1, 0)$ and $\Omega_2 = (0, 1)$ such that $\kappa|_{\Omega_1} = \alpha$ and $\kappa|_{\Omega_2} = 1$ with positive parameter α. The exact solution of (4.62) with $f \equiv 1$ is

$$u(x) = \begin{cases} a_1(1+x)^2 + b_1(1+x) & \text{if } x \in \Omega_1, \\ a_2(x-1)^2 + b_2(x-1) & \text{if } x \in \Omega_2, \end{cases}$$

where $a_1 = -\frac{1}{2\alpha}$, $a_2 = -\frac{1}{2}$, $b_1 = \frac{1+3\alpha}{2\alpha(1+\alpha)}$, and $b_2 = -\frac{\alpha+3}{2(1+\alpha)}$. Figure 4.4 presents the exact solutions obtained with $\alpha = 0.5$ (mild diffusion heterogeneity) and $\alpha = 0.01$ (strong diffusion heterogeneity). As expected, the exact solution is only continuous at $x = 0$, but not differentiable, and the jump in the derivative of the exact solution is more pronounced in the case of strong diffusion heterogeneity. Interestingly, the maximum value attained by the exact solution in Ω is substantially affected by the diffusion heterogeneity.

4.5.1.2 Two-Dimensional Example

In dimension $d \geq 2$, discontinuities in the diffusion coefficient can cause severe singularities in the exact solution. Exact solutions of two-dimensional heterogeneous diffusion problems with zero right-hand side are explicitly constructed by Kellogg [210]. A typical situation is the case where $\Omega = (-1, 1)^2$ is divided into four quadrants, and the diffusion coefficient takes the value κ_1 in the first and third quadrants and the value κ_2 in the second and fourth quadrants. Then, it is possible to construct an exact solution with zero source term and suitable nonhomogeneous Dirichlet boundary conditions such that, in polar coordinates, $u(r, \theta) = r^\gamma v(\theta)$ with a smooth function v. The exponent $\gamma > 0$ can be made as small as desired by taking large values of the ratio κ_1/κ_2. Figure 4.5 illustrates the exact solution for $\kappa_1/\kappa_2 = 5$ (left) and $\kappa_1/\kappa_2 = 100$ (right). In dimension 2, regularity results take the form $u \in H^{1+\epsilon}(\Omega)$ with $\epsilon > 0$ but arbitrary small. In dimension 3, regularity results have been obtained by Nicaise and Sändig [247] in some particular situations.

4.5. Heterogeneous Diffusion

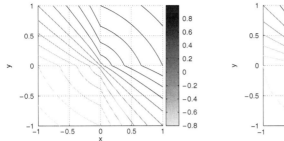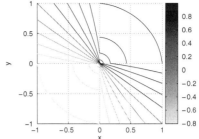

Fig. 4.5: Exact solution for heterogeneous diffusion problem; $\Omega = (-1, 1)^2$ is divided into four quadrants, and the diffusion coefficient takes the value κ_1 in the first and third quadrants and the value κ_2 in the second and fourth quadrants; *left*: $\kappa_1/\kappa_2 = 5$; *right*: $\kappa_1/\kappa_2 = 100$ (courtesy M. Vohralík)

4.5.2 Discretization

We aim at approximating the exact solution u of (4.62) by a dG method using the discrete space
$$V_h := \mathbb{P}_d^k(\mathcal{T}_h),$$
where $\mathbb{P}_d^k(\mathcal{T}_h)$ is defined by (1.15) with polynomial degree $k \geq 1$ and $\mathcal{T}_h$ belonging to an admissible mesh sequence. We consider the discrete problem:

$$\text{Find } u_h \in V_h \text{ s.t. } a_h^{\text{swip}}(u_h, v_h) = \int_\Omega f v_h \text{ for all } v_h \in V_h, \quad (4.63)$$

with the discrete bilinear form a_h^{swip} yet to be designed.

4.5.2.1 Mesh Compatibility

An important assumption on the mesh sequence $\mathcal{T}_\mathcal{H} := (\mathcal{T}_h)_{h \in \mathcal{H}}$ is its compatibility with the partition P_Ω.

Assumption 4.45 (Mesh compatibility). *We suppose that the admissible mesh sequence $\mathcal{T}_\mathcal{H}$ is such that, for each $h \in \mathcal{H}$, each $T \in \mathcal{T}_h$ is a subset of only one set Ω_i of the partition P_Ω. In this situation, the meshes are said to be* compatible *with the partition P_Ω.*

An example of compatible mesh is presented in Fig. 4.6. The motivation for the above assumption is to prevent jumps of the diffusion coefficient κ to occur inside mesh elements. Indeed, owing to Assumption 4.43, the diffusion coefficient is piecewise constant on each mesh $\mathcal{T}_h$. This fact is often used in what follows. The present setting can be enlarged, at the price of additional technicalities, by assuming that the diffusion coefficient is piecewise smooth (e.g., piecewise Lipschitz continuous). However, it is not reasonable to envisage a high-order dG method to approximate an heterogeneous diffusion problem if the mesh is not

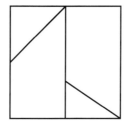 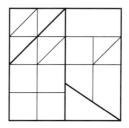

Fig. 4.6: Partition P_Ω (*left*) and compatible mesh (*right*)

compatible with the singularities of the diffusion coefficient. Indeed, the exact solution is not expected to be sufficiently smooth across these singularities to exploit the local degrees of freedom in the polynomial space.

4.5.2.2 Weighted Averages

While we keep Definitions 1.17 and 4.2 for interface and boundary jumps respectively, it is convenient to introduce weighted averages.

Definition 4.46 (Weighted averages)**.** To any interface $F \in \mathcal{F}_h^i$ with $F = \partial T_1 \cap \partial T_2$, we assign two nonnegative real numbers $\omega_{T_1,F}$ and $\omega_{T_2,F}$ such that

$$\omega_{T_1,F} + \omega_{T_2,F} = 1.$$

Then, for any scalar-valued function v defined on Ω that is smooth enough to admit a possibly two-valued trace on all $F \in \mathcal{F}_h^i$, we define its *weighted average* on F such that, for a.e. $x \in F$,

$$\{\!\{v\}\!\}_{\omega,F}(x) := \omega_{T_1,F} v|_{T_1}(x) + \omega_{T_2,F} v|_{T_2}(x).$$

On boundary faces $F \in \mathcal{F}_h^b$ with $F = \partial T \cap \partial \Omega$, we set $\{\!\{v\}\!\}_{\omega,F}(x) := v|_T(x)$. When v is vector-valued, the weighted average operator acts componentwise on the function v. Whenever no confusion can arise, the subscript F and the variable x are omitted and we simply write $\{\!\{v\}\!\}_\omega$.

Clearly, the usual (arithmetic) average of Definition 1.17 at interfaces corresponds to the particular choice $\omega_{T_1,F} = \omega_{T_2,F} = 1/2$. Henceforth, we consider a specific diffusion-dependent choice for the weights, namely, for all $F \in \mathcal{F}_h^i$, $F = \partial T_1 \cap \partial T_2$,

$$\omega_{T_1,F} := \frac{\kappa_2}{\kappa_1 + \kappa_2}, \qquad \omega_{T_2,F} := \frac{\kappa_1}{\kappa_1 + \kappa_2},$$

where $\kappa_i = \kappa|_{T_i}$, $i \in \{1, 2\}$. In particular, the case of homogeneous diffusion yields the usual (arithmetic) averages.

4.5.2.3 The SWIP Bilinear Form

In the context of heterogeneous diffusion problems, we modify the SIP bilinear form defined by (4.12) as follows: For all $(v_h, y_h) \in V_h \times V_h$,

$$a_h^{\text{swip}}(v_h, y_h) := \int_\Omega \kappa \nabla_h v_h \cdot \nabla_h y_h + \sum_{F \in \mathcal{F}_h} \eta \frac{\gamma_{\kappa,F}}{h_F} \int_F [\![v_h]\!][\![y_h]\!] \qquad (4.64)$$

$$- \sum_{F \in \mathcal{F}_h} \int_F \left(\{\!\{\kappa \nabla_h v_h\}\!\}_\omega \cdot \mathrm{n}_F [\![y_h]\!] + [\![v_h]\!] \{\!\{\kappa \nabla_h y_h\}\!\}_\omega \cdot \mathrm{n}_F \right).$$

The quantity $\eta > 0$ denotes a user-dependent penalty parameter which is independent of the diffusion coefficient, while the diffusion-dependent penalty parameter $\gamma_{\kappa,F}$ is such that for all $F \in \mathcal{F}_h^i$, $F = \partial T_1 \cap \partial T_2$,

$$\gamma_{\kappa,F} := \frac{2\kappa_1 \kappa_2}{\kappa_1 + \kappa_2},$$

where, as above, $\kappa_i = \kappa|_{T_i}$, $i \in \{1, 2\}$, while, for all $F \in \mathcal{F}_h^b$, $F = \partial T \cap \partial \Omega$,

$$\gamma_{\kappa,F} := \kappa|_T.$$

We notice that the above choice for the penalty parameter $\gamma_{\kappa,F}$ on interfaces corresponds to the *harmonic mean* of the values of the diffusion coefficient on either side of the interface. Furthermore, we observe that, for all $F \in \mathcal{F}_h^i$,

$$\gamma_{\kappa,F} \leq 2 \min(\kappa_1, \kappa_2). \qquad (4.65)$$

This property is used in the convergence analysis of Sect. 4.6.3 in the context of diffusion-advection-reaction problems; cf., in particular, Remark 4.65.

The bilinear form a_h^{swip} defined by (4.64) is termed the Symmetric Weighted Interior Penalty (SWIP) bilinear form. It has been introduced by Dryja [136] for heterogeneous diffusion problems and analyzed (in the more general context of diffusion-advection-reaction problems) by Di Pietro, Ern, and Guermond [133] and Ern, Stephansen, and Zunino [150]. The two differences with respect to the more usual SIP bilinear form are the use of (diffusion-dependent) weighted averages to formulate the consistency and symmetry terms and the presence of the diffusion-dependent penalty parameter. Whenever κ is constant in Ω, the usual (arithmetic) averages are recovered in the consistency and symmetry terms. The possibility of using non-arithmetic averages in dG methods has been pointed out and used in various contexts, e.g., by Stenberg [282], by Heinrich and co-workers [188–190], and by Hansbo and Hansbo [182] in the context of unfitted finite element methods based on Nitsche's method. The idea of connecting the actual value of the weights to the diffusion coefficient was also considered by Burman and Zunino [73] in the context of mortaring techniques for a singularly perturbed diffusion-advection equation.

Lemma 4.47 (Reformulation of SWIP bilinear form). *There holds, for all* $(v_h, y_h) \in V_h \times V_h$,

$$a_h^{\text{swip}}(v_h, y_h) = -\sum_{T \in \mathcal{T}_h} \int_T \nabla \cdot (\kappa \nabla v_h) y_h + \sum_{F \in \mathcal{F}_h} \eta \frac{\gamma_{\kappa,F}}{h_F} \int_F [\![v_h]\!][\![y_h]\!] \qquad (4.66)$$
$$+ \sum_{F \in \mathcal{F}_h^i} \int_F [\![\kappa \nabla_h v_h]\!] \cdot \mathbf{n}_F \{\!\{y_h\}\!\}_{\overline{\omega}} - \sum_{F \in \mathcal{F}_h} \int_F [\![v_h]\!]\{\!\{\kappa \nabla_h y_h\}\!\}_\omega \cdot \mathbf{n}_F,$$

where $\{\!\{y_h\}\!\}_{\overline{\omega}}$ *is the skew-weighted average value of* y_h *defined as*

$$\{\!\{y_h\}\!\}_{\overline{\omega}} := \omega_{T_2,F} y_h|_{T_1} + \omega_{T_1,F} y_h|_{T_2}.$$

Proof. Integrating by parts the first term in (4.64) yields

$$\int_\Omega \kappa \nabla_h v_h \cdot \nabla_h y_h = -\sum_{T \in \mathcal{T}_h} \int_T \nabla \cdot (\kappa \nabla v_h) y_h + \sum_{T \in \mathcal{T}_h} \int_{\partial T} \kappa (\nabla v_h \cdot \mathbf{n}_T) y_h. \qquad (4.67)$$

Rearranging the second term on the right-hand side as a sum over mesh faces leads to

$$\sum_{T \in \mathcal{T}_h} \int_{\partial T} \kappa(\nabla v_h \cdot \mathbf{n}_T) y_h = \sum_{F \in \mathcal{F}_h^i} \int_F [\![(\kappa \nabla_h v_h) y_h]\!] \cdot \mathbf{n}_F + \sum_{F \in \mathcal{F}_h^b} \int_F \kappa(\nabla v_h \cdot \mathbf{n}) y_h.$$

We now observe that, for all $F \in \mathcal{F}_h^i$,

$$[\![(\kappa \nabla_h v_h) y_h]\!] = \{\!\{\kappa \nabla_h v_h\}\!\}_\omega [\![y_h]\!] + [\![\kappa \nabla_h v_h]\!]\{\!\{y_h\}\!\}_{\overline{\omega}}.$$

To prove this identity, we set $a_i = (\kappa \nabla_h v_h)|_{T_i}$, $b_i = y_h|_{T_i}$, $\omega_i = \omega_{T_i,F}$, $i \in \{1,2\}$, so that

$$[\![(\kappa \nabla_h v_h) y_h]\!] = a_1 b_1 - a_2 b_2$$
$$= (\omega_1 a_1 + \omega_2 a_2)(b_1 - b_2) + (a_1 - a_2)(\omega_2 b_1 + \omega_1 b_2)$$
$$= \{\!\{\kappa \nabla_h v_h\}\!\}_\omega [\![y_h]\!] + [\![\kappa \nabla_h v_h]\!]\{\!\{y_h\}\!\}_{\overline{\omega}},$$

since $\omega_1 + \omega_2 = 1$. As a result and accounting for boundary faces,

$$\sum_{T \in \mathcal{T}_h} \int_{\partial T} \kappa(\nabla v_h \cdot \mathbf{n}_T) y_h = \sum_{F \in \mathcal{F}_h} \int_F \{\!\{\kappa \nabla_h v_h\}\!\}_\omega \cdot \mathbf{n}_F [\![y_h]\!] + \sum_{F \in \mathcal{F}_h^i} \int_F [\![\kappa \nabla_h v_h]\!] \cdot \mathbf{n}_F \{\!\{y_h\}\!\}_{\overline{\omega}}.$$

Combining this expression with (4.64) and (4.67) yields the assertion. □

4.5.3 Error Estimates for Smooth Solutions

In this section, we present the convergence analysis for the discrete problem (4.63) in the case where the exact solution is smooth enough to match the following assumption.

4.5. Heterogeneous Diffusion

Assumption 4.48 (Regularity of exact solution and space V_*). *We assume that the exact solution u is such that*
$$u \in V_* := V \cap H^2(P_\Omega).$$

In the spirit of Sect. 1.3, *we set* $V_{*h} := V_* + V_h$.

Assumption 4.48 implies that, for all $T \in \mathcal{T}_h$, letting $\sigma_T := -(\kappa \nabla u)|_T$ and $\sigma_{\partial T} = \sigma_T \cdot n_T$ on ∂T, the trace $\sigma_{\partial T}|_F$ is in $L^2(F)$ for all $F \in \mathcal{F}_T$. Using Lemma 1.23 for the jumps of the potential and proceeding as in the proof of Lemma 1.24 for the jumps of the diffusive flux, we infer that the exact solution satisfies

$$[\![u]\!] = 0 \qquad \forall F \in \mathcal{F}_h, \tag{4.68a}$$
$$[\![\kappa \nabla u]\!] \cdot n_F = 0 \qquad \forall F \in \mathcal{F}_h^i. \tag{4.68b}$$

The convergence analysis is performed in the spirit of Theorem 1.35 by establishing discrete coercivity, consistency, and boundedness for a_h^{swip}. The discrete bilinear form a_h^{swip} is extended to $V_{*h} \times V_h$.

Lemma 4.49 (Consistency). *Assume $u \in V_*$. Then, for all $v_h \in V_h$,*
$$a_h^{\text{swip}}(u, v_h) = \int_\Omega f v_h.$$

Proof. The result is a direct consequence of (4.66) and (4.68). □

To formulate discrete stability in the context of heterogeneous diffusion, we modify the $\|\cdot\|_{\text{sip}}$-norm considered for the Poisson problem (cf. (4.17)) as follows: For all $v \in V_{*h}$,

$$\|v\|_{\text{swip}} := \left(\|\kappa^{1/2} \nabla_h v\|_{[L^2(\Omega)]^d}^2 + |v|_{\text{J},\kappa}^2 \right)^{1/2}, \tag{4.69}$$

with the diffusion-dependent jump seminorm

$$|v|_{\text{J},\kappa} = \left(\sum_{F \in \mathcal{F}_h} \frac{\gamma_{\kappa,F}}{h_F} \|[\![v]\!]\|_{L^2(F)}^2 \right)^{1/2}. \tag{4.70}$$

Before addressing the discrete coercivity of the SWIP bilinear form, we derive a bound on the consistency term.

Lemma 4.50 (Bound on consistency term). *For all $(v, y_h) \in V_{*h} \times V_h$,*

$$\left| \sum_{F \in \mathcal{F}_h} \int_F \{\!\!\{\kappa \nabla_h v\}\!\!\}_\omega \cdot n_F [\![y_h]\!] \right| \leq \left(\sum_{T \in \mathcal{T}_h} \sum_{F \in \mathcal{F}_T} h_F \|\kappa^{1/2} \nabla v|_T \cdot n_F\|_{L^2(F)}^2 \right)^{1/2} |y_h|_{\text{J},\kappa}. \tag{4.71}$$

Proof. For all $F \in \mathcal{F}_h^i$ with $F = \partial T_1 \cap \partial T_2$, $\omega_i = \omega_{T_i, F}$, $\kappa_i = \kappa|_{T_i}$, and $a_i = \kappa_i^{1/2}(\nabla v)|_{T_i} \cdot \mathbf{n}_F$, $i \in \{1, 2\}$, the Cauchy–Schwarz inequality yields

$$\int_F \{\kappa \nabla_h v\}_\omega \cdot \mathbf{n}_F [\![y_h]\!] = \int_F (\omega_1 \kappa_1^{1/2} a_1 + \omega_2 \kappa_2^{1/2} a_2) [\![y_h]\!]$$

$$\leq \left(\frac{1}{2} h_F (\|a_1\|_{L^2(F)}^2 + \|a_2\|_{L^2(F)}^2)\right)^{1/2}$$

$$\times \left(2(\omega_1^2 \kappa_1 + \omega_2^2 \kappa_2) \frac{1}{h_F} \|[\![y_h]\!]\|_{L^2(F)}^2\right)^{1/2},$$

and since $2(\omega_1^2 \kappa_1 + \omega_2^2 \kappa_2) = \gamma_{\kappa, F}$, we infer

$$\int_F \{\kappa \nabla_h v\}_\omega \cdot \mathbf{n}_F [\![y_h]\!] \leq \left(\frac{1}{2} h_F (\|a_1\|_{L^2(F)}^2 + \|a_2\|_{L^2(F)}^2)\right)^{1/2}$$

$$\times \left(\frac{\gamma_{\kappa, F}}{h_F}\right)^{1/2} \|[\![y_h]\!]\|_{L^2(F)}.$$

Moreover, for all $F \in \mathcal{F}_h^b$ with $F = \partial T \cap \partial \Omega$,

$$\int_F \{\kappa \nabla_h v\}_\omega \cdot \mathbf{n}_F [\![y_h]\!] \leq h_F^{1/2} \|(\kappa^{1/2} \nabla v)|_T \cdot \mathbf{n}_F\|_{L^2(F)} \times \left(\frac{\gamma_{\kappa, F}}{h_F}\right)^{1/2} \|[\![y_h]\!]\|_{L^2(F)}.$$

Summing over mesh faces, using the Cauchy–Schwarz inequality, and regrouping the face contributions for each mesh element yields the assertion. □

We now establish the discrete coercivity of the SWIP bilinear form under the usual assumption that the penalty parameter η is large enough. An important point is that the minimal threshold on the penalty parameter is independent of the diffusion coefficient (it is actually the same as for the Poisson problem).

Lemma 4.51 (Discrete coercivity). *For all $\eta > \underline{\eta}$ with $\underline{\eta}$ defined in Lemma 4.12, the SWIP bilinear form defined by (4.64) is coercive on V_h with respect to the $\|\cdot\|_{\mathrm{swip}}$-norm, i.e.,*

$$\forall v_h \in V_h, \qquad a_h^{\mathrm{swip}}(v_h, v_h) \geq C_\eta \|v_h\|_{\mathrm{swip}}^2,$$

with C_η defined in Lemma 4.12.

Proof. Let $v_h \in V_h$. Owing to the discrete trace inequality (1.40), the fact that $h_F \leq h_T$ for all $T \in \mathcal{T}_h$ and for all $F \in \mathcal{F}_T$, and since κ is piecewise constant on $\mathcal{T}_h$, we infer from the bound (4.71) that

$$\left| \sum_{F \in \mathcal{F}_h} \int_F \{\kappa \nabla_h v_h\}_\omega \cdot \mathbf{n}_F [\![v_h]\!] \right| \leq C_{\mathrm{tr}} N_\partial^{1/2} \|\kappa^{1/2} \nabla_h v_h\|_{[L^2(\Omega)]^d} |v_h|_{\mathrm{J}, \kappa}.$$

We conclude as in the proof of Lemma 4.12. □

4.5. Heterogeneous Diffusion

A straightforward consequence of the Lax–Milgram Lemma is that the discrete problem (4.63) is well-posed.

Our last step in the convergence analysis is to prove the boundedness of the SWIP bilinear form. To formulate this result, we define on V_{*h} the norm

$$\|v\|_{\text{swip},*} := \left(\|v\|_{\text{swip}}^2 + \sum_{T \in \mathcal{T}_h} h_T \|\kappa^{1/2} \nabla v|_T \cdot \mathbf{n}_T\|_{L^2(\partial T)}^2 \right)^{1/2}.$$

Lemma 4.52 (Boundedness). *There is C_{bnd}, independent of h and κ, such that*

$$\forall (v, y_h) \in V_{*h} \times V_h, \qquad a_h^{\text{swip}}(v, y_h) \leq C_{\text{bnd}} \|v\|_{\text{swip},*} \|y_h\|_{\text{swip}}.$$

Proof. Let $(v, y_h) \in V_{*h} \times V_h$ and observe that

$$a_h^{\text{swip}}(v, y_h) := \int_\Omega \kappa \nabla_h v \cdot \nabla_h y_h + \sum_{F \in \mathcal{F}_h} \eta \frac{\gamma_{\kappa,F}}{h_F} \int_F [\![v]\!][\![y_h]\!]$$

$$- \sum_{F \in \mathcal{F}_h} \int_F \{\!\{\kappa \nabla_h v\}\!\}_\omega \cdot \mathbf{n}_F [\![y_h]\!] - \sum_{F \in \mathcal{F}_h} \int_F [\![v]\!] \{\!\{\kappa \nabla_h y_h\}\!\}_\omega \cdot \mathbf{n}_F$$

$$= \mathfrak{T}_1 + \mathfrak{T}_2 + \mathfrak{T}_3 + \mathfrak{T}_4. \tag{4.72}$$

Using the Cauchy–Schwarz inequality yields

$$|\mathfrak{T}_1 + \mathfrak{T}_2| \leq (1 + \eta)\|v\|_{\text{swip}}\|y_h\|_{\text{swip}}.$$

Moreover, owing to the bound (4.71),

$$|\mathfrak{T}_3| \leq \|v\|_{\text{swip},*}|y_h|_{J,\kappa} \leq \|v\|_{\text{swip},*}\|y_h\|_{\text{swip}}$$

by definition of the $\|\cdot\|_{\text{swip},*}$-norm. Finally, still owing to the bound (4.71) and proceeding as in the proof of Lemma 4.51 leads to

$$|\mathfrak{T}_4| \leq C_{\text{tr}} N_\partial^{1/2} |v|_{J,\kappa} \|\kappa^{1/2} \nabla_h y_h\|_{[L^2(\Omega)]^d} \leq C_{\text{tr}} N_\partial^{1/2} \|v\|_{\text{swip}} \|y_h\|_{\text{swip}}.$$

Collecting the above bounds yields the assertion with $C_{\text{bnd}} = 2 + \eta + C_{\text{tr}} N_\partial^{1/2}$. □

A straightforward consequence of Theorem 1.35, together with Lemmata 1.58 and 1.59, is the following convergence result.

Theorem 4.53 ($\|\cdot\|_{\text{swip}}$-norm error estimate and convergence rate). *Let $u \in V_*$ solve (4.62). Let u_h solve (4.63) with a_h^{swip} defined by (4.64) and penalty parameter as in Lemma 4.12. Then, there is C, independent of h and κ, such that*

$$\|u - u_h\|_{\text{swip}} \leq C \inf_{v_h \in V_h} \|u - v_h\|_{\text{swip},*}.$$

Moreover, if $u \in H^{k+1}(P_\Omega)$,

$$\|u - u_h\|_{\text{swip}} \leq C_u \|\kappa\|_{L^\infty(\Omega)}^{1/2} h^k,$$

with $C_u = C\|u\|_{H^{k+1}(P_\Omega)}$ and C independent of h and κ.

Since the quantity C in the error estimates is independent of the diffusion coefficient κ, the approximation method is robust with respect to diffusion heterogeneities (observing that the $\|\cdot\|_{\text{swip}}$-norm depends on κ). The convergence rate in the $\|\cdot\|_{\text{swip}}$-norm is optimal, both for the broken gradient and the jump seminorm.

4.5.4 Error Estimates for Low-Regularity Solutions

In this section, following [132], we present the convergence analysis for the discrete problem (4.63) for an exact solution with low-regularity.

Assumption 4.54 (Regularity of exact solution and space V_*). *We assume that $d \geq 2$ and that there is $p \in (\frac{2d}{d+2}, 2]$ such that, for the exact solution u,*

$$u \in V_* := V \cap W^{2,p}(P_\Omega).$$

In the spirit of Sect. 1.3, *we set* $V_{*h} := V_* + V_h$.

Assumption 4.48 implies that, for all $T \in \mathcal{T}_h$, letting $\sigma_T := -(\kappa \nabla u)|_T$ and $\sigma_{\partial T} = \sigma_T \cdot n_T$ on ∂T, the trace $\sigma_{\partial T}|_F$ is in $L^p(F)$ for all $F \in \mathcal{F}_T$. We adapt the analysis of Sect. 4.2.5 for the Poisson problem to the present setting with heterogeneous diffusion.

We already know that discrete coercivity holds true provided the penalty parameter is chosen as in Lemma 4.12. Moreover, since the jump conditions (4.68) still hold true, consistency can be asserted. Thus, it only remains to prove boundedness, which we do by redefining on V_{*h} the $\|\cdot\|_{\text{swip},*}$-norm as

$$\|v\|_{\text{swip},*} := \left(\|v\|_{\text{swip}}^p + \sum_{T \in \mathcal{T}_h} h_T^{1+\gamma_p} \|\kappa^{1/2} \nabla v|_T \cdot n_T\|_{L^p(\partial T)}^p \right)^{1/p}, \quad (4.73)$$

where $\gamma_p := \frac{1}{2} d(p-2)$. We observe that, for $p = 2$, we recover the previous definition of the $\|\cdot\|_{\text{swip},*}$-norm. The value for γ_p is motivated by the following boundedness result.

Lemma 4.55 (Boundedness). *There is C_{bnd}, independent of h and κ, such that*

$$\forall (v, w_h) \in V_{*h} \times V_h, \qquad a_h^{\text{swip}}(v, w_h) \leq C_{\text{bnd}} \|v\|_{\text{swip},*} \|w_h\|_{\text{swip}}.$$

Proof. Let $(v, w_h) \in V_{*h} \times V_h$. We need to bound the four terms $\mathfrak{T}_1, \ldots, \mathfrak{T}_4$ in (4.72). Proceeding as in the proof of Lemma 4.52, we obtain

$$|\mathfrak{T}_1 + \mathfrak{T}_3 + \mathfrak{T}_4| \leq C \|v\|_{\text{swip}} \|w_h\|_{\text{swip}},$$

with C independent of h and κ, so that it only remains to bound the consistency term $\mathfrak{T}_2$. For all $F \in \mathcal{F}_h^i$ with $F = \partial T_1 \cap \partial T_2$, and $a_i = (\kappa^{1/2} \nabla v)|_{T_i} \cdot n_F$, $i \in \{1, 2\}$,

4.5. Heterogeneous Diffusion

Hölder's inequality yields

$$\int_F \{\kappa \nabla_h v\}_\omega \cdot n_F [\![w_h]\!] = \int_F (\omega_1 \kappa_1^{1/2} a_1 + \omega_2 \kappa_2^{1/2} a_2)[\![w_h]\!]$$
$$\leq \left(\frac{1}{2} h_F^{1+\gamma_p}(\|a_1\|_{L^p(F)}^p + \|a_2\|_{L^p(F)}^p)\right)^{1/p}$$
$$\times 2^{1/p}\left((\omega_1^q \kappa_1^{q/2} + \omega_2^q \kappa_2^{q/2}) h_F^{-q\beta_p} \|[\![w_h]\!]\|_{L^q(F)}^q\right)^{1/q},$$

with $\beta_p = \frac{1+\gamma_p}{p}$ and $q = \frac{p}{p-1}$. Since $q \geq 2$, we obtain

$$(\omega_1^q \kappa_1^{q/2} + \omega_2^q \kappa_2^{q/2}) = \frac{(\kappa_1 \kappa_2)^{q/2}}{(\kappa_1 + \kappa_2)^q}(\kappa_1^{q/2} + \kappa_2^{q/2}) \leq \frac{(\kappa_1 \kappa_2)^{q/2}}{(\kappa_1 + \kappa_2)^q}(\kappa_1 + \kappa_2)^{q/2} = 2^{-q/2}\gamma_{\kappa,F}^{q/2}.$$

Hence, since $2^{1/p - 1/2} \leq 2$,

$$\int_F \{\kappa \nabla_h v\}_\omega \cdot n_F [\![w_h]\!] \leq \left(\frac{1}{2} h_F^{1+\gamma_p}(\|a_1\|_{L^p(F)}^p + \|a_2\|_{L^p(F)}^p)\right)^{1/p}$$
$$\times 2\gamma_{\kappa,F}^{1/2} h_F^{-\beta_p} \|[\![w_h]\!]\|_{L^q(F)}.$$

Moreover, for all $F \in \mathcal{F}_h^b$ with $F = \partial T \cap \partial \Omega$,

$$\int_F \{\kappa \nabla_h v\}_\omega \cdot n_F [\![w_h]\!] \leq \left(h_F^{1+\gamma_p} \|\kappa^{1/2} \nabla v|_T \cdot n_F\|_{L^p(F)}^p\right)^{1/p} \gamma_{\kappa,F}^{1/2} h_F^{-\beta_p} \|[\![w_h]\!]\|_{L^q(F)}.$$

We can now conclude as in the proof of Lemma 4.30. $\square$

A straightforward consequence of Theorem 1.35 is the following convergence result. The achieved convergence rates are optimal, both for the broken gradient and the jump seminorm.

Theorem 4.56 ($\|\cdot\|_{\text{swip}}$-norm error estimate and convergence rate). *Let $u \in V_*$ solve (4.62). Let u_h solve (4.63) with a_h^{swip} defined by (4.64) and penalty parameter as in Lemma 4.12. Then, there is C, independent of h and κ, such that*

$$\|u - u_h\|_{\text{swip}} \leq C \inf_{v_h \in V_h} \|u - v_h\|_{\text{swip},*},$$

where the $\|\cdot\|_{\text{swip},*}$-norm is defined by (4.73). Moreover, under Assumption 4.31, there holds

$$\|u - u_h\|_{\text{swip}} \leq C_u h^{\alpha_p},$$

with $C_u = C|u|_{W^{2,p}(P_\Omega)}$, C independent of h and κ, and $\alpha_p = \frac{d+2}{2} - \frac{d}{p} > 0$.

4.5.5 Numerical Fluxes

As for the Poisson problem in Sect. 4.3.4, it is possible to derive a local formulation of the discrete problem (4.63) by localizing test functions to mesh elements. To this purpose, we first modify the definition of the lifting operators and discrete gradients (cf. Sects. 4.3.1 and 4.3.2) to account for diffusion heterogeneities. For any face $F \in \mathcal{F}_h$ and for any integer $l \geq 0$, we define the (local) lifting operator $\mathrm{r}_{F,\kappa}^l : L^2(F) \to [\mathbb{P}_d^l(\mathcal{T}_h)]^d$ as follows: For all $\varphi \in L^2(F)$,

$$\int_\Omega \kappa \mathrm{r}_{F,\kappa}^l(\varphi) \cdot \tau_h = \int_F \{\!\{\kappa \tau_h\}\!\}_\omega \cdot \mathrm{n}_F \varphi \qquad \forall \tau_h \in [\mathbb{P}_d^l(\mathcal{T}_h)]^d. \tag{4.74}$$

Clearly, if κ does not jump across F (so that κ is constant in the support of $\mathrm{r}_{F,\kappa}^l(\varphi)$), definitions (4.37) and (4.74) produce the same result, but this is no longer the case in the presence of diffusion heterogeneities. Then, for any function $v \in H^1(\mathcal{T}_h)$, we define the (global) lifting of its interface and boundary jumps as

$$\mathrm{R}_{h,\kappa}^l([\![v]\!]) := \sum_{F \in \mathcal{F}_h} \mathrm{r}_{F,\kappa}^l([\![v]\!]) \in [\mathbb{P}_d^l(\mathcal{T}_h)]^d, \tag{4.75}$$

being implicitly understood that $\mathrm{r}_{F,\kappa}^l$ acts on the function $[\![v]\!]_F$ (which is in $L^2(F)$ since $v \in H^1(\mathcal{T}_h)$). If κ is constant in Ω, definitions (4.40) and (4.75) produce the same result. Finally, the definition (4.44) of the discrete gradient is extended to the heterogeneous diffusion case by setting, for all $v \in H^1(\mathcal{T}_h)$,

$$G_{h,\kappa}^l(v) := \nabla_h v - \mathrm{R}_{h,\kappa}^l([\![v]\!]) \in [L^2(\Omega)]^d.$$

Let $T \in \mathcal{T}_h$ and let $\xi \in \mathbb{P}_d^k(T)$. Then, using $v_h = \xi \chi_T$ as test function in (4.63) where χ_T is the characteristic function of T, proceeding as in Sect. 4.3.4, and using the above definitions, we infer

$$\int_T \kappa G_{h,\kappa}^l(u_h) \cdot \nabla \xi + \sum_{F \in \mathcal{F}_T} \epsilon_{T,F} \int_F \phi_F(u_h) \xi = \int_T f \xi,$$

with $l \in \{k-1, k\}$, $\epsilon_{T,F} = \mathrm{n}_T \cdot \mathrm{n}_F$, and the numerical flux $\phi_F(u_h)$ defined as

$$\phi_F(u_h) := -\{\!\{\kappa \nabla_h u_h\}\!\}_\omega \cdot \mathrm{n}_F + \eta \frac{\gamma_{\kappa,F}}{h_F} [\![u_h]\!].$$

Remark 4.57 (Harmonic means). For all $F \in \mathcal{F}_h^i$ with $F = \partial T_1 \cap \partial T_2$ and $\kappa_i = \kappa|_{T_i}$, $i \in \{1,2\}$, we observe that

$$-\{\!\{\kappa \nabla_h u_h\}\!\}_\omega \cdot \mathrm{n}_F = -\frac{\kappa_2}{\kappa_1 + \kappa_2} \kappa_1 (\nabla u_h)|_{T_1} \cdot \mathrm{n}_F - \frac{\kappa_1}{\kappa_1 + \kappa_2} \kappa_2 (\nabla u_h)|_{T_2} \cdot \mathrm{n}_F$$

$$= -\frac{2\kappa_1 \kappa_2}{\kappa_1 + \kappa_2} \{\!\{\nabla_h u_h\}\!\} \cdot \mathrm{n}_F.$$

Thus, recalling that the jump seminorm of u_h tends to zero as $h \to 0$, the leading-order term in the numerical flux $\phi_F(u_h)$ uses the harmonic mean of the diffusion coefficient. A motivation for using harmonic means can be given in the context of heat transfer where κ represents the thermal conductivity, u the temperature, and $-\kappa \nabla u$ the heat flux. Consider an interface between a poorly conductive medium (where κ is relatively small) and a highly conductive medium (where κ is much larger), so that, at this interface, the harmonic mean of κ is close to the value in the poorly conductive medium. Then, the heat transfer through this interface is essentially governed by the poorly conductive medium.

4.5.6 Anisotropy

The above developments can be extended to the anisotropic case, that is, when for a.e. $x \in \Omega$, $\kappa(x)$ is a symmetric tensor in $\mathbb{R}^{d,d}$. Assuming that the lowest eigenvalue of κ is uniformly bounded from below in Ω by a positive real number, the model problem (4.62) is well-posed.

The SWIP bilinear form defined by (4.64) can be used to approximate heterogeneous anisotropic diffusion problems. Specifically, the weights $\{\omega_{T_1,F}, \omega_{T_2,F}\}$ and the penalty parameter $\gamma_{\kappa,F}$ are evaluated on any interface $F \in \mathcal{F}_h^i$ by using the normal component of the diffusion tensor on both sides of that interface, that is, for all $F \in \mathcal{F}_h^i$, $F = \partial T_1 \cap \partial T_2$, we now let $\kappa_i := \mathrm{n}_F^t (\kappa|_{T_i}) \mathrm{n}_F$, $i \in \{1,2\}$, and we set as before

$$\omega_{T_1,F} := \frac{\kappa_2}{\kappa_1 + \kappa_2}, \qquad \omega_{T_2,F} := \frac{\kappa_1}{\kappa_1 + \kappa_2}, \qquad \gamma_{\kappa,F} := \frac{2\kappa_1 \kappa_2}{\kappa_1 + \kappa_2}.$$

Moreover, for all $F \in \mathcal{F}_h^b$, $F = \partial T \cap \partial \Omega$, we set $\gamma_{\kappa,F} := \mathrm{n}^t (\kappa|_T) \mathrm{n}$. With these modifications, the convergence analysis proceeds as in the isotropic case. Since κ takes symmetric positive definite values, it is in particular possible to define $\kappa^{1/2}$ as the symmetric positive definite matrix such that $\kappa^{1/2} \kappa^{1/2} = \kappa$. We refer the reader to [133, 150] for a detailed presentation of the convergence analysis.

4.6 Diffusion-Advection-Reaction

In this section, we consider a model diffusion-advection-reaction problem. The design and analysis of the dG approximation combine the ideas of Sect. 4.5 to handle the diffusion part and those of Sect. 2.3 to handle the advection-reaction part. One issue of particular interest is the robustness of the approximation in the singularly perturbed regime where advection-reaction effects dominate over diffusion effects. In particular, we address at the end of this section the situation where the diffusion coefficient can actually vanish locally, so that a first-order PDE in some part of the domain is coupled to an elliptic PDE in the remaining part.

4.6.1 The Continuous Setting

Let $\kappa \in L^\infty(\Omega)$ and assume that κ is uniformly bounded from below in Ω by a positive real number; the singular limit where κ can actually vanish locally in some parts of Ω is addressed in Sect. 4.6.4. Moreover, we keep Assumption 4.43 so as to localize possible jumps in the diffusion coefficient. Let $\beta \in [\mathrm{Lip}(\Omega)]^d$ be the advective velocity and let $\tilde{\mu} \in L^\infty(\Omega)$ be the reaction coefficient. We are interested in the problem:

$$\nabla \cdot (-\kappa \nabla u + \beta u) + \tilde{\mu} u = f \quad \text{in } \Omega,$$
$$u = 0 \quad \text{on } \partial\Omega,$$

with source term $f \in L^2(\Omega)$. The weak form of this problem reads

$$\text{Find } u \in V \text{ s.t. } a(u,v) = \int_\Omega fv \text{ for all } v \in V, \tag{4.76}$$

with energy space $V = H_0^1(\Omega)$ and bilinear form

$$a(u,v) := \int_\Omega \kappa \nabla u \cdot \nabla v - \int_\Omega u\beta \cdot \nabla v + \int_\Omega \tilde{\mu} uv.$$

We observe that the advective term is written in conservative form. The $\mathbb{R}^d$-valued function

$$\Phi(u) = -\kappa \nabla u + \beta u$$

is termed the diffusive-advective flux. By construction, $\Phi(u)$ is in $H(\mathrm{div}; \Omega)$. The diffusion-advection-reaction can be rewritten as

$$-\nabla \cdot \Phi(u) + \tilde{\mu} u = f,$$

and the bilinear form a as

$$a(u,v) = \int_\Omega -\Phi(u) \cdot \nabla v + \int_\Omega \tilde{\mu} uv. \tag{4.77}$$

Since $u \in H^1(\Omega)$ and β is smooth, it is equivalent to consider the advective term in its non-conservative form, i.e.,

$$-\nabla \cdot (\kappa \nabla u) + \beta \cdot \nabla u + \mu u = f,$$

with $\mu := \tilde{\mu} + \nabla \cdot \beta$. However, if κ vanishes locally, the exact solution can feature discontinuities, and the two forms are no longer equivalent. The conservative form is more natural from a physical viewpoint since it expresses a basic conservation principle. Indeed, integrating the diffusion-advection-reaction equation over a control volume $V \subset \Omega$, we obtain formally

$$\int_{\partial V} \Phi(u) \cdot n_V + \int_V \tilde{\mu} u = \int_V f,$$

where n_V denotes the outward normal to ∂V. This equation expresses the fact that the variation of u in the control volume V due to the diffusive and advective exchanges through ∂V plus the quantity of u generated/depleted by reaction over V is equal to the integral of the source term f over V.

4.6. Diffusion-Advection-Reaction

As in Sect. 2.1, we assume that there is a real number $\mu_0 > 0$ such that

$$\Lambda := \tilde{\mu} + \frac{1}{2}\nabla\cdot\beta = \mu - \frac{1}{2}\nabla\cdot\beta \geq \mu_0 \text{ a.e. in } \Omega.$$

Hence, using integration by parts, the bilinear form a is coercive on V,

$$\forall v \in V, \quad a(v,v) = \|\kappa^{1/2}\nabla v\|_{[L^2(\Omega)]^d}^2 + \|\Lambda^{1/2}v\|_{L^2(\Omega)}^2.$$

Owing to the Lax–Milgram Lemma, (4.76) is therefore well-posed.

4.6.2 Discretization

We aim at approximating the exact solution u of (4.76) by a dG method using the discrete space

$$V_h := \mathbb{P}_d^k(\mathcal{T}_h),$$

where $\mathbb{P}_d^k(\mathcal{T}_h)$ is defined by (1.15) with polynomial degree $k \geq 1$ and $\mathcal{T}_h$ belonging to an admissible mesh sequence. We keep Assumption 4.45 on the compatibility of the meshes with the partition P_Ω associated with the diffusion coefficient κ. Moreover, concerning the regularity of the exact solution, we assume that (cf. Assumption 4.48)

$$u \in V_* := V \cap H^2(P_\Omega),$$

and we set, as before, $V_{*h} = V_* + V_h$. It is also possible to analyze the dG approximation in the case of low-regularity exact solutions matching only Assumption 4.54.

The dG method considered herein combines the SWIP bilinear form of Sect. 4.5 to handle the diffusion term and the upwind dG method of Sect. 2.3 to handle the advection-reaction terms. Thus, we let, for all $(v,w_h) \in V_{*h} \times V_h$,

$$a_h^{\text{dar}}(v,w_h) = a_h^{\text{swip}}(v,w_h) + a_h^{\text{upw}}(v,w_h), \tag{4.78}$$

where (cf. (4.64))

$$a_h^{\text{swip}}(v,w_h) := \int_\Omega \kappa \nabla_h v \cdot \nabla_h w_h + \sum_{F \in \mathcal{F}_h} \eta \frac{\gamma_{\kappa,F}}{h_F} \int_F [\![v]\!][\![w_h]\!]$$
$$- \sum_{F \in \mathcal{F}_h} \int_F (\{\!\{\kappa \nabla_h v\}\!\}_\omega \cdot n_F [\![w_h]\!] + [\![v]\!]\{\!\{\kappa \nabla_h w_h\}\!\}_\omega \cdot n_F),$$

and (cf. (2.34))

$$a_h^{\text{upw}}(v,w_h) = \int_\Omega [\tilde{\mu}vw_h + \nabla_h\cdot(\beta v)w_h] + \int_{\partial\Omega} (\beta\cdot n)^\ominus v w_h$$
$$- \sum_{F \in \mathcal{F}_h^i} \int_F (\beta\cdot n_F)[\![v]\!]\{\!\{w_h\}\!\} + \sum_{F \in \mathcal{F}_h^i} \int_F \gamma_{\beta,F}[\![v]\!][\![w_h]\!],$$

or equivalently, after integrating by parts the advective derivative (cf. (2.35)),

$$a_h^{\text{upw}}(v, w_h) = \int_\Omega [\tilde{\mu} v w_h - v(\beta \cdot \nabla_h w_h)] + \int_{\partial\Omega} (\beta \cdot \mathbf{n})^\oplus v w_h$$
$$+ \sum_{F \in \mathcal{F}_h^i} \int_F (\beta \cdot \mathbf{n}_F) \{v\} [\![w_h]\!] + \sum_{F \in \mathcal{F}_h^i} \int_F \gamma_{\beta,F} [\![v]\!] [\![w_h]\!].$$

In what follows, we set

$$\gamma_{\beta,F} := \frac{1}{2} |\beta \cdot \mathbf{n}_F|.$$

It is also possible to multiply $\gamma_{\beta,F}$ by a positive user-dependent parameter as in Sect. 2.3 (cf., e.g., (2.33)), but the present choice is needed for consistency reasons in Sect. 4.6.4 in the singular limit of locally vanishing diffusion; cf. Remark 4.69. We also observe that the penalty terms can be grouped to obtain

$$\sum_{F \in \mathcal{F}_h^i} \left(\eta \frac{\gamma_{\kappa,F}}{h_F} + \frac{1}{2} |\beta \cdot \mathbf{n}_F| \right) \int_F [\![v]\!] [\![w_h]\!].$$

In the diffusion-dominated regime where $h_F |\beta \cdot \mathbf{n}_F| \lesssim \gamma_{\kappa,F}$, the amount of penalty introduced by the SWIP bilinear form suffices for discrete stability, and it is possible to drop upwinding for the advective terms (that is, to approximate the advective term using centered fluxes). The ratio $h_F |\beta \cdot \mathbf{n}_F| / \gamma_{\kappa,F}$ is termed a local Péclet number. In practice, local Péclet numbers are often large, generally because the diffusion coefficient is (locally) small, so that upwinding is necessary. In this situation, the exact solution features inner and outflow layers where it varies quite sharply, and practical meshes may not be fine enough to resolve these layers; we refer the reader, e.g., to Roos, Stynes, and Tobiska [274] for a general overview on singularly perturbed diffusion-advection-reaction problems and stabilized finite element approximations.

4.6.3 Error Estimates

To approximate the model problem (4.76), we consider the discrete problem:

$$\text{Find } u_h \in V_h \text{ s.t. } a_h^{\text{dar}}(u_h, v_h) = \int_\Omega f v_h \text{ for all } v_h \in V_h, \tag{4.79}$$

where a_h^{dar} is the discrete bilinear form defined by (4.78). The convergence analysis is performed by establishing discrete stability, consistency, and boundedness for a_h^{dar}. We begin with consistency.

Lemma 4.58 (Consistency). *Assume $u \in V_*$. Then, for all $w_h \in V_h$,*

$$a_h^{\text{dar}}(u, w_h) = \int_\Omega f w_h.$$

4.6. Diffusion-Advection-Reaction

Proof. The proof of Lemma 4.49 yields, for all $w_h \in V_h$,

$$a_h^{\mathrm{swip}}(u, w_h) = \sum_{T \in \mathcal{T}_h} \int_T \nabla\cdot(-\kappa \nabla u) w_h.$$

Moreover, adapting the proof of Lemma 2.27, we infer

$$a_h^{\mathrm{upw}}(u, w_h) = \sum_{T \in \mathcal{T}_h} \int_T \nabla\cdot(\beta u) w_h + \int_\Omega \tilde{\mu} u w_h.$$

Summing up and observing that $\nabla\cdot(-\kappa\nabla u + \beta u) + \tilde{\mu} u = f$ in all $T \in \mathcal{T}_h$ yields the assertion. $\square$

4.6.3.1 Analysis Based on Discrete Coercivity

The convergence analysis is performed in the spirit of Theorem 2.31 by combining consistency (cf. Lemma 4.58) with discrete coercivity and boundedness on orthogonal subscales for the discrete bilinear form a_h^{dar}. We recall that in the context of the advection-reaction equation, we introduced in Sect. 2.1 the reference time τ_c and the reference velocity β_c such that

$$\tau_c := \{\max(\|\mu\|_{L^\infty(\Omega)}, L_\beta)\}^{-1}, \qquad \beta_c := \|\beta\|_{[L^\infty(\Omega)]^d},$$

where L_β is the Lipschitz module of β (cf. (2.5)). We define on V_{*h} the norm

$$\|v\|_{\mathrm{dab}} := \left(\|v\|_{\mathrm{swip}}^2 + |v|_\beta^2 + \tau_c^{-1} \|v\|_{L^2(\Omega)}^2 \right)^{1/2}, \qquad (4.80)$$

where the $\|\cdot\|_{\mathrm{swip}}$-norm is defined by (4.69) and (4.70) while the $|\cdot|_\beta$-seminorm is defined as

$$|v|_\beta := \left(\int_{\partial\Omega} \frac{1}{2} |\beta\cdot n| v^2 + \sum_{F \in \mathcal{F}_h^i} \int_F \frac{1}{2} |\beta\cdot n_F| [\![v]\!]^2 \right)^{1/2}.$$

The two rightmost terms in (4.80) form the stability norm (cf. (2.37)) considered in Sect. 2.3.2 for the advection-reaction equation.

Lemma 4.59 (Discrete coercivity). *For all $\eta > \underline{\eta}$ with $\underline{\eta}$ defined in Lemma 4.12, the discrete bilinear form a_h^{dar} defined by (4.78) is coercive on V_h, i.e.,*

$$\forall v_h \in V_h, \qquad a_h^{\mathrm{dar}}(v_h, v_h) \geq \min(1, \tau_c \mu_0, C_\eta) \|v_h\|_{\mathrm{dab}}^2,$$

with C_η defined in Lemma 4.12.

Proof. Let $v_h \in V_h$. Lemma 4.51 yields

$$a_h^{\mathrm{swip}}(v_h, v_h) \geq C_\eta \|v_h\|_{\mathrm{swip}}^2.$$

Moreover, owing to Lemma 2.27,

$$a_h^{\mathrm{upw}}(v_h, v_h) \geq \min(1, \tau_c \mu_0) \left(|v_h|_\beta^2 + \tau_c^{-1} \|v_h\|_{L^2(\Omega)}^2 \right).$$

Combining these lower bounds yields the assertion. $\square$

A straightforward consequence of the Lax–Milgram Lemma is that the discrete problem (4.79) is well-posed.

The last ingredient is boundedness on orthogonal subscales for the discrete bilinear form a_h^{dar}. To this purpose, we define on V_{*h} the norm

$$\|v\|_{\mathrm{dab},*} := \left(\|v\|_{\mathrm{dab}}^2 + \sum_{T \in \mathcal{T}_h} \beta_{\mathrm{c}} \|v\|_{L^2(\partial T)}^2 + \sum_{T \in \mathcal{T}_h} h_T \|\kappa^{1/2} \nabla v \cdot \mathbf{n}_T\|_{L^2(\partial T)}^2 \right)^{1/2}.$$

Lemma 4.60 (Boundedness on orthogonal subscales). *There is C_{bnd}, independent of h and the data κ, β, and $\tilde{\mu}$, such that*

$$\forall (v, w_h) \in V_* \times V_h, \qquad a_h^{\mathrm{dar}}(v - \pi_h v, w_h) \le C_{\mathrm{bnd}} \|v - \pi_h v\|_{\mathrm{dab},*} \|w_h\|_{\mathrm{dab}},$$

where π_h denotes the L^2-orthogonal projection onto V_h.

Proof. Combine Lemma 4.52 with Lemma 2.30. (The fact that the first argument in a_h^{dar} is L^2-orthogonal to V_h is only needed to apply Lemma 2.30.) □

Proceeding as in the proof of Theorem 2.31 leads to the following error estimate.

Theorem 4.61 (Error estimate). *Let $u \in V_*$ solve (4.76). Let u_h solve (4.79) with a_h^{dar} defined by (4.78) and penalty parameter as in Lemma 4.12. Then, there is C, independent of h and the data κ, β, and $\tilde{\mu}$, such that*

$$\|u - u_h\|_{\mathrm{dab}} \le C \max(1, \tau_{\mathrm{c}}^{-1} \mu_0^{-1}, C_\eta^{-1}) \|u - \pi_h u\|_{\mathrm{dab},*}. \qquad (4.81)$$

A convergence rate can be inferred from (4.81) using Lemmata 1.58 and 1.59 if the exact solution is smooth enough. Namely, if $u \in H^{k+1}(\Omega)$, (4.81) yields

$$\|u - u_h\|_{\mathrm{dab}} \le C'_u \max(1, \tau_{\mathrm{c}}^{-1} \mu_0^{-1}, C_\eta^{-1})(\overline{\kappa}^{1/2} + \beta_{\mathrm{c}}^{1/2} h^{1/2} + \tau_{\mathrm{c}}^{-1/2} h) h^k, \qquad (4.82)$$

with $\overline{\kappa} := \|\kappa\|_{L^\infty(\Omega)}$, $C'_u = C' \|u\|_{H^{k+1}(\Omega)}$, and C' independent of h and the data κ, β, and $\tilde{\mu}$. The estimate can be simplified by dropping the last term under the reasonable assumption that $h \le \beta_{\mathrm{c}} \tau_{\mathrm{c}}$; cf. (2.41). Moreover, observing that $h\beta_{\mathrm{c}}/\overline{\kappa}$ represents a Péclet number and recalling the definition (4.80) of the $\|\cdot\|_{\mathrm{dab}}$-norm, we conclude that in the advection-dominated regime, the convergence rate of $|u - u_h|_\beta + \tau_{\mathrm{c}}^{-1/2} \|u - u_h\|_{L^2(\Omega)}$ is of order $h^{k+1/2}$ (as for the pure advection-reaction problem; cf. Sect. 2.3.2), while in the diffusion-dominated regime, the convergence rate of $\|u - u_h\|_{\mathrm{swip}}$ is of order h^k (as for the purely diffusive problem; cf. Sect. 4.5.3).

4.6.3.2 Analysis Based on Discrete Inf-Sup Condition

As shown in [133, 150], the above convergence analysis can be improved by including a bound on the advective derivative of the error. To this purpose, we need to tighten the discrete stability norm. Indeed, using the $\|\cdot\|_{\mathrm{swip}}$-norm contribution to the $\|\cdot\|_{\mathrm{dab}}$-norm to bound the advective derivative leads to an error bound

4.6. Diffusion-Advection-Reaction

that scales unfavorably with the Péclet number. Instead, we define on V_{*h} the norm

$$\|v\|_{\mathrm{da}\sharp} := \left(\|v\|_{\mathrm{da}\flat}^2 + \sum_{T\in\mathcal{T}_h} \beta_\mathrm{c}^{-1} h_T \|\beta\cdot\nabla v\|_{L^2(T)}^2 \right)^{1/2}.$$

As in Sect. 2.3.3, asserting discrete stability in the $\|\cdot\|_{\mathrm{da}\sharp}$-norm requires proving a discrete inf-sup condition.

Lemma 4.62 (Discrete inf-sup stability). *There is C_{sta}, independent of h and the data κ, β, and $\tilde{\mu}$, such that*

$$\forall v_h \in V_h, \qquad C_{\mathrm{sta}} \min(1, \tau_\mathrm{c}\mu_0, C_\eta)\|v_h\|_{\mathrm{da}\sharp} \leq \sup_{w_h\in V_h\setminus\{0\}} \frac{a_h^{\mathrm{dar}}(v_h, w_h)}{\|w_h\|_{\mathrm{da}\sharp}}.$$

Proof. The proof is similar to that of Lemma 2.35. Let $v_h \in V_h$ and set $\mathcal{S} = \sup_{w_h\in V_h\setminus\{0\}} \frac{a_h^{\mathrm{dar}}(v_h, w_h)}{\|w_h\|_{\mathrm{da}\sharp}}$. Lemma 4.59 implies that

$$\min(1, \tau_\mathrm{c}\mu_0, C_\eta)\|v_h\|_{\mathrm{da}\flat}^2 \leq a_h^{\mathrm{dar}}(v_h, v_h) \leq \mathcal{S}\|v_h\|_{\mathrm{da}\sharp}.$$

To bound the contribution of the advective derivative in the expression for $\|v_h\|_{\mathrm{da}\sharp}$, we consider the function $w_h \in V_h$ such that, for all $T\in\mathcal{T}_h$, $w_h|_T = \beta_\mathrm{c}^{-1} h_T \langle\beta\rangle_T\cdot\nabla v_h$ where $\langle\beta\rangle_T$ denotes the mean value of β over T. To alleviate the notation, we abbreviate as $a \lesssim b$ the inequality $a \leq Cb$ with positive C independent of h and the data κ, β, and $\tilde{\mu}$.
(i) Let us bound $\|w_h\|_{\mathrm{da}\sharp}$ by $\|v_h\|_{\mathrm{da}\sharp}$. As in the proof of Lemma 2.35, we obtain

$$|w_h|_\beta^2 + \tau_\mathrm{c}^{-1}\|w_h\|_{L^2(\Omega)}^2 + \sum_{T\in\mathcal{T}_h} \beta_\mathrm{c}^{-1} h_T \|\beta\cdot\nabla w_h\|_{L^2(T)}^2 \lesssim \|v_h\|_{\mathrm{da}\sharp}^2.$$

Moreover, owing to the inverse inequality (1.36) and the fact that $\kappa|_T$ and $\langle\beta\rangle_T$ are constant in any mesh element $T\in\mathcal{T}_h$,

$$\|\kappa^{1/2}\nabla_h w_h\|_{[L^2(\Omega)]^d}^2 = \sum_{T\in\mathcal{T}_h} \kappa|_T \beta_\mathrm{c}^{-2} h_T^2 \|\nabla(\langle\beta\rangle_T\cdot\nabla v_h)\|_{L^2(T)}^2$$

$$\lesssim \sum_{T\in\mathcal{T}_h} \kappa|_T \|\nabla v_h\|_{[L^2(T)]^d}^2 = \|\kappa^{1/2}\nabla_h v_h\|_{[L^2(\Omega)]^d}^2.$$

In addition, for all $F \in \mathcal{F}_h^i$ with $F = \partial T_1 \cap \partial T_2$,

$$\frac{\gamma_{\kappa,F}}{h_F}\|[\![w_h]\!]\|_{L^2(F)}^2 \leq 2\frac{\gamma_{\kappa,F}}{h_F}\beta_\mathrm{c}^{-2} \sum_{i\in\{1,2\}} h_{T_i}^2 \|\langle\beta\rangle_{T_i}\cdot(\nabla v_h)|_{T_i}\|_{L^2(F)}^2$$

$$\lesssim \sum_{i\in\{1,2\}} \kappa|_{T_i}\|\nabla v_h\|_{[L^2(T_i)]^d}^2,$$

where we have used the discrete trace inequality (1.37), the mesh regularity, and the bound (4.65) on $\gamma_{\kappa,F}$. Hence,

$$|w_h|_{\mathrm{J},\kappa} \lesssim \|\kappa^{1/2}\nabla_h v_h\|_{[L^2(\Omega)]^d},$$

and collecting the above bounds yields $\|w_h\|_{\mathrm{da}\sharp} \lesssim \|v_h\|_{\mathrm{da}\sharp}$.
(ii) Proceeding as in step (ii) of the proof of Lemma 2.35, we observe that

$$\sum_{T\in\mathcal{T}_h} \beta_{\mathrm{c}}^{-1} h_T \|\beta\cdot\nabla v_h\|_{L^2(T)}^2 = a_h^{\mathrm{dar}}(v_h,w_h) - a_h^{\mathrm{swip}}(v_h,w_h) - \int_\Omega \mu v_h w_h$$

$$+ \sum_{T\in\mathcal{T}_h} \beta_{\mathrm{c}}^{-1} h_T \int_T (\beta\cdot\nabla v_h)(\beta - \langle\beta\rangle_T)\cdot\nabla v_h$$

$$- \int_{\partial\Omega} (\beta\cdot\mathrm{n})^\ominus v_h w_h + \sum_{F\in\mathcal{F}_h^i} \int_F (\beta\cdot\mathrm{n}_F)\llbracket v_h\rrbracket\{\!\!\{w_h\}\!\!\}$$

$$- \sum_{F\in\mathcal{F}_h^i} \int_F \frac{1}{2} |\beta\cdot\mathrm{n}_F| \llbracket v_h\rrbracket\llbracket w_h\rrbracket = \mathfrak{T}_1 + \ldots + \mathfrak{T}_7.$$

Clearly, $|\mathfrak{T}_1| \leq \mathbb{S}\|w_h\|_{\mathrm{da}\sharp} \lesssim \mathbb{S}\|v_h\|_{\mathrm{da}\sharp}$ and

$$|\mathfrak{T}_2| = |a_h^{\mathrm{swip}}(v_h,w_h)| \lesssim \|v_h\|_{\mathrm{da}\flat}\|w_h\|_{\mathrm{da}\flat} \lesssim \|v_h\|_{\mathrm{da}\flat}\|v_h\|_{\mathrm{da}\sharp}.$$

Finally, the terms $\mathfrak{T}_3,\ldots,\mathfrak{T}_7$ are those already bounded in the proof of Lemma 2.35. As a result,

$$\sum_{T\in\mathcal{T}_h} \beta_{\mathrm{c}}^{-1} h_T \|\beta\cdot\nabla v_h\|_{L^2(T)}^2 \lesssim \mathbb{S}\|v_h\|_{\mathrm{da}\sharp} + \|v_h\|_{\mathrm{da}\flat}\|v_h\|_{\mathrm{da}\sharp} + \|v_h\|_{\mathrm{da}\flat}^2.$$

We conclude as in the proof of Lemma 2.35. □

To formulate a boundedness result, we define on V_{*h} the norm

$$\|v\|_{\mathrm{da}\sharp,*} := \left(\|v\|_{\mathrm{da}\sharp}^2 + \sum_{T\in\mathcal{T}_h} \beta_{\mathrm{c}} \left(h_T^{-1} \|v\|_{L^2(T)}^2 + \|v\|_{L^2(\partial T)}^2 \right) \right.$$

$$\left. + \sum_{T\in\mathcal{T}_h} h_T \|\kappa^{1/2}\nabla v\cdot\mathrm{n}_T\|_{L^2(\partial T)}^2 \right)^{1/2}.$$

Lemma 4.63 (Boundedness). *There is C_{bnd}, independent of h and the data κ, β, and $\tilde\mu$, such that*

$$\forall (v,w_h) \in V_{*h} \times V_h, \qquad a_h^{\mathrm{dar}}(v,w_h) \leq C_{\mathrm{bnd}} \|v\|_{\mathrm{da}\sharp,*} \|w_h\|_{\mathrm{da}\sharp}.$$

Proof. Combine Lemma 4.52 with Lemma 2.36. □

A straightforward consequence of Theorem 1.35 is the following error estimate.

Theorem 4.64 (Error estimate). *Under the hypotheses of Theorem 4.61, there is C, independent of h and the data κ, β, and $\tilde\mu$, such that*

$$\|u - u_h\|_{\mathrm{da}\sharp} \leq C \max(1, \tau_{\mathrm{c}}^{-1}\mu_0^{-1}, C_\eta^{-1}) \inf_{v_h\in V_h} \|u - v_h\|_{\mathrm{da}\sharp,*}. \qquad (4.83)$$

4.6. Diffusion-Advection-Reaction

Finally, a convergence rate can be inferred from (4.83) using Lemmata 1.58 and 1.59 if $u \in H^{k+1}(\Omega)$ since (4.83) yields an error estimate with the same upper bound as in (4.82), namely

$$\|u - u_h\|_{\mathrm{da}\sharp} \leq C'_u \max(1, \tau_c^{-1}\mu_0^{-1}, C_\eta^{-1})(\overline{\kappa}^{1/2} + \beta_c^{1/2} h^{1/2} + \tau_c^{-1/2} h) h^k.$$

Thus, in the advection-dominated regime, the convergence rate of $|u - u_h|_\beta + \tau_c^{-1/2}\|u - u_h\|_{L^2(\Omega)} + (\sum_{T \in \mathcal{T}_h} \beta_c^{-1} h_T \|\beta \cdot \nabla v\|_{L^2(T)}^2)^{1/2}$ is of order $h^{k+1/2}$ (as for the pure advection-reaction problem; cf. Sect. 2.3.3), while in the diffusion-dominated regime, the convergence rate of $\|u - u_h\|_{\mathrm{swip}}$ is of order h^k (as for the purely diffusive problem; cf. Sect. 4.5.3).

Remark 4.65 (Harmonic means in the penalty term). The bound (4.65) plays an important role in the proof of Lemma 4.62 since it allows one to bound the jump seminorm $|w_h|_{\mathrm{J},\kappa}$. We observe that this bound results from the fact that the harmonic mean of the diffusion coefficient is used to penalize jumps across interfaces in the SWIP bilinear form.

Remark 4.66 (Numerical fluxes). A local formulation using numerical fluxes can be derived for the discrete problem (4.79) by combining the results of Sect. 4.5.5 for the diffusion terms and those of Sect. 2.3.4 for the advection-reaction terms. Specifically, letting $T \in \mathcal{T}_h$ and $\xi \in \mathbb{P}_d^k(T)$, we infer (compare with (4.77))

$$\int_T (\kappa G_{h,\kappa}^l(u_h) - u_h \beta) \cdot \nabla \xi + \int_T \tilde{\mu} u_h \xi + \sum_{F \in \mathcal{F}_T} \epsilon_{T,F} \int_F \phi_F(u_h) \xi = \int_T f \xi,$$

with $l \in \{k-1, k\}$, $\epsilon_{T,F} = \mathrm{n}_T \cdot \mathrm{n}_F$, and the numerical flux $\phi_F(u_h)$ defined as

$$\phi_F(u_h) := \begin{cases} (-\{\!\!\{\kappa \nabla_h u_h\}\!\!\}_\omega + \beta\{\!\!\{u_h\}\!\!\}) \cdot \mathrm{n}_F + (\eta \frac{\gamma_{\kappa,F}}{h_F} + \frac{1}{2}|\beta \cdot \mathrm{n}_F|)[\![u_h]\!] & \text{if } F \in \mathcal{F}_h^i, \\ -\kappa \nabla_h u_h \cdot \mathrm{n} + (\beta \cdot \mathrm{n})^\oplus u_h + \eta \frac{\gamma_{\kappa,F}}{h_F} u_h & \text{if } F \in \mathcal{F}_h^b. \end{cases}$$

Remark 4.67 (Anisotropic diffusion). In the case of anisotropic diffusion, the SWIP bilinear form is modified as discussed in Sect. 4.5.6. The convergence analysis based on discrete coercivity can be extended to this case. However, it is not clear how to extend the proof of Lemma 4.62 since the bound on $\|\kappa^{1/2} \nabla_h w_h\|_{[L^2(\Omega)]^d}$ uses the assumption that κ is scalar-valued; see [150] for further discussion.

4.6.4 Locally Vanishing Diffusion

In this section, we are interested in the case where κ only takes nonnegative values in the domain Ω, a typical example being that κ vanishes in some parts of Ω. In the anisotropic case, a more complex situation is that where κ only takes symmetric semidefinite values, for instance because different eigenvalues of κ vanish in different parts of Ω.

4.6.4.1 The Continuous Setting

As before, we keep Assumption 4.43 so as to localize the jumps of κ, and we consider the resulting partition P_Ω. We say that I is a *partition interface* if:

(a) I has positive $(d-1)$-dimensional Hausdorff measure.

(b) I is part of a hyperplane, say H_I.

(c) There are distinct Ω_i and Ω_j belonging to P_Ω such that $I = H_I \cap \partial\Omega_i \cap \partial\Omega_j$.

Partition interfaces are collected into the set I_Ω and points in Ω belonging to partition interfaces are collected into the set $\mathcal{I}_\Omega$. Of particular interest are those partition interfaces for which the normal component of the diffusion tensor becomes singular on one of its sides, say Ω_j. Specifically, we set

$$I_{0,\Omega} := \{I \in I_\Omega \mid \mathrm{n}_I^t(\kappa|_{\Omega_i})\mathrm{n}_I > \mathrm{n}_I^t(\kappa|_{\Omega_j})\mathrm{n}_I = 0\},$$

where n_I denotes a unit normal vector to I, and without loss of generality we assume that n_I points from Ω_i toward Ω_j. On a (partition) interface $I \in I_{0,\Omega}$, we loosely say that the subdomain Ω_i is the *diffusive side* and the subdomain Ω_j the *nondiffusive side*. Points in Ω belonging to (partition) interfaces in $I_{0,\Omega}$ are collected into the set $\mathcal{I}_{0,\Omega}$. It is important to identify those points in $\mathcal{I}_{0,\Omega}$ where the advective field flows from the diffusive side to the nondiffusive side and to distinguish them from the remaining points, namely

$$\mathcal{I}_{0,\Omega}^+ := \{x \in \mathcal{I}_{0,\Omega} \mid (\beta \cdot \mathrm{n}_I)(x) > 0\},$$
$$\mathcal{I}_{0,\Omega}^- := \{x \in \mathcal{I}_{0,\Omega} \mid (\beta \cdot \mathrm{n}_I)(x) < 0\},$$

and we assume that $(\beta \cdot \mathrm{n}_I)(x) \neq 0$ for a.e. $x \in \mathcal{I}_{0,\Omega}$. Following Di Pietro, Ern, and Guermond [133], we consider the following diffusion-advection-reaction problem with locally vanishing diffusion

$$\nabla \cdot (-\kappa \nabla u + \beta u) + \tilde{\mu} u = f \quad \text{in } \Omega \setminus \mathcal{I}_{0,\Omega}, \quad (4.84a)$$
$$u = 0 \quad \text{on } \partial\Omega_{\kappa,\beta}, \quad (4.84b)$$

where

$$\partial\Omega_{\kappa,\beta} := \{x \in \partial\Omega \mid \mathrm{n}^t \kappa \mathrm{n} > 0 \text{ or } \beta \cdot \mathrm{n} < 0\},$$

and supplemented with the following conditions on $\mathcal{I}_{0,\Omega}$:

$$[\![-\kappa \nabla u + \beta u]\!] \cdot \mathrm{n}_I = 0 \quad \text{on } \mathcal{I}_{0,\Omega}, \quad (4.85a)$$
$$[\![u]\!] = 0 \quad \text{on } \mathcal{I}_{0,\Omega}^+. \quad (4.85b)$$

We observe that (4.84b) enforces a homogeneous Dirichlet condition if $\mathrm{n}^t \kappa \mathrm{n} > 0$ (as for pure diffusion problems) or if $\beta \cdot \mathrm{n} < 0$ (as for advection-reaction problems). Moreover, (4.85a) enforces the continuity of the normal component of the diffusive-advective flux on the whole partition interface $\mathcal{I}_{0,\Omega}$, whereas (4.85b) enforces the continuity of the exact solution only on $\mathcal{I}_{0,\Omega}^+$, that is, where the

4.6. Diffusion-Advection-Reaction

advection field flows from the diffusive side toward the nondiffusive side, while the exact solution can jump across $\mathcal{I}_{0,\Omega}^-$. We also notice that combining (4.85a) and (4.85b) yields $[\![\kappa \nabla u]\!]\cdot n_I = 0$ across $\mathcal{I}_{0,\Omega}^+$ (recall that β is smooth so that its normal component is continuous across partition interfaces). Since $\kappa|_{\Omega_j}\cdot n_I = 0$ on the nondiffusive side, this yields the homogeneous Neumann condition $(\kappa \nabla u)|_{\Omega_i}\cdot n_I = 0$ on the diffusive side.

The mathematical analysis of the model problem (4.84) with conditions (4.85) can be found in [133]. We only give here a brief motivation for condition (4.85b). In one space dimension, these conditions were derived by Gastaldi and Quarteroni [165], where it is proven that the solution u_ϵ of the following regularized problem with suitable boundary conditions:

$$(-\kappa u'_\epsilon + \beta u_\epsilon)' + \tilde{\mu} u_\epsilon - \epsilon u''_\epsilon = f, \qquad (4.86)$$

converges in $L^2(\Omega)$, as $\epsilon \to 0$, to the so-called viscosity solution of (4.84a) which satisfies conditions (4.85). As an example, let $\Omega = (0,1)$ be partitioned into $\Omega_1 = \left(0, \frac{1}{3}\right)$, $\Omega_2 = \left(\frac{1}{3}, \frac{2}{3}\right)$, and $\Omega_3 = \left(\frac{2}{3}, 1\right)$ and set $f = 0$, $\mu = 0$, $\beta = 1$, $\kappa|_{\Omega_1 \cup \Omega_3} = 1$, and $\kappa|_{\Omega_2} = 0$. Then, $\mathcal{I}_{0,\Omega} = \left\{\frac{1}{3}, \frac{2}{3}\right\}$ with $\mathcal{I}_{0,\Omega}^+ = \left\{\frac{1}{3}\right\}$ and $\mathcal{I}_{0,\Omega}^- = \left\{\frac{2}{3}\right\}$. The viscosity solution to (4.86) with the boundary conditions $u(0) = 1$ and $u(1) = 0$ is (cf. Fig. 4.7)

$$u|_{\Omega_1} = u|_{\Omega_2} = 1, \qquad u|_{\Omega_3} = 1 - e^{(x-1)}.$$

This solution satisfies (4.85).

4.6.4.2 Discretization

We set $V_h = \mathbb{P}_d^k(\mathcal{T}_h)$ with $k \geq 1$ and $\mathcal{T}_h$ belonging to an admissible mesh sequence satisfying Assumption 4.45. In addition, we assume that each (mesh) interface

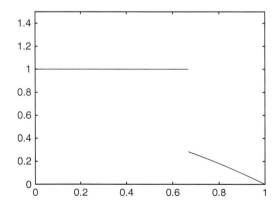

Fig. 4.7: Exact solution with vanishing diffusion

$F \in \mathcal{F}_h^i$ such that $F \cap \mathcal{I}_{0,\Omega}$ has positive $(d-1)$-dimensional Hausdorff measure is either a subset of $\mathcal{I}_{0,\Omega}^-$ or of $\mathcal{I}_{0,\Omega}^+$. We define $\mathcal{F}_h^{i*}$ as the set of (mesh) interfaces such that $F \cap \mathcal{I}_{0,\Omega}$ is a subset of $\mathcal{I}_{0,\Omega}^-$. Without loss of generality, we assume that the normal n_F to each $F \in \mathcal{F}_h^{i*}$ points from the diffusive side, say Ω_1, to the nondiffusive side, say Ω_2. As a result, the weights at F are such that $\omega_{T_1,F} = 0$ and $\omega_{T_2,F} = 1$. We also assume that each boundary face $F \in \mathcal{F}_h^b$ is either a subset of $\partial\Omega_{\kappa,\beta}$ or of $\partial\Omega \setminus \partial\Omega_{\kappa,\beta}$, and we define $\mathcal{F}_h^{b*}$ as the set of boundary faces that are a subset of $\partial\Omega \setminus \partial\Omega_{\kappa,\beta}$. With these definitions, we obtain

$$[\![u]\!] = 0 \quad \forall F \in \mathcal{F}_h \setminus (\mathcal{F}_h^{i*} \cup \mathcal{F}_h^{b*}). \tag{4.87}$$

The key property is that the discrete bilinear form a_h^{dar} defined by (4.78) remains consistent even in the singular limit of vanishing diffusion.

Lemma 4.68 (Consistency). *For all $v_h \in V_h$,*

$$a_h^{\mathrm{dar}}(u, v_h) = \int_\Omega f v_h.$$

Proof. Let $v_h \in V_h$. Consider first the contribution of the SWIP bilinear form. Owing to (4.66) and (4.87),

$$a_h^{\mathrm{swip}}(u, v_h) = -\sum_{T \in \mathcal{T}_h} \int_T \nabla \cdot (\kappa \nabla u) v_h + \sum_{F \in \mathcal{F}_h} \eta \frac{\gamma_{\kappa,F}}{h_F} \int_F [\![u]\!][\![v_h]\!]$$

$$+ \sum_{F \in \mathcal{F}_h^i} \int_F [\![\kappa \nabla_h u]\!] \cdot n_F \{\!\!\{v_h\}\!\!\}_{\overline{\omega}} - \sum_{F \in \mathcal{F}_h} \int_F [\![u]\!] \{\!\!\{\kappa \nabla_h v_h\}\!\!\}_\omega \cdot n_F$$

$$= -\sum_{T \in \mathcal{T}_h} \int_T \nabla \cdot (\kappa \nabla u) v_h + \sum_{F \in \mathcal{F}_h^{i*} \cup \mathcal{F}_h^{b*}} \eta \frac{\gamma_{\kappa,F}}{h_F} \int_F [\![u]\!][\![v_h]\!]$$

$$+ \sum_{F \in \mathcal{F}_h^{i*}} \int_F [\![\kappa \nabla_h u]\!] \cdot n_F \{\!\!\{v_h\}\!\!\}_{\overline{\omega}} - \sum_{F \in \mathcal{F}_h^{i*} \cup \mathcal{F}_h^{b*}} \int_F [\![u]\!] \{\!\!\{\kappa \nabla_h v_h\}\!\!\}_\omega \cdot n_F,$$

where we have used the fact that $[\![\kappa \nabla u]\!] \cdot n_F = 0$ on all $F \in \mathcal{F}_h^i \setminus \mathcal{F}_h^{i*}$ owing to (4.85). Moreover, for all $F \in \mathcal{F}_h^{i*}$, $\gamma_{\kappa,F} = 0$ and $\{\!\!\{\kappa \nabla_h v_h\}\!\!\}_\omega \cdot n_F = 0$ owing to the definition of the penalty parameter and the weights. Similarly, owing to the boundary condition (4.84b), $n^t \kappa n = 0$ for all $F \in \mathcal{F}_h^{b*}$. As a result,

$$a_h^{\mathrm{swip}}(u, v_h) = -\sum_{T \in \mathcal{T}_h} \int_T \nabla \cdot (\kappa \nabla u) v_h + \sum_{F \in \mathcal{F}_h^{i*}} \int_F [\![\kappa \nabla_h u]\!] \cdot n_F v_h|_{\Omega_1},$$

where we have used the fact that $\{\!\!\{v_h\}\!\!\}_{\overline{\omega}} = v_h|_{\Omega_1}$ since $\omega_{T_1,F} = 0$ and $\omega_{T_2,F} = 1$.

4.6. Diffusion-Advection-Reaction

Consider now the contribution of the upwind bilinear form, namely

$$a_h^{\text{upw}}(u, v_h) = \sum_{T \in \mathcal{T}_h} \int_T [\tilde{\mu} u v_h + \nabla \cdot (\beta u) v_h] + \int_{\partial \Omega} (\beta \cdot n)^{\ominus} u v_h$$
$$- \sum_{F \in \mathcal{F}_h^i} \int_F (\beta \cdot n_F) \llbracket u \rrbracket \{v_h\} + \sum_{F \in \mathcal{F}_h^i} \int_F \frac{1}{2} |\beta \cdot n_F| \llbracket u \rrbracket \llbracket v_h \rrbracket$$
$$= \sum_{T \in \mathcal{T}_h} \int_T [\tilde{\mu} u v_h + \nabla \cdot (\beta u) v_h] - \sum_{F \in \mathcal{F}_h^{i*}} \int_F (\beta \cdot n_F) \llbracket u \rrbracket v_h|_{\Omega_1},$$

where we have used (4.87), $(\beta \cdot n)^{\ominus} = 0$ on all $F \in \mathcal{F}_h^{b*}$, and that the upwind side is the nondiffusive side on all $F \in \mathcal{F}_h^{i*}$ so that

$$-(\beta \cdot n_F) \llbracket u \rrbracket \{v_h\} + \frac{1}{2} |\beta \cdot n_F| \llbracket u \rrbracket \llbracket v_h \rrbracket = -(\beta \cdot n_F) \llbracket u \rrbracket \left(\{v_h\} + \frac{1}{2} \llbracket v_h \rrbracket \right)$$
$$= -(\beta \cdot n_F) \llbracket u \rrbracket v_h|_{\Omega_1}.$$

Summing up yields

$$a_h^{\text{dar}}(u, v_h) = \sum_{T \in \mathcal{T}_h} \int_T (\nabla \cdot (-\kappa \nabla u + \beta u) + \tilde{\mu} u) v_h + \sum_{F \in \mathcal{F}_h^{i*}} \int_F \llbracket \kappa \nabla u - \beta u \rrbracket \cdot n_F v_h|_{\Omega_1}$$
$$= \sum_{T \in \mathcal{T}_h} \int_T (\nabla \cdot (-\kappa \nabla u + \beta u) + \tilde{\mu} u) v_h,$$

owing to (4.85a). The assertion follows. □

Remark 4.69 (Amount of upwinding). The choice $\gamma_{\beta,F} = \frac{1}{2} |\beta \cdot n_F|$, corresponding to the usual amount of upwinding, is instrumental in the above proof so as to combine the two terms multiplying $v_h|_{\Omega_1}$ and recover the jump of the total diffusive-advective flux.

The rest of the convergence analysis proceeds as in Sect. 4.6.3 yielding the error estimates (4.81) and (4.83). We observe that the approximate solution exhibits, like the exact solution, a finite jump across $\mathcal{I}_{0,\Omega}^-$. The approximation error on this jump is controlled via the $|\cdot|_\beta$-seminorm present in the error estimates.

To illustrate with a two-dimensional example, we consider $\Omega = (0,1)^2$ partitioned into the two subdomains depicted in the left panel of Fig. 4.8. The subdomain Ω_1 is a trapezoidal inclusion. The diffusion is anisotropic and such that

$$\kappa|_{\Omega_1} = \begin{bmatrix} 1 & 0 \\ 0 & 0.5 \end{bmatrix}, \quad \kappa|_{\Omega_2} = \begin{bmatrix} 0 & 0 \\ 0 & 1 \end{bmatrix}.$$

The advection field is horizontal and uniform with $\beta = (-5, 0)$, and the reaction coefficient is uniform with $\tilde{\mu} = 1$. The partition interface $\mathcal{I}_{0,\Omega}$ consists of the two vertical sides of Ω_1, with $\mathcal{I}_{0,\Omega}^+$ equal to the left side and $\mathcal{I}_{0,\Omega}^-$ to the right side. The approximate solution obtained with the above dG method and polynomial degree $k = 1$ is shown in the right panel of Fig. 4.8, showing that the expected behavior of the exact solution is captured accurately. In particular, the jump across $\mathcal{I}_{0,\Omega}^-$ is clearly visible.

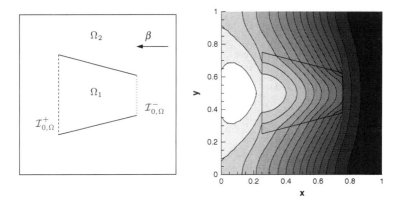

Fig. 4.8: Two-dimensional example of heterogeneous and anisotropic diffusion-advection-reaction problem: problem setting (*left*) and approximate dG solution (*right*)

4.7 An Unsteady Example: The Heat Equation

To illustrate the approximation of unsteady scalar PDEs with diffusion, we consider the heat equation, which we approximate in space using the SIP scheme of Sect. 4.2 and in time using (implicit) A-stable finite difference schemes, e.g., the backward Euler and BDF2 schemes. Implicit time-marching is usually preferred for parabolic problems because explicit schemes lead to the stringent parabolic CFL condition $\delta t \leq Ch^2$ where δt is the time step and h the meshsize.

4.7.1 The Continuous Setting

For given finite time $t_F > 0$, source term f, and initial datum u_0, we consider the unsteady version of the Poisson problem (4.1), namely,

$$\partial_t u - \triangle u = f \quad \text{in } \Omega \times (0, t_F), \qquad (4.88\text{a})$$
$$u = 0 \quad \text{on } \partial\Omega \times (0, t_F), \qquad (4.88\text{b})$$
$$u(\cdot, t = 0) = u_0 \quad \text{in } \Omega. \qquad (4.88\text{c})$$

Problem (4.88) is termed the *heat equation*.

We recall (cf. Sect. 3.1.1) that, for a function ψ defined on the space-time cylinder $\Omega \times (0, t_F)$, we consider ψ as a function of the time variable with values in a Hilbert space, say V, spanned by functions of the space variable, in such a way that

$$\psi : (0, t_F) \ni t \longmapsto \psi(t) \equiv \psi(\cdot, t) \in V.$$

We also recall that, for an integer $l \geq 0$, $C^l(V)$ denotes the space of V-valued functions that are l times continuously differentiable in the interval $[0, t_F]$.

4.7. An Unsteady Example: The Heat Equation

We take the source term f in $C^0(L^2(\Omega))$. Moreover, we are interested in smooth solutions such that

$$u \in C^0(H_0^1(\Omega)) \cap C^1(L^2(\Omega)).$$

This implies, in particular, that the initial datum u_0 is in the energy space $H_0^1(\Omega)$. In addition, since $u \in C^0(H_0^1(\Omega)) \cap C^1(L^2(\Omega))$, $d_t u \in L^2(\Omega)$ for all $t \in (0, t_F)$, so that we consider the following weak formulation of (4.88): For all $t \in [0, t_F]$,

$$(d_t u, v)_{L^2(\Omega)} + a(u, v) = (f, v)_{L^2(\Omega)} \quad \forall v \in H_0^1(\Omega), \tag{4.89}$$

where, as in the steady case, $a(u, v) = \int_\Omega \nabla u \cdot \nabla v$.

We now establish the basic stability result for (4.89).

Lemma 4.70 (Stability). *Let* $u \in C^0(H_0^1(\Omega)) \cap C^1(L^2(\Omega))$ *solve* (4.89). *Then, for all* $t \in [0, t_F]$,

$$\|u(t)\|_{L^2(\Omega)}^2 + \int_0^t \|\nabla u(s)\|_{[L^2(\Omega)]^d}^2 \, ds \leq \|u_0\|_{L^2(\Omega)}^2 + C_\Omega^2 \int_0^t \|f(s)\|_{L^2(\Omega)}^2 \, ds,$$

where C_Ω *results from the Poincaré inequality* (4.4).

Proof. For a fixed $t \in (0, t_F)$, selecting $u(t)$ as a test function in (4.89) and using the Cauchy–Schwarz inequality followed by the Poincaré inequality (4.4), we infer

$$\frac{1}{2} d_t \|u(t)\|_{L^2(\Omega)}^2 + \|\nabla u(t)\|_{[L^2(\Omega)]^d}^2 = (f(t), u(t))_{L^2(\Omega)}$$

$$\leq C_\Omega \|f(t)\|_{L^2(\Omega)} \|\nabla u(t)\|_{[L^2(\Omega)]^d}$$

$$\leq \frac{C_\Omega^2}{2} \|f(t)\|_{L^2(\Omega)}^2 + \frac{1}{2} \|\nabla u(t)\|_{[L^2(\Omega)]^d}^2.$$

Rearranging terms and integrating in time yields the assertion. $\square$

4.7.2 Discretization

As in Chap. 3, we focus on the method of lines in which the time evolution problem (4.89) is first semidiscretized in space yielding a system of coupled ODEs which is then discretized in time. Specifically, we consider space semidiscretization by the SIP dG method of Sect. 4.2 together with a *backward* (also called *implicit*) *Euler scheme* for time discretization. The BDF2 scheme to discretize in time is addressed in Sect. 4.7.4.

4.7.2.1 Space Semidiscretization

Let $V_h = \mathbb{P}_d^k(\mathcal{T}_h)$ with polynomial degree $k \geq 1$ and $\mathcal{T}_h$ belonging to an admissible mesh sequence. The spaces V_* and V_{*h} are defined in Assumption 4.4.

The discrete problem is formulated as follows: For all $t \in [0, t_F]$,
$$(d_t u_h, v_h)_{L^2(\Omega)} + a_h^{\text{sip}}(u_h, v_h) = (f, v_h) \qquad \forall v_h \in V_h,$$
with bilinear form a_h^{sip} defined by (4.10). We introduce the discrete differential operator $A_h^{\text{sip}} : V_{*h} \to V_h$ such that, for all $(v, w_h) \in V_{*h} \times V_h$,
$$(A_h^{\text{sip}} v, w_h)_{L^2(\Omega)} = a_h^{\text{sip}}(v, w_h).$$

The discrete operator A_h^{sip} can be used to formulate the space semidiscrete problem in the following form: For all $t \in [0, t_F]$,
$$d_t u_h(t) + A_h^{\text{sip}} u_h(t) = f_h(t), \qquad (4.90)$$
with initial condition $u_h(0) = \pi_h u_0$ and source term
$$f_h(t) = \pi_h f(t) \qquad \forall t \in [0, t_F],$$
where π_h denotes, as usual, the L^2-orthogonal projection onto V_h. Choosing a basis for V_h, the space semidiscrete evolution problem (4.90) can be transformed into a system of coupled ODEs for the time-dependent components of $u_h(t)$ on the selected basis. Written in component form, (4.90) leads to the appearance of the mass matrix in front of the time derivative. In the context of dG methods, the mass matrix is block-diagonal; cf. Sect. A.1.2.

We now state the important properties of the discrete operator A_h^{sip}. These properties result from the consistency, discrete coercivity, and boundedness of the SIP bilinear form.

Lemma 4.71 (Properties of A_h^{sip}). *The discrete operator A_h^{sip} satisfies the following properties:*

(i) *Consistency:* For the exact solution u, assuming $u \in C^0(V_*)$,
$$\pi_h d_t u(t) + A_h^{\text{sip}} u(t) = f_h(t) \qquad \forall t \in [0, t_F].$$

(ii) *Discrete coercivity:* For all $v_h \in V_h$,
$$(A_h^{\text{sip}} v_h, v_h)_{L^2(\Omega)} \geq C_{\text{sta}} \|v_h\|_{\text{sip}}^2.$$

(iii) *Boundedness:* For all $(v, w_h) \in V_{*h} \times V_h$,
$$(A_h^{\text{sip}} v, w_h)_{L^2(\Omega)} \leq C_{\text{bnd}} \|v\|_{\text{sip},*} \|w_h\|_{\text{sip}}.$$

Here, C_{sta} and C_{bnd} are independent of h and δt, the $\|\cdot\|_{\text{sip}}$-norm is defined by (4.17), and the $\|\cdot\|_{\text{sip},*}$-norm by (4.22).

4.7.2.2 Time Discretization

To discretize in time the space semidiscrete problem (4.90), we introduce a partition of $(0, t_F)$ into N intervals of length δt (the time step) so that $N\delta t = t_F$; more generally, a variable time step can be considered. For $n \in \{0, \ldots, N\}$, a superscript n indicates the value of a function at the discrete time $t^n := n\delta t$.

For a function $v \in C^1(V)$ with some function space V of the space variable, we introduce the backward Euler operator $\delta_t^{(1)}$ such that, for all $n \in \{1, \ldots, N\}$,

$$\delta_t^{(1)} v^n := \frac{v^n - v^{n-1}}{\delta t} \in V, \tag{4.91}$$

thereby providing a first-order finite difference approximation of the time derivative $d_t v^n$. The discrete solution is obtained from the backward Euler scheme,

$$\delta_t^{(1)} u_h^{n+1} + A_h^{\text{sip}} u_h^{n+1} = f_h^{n+1}, \tag{4.92}$$

with the initial condition $u_h^0 = \pi_h u_0$ and the source term $f_h^{n+1} = \pi_h f^{n+1}$ for all $n \in \{0, \ldots, N-1\}$. Problem (4.92) can be equivalently rewritten as

$$u_h^{n+1} + \delta t A_h^{\text{sip}} u_h^{n+1} = u_h^n + \delta t f_h^{n+1},$$

thus highlighting the fact that u_h^{n+1} is obtained from u_h^n by solving a linear problem. In what follows, we abbreviate as $a \lesssim b$ the inequality $a \leq Cb$ with positive C independent of h, δt, and f.

4.7.3 Error Estimates

The analysis follows a path similar to that deployed in Sect. 3.1 for the unsteady advection-reaction equation. We first derive the error equation, then establish an energy estimate and finally infer the convergence result. The analysis is much simpler than in Sect. 3.1 since we use an implicit scheme to march in time. Indeed, contrary to explicit schemes, implicit schemes are dissipative.

Letting

$$\xi_h^n := u_h^n - \pi_h u^n, \qquad \xi_\pi^n := u^n - \pi_h u^n, \tag{4.93}$$

the approximation error at the discrete time t^n is decomposed as

$$u^n - u_h^n = \xi_\pi^n - \xi_h^n.$$

Lemma 4.72 (Error equation). *Assume $u \in C^0(V_*) \cap C^2(L^2(\Omega))$. Then,*

$$\delta_t^{(1)} \xi_h^{n+1} = -A_h^{\text{sip}} \xi_h^{n+1} + \alpha_h^{n+1}, \tag{4.94}$$

with $\alpha_h^{n+1} := A_h^{\text{sip}} \xi_\pi^{n+1} - \pi_h \theta^{n+1}$ and $\theta^{n+1} := \delta t^{-1} \int_{t^n}^{t^{n+1}} (t^n - t) d_t^2 u(t)\, dt$.

Proof. A Taylor expansion in time yields

$$u^n = u^{n+1} - \delta t\, d_t u^{n+1} - \int_{t^n}^{t^{n+1}} (t^n - t) d_t^2 u(t)\, dt, \tag{4.95}$$

i.e.,
$$\delta_t^{(1)} u^{n+1} = d_t u^{n+1} + \theta^{n+1}.$$

Projecting this equation onto V_h and replacing $\pi_h d_t u^{n+1}$ by $f_h^{n+1} - A_h^{\mathrm{sip}} u^{n+1}$ owing to consistency, we obtain

$$\delta_t^{(1)} \pi_h u^{n+1} + A_h^{\mathrm{sip}} u^{n+1} = f_h^{n+1} + \pi_h \theta^{n+1}. \qquad (4.96)$$

Subtracting (4.96) from (4.92) yields the assertion. $\square$

We now derive an energy estimate for the discrete scheme.

Lemma 4.73 (Energy estimate). *Assume $u \in C^0(V_*) \cap C^2(L^2(\Omega))$. Then, for all $n \in \{0, \ldots, N-1\}$, there holds*

$$\|\xi_h^{n+1}\|_{L^2(\Omega)}^2 \quad \|\xi_h^n\|_{L^2(\Omega)}^2 + \|\xi_h^{n+1} - \xi_h^n\|_{L^2(\Omega)}^2 + \delta t C_{\mathrm{sta}} \|\xi_h^{n+1}\|_{\mathrm{sip}}^2$$
$$\lesssim \delta t (\|\xi_\pi^{n+1}\|_{\mathrm{sip},*}^2 + C_u^2 \delta t^2), \qquad (4.97)$$

with $C_u := \|d_t^2 u\|_{C^0(L^2(\Omega))}$.

Proof. Testing (4.94) with $\delta t \xi_h^{n+1}$, we obtain

$$\|\xi_h^{n+1}\|_{L^2(\Omega)}^2 + \delta t (A_h^{\mathrm{sip}} \xi_h^{n+1}, \xi_h^{n+1})_{L^2(\Omega)}$$
$$= (\xi_h^n, \xi_h^{n+1})_{L^2(\Omega)} + \delta t (A_h^{\mathrm{sip}} \xi_\pi^{n+1}, \xi_h^{n+1})_{L^2(\Omega)} - \delta t (\theta^{n+1}, \xi_h^{n+1})_{L^2(\Omega)},$$

since $(\pi_h \theta^{n+1}, \xi_h^{n+1})_{L^2(\Omega)} = (\theta^{n+1}, \xi_h^{n+1})_{L^2(\Omega)}$. Using the algebraic relation $ab = \frac{1}{2} a^2 + \frac{1}{2} b^2 - \frac{1}{2}(a-b)^2$ for the first term on the right-hand side, the above energy equality becomes

$$\|\xi_h^{n+1}\|_{L^2(\Omega)}^2 + \|\xi_h^{n+1} - \xi_h^n\|_{L^2(\Omega)}^2 + 2\delta t (A_h^{\mathrm{sip}} \xi_h^{n+1}, \xi_h^{n+1})_{L^2(\Omega)}$$
$$= \|\xi_h^n\|_{L^2(\Omega)}^2 + 2\delta t (A_h^{\mathrm{sip}} \xi_\pi^{n+1}, \xi_h^{n+1})_{L^2(\Omega)} - 2\delta t (\theta^{n+1}, \xi_h^{n+1})_{L^2(\Omega)}.$$

Using discrete coercivity and boundedness of A_h^{sip} together with the Cauchy–Schwarz inequality, we obtain

$$\|\xi_h^{n+1}\|_{L^2(\Omega)}^2 + \|\xi_h^{n+1} - \xi_h^n\|_{L^2(\Omega)}^2 + 2\delta t C_{\mathrm{sta}} \|\xi_h^{n+1}\|_{\mathrm{sip}}^2$$
$$\leq \|\xi_h^n\|_{L^2(\Omega)}^2 + 2C_{\mathrm{bnd}} \delta t \|\xi_\pi^{n+1}\|_{\mathrm{sip},*} \|\xi_h^{n+1}\|_{\mathrm{sip}} + 2\delta t \|\theta^{n+1}\|_{L^2(\Omega)} \|\xi_h^{n+1}\|_{L^2(\Omega)}.$$

Owing to the discrete Poincaré inequality (4.20),

$$\|\xi_h^{n+1}\|_{L^2(\Omega)}^2 + \|\xi_h^{n+1} - \xi_h^n\|_{L^2(\Omega)}^2 + 2\delta t C_{\mathrm{sta}} \|\xi_h^{n+1}\|_{\mathrm{sip}}^2$$
$$\leq \|\xi_h^n\|_{L^2(\Omega)}^2 + 2C_{\mathrm{bnd}} \delta t \|\xi_\pi^{n+1}\|_{\mathrm{sip},*} \|\xi_h^{n+1}\|_{\mathrm{sip}} + 2\sigma_2 \delta t \|\theta^{n+1}\|_{L^2(\Omega)} \|\xi_h^{n+1}\|_{\mathrm{sip}}.$$

Using Young's inequality for the two rightmost terms yields

$$\|\xi_h^{n+1}\|_{L^2(\Omega)}^2 + \|\xi_h^{n+1} - \xi_h^n\|_{L^2(\Omega)}^2 + \delta t C_{\mathrm{sta}} \|\xi_h^{n+1}\|_{\mathrm{sip}}^2$$
$$\leq \|\xi_h^n\|_{L^2(\Omega)}^2 + C \delta t (\|\xi_\pi^{n+1}\|_{\mathrm{sip},*}^2 + \|\theta^{n+1}\|_{L^2(\Omega)}^2).$$

4.7. An Unsteady Example: The Heat Equation

Finally, proceeding as in the proof of Lemma 3.20, we infer
$$\|\theta^{n+1}\|_{L^2(\Omega)} \lesssim C_u \delta t,$$
whence the assertion. □

Remark 4.74 (Dissipation in backward Euler scheme). The dissipative nature of the backward Euler scheme is reflected by the presence of the time increment $\|\xi_h^{n+1} - \xi_h^n\|_{L^2(\Omega)}^2$ on the left-hand side of the energy estimate (4.97). We observe that, up to a factor δt, the increment $\xi_h^{n+1} - \xi_h^n$ can be interpreted as a first-order finite difference approximation of the time derivative of the error component in V_h.

Finally, we arrive at our main convergence result.

Theorem 4.75 (Convergence). *Let $u \in C^0(V_*) \cap C^2(L^2(\Omega))$ solve (4.89) and let $(u_h^n)_{1 \le n \le N}$ solve (4.92) with $u_h^0 = \pi_h u_0$. Assume $u \in C^0(H^{k+1}(\Omega))$. Then, there holds*

$$\|u^N - u_h^N\|_{L^2(\Omega)} + \left(C_{\text{sta}} \sum_{n=1}^N \delta t \|u^n - u_h^n\|_{\text{sip}}^2\right)^{1/2} \lesssim \chi_1 \delta t + \chi_2 h^k, \tag{4.98}$$

with $\chi_1 = t_F^{1/2} \|d_t^2 u\|_{C^0(L^2(\Omega))}$ and $\chi_2 = \|u\|_{C^0(H^{k+1}(\Omega))}$.

Proof. Summing (4.97) for $n \in \{0, \ldots, N-1\}$, dropping the nonnegative contribution $\|\xi_h^{n+1} - \xi_h^n\|_{L^2(\Omega)}^2$, and observing that $\xi_h^0 = 0$, we obtain

$$\|\xi_h^N\|_{L^2(\Omega)}^2 + \delta t C_{\text{sta}} \sum_{n=1}^N \|\xi_h^n\|_{\text{sip}}^2 \lesssim \delta t \sum_{n=1}^N \|\xi_\pi^n\|_{\text{sip},*}^2 + t_F C_u^2 \delta t^2.$$

Recalling the results of Sect. 4.2.3, we infer that, for all $n \in \{1, \ldots, N\}$, $\|\xi_\pi^n\|_{\text{sip},*} \lesssim h^k \|u^n\|_{H^{k+1}(\Omega)}$. Hence,

$$\|\xi_h^N\|_{L^2(\Omega)}^2 + \delta t C_{\text{sta}} \sum_{n=1}^N \|\xi_h^n\|_{\text{sip}}^2 \lesssim (\chi_1 \delta t + \chi_2 h^k)^2.$$

Using the triangle inequality $\|u^N - u_h^N\|_{L^2(\Omega)} \le \|\xi_\pi^N\|_{L^2(\Omega)} + \|\xi_h^N\|_{L^2(\Omega)}$ together with

$$\sum_{n=1}^N \delta t \|u^n - u_h^n\|_{\text{sip}}^2 \le 2 \sum_{n=1}^N \delta t \left(\|\xi_h^n\|_{\text{sip}}^2 + \|\xi_\pi^n\|_{\text{sip}}^2\right),$$

and observing that

$$\|\xi_\pi^N\|_{L^2(\Omega)}^2 + \delta t C_{\text{sta}} \sum_{n=1}^N \|\xi_\pi^n\|_{\text{sip}}^2 \lesssim (\chi_2 h^k)^2,$$

yields the assertion. □

4.7.4 BDF2 Time Discretization

To improve the convergence rate in time, we can consider higher-order backward approximations of the time derivative $d_t u$. In this section, we briefly examine time discretization using the *second-order backward difference formula (BDF2)*. We show, in particular, that also in this case, stability is related to the dissipative nature of the scheme. We proceed as in Sect. 4.7.3, whereby we derive the error equation, establish an energy estimate, and finally infer the convergence result.

For a function $v \in C^1(V)$, we introduce the BDF2 operator $\delta_t^{(2)}$ such that, for all $n \in \{2, \ldots, N\}$,

$$\delta_t^{(2)} v^n := \frac{3v^n - 4v^{n-1} + v^{n-2}}{2\delta t} \in V,$$

thereby providing a second-order finite difference approximation of the time derivative $d_t v^n$. Then, the discrete solution is obtained from

$$\delta_t^{(1)} u_h^1 + A_h^{\text{sip}} \frac{u_h^0 + u_h^1}{2} = \frac{f_h^0 + f_h^1}{2}, \tag{4.99a}$$

$$\delta_t^{(2)} u_h^{n+1} + A_h^{\text{sip}} u_h^{n+1} = f_h^{n+1} \quad \text{for } n \in \{1, \ldots, N-1\}, \tag{4.99b}$$

with $u_h^0 = \pi_h u_0$. The operator $\delta_t^{(2)}$ cannot be used for the first time step $n = 1$, since only the initial value is available. In (4.99a), the value u_h^1 is computed from u_h^0 using the Crank–Nicolson scheme which is also second-order accurate in time.

We first derive the error equation, recalling that the components ξ_h^n and ξ_π^n of the approximation errors are defined by (4.93).

Lemma 4.76 (Error equation). *Assume $u \in C^0(V_*) \cap C^3(L^2(\Omega))$. Then,*

$$\delta_t^{(1)} \xi_h^1 + A_h^{\text{sip}} \frac{\xi_h^0 + \xi_h^1}{2} = \alpha_h^1, \tag{4.100}$$

where $\alpha_h^1 := A_h^{\text{sip}} \frac{\xi_\pi^0 + \xi_\pi^1}{2} - \pi_h \theta^1$, $\theta^1 := -\frac{1}{2} \delta t^{-1} \int_0^{\delta t} t(\delta t - t) d_t^3 u(t) \, dt$, and, for all $n \in \{1, \ldots, N-1\}$,

$$\delta_t^{(2)} \xi_h^{n+1} + A_h^{\text{sip}} \xi_h^{n+1} = \alpha_h^{n+1}, \tag{4.101}$$

where $\alpha_h^{n+1} := A_h^{\text{sip}} \xi_\pi^{n+1} - \pi_h \theta^{n+1}$ and

$$\theta^{n+1} := \delta t^{-1} \int_{t^n}^{t^{n+1}} (t^n - t)^2 d_t^3 u(t) \, dt - \frac{1}{4} \delta t^{-1} \int_{t^{n-1}}^{t^{n+1}} (t^{n-1} - t)^2 d_t^3 u(t) \, dt.$$

Proof. We observe that

$$\delta_t^{(1)} u^1 = d_t u^1 - \delta t^{-1} \int_0^{\delta t} t \, d_t^2 u(t) \, dt,$$

$$\delta_t^{(1)} u^1 = d_t u^0 + \delta t^{-1} \int_0^{\delta t} (\delta t - t) d_t^2 u(t) \, dt,$$

4.7. An Unsteady Example: The Heat Equation

so that, integrating by parts in time,

$$\delta_t^{(1)} u^1 = d_t \frac{u^0 + u^1}{2} + \frac{1}{2}\delta t^{-1} \int_0^{\delta t} (\delta t - 2t) d_t^2 u(t) \, dt = d_t \frac{u^0 + u^1}{2} + \theta^1.$$

Proceeding as in the proof of Lemma 4.72 yields (4.100). Furthermore, for all $n \in \{2, \ldots, N\}$, a direct calculation shows that

$$\delta_t^{(2)} u^{n+1} = d_t u^{n+1} + \theta^{n+1},$$

and proceeding again as in the proof of Lemma 4.72 yields (4.101). □

We can now derive the energy estimate.

Lemma 4.77 (Energy estimate). *Assume* $u \in C^0(V_*) \cap C^3(L^2(\Omega))$. *Then, there holds*

$$\|\xi_h^1\|_{L^2(\Omega)}^2 + \delta t C_{\text{sta}} \|\xi_h^1\|_{\text{sip}}^2 \lesssim \delta t \left(\|\xi_\pi^1\|_{\text{sip},*}^2 + \|\xi_h^0\|_{\text{sip},*}^2 + C_u^2 \delta t^4 \right), \quad (4.102)$$

and, for all $n \in \{1, \ldots, N-1\}$,

$$\|\xi_h^{n+1}\|_{L^2(\Omega)}^2 - \|\xi_h^n\|_{L^2(\Omega)}^2 + \|2\xi_h^{n+1} - \xi_h^n\|_{L^2(\Omega)}^2 - \|2\xi_h^n - \xi_h^{n-1}\|_{L^2(\Omega)}^2$$
$$+ \|\delta_{tt} \xi_h^{n+1}\|_{L^2(\Omega)}^2 + \delta t C_{\text{sta}} \|\xi_h^{n+1}\|_{\text{sip}}^2 \lesssim \delta t \left(\|\xi_\pi^{n+1}\|_{\text{sip},*}^2 + C_u^2 \delta t^4 \right), \quad (4.103)$$

with $C_u := \|d_t^3 u\|_{C^0(L^2(\Omega))}$ *and* $\delta_{tt} \xi_h^{n+1} := \xi_h^{n+1} - 2\xi_h^n + \xi_h^{n-1}$.

Proof. Testing (4.100) with $\delta t \xi_h^1$, observing that $\xi_h^0 = 0$, rearranging terms, and using $\|\theta^1\|_{L^2(\Omega)} \lesssim C_u \delta t^2$ yields (4.102). Furthermore, testing (4.101) with $4\delta t \xi_h^{n+1}$, we infer

$$4\delta t (\delta_t^{(2)} \xi_h^{n+1}, \xi_h^{n+1})_{L^2(\Omega)} + 4\delta t (A_h \xi_h^{n+1}, \xi_h^{n+1})_{L^2(\Omega)} = 4\delta t (\alpha_h^{n+1}, \xi_h^{n+1})_{L^2(\Omega)}.$$

We observe that

$$4\delta t (\delta_t^{(2)} \xi_h^{n+1}, \xi_h^{n+1})_{L^2(\Omega)} = \|\xi_h^{n+1}\|_{L^2(\Omega)}^2 - \|\xi_h^n\|_{L^2(\Omega)}^2$$
$$+ \|2\xi_h^{n+1} - \xi_h^n\|_{L^2(\Omega)}^2 - \|2\xi_h^n - \xi_h^{n-1}\|_{L^2(\Omega)}^2$$
$$+ \|\delta_{tt} \xi_h^{n+1}\|_{L^2(\Omega)}^2.$$

Finally, we bound $(\alpha_h^{n+1}, \xi_h^{n+1})_{L^2(\Omega)}$ by proceeding as in the proof of Lemma 4.73 using $\|\theta^{n+1}\|_{L^2(\Omega)} \lesssim C_u \delta t^2$ for all $n \in \{1, \ldots, N-1\}$. □

Remark 4.78 (Dissipation in BDF2 scheme). The dissipative nature of the BDF2 scheme is reflected by the term $\|\delta_{tt} \xi_h^{n+1}\|_{L^2(\Omega)}^2$ on the left-hand side of (4.103). We observe that, up to a factor δt^2, $\delta_{tt} \xi_h^{n+1}$ can be interpreted as a second-order finite difference approximation of the second-order time derivative of the error component in V_h.

Finally, we arrive at our main convergence result. The proof is skipped since it is similar to that of Theorem 4.75.

Theorem 4.79 (Convergence). *Let $u \in C^0(V_*) \cap C^3(L^2(\Omega))$ solve (4.89) and let $(u_h^n)_{1 \le n \le N}$ solve (4.99) with $u_h^0 = \pi_h u_0$. Assume $u \in C^0(H^{k+1}(\Omega))$. Then, there holds*

$$\|u^N - u_h^N\|_{L^2(\Omega)} + \left(C_{\text{sta}} \sum_{n=1}^{N} \delta t \|u^n - u_h^n\|_{\text{sip}}^2\right)^{1/2} \lesssim \chi_1 \delta t^2 + \chi_2 h^k,$$

with $\chi_1 = t_{\text{F}}^{1/2} \|d_t^3 u\|_{C^0(L^2(\Omega))}$ and $\chi_2 = \|u\|_{C^0(H^{k+1}(\Omega))}$.

4.7.5 Improved $C^0(L^2(\Omega))$-Error Estimate

The error estimate (4.98) is suboptimal for the term $\|u^N - u_h^N\|_{L^2(\Omega)}$ by one power in h. Following the ideas of Wheeler [305], a sharper result can be obtained by replacing the L^2-orthogonal projector in (4.93) by the so-called *elliptic projector* $\pi_{\text{ell}} \in \mathcal{L}(V_*, V_h)$ such that, for all $w \in V_*$, $\pi_{\text{ell}} w \in V_h$ solves

$$a_h^{\text{sip}}(\pi_{\text{ell}} w, v_h) = a_h^{\text{sip}}(w, v_h) \qquad \forall v_h \in V_h,$$

or, equivalently,

$$A_h^{\text{sip}} \pi_{\text{ell}} w = A_h^{\text{sip}} w.$$

If elliptic regularity holds (cf. Definition 4.24), there is C, independent of h, such that, for all $w \in V_* \cap H^{k+1}(\Omega)$,

$$\|w - \pi_{\text{ell}} w\|_{L^2(\Omega)} \le C h^{k+1} \|w\|_{H^{k+1}(\Omega)}. \tag{4.104}$$

Redefining the components of the approximation error as

$$\zeta_\pi^n := u^n - \pi_{\text{ell}} u^n, \qquad \zeta_h^n := \pi_{\text{ell}} u^n - u_h^n,$$

the approximation error at the discrete time t^n is now decomposed as

$$u^n - u_h^n = \zeta_\pi^n + \zeta_h^n.$$

We consider again the backward Euler operator $\delta_t^{(1)}$ defined by (4.91).

Lemma 4.80 (Error equation). *Assume $u \in C^0(V_*) \cap C^2(L^2(\Omega))$. Then,*

$$\delta_t^{(1)} \zeta_h^{n+1} + A_h^{\text{sip}} \zeta_h^{n+1} = \alpha_h^{n+1}, \tag{4.105}$$

with $\alpha_h^{n+1} := \pi_h \delta_t^{(1)} \zeta_\pi^{n+1} - \pi_h \theta^{n+1}$ and θ^{n+1} defined in Lemma 4.72.

Proof. Recalling that $\delta_t^{(1)} u^{n+1} = d_t u^{n+1} + \theta^{n+1}$ and observing that $\delta_t^{(1)} u^{n+1} = \delta_t^{(1)} \pi_{\text{ell}} u^{n+1} + \delta_t^{(1)} \zeta_\pi^{n+1}$, we obtain

$$\delta_t^{(1)} \pi_{\text{ell}} u^{n+1} = d_t u^{n+1} + \theta^{n+1} - \delta_t^{(1)} \zeta_\pi^{n+1}.$$

4.7. An Unsteady Example: The Heat Equation

Projecting this equation onto V_h, replacing $\pi_h d_t u^{n+1}$ by $f_h^{n+1} - A_h^{\text{sip}} u^{n+1}$, and observing that $A_h^{\text{sip}} \pi_{\text{ell}} u^{n+1} = A_h^{\text{sip}} u^{n+1}$, we infer

$$\delta_t^{(1)} \pi_{\text{ell}} u^{n+1} = f_h^{n+1} - A_h^{\text{sip}} \pi_{\text{ell}} u^{n+1} + \pi_h \theta^{n+1} - \pi_h \delta_t^{(1)} \zeta_\pi^{n+1}.$$

Subtracting this relation from (4.92) yields the assertion. □

The difference between (4.105) and (4.94) lies in the residual α_h^{n+1}. When using the elliptic projector, the term $A_h^{\text{sip}} \zeta_\pi^{n+1}$ in α_h^{n+1} is replaced $\pi_h \delta_t^{(1)} \zeta_\pi^{n+1}$. This is a key point, since the latter scales in space as h^{k+1} (cf. the proof of Theorem 4.82 below), while the former only scales as h^k.

The next step is to derive an energy estimate. The proof is skipped since it is similar to that of Lemma 4.73.

Lemma 4.81 (Energy estimate). *For all $n \in \{0,\ldots,N-1\}$, there holds*

$$\|\zeta_h^{n+1}\|_{L^2(\Omega)}^2 - \|\zeta_h^n\|_{L^2(\Omega)}^2 + \|\zeta_h^{n+1} - \zeta_h^n\|_{L^2(\Omega)}^2 + \delta t C_{\text{sta}} \|\zeta_h^{n+1}\|_{\text{sip}}^2$$
$$\lesssim \delta t (\|\delta_t^{(1)} \zeta_\pi^{n+1}\|_{L^2(\Omega)}^2 + C_u^2 \delta t^2), \quad (4.106)$$

with $C_u := \|d_t^2 u\|_{C^0(L^2(\Omega))}$.

Finally, we arrive at our improved convergence result.

Theorem 4.82 (Convergence). *Let $u \in C^0(V_*) \cap C^2(L^2(\Omega))$ solve (4.89) and additionally assume that $u \in C^1(H^{k+1}(\Omega))$. Then,*

$$\|u^N - u_h^N\|_{L^2(\Omega)} \lesssim \chi_1 \delta t + \chi_2 h^{k+1},$$

with $\chi_1 = t_F^{1/2} \|d_t^2 u\|_{C^0(L^2(\Omega))}$ and $\chi_2 = t_F^{1/2} \|u\|_{C^1(H^{k+1}(\Omega))}$.

Proof. We first observe that

$$\delta_t^{(1)} \zeta_\pi^{n+1} = \pi_{\text{ell}} \delta_t^{(1)} u^{n+1} - \delta_t^{(1)} u^{n+1},$$

so that, owing to (4.104),

$$\|\delta_t^{(1)} \zeta_\pi^{n+1}\|_{L^2(\Omega)} \lesssim h^{k+1} \|\delta_t^{(1)} u^{n+1}\|_{H^{k+1}(\Omega)}.$$

Moreover,

$$\|\delta_t^{(1)} u^{n+1}\|_{H^{k+1}(\Omega)}^2 = \sum_{|\alpha| \leq k+1} \int_\Omega \frac{1}{\delta t^2} \left| \int_{t^n}^{t^{n+1}} \partial^\alpha d_t u(s) \, ds \right|^2$$
$$\leq \sum_{|\alpha| \leq k+1} \int_\Omega \frac{1}{\delta t} \int_{t^n}^{t^{n+1}} |\partial^\alpha d_t u(s)|^2 \, ds$$
$$= \frac{1}{\delta t} \int_{t^n}^{t^{n+1}} \|d_t u(s)\|_{H^{k+1}(\Omega)}^2 \, ds$$
$$\leq \|d_t u\|_{C^0(H^{k+1}(\Omega))}^2 \leq \|u\|_{C^1(H^{k+1}(\Omega))}^2.$$

Then, summing (4.106) for $n \in \{0, \ldots, N-1\}$, dropping the nonnegative contribution $\|\zeta_h^{n+1} - \zeta_h^n\|_{L^2(\Omega)}^2$, and observing that $\zeta_h^0 = 0$ we obtain

$$\|\zeta_h^N\|_{L^2(\Omega)}^2 + \delta t C_{\text{sta}} \sum_{n=1}^{N} \|\zeta_h^n\|_{\text{sip}}^2 \lesssim (\chi_1 \delta t + \chi_2 h^{k+1})^2. \qquad (4.107)$$

Owing to the triangle inequality, $\|u^N - u_h^N\|_{L^2(\Omega)} \leq \|\zeta_\pi^N\|_{L^2(\Omega)} + \|\zeta_h^N\|_{L^2(\Omega)}$ and we conclude using (4.104). $\square$

Remark 4.83 (Superconvergence of $\delta t \sum_{n=1}^{N} \|\zeta_h^n\|_{\text{sip}}^2$). The error decomposition based on ζ_h^n and ζ_π^n can also be used to derive an energy-error estimate of order h^k in space. The bound (4.107) shows that $\left(\sum_{n=1}^{N} \delta t \|\zeta_h^n\|_{\text{sip}}^2\right)^{1/2}$ scales as h^{k+1}, and, therefore, superconverges. The leading order term in the energy-error estimate is the projection term $\left(\sum_{n=1}^{N} \delta t \|\zeta_\pi^n\|_{\text{sip}}\right)^{1/2}$ which scales as h^k.

Chapter 5
Additional Topics on Pure Diffusion

In this chapter, we discuss some advanced topics focusing on the Poisson problem with homogeneous Dirichlet boundary conditions (4.1). Most of the results can be extended to more general scalar PDEs with diffusion. In Sect. 5.1, using basic properties of the space of functions with bounded variation (the so-called BV space), we derive discrete Sobolev embeddings for broken polynomial spaces equipped with the $\|\cdot\|_{\mathrm{dG},p}$–norms (5.1), and a discrete Rellich–Kondrachov compactness theorem for sequences bounded in the $\|\cdot\|_{\mathrm{dG},p}$-norm. We also prove an important result concerning the (weak) asymptotic consistency of the discrete gradients introduced in Sect. 4.3.2. In Sect. 5.2, we revisit the approximation of the Poisson problem by the SIP method (cf. Sect. 4.2) and use the results of Sect. 5.1 to prove the convergence of the sequence of discrete solutions to minimal regularity solutions. The results of Sects. 5.1 and 5.2 lay the basis for the convergence analysis of the dG method for the Navier–Stokes equations presented in Chap. 6. In Sect. 5.3, we briefly review some possible variations on penalty and symmetry that can be envisaged for the SIP bilinear form. A different extension is considered in Sect. 5.4, where we introduce a consistent discretization featuring one degree of freedom per cell in the spirit of cell-centered Galerkin methods. Such methods bridge somewhat the gap between dG and finite volume methods, while at the same time reducing computational costs with respect to the standard dG approach of Chap. 4 that requires at least piecewise affine polynomials.

In Sect. 5.5, we return to the SIP method (variations on symmetry and penalty can be easily accommodated). Referring to Definition 4.1, our goal is to postprocess the discrete solution so as to reconstruct a discrete potential that is $H_0^1(\Omega)$-conforming and a diffusive flux that is $H(\mathrm{div};\Omega)$-conforming. Finally, in Sect. 5.6, we present a posteriori error estimates for the Poisson problem. Such estimates deliver upper and lower bounds on the approximation error. The upper bound is fully computable and, as such, can be used to certify the accuracy of the discrete solution. Moreover, the estimates can be decomposed as elementwise sums of error indicators that, in turn, can be used to adapt the mesh by concentrating degrees of freedom where they are actually needed to reduce the error most substantially.

5.1 Discrete Functional Analysis

This section contains important results of discrete functional analysis. These results are used in Sect. 5.2 (and in Chap. 6) to analyze the convergence, as the meshsize goes to zero, of dG approximations when the exact solution only satisfies minimal regularity assumptions. The results presented in this section are discrete Sobolev embeddings and a discrete Rellich–Kondrachov compactness theorem in broken polynomial spaces, together with the (weak) asymptotic consistency of discrete gradients. Such results, which hinge on BV-norm estimates, have been derived by the authors [131] in the context of dG methods, taking inspiration from the penetrating work of Eymard, Gallouët, and Herbin [159] in the context of finite volume methods. These results have also been obtained in a similar form independently by Buffa and Ortner [61].

Let $\mathcal{T}_h$ be a mesh of Ω belonging to an admissible mesh sequence with mesh regularity parameters denoted by ϱ. As in Chap. 4 (cf. Definition 4.5), in dimension $d \geq 2$, h_F denotes the diameter of the face F (other local length scales can be chosen as discussed in Remark 4.6), while, in dimension 1, we set $h_F := \min(h_{T_1}, h_{T_2})$ if $F \in \mathcal{F}_h^i$ with $F = \partial T_1 \cap \partial T_2$ and $h_F := h_T$ if $F \in \mathcal{F}_h^b$ with $F = \partial T \cap \partial \Omega$.

5.1.1 The BV Space and the $\|\cdot\|_{\mathrm{dG},p}$-Norms

For $v \in L^1(\mathbb{R}^d)$, we introduce its $\|\cdot\|_{\mathrm{BV}}$-norm

$$\|v\|_{\mathrm{BV}} := \sum_{i=1}^d \sup\left\{ \int_{\mathbb{R}^d} v \partial_i \varphi \mid \varphi \in C_0^\infty(\mathbb{R}^d),\ \|\varphi\|_{L^\infty(\mathbb{R}^d)} \leq 1 \right\},$$

and we define the space of integrable functions with bounded variation in $\mathbb{R}^d$ as

$$\mathrm{BV}(\mathbb{R}^d) := \{ v \in L^1(\mathbb{R}^d) \mid \|v\|_{\mathrm{BV}} < \infty \}.$$

For a real number $1 \leq p < \infty$, let $W^{1,p}(\mathcal{T}_h)$ be the broken Sobolev space defined by (1.17). We introduce the following norms: For all $v \in W^{1,p}(\mathcal{T}_h)$,

$$\begin{aligned}
\|v\|_{\mathrm{dG},p} &:= \left(\|\nabla_h v\|_{[L^p(\Omega)]^d}^p + \sum_{F \in \mathcal{F}_h} \frac{1}{h_F^{p-1}} \|\llbracket v \rrbracket\|_{L^p(F)}^p \right)^{1/p} \\
&= \left(\sum_{T \in \mathcal{T}_h} \int_T |\nabla v|_{\ell^p}^p + \sum_{F \in \mathcal{F}_h} \frac{1}{h_F^{p-1}} \int_F |\llbracket v \rrbracket|^p \right)^{1/p}, \qquad (5.1)
\end{aligned}$$

5.1. Discrete Functional Analysis

where we recall that $|\nabla v|_{\ell^p} = (\sum_{i=1}^d |\partial_i v|^p)^{1/p}$. As the case $p = 2$ is particularly important in the context of linear diffusive PDEs, we omit the subscript in this case and simply write, for all $v \in H^1(\mathcal{T}_h)$,

$$\|v\|_{\mathrm{dG}} := \left(\|\nabla_h v\|_{[L^2(\Omega)]^d}^2 + \sum_{F \in \mathcal{F}_h} \frac{1}{h_F} \|[\![v]\!]\|_{L^2(F)}^2 \right)^{1/2}$$

$$= \left(\sum_{T \in \mathcal{T}_h} \int_T |\nabla v|^2 + \sum_{F \in \mathcal{F}_h} \frac{1}{h_F} \int_F |[\![v]\!]|^2 \right)^{1/2}, \qquad (5.2)$$

where $|\nabla v|$ denotes the Euclidean norm of ∇v. We observe that the $\|\cdot\|_{\mathrm{dG}}$-norm coincides with the $\|\cdot\|_{\mathrm{sip}}$-norm defined by (4.17) and extensively used in Chap. 4.

Our first preliminary result concerns the comparison of $\|\cdot\|_{\mathrm{dG},p}$-norms.

Lemma 5.1 (Comparison of $\|\cdot\|_{\mathrm{dG},p}$-norms). *Let $1 \leq s < t < \infty$. Then,*

$$\forall v \in W^{1,t}(\mathcal{T}_h), \qquad \|v\|_{\mathrm{dG},s} \leq C_{\mathrm{dG}} \|v\|_{\mathrm{dG},t},$$

with $C_{\mathrm{dG}} := \max(1, d|\Omega|_d (1 + (1+d)(\varrho_1 \varrho_2)^{-1} N_\partial))$.

Proof. Let $v \in W^{1,t}(\mathcal{T}_h)$. Observing that, for all $x \in \mathbb{R}^d$, $|x|_{\ell^s} \leq d^{1/s - 1/t} |x|_{\ell^t}$ and using Hölder's inequality with $\lambda = t/s > 1$ and $\lambda' = \lambda/(\lambda - 1)$ so that $1/\lambda + 1/\lambda' = 1$, we infer

$$\|v\|_{\mathrm{dG},s}^s = \sum_{T \in \mathcal{T}_h} \int_T |\nabla v|_{\ell^s}^s + \sum_{F \in \mathcal{F}_h} \frac{1}{h_F^{s-1}} \int_F |[\![v]\!]|^s$$

$$\leq \sum_{T \in \mathcal{T}_h} \int_T d^{\frac{1}{\lambda'}} |\nabla v|_{\ell^t}^s + \sum_{F \in \mathcal{F}_h} \int_F h_F^{\frac{1}{\lambda'}} h_F^{\frac{1-t}{\lambda}} |[\![v]\!]|^s$$

$$\leq \left(\sum_{T \in \mathcal{T}_h} d \int_T 1^{\lambda'} \right)^{1/\lambda'} \times \left(\sum_{T \in \mathcal{T}_h} \int_T |\nabla v|_{\ell^t}^t \right)^{1/\lambda}$$

$$+ \left(\sum_{F \in \mathcal{F}_h} h_F \int_F 1^{\lambda'} \right)^{1/\lambda'} \times \left(\sum_{F \in \mathcal{F}_h} \frac{1}{h_F^{t-1}} \int_F |[\![v]\!]|^t \right)^{1/\lambda}$$

$$\leq \left(d|\Omega|_d + \sum_{F \in \mathcal{F}_h} h_F |F|_{d-1} \right)^{1/\lambda'} \|v\|_{\mathrm{dG},t}^s.$$

For all $F \in \mathcal{F}_h$, we pick $T \in \mathcal{T}_h$ such that $F \in \mathcal{F}_T$ so that $h_F \leq h_T$. Using the continuous trace inequality (1.41) with $v \equiv 1$ on T leads to

$$|F|_{d-1} \leq dC_{\mathrm{cti}} h_T^{-1} |T|_d \leq dC_{\mathrm{cti}} h_F^{-1} |T|_d,$$

with $C_{\mathrm{cti}} = (1+d)(\varrho_1 \varrho_2)^{-1}$, so that $\sum_{F \in \mathcal{F}_h} h_F |F|_{d-1} \leq dC_{\mathrm{cti}} N_\partial |\Omega|_d$, since each mesh element is counted at most N_∂ times. Hence,

$$\|v\|_{\mathrm{dG},s}^s \leq (d|\Omega|_d (1 + C_{\mathrm{cti}} N_\partial))^{1/\lambda'} \|v\|_{\mathrm{dG},t}^s,$$

and we conclude observing that $0 \leq \frac{1}{\lambda's} = \frac{1}{s} - \frac{1}{t} \leq 1$ so that, for any positive real number x, $x^{1/\lambda's} \leq \max(1, x)$. □

Our second preliminary result allows us to bound the $\|\cdot\|_{\mathrm{BV}}$-norm by the $\|\cdot\|_{\mathrm{dG},p}$-norms. The observation that, for $p = 2$, the $\|\cdot\|_{\mathrm{dG}}$-norm controls the BV norm can also be found in the work of Lew, Neff, Sulsky, and Ortiz [232].

Lemma 5.2 (Bound on $\|\cdot\|_{\mathrm{BV}}$-norm). Let $1 \leq p < \infty$. Let $v \in W^{1,1}(\mathcal{T}_h)$ and extend v by zero outside Ω. Then, $v \in \mathrm{BV}(\mathbb{R}^d)$ and there holds

$$\|v\|_{\mathrm{BV}} \leq C_{\mathrm{BV}} \|v\|_{\mathrm{dG},p}, \qquad (5.3)$$

with $C_{\mathrm{BV}} = d^{1/2} \max(1, C_{\mathrm{dG}})$ and C_{dG} defined in Lemma 5.1.

Proof. Let $v \in W^{1,1}(\mathcal{T}_h)$ and extend v by zero outside Ω. It is clear that $v \in L^1(\mathbb{R}^d)$. Let now $\varphi \in C_0^\infty(\mathbb{R}^d)$ with $\|\varphi\|_{L^\infty(\mathbb{R}^d)} \leq 1$. Integrating by parts yields

$$\int_{\mathbb{R}^d} v \partial_i \varphi = -\int_\Omega (\nabla_h v \cdot e_i) \varphi + \sum_{F \in \mathcal{F}_h} \int_F \llbracket v \rrbracket (e_i \cdot n_F) \varphi \qquad \forall i \in \{1, \ldots, d\},$$

where $(e_1, \ldots, e_d)$ denotes the Cartesian basis of $\mathbb{R}^d$. Using $\|\varphi\|_{L^\infty(\mathbb{R}^d)} \leq 1$ and summing over $i \in \{1, \ldots, d\}$, we obtain

$$\|v\|_{\mathrm{BV}} \leq \|\nabla_h v\|_{[L^1(\Omega)]^d} + d^{1/2} \sum_{F \in \mathcal{F}_h} \|\llbracket v \rrbracket\|_{L^1(F)}.$$

This yields the assertion for $p = 1$. The assertion for $p > 1$ then follows from Lemma 5.1. □

5.1.2 Discrete Sobolev Embeddings

The usual Sobolev embeddings allow one to control the L^q-norm of a function in $W^{1,p}(\Omega)$ by the L^p-norm of its gradient (the maximum value of q for which control is achieved depends on both p and the space dimension d); see Evans [153, Sect. 5.6] or Brézis [55, Sect. IX.3]. The aim of this section is to establish the counterpart of these embeddings for functions in the broken polynomial space $\mathbb{P}_d^k(\mathcal{T}_h)$ with polynomial degree $k \geq 0$ (other broken polynomial spaces can be considered), so as to achieve control, uniformly in h, of the L^q-norm of a function in $\mathbb{P}_d^k(\mathcal{T}_h)$ by its $\|\cdot\|_{\mathrm{dG},p}$-norm with the same maximal value for q as in the continuous case. Obviously, the difficulty is that the broken polynomial space $\mathbb{P}_d^k(\mathcal{T}_h)$ is nonconforming in $W^{1,p}(\Omega)$. As mentioned above, the control of the $\|\cdot\|_{\mathrm{BV}}$-norm by the $\|\cdot\|_{\mathrm{dG},p}$-norm (cf. Lemma 5.2) plays a central role in the proof.

Theorem 5.3 (Discrete Sobolev embeddings). Let $1 \leq p < \infty$. Let $k \geq 0$. Then, for all q satisfying

(i) $1 \leq q \leq p^* := \frac{pd}{d-p}$ if $1 \leq p < d$;

5.1. Discrete Functional Analysis

(ii) $1 \le q < \infty$ if $d \le p < \infty$ (and $1 \le q \le \infty$ if $d = 1$);

there is $\sigma_{p,q}$ such that

$$\forall v_h \in \mathbb{P}_d^k(\mathcal{T}_h), \qquad \|v_h\|_{L^q(\Omega)} \le \sigma_{p,q} \|v_h\|_{\mathrm{dG},p}. \tag{5.4}$$

The quantity $\sigma_{p,q}$ additionally depends on $|\Omega|_d$, k, and ϱ.

Proof. For brevity of notation, we abbreviate as $a \lesssim b$ the inequality $a \le Cb$ with positive C having the same dependencies as $\sigma_{p,q}$.
(i) *The case $p = 1$.* Set $1^* := \frac{d}{d-1}$ (conventionally, $1^* = \infty$ if $d = 1$). A classical result (see, e.g., Eymard, Gallouët, and Herbin [159]) states that, for all $v \in \mathrm{BV}(\mathbb{R}^d)$,

$$\|v\|_{L^{1^*}(\mathbb{R}^d)} \le \frac{1}{2d}\|v\|_{\mathrm{BV}}. \tag{5.5}$$

Let now $v_h \in \mathbb{P}_d^k(\mathcal{T}_h)$ and extend v_h by zero outside Ω. Then, owing to (5.5) and Lemma 5.2, we infer

$$\|v_h\|_{L^{1^*}(\Omega)} \lesssim \|v_h\|_{\mathrm{dG},1},$$

yielding (5.4) for $p = 1$ and $q = 1^*$ (and, hence, for all $1 \le q \le 1^*$ since Ω is bounded). This also proves (5.4) for $d = 1$.
(ii) *The case $1 < p < d$.* It is sufficient to establish (5.4) for $q = p^*$ since Ω is bounded. Set $\lambda := \frac{p(d-1)}{d-p} > 1$ and notice that $p^* = \lambda 1^*$. Let $v_h \in \mathbb{P}_d^k(\mathcal{T}_h)$. Considering the function $\xi := |v_h|^\lambda$ (extended by zero outside Ω) and using (5.5) yields

$$\left(\int_\Omega |v_h|^{p^*}\right)^{\frac{d-1}{d}} = \|\xi\|_{L^{1^*}(\Omega)} = \|\xi\|_{L^{1^*}(\mathbb{R}^d)} \lesssim \|\xi\|_{\mathrm{BV}} \lesssim \|\xi\|_{\mathrm{dG},1}$$

$$= \sum_{T \in \mathcal{T}_h} \int_T |\nabla |v_h|^\lambda|_{\ell^1} + \sum_{F \in \mathcal{F}_h} \int_F |[\![|v_h|^\lambda]\!]| := \mathfrak{T}_1 + \mathfrak{T}_2.$$

We observe that, a.e. in each $T \in \mathcal{T}_h$, $|\partial_i |v_h|^\lambda| = \lambda|v_h|^{\lambda-1}|\partial_i v_h|$ for all $i \in \{1,\ldots,d\}$, so that $|\nabla |v_h|^\lambda|_{\ell^1} = \lambda|v_h|^{\lambda-1}|\nabla v_h|_{\ell^1}$. Using Hölder's inequality with p and $r = \frac{p}{p-1}$, the term $\mathfrak{T}_1$ is then bounded as

$$|\mathfrak{T}_1| \le \lambda \left(\sum_{T \in \mathcal{T}_h} \int_T |v_h|^{r(\lambda-1)}\right)^{\frac{1}{r}} \left(\sum_{T \in \mathcal{T}_h} \int_T |\nabla v_h|^p_{\ell^1}\right)^{\frac{1}{p}}$$

$$\lesssim \left(\int_\Omega |v_h|^{p^*}\right)^{\frac{1}{r}} \left(\sum_{T \in \mathcal{T}_h} \int_T |\nabla v_h|^p_{\ell^p}\right)^{\frac{1}{p}},$$

since $r(\lambda-1) = \frac{p}{p-1}\frac{d(p-1)}{d-p} = \frac{pd}{d-p} = p^*$ and $|\nabla v_h|_{\ell^1} \le d^{1/r}|\nabla v_h|_{\ell^p}$. Furthermore, concerning $\mathfrak{T}_2$, we observe that $|[\![|v_h|^\lambda]\!]| \le 2\lambda \{\!\!\{|v_h|^{\lambda-1}\}\!\!\}|[\![v_h]\!]|$. Using again

Hölder's inequality, we then infer

$$|\mathfrak{T}_2| \leq \lambda \sum_{T \in \mathcal{T}_h} \sum_{F \in \mathcal{F}_T} \int_F h_F^{\frac{1}{r}} |v_h|_T|^{\lambda-1} h_F^{-\frac{1}{r}} |[\![v_h]\!]|$$

$$\leq \lambda \left(\sum_{T \in \mathcal{T}_h} \sum_{F \in \mathcal{F}_T} \int_F h_F |v_h|_T|^{p^*} \right)^{\frac{1}{r}} \left(\sum_{T \in \mathcal{T}_h} \sum_{F \in \mathcal{F}_T} \frac{1}{h_F^{p-1}} \int_F |[\![v_h]\!]|^p \right)^{\frac{1}{p}}.$$

To bound the first factor on the right-hand side, we use the discrete trace inequality (1.44) in $L^{p^*}(F)$ for all $F \in \mathcal{F}_T$ and the fact that N_∂ is uniformly bounded in h. We obtain

$$|\mathfrak{T}_2| \lesssim \left(\int_\Omega |v_h|^{p^*} \right)^{\frac{1}{r}} \left(\sum_{F \in \mathcal{F}_h} \frac{1}{h_F^{p-1}} \int_F |[\![v_h]\!]|^p \right)^{\frac{1}{p}}.$$

Collecting the above bounds for $\mathfrak{T}_1$ and $\mathfrak{T}_2$, we infer

$$\left(\int_\Omega |v_h|^{p^*} \right)^{\frac{d-1}{d}} \lesssim \left(\int_\Omega |v_h|^{p^*} \right)^{\frac{1}{r}} \|v_h\|_{\mathrm{dG},p},$$

whence (5.4) for $q = p^*$ since $\frac{d-1}{d} - \frac{1}{r} = \frac{1}{p^*}$.

(iii) *The case $d \leq p < \infty$.* Let $q_1 > d$ and set $p_1 := \frac{dq_1}{d+q_1}$ so that $1 \leq p_1 < d$ and $p_1^* = q_1$. Then, owing to point (ii) in this proof, we infer for all $v_h \in \mathbb{P}_d^k(\mathcal{T}_h)$,

$$\|v_h\|_{L^{q_1}(\Omega)} \leq \sigma_{p_1,q_1} \|v_h\|_{\mathrm{dG},p_1},$$

and the conclusion follows from Lemma 5.1 since $p_1 < d \leq p$ and the fact that Ω is bounded. □

Because of its importance in the context of linear diffusive PDEs, we state explicitly Theorem 5.3 in the special case $q = p = 2$ yielding the well-known discrete Poincaré inequality (already stated in Chap. 4; cf. (4.20)).

Corollary 5.4 (Discrete Poincaré inequality). *Let $k \geq 0$. There is σ_2, only depending on $|\Omega|_d$, k, and ϱ, such that*

$$\forall v_h \in \mathbb{P}_d^k(\mathcal{T}_h), \qquad \|v_h\|_{L^2(\Omega)} \leq \sigma_2 \|v_h\|_{\mathrm{dG}}. \qquad (5.6)$$

Remark 5.5 (Embeddings in broken Sobolev spaces). We established Theorem 5.3 in broken polynomial spaces since this result is sufficient for our purposes, that is, to analyze the convergence of dG methods for exact solutions with minimal regularity. Incidentally, we notice that the above proof cannot be extended to broken Sobolev spaces since we used a discrete trace inequality. Yet, by proceeding differently, it is sometimes possible to extend these embeddings to broken Sobolev spaces; we refer the reader to Brenner [51] for the case $p = q = 2$ and to Lasis and Süli [221] for the general case in a Hilbertian setting ($p = 2$).

5.1.3 Discrete Compactness

The Rellich–Kondrachov Theorem (see, e.g., Evans [153, p. 272] or Brézis [55, p. 169]) states that bounded sequences in $W^{1,p}(\Omega)$ are relatively compact in $L^q(\Omega)$ for suitable q depending on both p and the space dimension d. The aim of this section is to establish the counterpart of this result for sequences

$$v_{\mathcal{H}} := (v_h)_{h \in \mathcal{H}}$$

that are bounded in the $\|\cdot\|_{\mathrm{dG},p}$-norm and such that, for all $h \in \mathcal{H}$, v_h is in the broken polynomial space $\mathbb{P}_d^k(\mathcal{T}_h)$ for a fixed $k \geq 0$. Moreover, we identify the regularity of the limit and prove a weak asymptotic consistency result for the discrete gradients introduced in Sect. 4.3.2. In what follows, to alleviate the notation, subsequences are not renumbered.

We begin with the discrete Rellich–Kondrachov theorem. As in the previous section, the difficulty is that the broken polynomial space $\mathbb{P}_d^k(\mathcal{T}_h)$ is nonconforming in $W^{1,p}(\Omega)$. Also in this case, the control of the $\|\cdot\|_{\mathrm{BV}}$-norm by the $\|\cdot\|_{\mathrm{dG},p}$-norm (cf. Lemma 5.2) plays a central role in the proof. We mention the early work of Stummel [286] in the case of nonconforming finite element methods, and in the context of Maxwell's equations, the work of Kikuchi [211], Boffi [44], and Monk and Demkowicz [237] on edge elements and that of Creusé and Nicaise [116] on dG approximations.

Theorem 5.6 (Discrete Rellich–Kondrachov theorem). *Let $1 \leq p < \infty$. Let $k \geq 0$. Let $v_{\mathcal{H}}$ be a sequence in $\mathbb{P}_d^k(\mathcal{T}_{\mathcal{H}}) := (\mathbb{P}_d^k(\mathcal{T}_h))_{h \in \mathcal{H}}$ bounded in the $\|\cdot\|_{\mathrm{dG},p}$-norm. Then, for all q such that $1 \leq q < p^*$ if $1 \leq p < d$ or $1 \leq q < \infty$ if $d \leq p < \infty$, the sequence $v_{\mathcal{H}}$ is relatively compact in $L^q(\Omega)$.*

Proof. Let $v_{\mathcal{H}}$ be a sequence in $\mathbb{P}_d^k(\mathcal{T}_{\mathcal{H}})$ bounded in the $\|\cdot\|_{\mathrm{dG},p}$-norm. As above, functions are extended by zero outside Ω, and we use the symbol $\lesssim$ as in the proof of Theorem 5.3. Following Eymard, Gallouët, and Herbin [159, Lemma 5.4], we observe that, for all $\xi \in \mathbb{R}^d$,

$$\|v_h(\cdot + \xi) - v_h\|_{L^1(\mathbb{R}^d)} \lesssim |\xi|_{\ell^1} \|v_h\|_{\mathrm{BV}}.$$

As a result, using Lemma 5.2,

$$\|v_h(\cdot + \xi) - v_h\|_{L^1(\mathbb{R}^d)} \lesssim |\xi|_{\ell^1} \|v_h\|_{\mathrm{dG},p} \lesssim |\xi|_{\ell^1},$$

owing to the boundedness of the sequence $v_{\mathcal{H}}$ in the $\|\cdot\|_{\mathrm{dG},p}$-norm. Hence, owing to Kolmogorov's Compactness Criterion, the sequence $v_{\mathcal{H}}$ is relatively compact in $L^1(\mathbb{R}^d)$, and thus in $L^1(\Omega)$. Let now q be such that $1 < q < p^*$ if $1 \leq p < d$ or $1 < q < \infty$ if $d \leq p < \infty$. Then, owing to Theorem 5.3, there is $r > q$ such that the sequence $v_{\mathcal{H}}$ is bounded in $L^r(\Omega)$. We now make use of the following interpolation inequality between $L^1(\Omega)$ and $L^r(\Omega)$ (see Evans [153, p. 623] or Brezis [55, p. 57]): For all $w \in L^r(\Omega)$,

$$\|w\|_{L^q(\Omega)} \leq \|w\|_{L^1(\Omega)}^{\theta} \|w\|_{L^r(\Omega)}^{1-\theta},$$

where $\theta = \frac{r-q}{q(r-1)} \in (0,1)$. Hence, $v_\mathcal{H}$ is a Cauchy sequence in $L^q(\Omega)$ since, for all $h, h' \in \mathcal{H}$, taking $w = v_h - v_{h'} \in L^r(\Omega)$ in the above inequality yields

$$\|v_h - v_{h'}\|_{L^q(\Omega)} \leq \|v_h - v_{h'}\|_{L^1(\Omega)}^\theta \|v_h - v_{h'}\|_{L^r(\Omega)}^{1-\theta} \lesssim \|v_h - v_{h'}\|_{L^1(\Omega)}^\theta.$$

This completes the proof. □

For simplicity, we now specialize the setting to the Hilbertian case $p = 2$; cf. Remark 5.8 for the general case $p \neq 2$. The compactness property stated in Theorem 5.6 allows us to infer, in particular, the existence of a function $v \in L^2(\Omega)$ such that, up to a subsequence, $v_h \to v$ in $L^2(\Omega)$. However, the regularity of the limit is still insufficient for use in second-order elliptic problems where the natural space for the solution is $H_0^1(\Omega)$. We now prove that indeed $v \in H_0^1(\Omega)$. The tool to prove this result is the weak asymptotic consistency for the discrete gradients, and this property turns out to be of independent interest.

Theorem 5.7 (Regularity of the limit and weak asymptotic consistency of discrete gradients). *Let $k \geq 0$. Let $v_\mathcal{H}$ be a sequence in $\mathbb{P}_d^k(\mathcal{T}_\mathcal{H})$ bounded in the $\|\cdot\|_{\mathrm{dG}}$-norm. Then, there is a function $v \in H_0^1(\Omega)$ such that as $h \to 0$, up to a subsequence,*

$$v_h \to v \quad \text{strongly in } L^2(\Omega),$$

and, for all $l \geq 0$, the discrete gradients defined by (4.44) are such that

$$G_h^l(v_h) \rightharpoonup \nabla v \qquad \text{weakly in } [L^2(\Omega)]^d. \tag{5.7}$$

Proof. Let $v_\mathcal{H}$ be a sequence in $\mathbb{P}_d^k(\mathcal{T}_\mathcal{H})$ bounded in the $\|\cdot\|_{\mathrm{dG}}$-norm. The existence of $v \in L^2(\Omega)$ such that, up to a subsequence, $v_h \to v$ strongly in $L^2(\Omega)$ is a direct consequence of Theorem 5.6. To prove that $v \in H_0^1(\Omega)$, we first establish (5.7). Let $l \geq 0$. We extend functions by zero outside Ω. Owing to Proposition 4.35,

$$\|G_h^l(v_h)\|_{[L^2(\mathbb{R}^d)]^d} = \|G_h^l(v_h)\|_{[L^2(\Omega)]^d} \lesssim \|v_h\|_{\mathrm{sip}} = \|v_h\|_{\mathrm{dG}},$$

so that the sequence $(G_h^l(v_h))_{h \in \mathcal{H}}$ is bounded in $[L^2(\mathbb{R}^d)]^d$. Hence, up to a subsequence, there is $w \in [L^2(\mathbb{R}^d)]^d$ such that $G_h^l(v_h) \rightharpoonup w$ weakly in $[L^2(\mathbb{R}^d)]^d$. To prove that $w = \nabla v$, let $\Phi \in [C_0^\infty(\mathbb{R}^d)]^d$ and observe that

$$\int_{\mathbb{R}^d} G_h^l(v_h) \cdot \Phi = \int_{\mathbb{R}^d} \nabla_h v_h \cdot \Phi - \int_{\mathbb{R}^d} R_h^l(\llbracket v_h \rrbracket) \cdot \pi_h^l \Phi$$

$$= -\int_{\mathbb{R}^d} v_h \nabla \cdot \Phi + \sum_{F \in \mathcal{F}_h} \int_F \{\!\!\{\Phi - \pi_h^l \Phi\}\!\!\} \cdot n_F \llbracket v_h \rrbracket := \mathfrak{T}_1 + \mathfrak{T}_2.$$

where π_h^l denotes the L^2-orthogonal projection onto $\mathbb{P}_d^l(\mathcal{T}_h)$. Letting $h \to 0$, we observe that $\mathfrak{T}_1 \to -\int_{\mathbb{R}^d} v(\nabla \cdot \Phi)$ since $v_h \to v$ strongly in $L^2(\Omega)$. Furthermore, using the Cauchy–Schwarz inequality and recalling the definition (4.18) of the $|\cdot|_\mathrm{J}$-seminorm, we infer

$$|\mathfrak{T}_2| \leq \left(\sum_{F \in \mathcal{F}_h} h_F \|\{\!\!\{\Phi - \pi_h^l \Phi\}\!\!\}\|_{L^2(F)}^2 \right)^{1/2} |v_h|_\mathrm{J} \lesssim h^{l+1} \|\Phi\|_{H^{l+1}(\Omega)} |v_h|_\mathrm{J}.$$

Since $|v_h|_J$ is uniformly bounded, we infer that $\mathfrak{T}_2 \to 0$ as $h \to 0$. Collecting the above limits yields

$$\int_{\mathbb{R}^d} w \cdot \Phi = \lim_{h \to 0} \int_{\mathbb{R}^d} G_h^l(v_h) \cdot \Phi = -\int_{\mathbb{R}^d} v(\nabla \cdot \Phi),$$

implying that $w = \nabla v$. This yields (5.7) as well as $v \in H^1(\mathbb{R}^d)$. Moreover, since v is zero outside Ω, we conclude that $v \in H_0^1(\Omega)$. □

Remark 5.8 (The non-Hilbertian case). Let $1 < p < \infty$. Let $k \geq 0$. Let $v_{\mathcal{H}}$ be a sequence in $\mathbb{P}_d^k(\mathcal{T}_{\mathcal{H}})$ bounded in the $\|\cdot\|_{\mathrm{dG},p}$-norm. Theorem 5.6 allows us to infer, in particular, the existence of a function $v \in L^p(\Omega)$ such that, up to a subsequence, $v_h \to v$ in $L^p(\Omega)$. Moreover, it can be shown (see [131]) that $v \in W_0^{1,p}(\Omega)$. The proof follows a similar path to that deployed above by establishing preliminarily that $G_h^0(v_h) \rightharpoonup \nabla v$ weakly in $[L^p(\Omega)]^d$.

5.2 Convergence to Minimal Regularity Solutions

The aim of this section is to investigate the convergence, as the meshsize goes to zero, of the sequence of solutions to the discrete Poisson problem

$$\text{Find } u_h \in V_h \text{ s.t. } a_h^{\mathrm{sip}}(u_h, v_h) = \int_\Omega f v_h \text{ for all } v_h \in V_h, \qquad (5.8)$$

when the exact solution u has only the minimal regularity $u \in H_0^1(\Omega)$. For the Poisson problem, the regularity theory for the Laplace operator can in general be invoked to infer stronger regularity for the exact solution. However, this is no longer the case when working with more complex problems, such as diffusion problems with rough coefficients or nonlinear problems. In particular, the results presented in this section constitute the first stone upon which the convergence analysis of dG approximations for the incompressible Navier–Stokes equations is built in Chap. 6.

The key idea is to revisit the concept of consistency and introduce a new point of view based on asymptotic consistency. This new form of consistency, together with the usual stability of the discrete bilinear form, are the two main ingredients for asserting convergence to minimal regularity solutions. In what follows, we consider sequences in the discrete spaces $V_{\mathcal{H}} := (V_h)_{h \in \mathcal{H}}$ where $V_h := \mathbb{P}_d^k(\mathcal{T}_h)$ with fixed polynomial degree $k \geq 1$ and $\mathcal{T}_h$ belonging to an admissible mesh sequence.

5.2.1 Consistency Revisited

The convergence analysis to minimal regularity solutions does not require to plug in the exact solution into the first argument of the discrete bilinear form. In fact, the latter is only employed using discrete functions as arguments. This leads us to redefine the concept of consistency. We first present a general definition and then verify it on the SIP bilinear form.

Definition 5.9 (Asymptotic consistency). We say that the discrete bilinear form a_h is *asymptotically consistent* on $V_{\mathcal{H}}$ with the exact bilinear form a defined by (4.3) if, for any sequence $v_{\mathcal{H}}$ in $V_{\mathcal{H}}$ bounded in the $\|\cdot\|_{\mathrm{dG}}$-norm and for any smooth function $\varphi \in C_0^\infty(\Omega)$, there is a sequence $\varphi_{\mathcal{H}}$ in $V_{\mathcal{H}}$ converging to φ in the $\|\cdot\|_{\mathrm{dG}}$-norm and such that, up to a subsequence,

$$\lim_{h \to 0} a_h(v_h, \varphi_h) = a(v, \varphi) = \int_\Omega \nabla v \cdot \nabla \varphi,$$

where $v \in H_0^1(\Omega)$ is the limit of the subsequence identified in Theorem 5.7.

We now focus on the SIP bilinear form a_h^{sip} defined by (4.12) or, equivalently at the discrete level, by (4.46). Here, it is more convenient to consider the latter expression, namely

$$a_h^{\mathrm{sip}}(v_h, w_h) = \int_\Omega G_h^l(v_h) \cdot G_h^l(w_h) + \hat{s}_h^{\mathrm{sip}}(v_h, w_h), \qquad (5.9)$$

with $l \in \{k-1, k\}$ and the stabilization bilinear form

$$\hat{s}_h^{\mathrm{sip}}(v_h, w_h) := \sum_{F \in \mathcal{F}_h} \frac{\eta}{h_F} \int_F [\![v_h]\!] [\![w_h]\!] - \int_\Omega \mathrm{R}_h^l([\![v_h]\!]) \cdot \mathrm{R}_h^l([\![w_h]\!]).$$

We stress that, in this section, the SIP bilinear form is not extended to larger (e.g., broken Sobolev) spaces and that its arguments are always discrete functions.

Lemma 5.10 (Asymptotic consistency of the SIP bilinear form). *Let $k \geq 1$. The SIP bilinear form is asymptotically consistent with the exact bilinear form a on $V_{\mathcal{H}}$.*

Proof. Let $v_{\mathcal{H}}$ be a sequence in $V_{\mathcal{H}}$ bounded in the $\|\cdot\|_{\mathrm{dG}}$-norm and let $\varphi \in C_0^\infty(\Omega)$. For all $h \in \mathcal{H}$, we set $\varphi_h = \pi_h \varphi$ where π_h denotes the L^2-orthogonal projection onto V_h. Since $k \geq 1$, we infer $\|\varphi - \pi_h \varphi\|_{\mathrm{dG}} \to 0$ as $h \to 0$. Owing to Proposition 4.35 and since $G_h^l(\varphi) = \nabla \varphi$, we obtain, for all $l \geq 0$,

$$G_h^l(\pi_h \varphi) \to \nabla \varphi \qquad \text{strongly in } [L^2(\Omega)]^d. \qquad (5.10)$$

We now observe that

$$a_h^{\mathrm{sip}}(v_h, \pi_h \varphi) = \int_\Omega G_h^l(v_h) \cdot G_h^l(\pi_h \varphi) + \hat{s}_h^{\mathrm{sip}}(v_h, \pi_h \varphi) := \mathfrak{T}_1 + \mathfrak{T}_2.$$

Clearly, as $h \to 0$, $\mathfrak{T}_1 \to \int_\Omega \nabla v \cdot \nabla \varphi$ owing to the weak convergence of $G_h^l(v_h)$ to ∇v (cf. (5.7)) and to the strong convergence of $G_h^l(\pi_h \varphi)$ to $\nabla \varphi$. Furthermore, using the Cauchy–Schwarz inequality together with (4.42) yields

$$|\mathfrak{T}_2| = |\hat{s}_h^{\mathrm{sip}}(v_h, \pi_h \varphi)| \leq (\eta + C_{\mathrm{tr}}^2 N_\partial) |v_h|_{\mathrm{J}} |\pi_h \varphi|_{\mathrm{J}}.$$

Since $|v_h|_{\mathrm{J}}$ is bounded by assumption and since $|\pi_h \varphi|_{\mathrm{J}} = |\varphi - \pi_h \varphi|_{\mathrm{J}}$ tends to zero as $h \to 0$, we infer $\mathfrak{T}_2 \to 0$. The proof is complete. □

Remark 5.11 (Strong asymptotic consistency). Property (5.10) can be interpreted as a strong asymptotic consistency of discrete gradients for approximations of smooth functions. It has to be compared with the weaker property (5.7).

5.2.2 Convergence

This section contains our main convergence result. The ideas in the proof can be summarized as follows:

(a) An a priori estimate on the discrete solution is proven using discrete coercivity in V_h with respect to the $\|\cdot\|_{\mathrm{dG}}$-norm.

(b) Discrete compactness (Theorem 5.7) yields the existence of a function $v \in H_0^1(\Omega)$ such that, up to a subsequence, the discrete solutions strongly converge to v in $L^2(\Omega)$, while the discrete gradients weakly converge to ∇v in $[L^2(\Omega)]^d$.

(c) Using asymptotic consistency and a density argument allows us to conclude that the limit function v solves the exact problem.

(d) Finally, additional properties of the exact problem (uniqueness of the exact solution and an energy estimate) are used to tighten the convergence result by showing that the whole sequence of discrete solutions converges and that the discrete gradients (as well as the broken gradients) strongly converge in $[L^2(\Omega)]^d$ to the correct limit, while the jump seminorm converges to zero.

As in Chap. 4, we assume that the penalty parameter is such that $\eta > C_{\mathrm{tr}}^2 N_\partial$ so that (cf. Lemma 4.12), for all $v_h \in V_h$,

$$a_h^{\mathrm{sip}}(v_h, v_h) \geq C_\eta \|v_h\|_{\mathrm{dG}}^2,$$

with $C_\eta = (\eta - C_{\mathrm{tr}}^2 N_\partial)(1+\eta)^{-1}$. We also recall the result of Proposition 4.36, namely, for all $v_h \in V_h$,

$$a_h^{\mathrm{sip}}(v_h, v_h) \geq \|G_h^l(v_h)\|_{[L^2(\Omega)]^d}^2 + (\eta - C_{\mathrm{tr}}^2 N_\partial)|v_h|_{\mathrm{J}}^2. \qquad (5.11)$$

Theorem 5.12 (Convergence to minimal regularity solutions). *Let $k \geq 1$. Let $u_\mathcal{H}$ be the sequence of approximate solutions generated by solving the discrete problems (5.8) with discrete bilinear form a_h^{sip} defined on $V_h \times V_h$ by (5.9). Then, as $h \to 0$,*

$$u_h \to u \qquad \text{strongly in } L^2(\Omega),$$
$$\nabla_h u_h \to \nabla u \qquad \text{strongly in } [L^2(\Omega)]^d,$$
$$|u_h|_{\mathrm{J}} \to 0,$$

where $u \in H_0^1(\Omega)$ is the unique solution of (4.2).

Proof. We follow the four steps outlined above.
(i) Owing to discrete coercivity and to the discrete Poincaré inequality (5.6),

$$C_\eta \|u_h\|_{\mathrm{dG}}^2 \leq a_h^{\mathrm{sip}}(u_h, u_h) = \int_\Omega f u_h$$
$$\leq \|f\|_{L^2(\Omega)} \|u_h\|_{L^2(\Omega)} \leq \sigma_2 \|f\|_{L^2(\Omega)} \|u_h\|_{\mathrm{dG}}.$$

Hence, the sequence of discrete solutions $u_\mathcal{H}$ is bounded in the $\|\cdot\|_{dG}$-norm.
(ii) Owing to Theorem 5.7, there exists $v \in H_0^1(\Omega)$ such that, as $h \to 0$, up to a subsequence, $u_h \to v$ strongly in $L^2(\Omega)$ and, for all $l \geq 0$, $G_h^l(u_h) \rightharpoonup \nabla v$ weakly in $[L^2(\Omega)]^d$.
(iii) Owing to asymptotic consistency (cf. Lemma 5.10), for all $\varphi \in C_0^\infty(\Omega)$,

$$\int_\Omega f\varphi \leftarrow \int_\Omega f\pi_h\varphi = a_h^{\mathrm{sip}}(u_h, \pi_h\varphi) \to \int_\Omega \nabla v \cdot \nabla \varphi,$$

i.e., v solves the Poisson problem by density of $C_0^\infty(\Omega)$ in $H_0^1(\Omega)$.
(iv) Since the solution u to the Poisson problem is unique, the whole sequence $u_\mathcal{H}$ strongly converges to u in $L^2(\Omega)$ and, for all $l \geq 0$, the sequence $(G_h^l(u_h))_{h \in \mathcal{H}}$ weakly converges to ∇u in $[L^2(\Omega)]^d$. Let $l \in \{k-1, k\}$. Owing to (5.11) and to weak convergence,

$$\liminf_{h \to 0} a_h^{\mathrm{sip}}(u_h, u_h) \geq \liminf_{h \to 0} \|G_h^l(u_h)\|_{[L^2(\Omega)]^d}^2 \geq \|\nabla u\|_{[L^2(\Omega)]^d}^2.$$

Furthermore, still owing to (5.11),

$$\|G_h^l(u_h)\|_{[L^2(\Omega)]^d}^2 \leq a_h^{\mathrm{sip}}(u_h, u_h) = \int_\Omega fu_h,$$

yielding

$$\limsup_{h \to 0} \|G_h^l(u_h)\|_{[L^2(\Omega)]^d}^2 \leq \limsup_{h \to 0} a_h^{\mathrm{sip}}(u_h, u_h)$$

$$= \limsup_{h \to 0} \int_\Omega fu_h = \int_\Omega fu = \|\nabla u\|_{[L^2(\Omega)]^d}^2.$$

Thus, $\|G_h^l(u_h)\|_{[L^2(\Omega)]^d} \to \|\nabla u\|_{[L^2(\Omega)]^d}$, classically yielding the strong convergence of the discrete gradient in $[L^2(\Omega)]^d$. Moreover, $a_h^{\mathrm{sip}}(u_h, u_h) \to \|\nabla u\|_{[L^2(\Omega)]^d}^2$. Owing to (5.11), we infer

$$(\eta - C_{\mathrm{tr}}^2 N_\partial)|u_h|_J^2 \leq a_h^{\mathrm{sip}}(u_h, u_h) - \|G_h^l(u_h)\|_{[L^2(\Omega)]^d}^2,$$

and, since $\eta > C_{\mathrm{tr}}^2 N_\partial$ and the right-hand side tends to zero, $|u_h|_J \to 0$. Finally, using the triangle inequality yields

$$\|\nabla_h u_h - \nabla u\|_{[L^2(\Omega)]^d} \leq \|G_h^l(u_h) - \nabla u\|_{[L^2(\Omega)]^d} + \|R_h^l(\llbracket u_h \rrbracket)\|_{[L^2(\Omega)]^d} \to 0,$$

as $h \to 0$, concluding the proof. □

5.3 Variations on Symmetry and Penalty

This section reviews variations on symmetry and penalty that can be envisaged to modify the SIP method analyzed in Sects. 4.2 and 5.2.

5.3.1 Variations on Symmetry

The interest in nonsymmetric dG approximations for the Poisson problem was prompted by the work of Oden, Babuška, and Baumann [250] who considered the discrete bilinear form

$$a_h^{\text{obb}}(v_h, w_h) := \int_\Omega \nabla_h v_h \cdot \nabla_h w_h - \sum_{F \in \mathcal{F}_h} \int_F \left(\{\!\{\nabla_h v_h\}\!\} \cdot \mathbf{n}_F [\![w_h]\!] - [\![v_h]\!] \{\!\{\nabla_h w_h\}\!\} \cdot \mathbf{n}_F \right).$$

There are two important differences with respect to the SIP bilinear form: the contribution of mesh faces is skew-symmetric and there is no penalty term. The skew-symmetry of the face contribution has the consequence that

$$\forall v_h \in V_h, \qquad a_h^{\text{obb}}(v_h, v_h) = \|\nabla_h v_h\|_{[L^2(\Omega)]^d}^2 \geq 0.$$

Various numerical results were presented in [250] indicating that despite the absence of a penalty term, the method behaves fairly well for quadratic and higher-order polynomial orders in one and two space dimensions. Further theoretical insight was gained by Larson and Niklasson [220] who showed discrete inf-sup stability for such polynomials on unstructured triangulations in two space dimensions. Alternatively, a penalty term of a form similar to (4.11) can be added to the OBB bilinear form leading to the *Nonsymmetric Interior Penalty* (NIP) method analyzed by Rivière, Wheeler, and Girault [271, 272]. The resulting discrete bilinear form is

$$a_h^{\text{nip}}(v_h, w_h) = a_h^{\text{obb}}(v_h, w_h) + \sum_{F \in \mathcal{F}_h} \frac{\eta}{h_F^\beta} \int_F [\![v_h]\!][\![w_h]\!]. \tag{5.12}$$

Taking $\beta = 1$ and $\eta > 0$ leads to optimal $\|\cdot\|_{\text{dG}}$-norm error estimates, as for the SIP method. Because of the lack of symmetry, the duality argument deployed in Sect. 4.2.4 for the SIP method cannot be used for the NIP method. Optimal L^2-norm error estimates can be derived by resorting to overpenalty, that is, by taking the exponent β larger than one; specifically, $\beta \geq 3$ for $d = 2$ and $\beta \geq 3/2$ for $d = 3$. A variant called Weakly Overpenalized NIP has been analyzed recently by Brenner and Owens [53], the idea being to overpenalize the jumps of the mean values only.

Finally, we mention the *Incomplete Interior Penalty* (IIP) method introduced by Dawson, Sun, and Wheeler [121] in the context of coupled porous media flow and contaminant transport. The discrete bilinear form reads

$$a_h^{\text{iip}}(v_h, w_h) := \int_\Omega \nabla_h v_h \cdot \nabla_h w_h - \sum_{F \in \mathcal{F}_h} \int_F \{\!\{\nabla_h v_h\}\!\} \cdot \mathbf{n}_F [\![w_h]\!] + \sum_{F \in \mathcal{F}_h} \frac{\eta}{h_F} \int_F [\![v_h]\!][\![w_h]\!],$$

the difference with SIP being that the symmetry term has been dropped.

In all cases, letting $V_h = \mathbb{P}_d^k(\mathcal{T}_h)$ with polynomial degree $k \geq 1$ and $\mathcal{T}_h$ belonging to an admissible mesh sequence, the discrete problem takes the form:

$$\text{Find } u_h \in V_h \text{ s.t. } a_h^*(u_h, v_h) = \int_\Omega f v_h \text{ for all } v_h \in V_h, \tag{5.13}$$

where a_h^* can be the OBB, NIP, or IIP discrete bilinear form. For NIP (with $\beta = 1$), discrete coercivity is achieved in the $\|\cdot\|_{\mathrm{dG}}$-norm as soon as $\eta > 0$, while for IIP, proceeding as in the proof of Lemma 4.12 shows that discrete coercivity is achieved in the $\|\cdot\|_{\mathrm{dG}}$-norm as soon as $\eta > \frac{1}{4}C_{\mathrm{tr}}^2 N_\partial$. Once discrete coercivity is asserted, the convergence analysis to smooth solutions in the $\|\cdot\|_{\mathrm{dG}}$-norm proceeds similarly to Sect. 4.2.3; details are omitted for brevity.

5.3.1.1 Local Conservation

Similarly to the SIP method (cf. Sect. 4.3.4), the OBB, NIP, and IIP methods satisfy a local conservation property. Specifically, for all $T \in \mathcal{T}_h$ and for all $\xi \in \mathbb{P}_d^k(T)$, there holds

$$\int_T (\nabla_h u_h - \alpha \, \mathrm{R}_h^{k-1}([\![u_h]\!])) \cdot \nabla \xi + \sum_{F \in \mathcal{F}_T} \epsilon_{T,F} \int_F \tilde{\phi}_F(u_h) \xi = \int_T f\xi,$$

where $\alpha = -1$ for OBB and NIP, while $\alpha = 0$ for IIP; furthermore, $\tilde{\phi}_F(u_h) = \phi_F(u_h) = -\{\!\{\nabla_h u_h\}\!\}\cdot \mathrm{n}_F + \eta h_F^{-1}[\![u_h]\!]$ for NIP (with $\beta = 1$) and IIP, while for OBB, $\tilde{\phi}_F(u_h) = -\{\!\{\nabla_h u_h\}\!\}\cdot \mathrm{n}_F$.

5.3.1.2 Convergence to Minimal Regularity Solutions

The NIP and IIP bilinear forms can be cast into the form (compare with (5.9))

$$a_h^*(v_h, w_h) = \int_\Omega \widehat{G}_h^l(v_h)\cdot G_h^l(w_h) + \hat{s}_h^*(v_h, w_h), \tag{5.14}$$

with $l \in \{k-1, k\}$ and where for NIP (with $\beta = 1$),

$$\widehat{G}_h^l(v_h) = \nabla_h v_h + \mathrm{R}_h^l([\![v_h]\!]), \tag{5.15}$$

$$\hat{s}_h^{\mathrm{nip}}(v_h, w_h) = \sum_{F \in \mathcal{F}_h} \frac{\eta}{h_F} \int_F [\![v_h]\!][\![w_h]\!] + \int_\Omega \mathrm{R}_h^l([\![v_h]\!]) \cdot \mathrm{R}_h^l([\![w_h]\!]), \tag{5.16}$$

while for IIP,

$$\widehat{G}_h^l(v_h) = \nabla_h v_h, \tag{5.17}$$

$$\hat{s}_h^{\mathrm{iip}}(v_h, w_h) = \sum_{F \in \mathcal{F}_h} \frac{\eta}{h_F} \int_F [\![v_h]\!][\![w_h]\!]. \tag{5.18}$$

We assume that the penalty parameter η is chosen such that $\eta > 0$ for NIP and $\eta > \frac{1}{4}C_{\mathrm{tr}}^2 N_\partial$ for IIP; this implies discrete coercivity in the $\|\cdot\|_{\mathrm{dG}}$-norm, and hence well-posedness of the discrete problem (5.13). Specifically, there is C_{sta}, independent of h, such that

$$\forall v_h \in V_h, \qquad C_{\mathrm{sta}} \|v_h\|_{\mathrm{dG}}^2 \leq a_h^*(v_h, v_h). \tag{5.19}$$

To analyze the convergence to minimal regularity solutions of the sequence of discrete solutions $u_\mathcal{H}$ solving (5.13), we adapt the ideas of Sect. 5.2.

5.3. Variations on Symmetry and Penalty

The difficulty is that the discrete bilinear form a_h^* defined by (5.14) does not match Definition 5.9 regarding asymptotic consistency. Instead, we consider asymptotic adjoint consistency.

Definition 5.13 (Asymptotic adjoint consistency). We say that the discrete bilinear form a_h^* is *asymptotically adjoint consistent* with the exact bilinear form a on $V_{\mathcal{H}}$ if for any sequence $v_{\mathcal{H}}$ in $V_{\mathcal{H}}$ bounded in the $\|\cdot\|_{\mathrm{dG}}$-norm and for any smooth function $\varphi \in C_0^\infty(\Omega)$, there is a sequence $\varphi_{\mathcal{H}}$ in $V_{\mathcal{H}}$ converging to φ in the $\|\cdot\|_{\mathrm{dG}}$-norm and such that, up to a subsequence,

$$\lim_{h \to 0} a_h(\varphi_h, v_h) = a(\varphi, v) = \int_\Omega \nabla\varphi \cdot \nabla v,$$

where $v \in H_0^1(\Omega)$ is the limit of the subsequence identified in Theorem 5.7.

Lemma 5.14 (Asymptotic adjoint consistency of NIP and IIP bilinear forms). *The discrete bilinear form a_h^* defined by (5.14) is asymptotically adjoint consistent with the exact bilinear form a on $V_{\mathcal{H}}$.*

Proof. We proceed as in the proof of Lemma 5.10 observing that, for both NIP and IIP, $\widehat{G}_h^l(\pi_h\varphi)$ strongly converges to $\nabla\varphi$ in $[L^2(\Omega)]^d$ where π_h denotes the L^2-orthogonal projection onto V_h. □

We can now state our convergence result to minimal regularity solutions. The general idea of the proof is inspired from Agélas, Di Pietro, Eymard, and Masson [6, Lemma 2.3 and Theorem 2.2].

Theorem 5.15 (Convergence to minimal regularity solutions). *Let $k \geq 1$. Let $u_{\mathcal{H}}$ be the sequence of approximate solutions generated by solving the discrete problems (5.13) with a_h^* defined by (5.14) and with penalty parameter ensuring coercivity. Then, as $h \to 0$,*

$$u_h \to u \qquad \text{strongly in } L^2(\Omega),$$
$$\nabla_h u_h \to \nabla u \qquad \text{strongly in } [L^2(\Omega)]^d,$$
$$|u_h|_J \to 0,$$

where $u \in H_0^1(\Omega)$ is the unique solution of (4.2).

Proof. Owing to the discrete coercivity of a_h^*, the sequence $u_{\mathcal{H}}$ is bounded in the $\|\cdot\|_{\mathrm{dG}}$-norm. Theorem 5.7 implies that there is $v \in H_0^1(\Omega)$ such that, up to a subsequence, $u_h \to v$ in $L^2(\Omega)$ and, for all $l \geq 0$, $G_h^l(u_h) \rightharpoonup \nabla v$ weakly in $[L^2(\Omega)]^d$ as $h \to 0$. Let $\varphi \in C_0^\infty(\Omega)$. Owing to Lemma 5.14, $a_h^*(\pi_h\varphi, u_h) \to a(\varphi, v)$ as $h \to 0$. Since u_h solves the discrete problem (5.13), we infer, as $h \to 0$,

$$a_h^*(u_h - \pi_h\varphi, u_h - \pi_h\varphi) \to \int_\Omega f(v - \varphi) - \int_\Omega \nabla\varphi \cdot \nabla(v - \varphi).$$

Hence, using (5.19)

$$C_{\mathrm{sta}} \limsup_{h \to 0} \|u_h - \pi_h\varphi\|_{\mathrm{dG}}^2 \leq \limsup_{h \to 0} a_h^*(u_h - \pi_h\varphi, u_h - \pi_h\varphi) \leq C_{f,\varphi} \|v - \varphi\|_{H^1(\Omega)},$$

with $C_{f,\varphi} = (\|f\|^2_{L^2(\Omega)} + \|\nabla\varphi\|^2_{[L^2(\Omega)]^d})^{1/2}$. As a consequence,

$$\limsup_{h\to 0} \|u_h - \pi_h\varphi\|^2_{dG} \leq C_{sta}^{-1} C_{f,\varphi} \|v - \varphi\|_{H^1(\Omega)}.$$

We now observe that both of the choices (5.15) and (5.17) for $\widehat{G}^l_h$ satisfy the stability property:

$$\forall v_h \in V_h, \qquad \|\widehat{G}^l_h(v_h)\|_{[L^2(\Omega)]^d} \leq \widehat{C}\|v_h\|_{dG},$$

for $\widehat{C}$ independent of h (the assertion is straightforward for IIP, whereas the bound (4.42) is used for NIP). As a result,

$$\limsup_{h\to 0} \|\widehat{G}^l_h(u_h) - \widehat{G}^l_h(\pi_h\varphi)\|^2_{[L^2(\Omega)]^d} \leq \widehat{C} C_{sta}^{-1} C_{f,\varphi} \|v - \varphi\|_{H^1(\Omega)},$$

and since $\widehat{G}^l_h(\pi_h\varphi)$ strongly converges to $\nabla\varphi$ in $[L^2(\Omega)]^d$, this yields

$$\limsup_{h\to 0} \|\widehat{G}^l_h(u_h) - \nabla\varphi\|^2_{[L^2(\Omega)]^d} \leq \widehat{C} C_{sta}^{-1} C_{f,\varphi} \|v - \varphi\|_{H^1(\Omega)}.$$

Since φ is arbitrary in $C_0^\infty(\Omega)$, and since this space is dense in $H_0^1(\Omega)$, the term on the right-hand side can be made as small as desired. Hence, we infer

$$\widehat{G}^l_h(u_h) \to \nabla v \qquad \text{strongly in } [L^2(\Omega)]^d.$$

As a result, taking again φ arbitrary in $C_0^\infty(\Omega)$ yields

$$\int_\Omega f\varphi \leftarrow \int_\Omega f\pi_h\varphi = a_h^*(u_h, \pi_h\varphi) \to \int_\Omega \nabla v \cdot \nabla\varphi.$$

The proof can now be concluded as in the symmetric case. $\square$

Remark 5.16 (Further variation on symmetry (with crime)). It is also possible to consider the discrete bilinear form (compare with (5.14))

$$a_h^{**}(v_h, w_h) = \int_\Omega G^l_h(v_h) \cdot \widehat{G}^l_h(w_h) + \hat{s}_h^*(v_h, w_h),$$

with corresponding sequence of discrete solutions $u_\mathcal{H}$. The asymptotic consistency of a_h^{**} is an immediate consequence of the discrete Rellich–Kondrachov theorem and the strong consistency of $\widehat{G}^l_h$ for smooth functions, so that proving the strong convergence in L^2 of the discrete solutions $u_\mathcal{H}$ is straightforward. However, the present techniques do not provide a convergence proof for the discrete gradients. This difficulty is also reflected by the lack of the consistency term (cf. Definition 4.7).

5.3.2 Variations on Penalty

Following an idea originally proposed by Bassi, Rebay, Mariotti, Pedinotti, and Savini [35], we analyze in this section a modification of the penalty strategy whereby the liftings of the interface and boundary jumps are penalized instead of the actual values of these quantities. Specifically, we define the so-called BRMPS discrete bilinear form

$$a_h^{\mathrm{brmps}}(v_h, w_h) := a_h^{\mathrm{cs}}(v_h, w_h) + s_h^{\mathrm{brmps}}(v_h, w_h), \qquad (5.20)$$

with a_h^{cs} defined by (4.8) and with the new stabilization bilinear form

$$s_h^{\mathrm{brmps}}(v_h, w_h) := \sum_{F \in \mathcal{F}_h} \eta \int_\Omega \mathrm{r}_F^l(\llbracket v_h \rrbracket) \cdot \mathrm{r}_F^l(\llbracket w_h \rrbracket),$$

where $\eta > 0$ is a user-dependent parameter and $l \in \{k-1, k\}$. The main advantage of this new penalty strategy is that coercivity holds provided $\eta > N_\partial$, thereby circumventing the presence of C_{tr} in the minimal threshold for the penalty parameter.

The natural norm to assert the discrete coercivity of a_h^{brmps} is

$$\|v_h\|_{\mathrm{brmps}} := \left(\|\nabla_h v_h\|_{[L^2(\Omega)]^d}^2 + \sum_{F \in \mathcal{F}_h} \|\mathrm{r}_F^l(\llbracket v_h \rrbracket)\|_{[L^2(\Omega)]^d}^2 \right)^{1/2}.$$

We observe that $\|\cdot\|_{\mathrm{brmps}}$ is a norm on $H^1(\mathcal{T}_h)$. Indeed, if $v \in H^1(\mathcal{T}_h)$ is such that $\|v\|_{\mathrm{brmps}} = 0$, then $\nabla_h v = 0$ and $\mathrm{r}_F^l(\llbracket v \rrbracket) = 0$ for all $F \in \mathcal{F}_h$. The first property implies that v is piecewise constant, and the second property yields that $\llbracket v \rrbracket = 0$ across all $F \in \mathcal{F}_h$; hence, $v = 0$.

Lemma 5.17 (Discrete coercivity). *Assume $\eta > N_\partial$. Then, the discrete bilinear form a_h^{brmps} defined by (5.20) is coercive on V_h with respect to the $\|\cdot\|_{\mathrm{brmps}}$-norm, i.e.,*

$$\forall v_h \in V_h, \qquad a_h^{\mathrm{brmps}}(v_h, v_h) \geq C_\eta \|v_h\|_{\mathrm{brmps}}^2,$$

with $C_\eta := (\eta - N_\partial)(1+\eta)^{-1}$.

Proof. Let $v_h \in V_h$. Observing that (cf. (4.45))

$$a_h^{\mathrm{cs}}(v_h, v_h) = \int_\Omega |\nabla_h v_h|^2 - 2 \int_\Omega \nabla_h v_h \cdot \mathrm{R}_h^l(\llbracket v_h \rrbracket),$$

and using (4.41) to bound $\|\mathrm{R}_h^l(\llbracket v_h \rrbracket)\|_{[L^2(\Omega)]^d}$, we obtain

$$a_h^{\mathrm{brmps}}(v_h, v_h) \geq \|\nabla_h v_h\|_{[L^2(\Omega)]^d}^2 + \eta \sum_{F \in \mathcal{F}_h} \|\mathrm{r}_F^l(\llbracket v_h \rrbracket)\|_{[L^2(\Omega)]^d}^2$$

$$- 2 N_\partial^{1/2} \|\nabla_h v_h\|_{[L^2(\Omega)]^d} \left(\sum_{F \in \mathcal{F}_h} \|\mathrm{r}_F^l(\llbracket v_h \rrbracket)\|_{[L^2(\Omega)]^d}^2 \right)^{1/2}.$$

The right-hand side takes the form $x^2 - 2\beta xy + \eta y^2$ with $x = \|\nabla_h v_h\|_{[L^2(\Omega)]^d}$ and $y = (\sum_{F\in\mathcal{F}_h} \|r_F^l(\llbracket v_h\rrbracket)\|_{[L^2(\Omega)]^d}^2)^{1/2}$. Proceeding as in Lemma 4.12 then yields the assertion. $\square$

We now establish the discrete coercivity of a_h^{brmps} on V_h using the $\|\cdot\|_{\mathrm{dG}}$-norm defined by (5.2) (or, equivalently, the $\|\cdot\|_{\mathrm{sip}}$-norm defined by (4.17)). This results from Lemma 5.17 and the following uniform equivalence of the $\|\cdot\|_{\mathrm{brmps}}$- and $\|\cdot\|_{\mathrm{dG}}$-norms on V_h.

Lemma 5.18 (Uniform norm equivalence). *Let $k \geq 1$ and let $l \geq 0$. There is $C_r > 0$, independent of h, such that, for all $v_h \in V_h$ and all $F \in \mathcal{F}_h$,*

$$C_r h_F^{-1/2} \|\llbracket v_h\rrbracket\|_{L^2(F)} \leq \left(\|r_F^l(\llbracket v_h\rrbracket)\|_{[L^2(\mathcal{T}_F)]^d}^2 + \|\nabla_h v_h\|_{[L^2(\mathcal{T}_F)]^d}^2 \right)^{1/2}, \quad (5.21)$$

where $\mathcal{T}_F$ is defined by (1.13) and $\|\cdot\|_{[L^2(\mathcal{T}_F)]^d} = (\sum_{T\in\mathcal{T}_F}\|\cdot\|_{[L^2(T)]^d}^2)^{1/2}$. As a result, the $\|\cdot\|_{\mathrm{dG}}$- and $\|\cdot\|_{\mathrm{brmps}}$-norms are uniformly equivalent on V_h: For all $v_h \in V_h$,

$$(1 + 2C_r^{-2})^{-1/2} \|v_h\|_{\mathrm{dG}} \leq \|v_h\|_{\mathrm{brmps}} \leq \max(1, C_{\mathrm{tr}}^2)^{1/2} \|v_h\|_{\mathrm{dG}}. \quad (5.22)$$

Proof. To alleviate the notation, we abbreviate as $a \lesssim b$ the inequality $a \leq Cb$ with positive C independent of h. Let $F \in \mathcal{F}_h$ and let $v_h \in V_h$. Let $y_h := \langle \llbracket v_h\rrbracket\rangle_F$ denote the mean value of $\llbracket v_h\rrbracket$ on F. The triangle inequality yields

$$\|\llbracket v_h\rrbracket\|_{L^2(F)} \leq \|y_h\|_{L^2(F)} + \|\llbracket v_h\rrbracket - y_h\|_{L^2(F)}. \quad (5.23)$$

Owing to the generalized Friedrichs inequality (see, e.g., Vohralík [300]), there holds

$$\|\llbracket v_h\rrbracket - y_h\|_{L^2(F)} \lesssim h_F^{1/2}\|\nabla_h v_h\|_{[L^2(\mathcal{T}_F)]^d}. \quad (5.24)$$

To bound the first term on the right-hand side of (5.23), let $n_{F,0}$ denote the mean value of n_F on F (we recall that mesh faces can be composed of several portions of hyperplanes). Then, since $n_F \cdot n_F = 1$ on F and since y_h is constant on F,

$$\|y_h\|_{L^2(F)}^2 = \int_F |y_h|^2 = \int_F y_h n_F \cdot n_F y_h = \int_F y_h n_F \cdot n_{F,0} y_h.$$

Therefore, setting $\Phi_h = y_h n_{F,0}$, extending Φ_h by its constant value on the elements in $\mathcal{T}_F$ so that $\Phi \in [\mathbb{P}_d^0(\mathcal{T}_h)]^d$ and $\{\!\{\Phi_h\}\!\} = \Phi_h$ on F, and using the definition (4.37) of r_F^l yields, for all $l \geq 0$,

$$\|y_h\|_{L^2(F)}^2 = \int_F y_h n_F \cdot \Phi_h = \int_\Omega r_F^l(y_h)\cdot\Phi_h \leq \|r_F^l(y_h)\|_{[L^2(\mathcal{T}_F)]^d}\|\Phi_h\|_{[L^2(\mathcal{T}_F)]^d},$$

and since $\|\Phi_h\|_{[L^2(\mathcal{T}_F)]^d} \lesssim h_F^{1/2}\|y_h\|_{L^2(F)}$, we conclude that

$$\|y_h\|_{L^2(F)} \lesssim h_F^{1/2}\|r_F^l(y_h)\|_{[L^2(\mathcal{T}_F)]^d}.$$

5.3. Variations on Symmetry and Penalty

Using the linearity of r_F^l, the triangle inequality, the bound (4.39) on $r_F^l(\cdot)$, and (5.24) leads to

$$\|y_h\|_{L^2(F)} \lesssim h_F^{1/2} \|r_F^l(\llbracket v_h \rrbracket)\|_{[L^2(\Omega)]^d} + h_F^{1/2}\|r_F^l(y_h - \llbracket v_h \rrbracket)\|_{[L^2(\Omega)]^d}$$
$$\lesssim h_F^{1/2} \|r_F^l(\llbracket v_h \rrbracket)\|_{[L^2(\Omega)]^d} + \|y_h - \llbracket v_h \rrbracket\|_{L^2(F)}$$
$$\lesssim h_F^{1/2} \|r_F^l(\llbracket v_h \rrbracket)\|_{[L^2(\Omega)]^d} + h_F^{1/2}\|\nabla_h v_h\|_{[L^2(T_F)]^d}.$$

This yields (5.21). Finally, the lower bound in (5.22) results from (5.21) after a summation over $F \in \mathcal{F}_h$, while the upper bound results from the bound (4.39) on each $r_F^l(\cdot)$ for all $F \in \mathcal{F}_h$. □

We can now reformulate Lemma 5.17.

Lemma 5.19 (Discrete coercivity). *Assume $\eta > N_\partial$. Then, the discrete bilinear form a_h^{brmps} defined by (5.20) is coercive on V_h with respect to the $\|\cdot\|_{\mathrm{dG}}$-norm, i.e.,*

$$\forall v_h \in V_h, \qquad a_h^{\mathrm{brmps}}(v_h, v_h) \geq C_{\mathrm{sta}} \|v_h\|_{\mathrm{dG}}^2,$$

with $C_{\mathrm{sta}} := C_\eta(1+2C_{\mathrm{r}}^{-2})^{-1}$, C_η defined in Lemma 5.17, and C_{r} in Lemma 5.18.

We consider the discrete problem

$$\text{Find } u_h \in V_h \text{ s.t. } a_h^{\mathrm{brmps}}(u_h, v_h) = \int_\Omega f v_h \text{ for all } v_h \in V_h. \tag{5.25}$$

For smooth solutions, the error analysis proceeds as in Sects. 4.2.3 and 4.2.4, leading to optimal error estimates in the $\|\cdot\|_{\mathrm{dG}}$- and L^2-norms respectively; details are omitted for brevity.

5.3.2.1 Local Formulation and Stencil

It is easily verified that the local formulation of the discrete problem (5.25) has the same form as (4.49) for the SIP method, except that the numerical flux $\phi_F(u_h)$ is now given by

$$\phi_F(u_h) = -\{\!\!\{\nabla_h u_h\}\!\!\}\cdot n_F + \eta\{\!\!\{r_F^l(\llbracket u_h \rrbracket)\}\!\!\}\cdot n_F.$$

Furthermore, the elementary stencil of the BRMPS bilinear form is the same as that of the SIP bilinear form; cf. Fig. 4.1.

5.3.2.2 Convergence to Minimal Regularity Solutions

In the spirit of Sect. 4.3.3, the discrete bilinear form a_h^{brmps} can be equivalently reformulated on $V_h \times V_h$ as

$$a_h^{\mathrm{brmps}}(v_h, w_h) = \int_\Omega G_h^l(v_h) \cdot G_h^l(w_h) + \hat{s}_h^{\mathrm{brmps}}(v_h, w_h),$$

with $l \in \{k-1, k\}$ and the stabilization bilinear form

$$\hat{s}_h^{\mathrm{brmps}}(v_h, w_h) = \sum_{F \in \mathcal{F}_h} \eta \int_\Omega r_F^l(\llbracket v_h \rrbracket) \cdot r_F^l(\llbracket w_h \rrbracket) - \int_\Omega R_h^l(\llbracket v_h \rrbracket) \cdot R_h^l(\llbracket w_h \rrbracket).$$

Theorem 5.20 (Convergence to minimal regularity solutions). *Let $k \geq 1$. Let $u_\mathcal{H}$ be the sequence of approximate solutions generated by solving the discrete problems (5.25) with a_h^{brmps} defined by (5.20). Then, as $h \to 0$,*

$$u_h \to u \quad \text{strongly in } L^2(\Omega),$$
$$\nabla_h u_h \to \nabla u \quad \text{strongly in } [L^2(\Omega)]^d,$$
$$|u_h|_{\mathrm{J}} \to 0,$$

where $u \in H_0^1(\Omega)$ is the unique solution of (4.2).

Proof. Owing to discrete coercivity and symmetry, we can proceed as in the proof of Theorem 5.12 to infer that $u_h \to u$ in $L^2(\Omega)$ and, for all $l \geq 0$, $G_h^l(u_h) \to \nabla u$ in $[L^2(\Omega)]^d$ where $u \in H_0^1(\Omega)$ is the unique solution of (4.2). Moreover, we obtain $a_h^{\mathrm{brmps}}(u_h, u_h) \to \|\nabla u\|_{[L^2(\Omega)]^d}^2$, and since

$$(\eta - N_\partial) \sum_{F \in \mathcal{F}_h} \|\mathrm{r}_F^l(\llbracket u_h \rrbracket)\|_{[L^2(\Omega)]^d}^2 \leq a_h^{\mathrm{brmps}}(u_h, u_h) - \|G_h^l(u_h)\|_{[L^2(\Omega)]^d}^2,$$

we infer that $\sum_{F \in \mathcal{F}_h} \|\mathrm{r}_F^l(\llbracket u_h \rrbracket)\|_{[L^2(\Omega)]^d}^2 \to 0$. It remains to prove that $|u_h|_{\mathrm{J}} \to 0$. To this purpose, let $\varphi \in C_0^\infty(\Omega)$ and set $\varphi_h = \pi_h \varphi$. Using Lemma 5.19 yields

$$C_{\mathrm{sta}} |u_h - \varphi_h|_{\mathrm{J}}^2 \leq C_{\mathrm{sta}} \|u_h - \varphi_h\|_{\mathrm{dG}}^2 \leq a_h^{\mathrm{brmps}}(u_h - \varphi_h, u_h - \varphi_h).$$

Letting $h \to 0$ yields

$$\limsup_{h \to 0} |u_h|_{\mathrm{J}}^2 \leq C_{\mathrm{sta}}^{-1} \|\nabla(u - \varphi)\|_{[L^2(\Omega)]^d}^2,$$

and the upper bound can be made as small as desired owing to the density of $C_0^\infty(\Omega)$ in $H_0^1(\Omega)$. □

5.3.3 Synopsis

All the variants considered in this section (including the SIP method of Sect. 4.2 and LDG method of Sect. 4.4.2) can be recast in the general form

$$a_h(v_h, w_h) := \int_\Omega \widehat{G}_{h,\alpha}^l(v_h) \cdot G_h^l(w_h) + \hat{s}_h^{\beta,\gamma}(v_h, w_h),$$

with $l \in \{k-1, k\}$, the discrete gradient $G_h^l(w_h)$ defined as usual by (4.44),

$$\widehat{G}_{h,\alpha}^l(v_h) := \nabla_h v_h - \alpha \, \mathrm{R}_h^l(\llbracket v_h \rrbracket) \qquad \alpha \in \{-1, 0, 1\},$$

and the stabilization bilinear form

$$\hat{s}_h^{\beta,\gamma}(v_h, w_h) := -\beta \int_\Omega \mathrm{R}_h(\llbracket v_h \rrbracket) \cdot \mathrm{R}_h(\llbracket w_h \rrbracket) + \gamma \sum_{F \in \mathcal{F}_h} \frac{\eta}{h_F} \int_F \llbracket v_h \rrbracket \llbracket w_h \rrbracket$$
$$+ (1-\gamma) \sum_{F \in \mathcal{F}_h} \eta \int_\Omega \mathrm{r}_F^l(\llbracket v_h \rrbracket) \cdot \mathrm{r}_F^l(\llbracket w_h \rrbracket),$$

5.4. Cell-Centered Galerkin

Table 5.1: Some common dG methods for the Poisson problem

α	β	$\zeta_{\alpha,\beta}$	Stencil	Name
1	1	1	N	SIP ($\gamma = 1$)/BRMPS ($\gamma = 0$)
1	0	0	NN	LDG-type
0	0	$1/4$	N	IIP
-1	-1	0	N	NIP

for parameters $\beta \in \{0, \alpha\}$ and $\gamma \in \{0, 1\}$ (taking $\beta = \alpha$ yields a more compact stencil, while the parameter γ serves to modify the penalty strategy). For given values of α, β, and γ, discrete coercivity is achieved for $\eta > \underline{\eta}_{\alpha,\beta,\gamma}$ where

$$\underline{\eta}_{\alpha,\beta,\gamma} = \zeta_{\alpha,\beta} N_\partial^{1/2}[1 + \gamma(C_{\mathrm{tr}} - 1)],$$

for some positive parameter $\zeta_{\alpha,\beta}$. A synopsis is presented in Table 5.1.

5.4 Cell-Centered Galerkin

The convergence analysis of Sects. 4.2.3 and 5.2 breaks down when using the SIP bilinear form with piecewise constant functions. In this section, we outline a different approach to derive a cost-effective, low-order dG method for the Poisson problem (4.2) with one degree of freedom per mesh cell. More details can be found in Di Pietro [128–130] and Eymard, Gallouët, and Herbin [159]; see also Agélas, Di Pietro, Eymard, and Masson [6].

Let $V_h := \mathbb{P}_d^0(\mathcal{T}_h)$ where $\mathcal{T}_h$ belongs to an admissible mesh sequence, so that the space V_h is spanned by piecewise constant functions. Our first ingredient is a discrete gradient reconstruction, i.e., a linear operator

$$G_h^{(0)} : V_h \ni v_h \longmapsto G_h^{(0)}(v_h) \in [\mathbb{P}_d^0(\mathcal{T}_h)]^d.$$

This operator is defined by (5.28) below. We can then consider the linear reconstruction operator $\mathcal{A}_h^{(1)}$ mapping piecewise constant functions onto piecewise affine functions,

$$\mathcal{A}_h^{(1)} : V_h \ni v_h \longmapsto \mathcal{A}_h^{(1)}(v_h) \in \mathbb{P}_d^1(\mathcal{T}_h),$$

such that, for all $v_h \in V_h$ and all $T \in \mathcal{T}_h$,

$$\mathcal{A}_h^{(1)}(v_h)|_T := v_T + G_h^{(0)}(v_h)|_T \cdot (x - x_T), \qquad v_T := v_h|_T,$$

where x_T denotes the center of gravity of T (in short, the cell center). We set

$$V_h^{\mathrm{cc}} := \mathcal{A}_h^{(1)}(V_h) \subset \mathbb{P}_d^1(\mathcal{T}_h).$$

The space V_h^{cc} has dimension equal to the number of mesh elements in $\mathcal{T}_h$ since the map $\mathcal{A}_h^{(1)}$ is injective. The key property required for the discrete gradient operator $G_h^{(0)}$ is strong consistency on smooth functions in the following sense.

Definition 5.21 (Strong consistency on smooth functions)**.** We say that the discrete gradient operator $G_h^{(0)}$ is *strongly consistent on smooth functions* if, for all $\varphi \in C_0^\infty(\Omega)$, letting

$$\Pi_h \varphi := \sum_{T \in \mathcal{T}_h} \varphi(x_T) \chi_T \in \mathbb{P}_d^0(\mathcal{T}_h), \qquad (5.26)$$

where χ_T denotes the characteristic function of T, there holds

$$\lim_{h \to 0} \|\varphi - \mathcal{A}_h^{(1)}(\Pi_h \varphi)\|_{\mathrm{dG}} = 0. \qquad (5.27)$$

A practical means to construct a discrete gradient operator $G_h^{(0)}$ matching property (5.27) has been proposed recently by Eymard, Gallouët, and Herbin [159] in the context of finite volume methods using the so-called barycentric trace interpolator. To use this construction, we make the following assumption on the mesh.

Assumption 5.22 (Assumption on the mesh)**.** *We assume that, for each mesh face $F \in \mathcal{F}_h^i$, there exists a set $\mathcal{B}_h^F \subset \mathcal{T}_h$ with $\mathrm{card}(\mathcal{B}_h^F) = (d+1)$ and such that the cell centers $\{x_T\}_{T \in \mathcal{B}_h^F}$ form a non-degenerate simplex S_F of $\mathbb{R}^d$. Moreover, for all $T \in \mathcal{T}_h$ and for all $F \in \mathcal{F}_T$, we also assume that $d_{T,F} := \mathrm{dist}(x_T, F)$ is uniformly comparable with the local length scale h_F (cf. Definition 4.5).*

The barycentric trace interpolator is conveniently defined using barycentric coordinates.

Definition 5.23 (Barycentric coordinates)**.** On a non-degenerate simplex S with vertices $\{a_0, \ldots, a_d\}$, the *barycentric coordinate* λ_i is defined, for all $i \in \{0, \ldots, d\}$, as the unique affine function in $\mathbb{P}_d^1(S)$ such that $\lambda_i(a_j) = \delta_{ij}$ for all $j \in \{0, \ldots, d\}$, where δ_{ij} is the Kronecker symbol.

For a point x belonging to a simplex S, the barycentric coordinate $\lambda_i(x)$, $i \in \{0, \ldots, d\}$, is equal to the ratio between the d-dimensional Hausdorff measure of the simplex obtained by joining the face of S opposite to the vertex a_i with the point x and the d-dimensional Hausdorff measure of S. The discrete gradient operator $G_h^{(0)}$ is then defined as follows (compare with the definition (4.43) of $\mathrm{R}_h^0(\llbracket v \rrbracket)|_T$ on broken polynomial spaces with $k \geq 1$): For all $v_h \in V_h$ and all $T \in \mathcal{T}_h$,

$$G_h^{(0)}(v_h)|_T := \sum_{F \in \mathcal{F}_T} \frac{|F|_{d-1}}{|T|_d} (I_F(v_h) - v_T) \mathrm{n}_{T,F}, \qquad (5.28)$$

where

$$I_F(v_h) := \begin{cases} \sum_{T \in \mathcal{B}_h^F} \lambda_{T,F} v_T & \text{if } F \in \mathcal{F}_h^i, \\ 0 & \text{if } F \in \mathcal{F}_h^b, \end{cases}$$

while $\lambda_{T,F}$ is the barycentric coordinate in the simplex S_F of Assumption 5.22 such that $\lambda_{T,F}$ is equal to one at the vertex x_T and to zero at the other vertices, while $\mathrm{n}_{T,F}$ denotes the outward normal to T on F; cf. Fig. 5.1.

5.4. Cell-Centered Galerkin

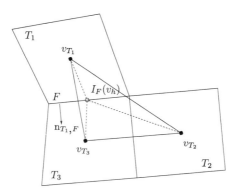

Fig. 5.1: Barycentric trace interpolator; here, $\mathcal{B}_h^F := \{T_1, T_2, T_3\}$

Lemma 5.24 (Strong consistency on smooth functions). *The discrete gradient operator $G_h^{(0)}$ defined by (5.28) is strongly consistent on smooth functions.*

Proof. Let $\varphi \in C_0^\infty(\Omega)$. We need to prove (5.27), and, to this purpose, we treat separately the two terms composing the $\|\cdot\|_{\mathrm{dG}}$-norm. To alleviate the notation, we abbreviate as $a \lesssim b$ the inequality $a \leq Cb$ with positive C that can depend on the derivatives of φ and mesh regularity, but not on the meshsize h. We observe preliminarily that it is proven in [159, Theorem 4.1] that, for all $F \in \mathcal{F}_h^i$,

$$|\varphi(x_F) - I_F(\Pi_h \varphi)| \lesssim h_F^2, \tag{5.29}$$

where x_F denotes the center of gravity of F.
(i) Let us prove that

$$\|\nabla_h \mathcal{A}_h^{(1)}(\Pi_h \varphi) - \nabla \varphi\|_{[L^2(\Omega)]^d} \lesssim h. \tag{5.30}$$

For all $T \in \mathcal{T}_h$, there holds, by construction, $\nabla_h \mathcal{A}_h^{(1)}(\Pi_h \varphi)|_T = G_h^{(0)}(\Pi_h \varphi)$. Moreover, for all $x \in T$, we observe that

$$G_h^{(0)}(\Pi_h \varphi) - \nabla \varphi(x) = \sum_{F \in \mathcal{F}_T} \frac{|F|_{d-1}}{|T|_d} (I_F(\Pi_h \varphi) - \varphi(x_F)) \, \mathrm{n}_{T,F}$$
$$+ (\nabla \varphi(x_T) - \nabla \varphi(x))$$
$$+ \left(\sum_{F \in \mathcal{F}_T} \frac{|F|_{d-1}}{|T|_d} (\varphi(x_F) - \varphi(x_T)) \, \mathrm{n}_{T,F} - \nabla \varphi(x_T) \right).$$

Let $\mathfrak{T}_1$, $\mathfrak{T}_2$, and $\mathfrak{T}_3$ denote the three terms on the right-hand side. Property (5.29), the geometric relation $\sum_{F \in \mathcal{F}_T} \frac{|F|_{d-1} d_{T,F}}{|T|_d} = d$, and Assumption 5.22 on $d_{T,F}$ yield

$$|\mathfrak{T}_1| \leq \sum_{F \in \mathcal{F}_T} \frac{|F|_{d-1} d_{T,F}}{|T|_d} \frac{|I_F(\Pi_h \varphi) - \varphi(x_F)|}{d_{T,F}} \lesssim h_F.$$

As a consequence, $\|\mathfrak{T}_1\|_{[L^2(T)]^d} \lesssim |T|_d^{1/2} h_T$. For the second term, since φ is smooth, we obtain $\|\mathfrak{T}_2\|_{[L^2(T)]^d} \lesssim |T|_d^{1/2} h_T$. Finally, owing to [159, Lemma 4.3], $\|\mathfrak{T}_3\|_{[L^2(T)]^d} \lesssim |T|_d^{1/2} h_T$. Summing the above estimates over $T \in \mathcal{T}_h$ yields (5.30).

(ii) Let us now show that $\mathfrak{T}_4 := |\varphi - \mathcal{A}_h^{(1)}(\Pi_h \varphi)|_{\mathrm{J}} \lesssim h$. Using the continuous trace inequality (1.41) and mesh regularity, we infer

$$\mathfrak{T}_4^2 \lesssim \sum_{F \in \mathcal{F}_h} \sum_{T \in \mathcal{T}_F} \left(h_T^{-2} \|\varphi - \mathcal{A}_h^{(1)}(\Pi_h \varphi)\|_{L^2(T)}^2 + \|\nabla \varphi - \nabla_h \mathcal{A}_h^{(1)}(\Pi_h \varphi)\|_{[L^2(T)]^d}^2 \right)$$

$$\lesssim \sum_{T \in \mathcal{T}_h} h_T^{-2} \|\varphi - \mathcal{A}_h^{(1)}(\Pi_h \varphi)\|_{L^2(T)}^2 + \|\nabla \varphi - \nabla_h \mathcal{A}_h^{(1)}(\Pi_h \varphi)\|_{[L^2(\Omega)]^d}^2.$$

To estimate $\|\varphi - \mathcal{A}_h^{(1)}(\Pi_h \varphi)\|_{L^2(T)}$, we expand φ to second-order at x_T and use $\mathcal{A}_h^{(1)}(\Pi_h \varphi)(x_T) = \varphi(x_T)$ to obtain

$$\|\varphi - \mathcal{A}_h^{(1)}(\Pi_h \varphi)\|_{L^2(T)} \lesssim h_T \|\nabla \varphi - \nabla_h \mathcal{A}_h^{(1)}(\Pi_h \varphi)\|_{[L^2(T)]^d} + h_T^2 \|\varphi\|_{H^2(T)}^2.$$

Combined with the above bound on $\mathfrak{T}_4^2$, this yields

$$|\varphi - \mathcal{A}_h^{(1)}(\Pi_h \varphi)|_{\mathrm{J}} \lesssim \|\nabla \varphi - \nabla_h \mathcal{A}_h^{(1)}(\Pi_h \varphi)\|_{[L^2(\Omega)]^d} + h \|\varphi\|_{H^2(\Omega)}.$$

Using (5.30) to bound the first term finally leads to

$$|\varphi - \mathcal{A}_h^{(1)}(\Pi_h \varphi)|_{\mathrm{J}} \lesssim h,$$

whence the convergence result (5.27). □

We consider the discrete problem

$$\text{Find } u_h \in V_h^{\mathrm{cc}} \text{ s.t. } a_h^{\mathrm{sip}}(u_h, v_h) = \int_\Omega f v_h \text{ for all } v_h \in V_h^{\mathrm{cc}}, \qquad (5.31)$$

with the SIP bilinear form a_h^{sip} defined by (4.46). An important consequence of Lemma 5.24 is the asymptotic consistency of the SIP bilinear form a_h^{sip} with the exact bilinear form a on $V_{\mathcal{H}}^{\mathrm{cc}}$.

Lemma 5.25 (Asymptotic consistency of SIP bilinear form). *The SIP bilinear form is asymptotically consistent with the exact bilinear form a on $V_{\mathcal{H}}^{\mathrm{cc}}$.*

Proof. We proceed as in Lemma 5.10 using the sequence $(\mathcal{A}_h^{(1)}(\Pi_h \varphi))_{h \in \mathcal{H}}$ with converges to φ in the $\|\cdot\|_{\mathrm{dG}}$-norm owing to Lemma 5.24. □

We can now state our convergence result to minimal regularity solutions. The proof, which proceeds exactly as that of Theorem 5.12 replacing the L^2-orthogonal projector π_h by the interpolator Π_h defined by (5.26), is omitted.

Theorem 5.26 (Convergence to minimal regularity solutions). *Let $u_\mathcal{H}$ be the sequence of approximate solutions generated by solving the discrete problems (5.31). Then, as $h \to 0$,*

$$u_h \to u \quad \text{strongly in } L^2(\Omega),$$
$$\nabla_h u_h \to \nabla u \quad \text{strongly in } [L^2(\Omega)]^d,$$
$$|u_h|_\mathrm{J} \to 0,$$

where $u \in H_0^1(\Omega)$ is the unique solution of (4.2).

5.5 Local Postprocessing

In this section, we consider the discrete Poisson problem (5.8) (variations in symmetry and penalty can be easily accommodated). We recall that the solution u_h is in $V_h := \mathbb{P}_d^k(\mathcal{T}_h)$ with polynomial degree $k \geq 1$ and $\mathcal{T}_h$ belonging to an admissible mesh sequence. Recalling Definition 4.38, the scalar-valued function u_h is termed the discrete potential. The aim of this section is to postprocess locally the discrete potential u_h so as to reconstruct an $H_0^1(\Omega)$-conforming potential u_h^* and an $H(\mathrm{div};\Omega)$-conforming diffusive flux σ_h^* that accurately approximate the exact potential u and the exact diffusive flux $\sigma = -\nabla u$. Incidentally, we notice that the discrete diffusive flux $\sigma_h \in [\mathbb{P}_d^l(\mathcal{T}_h)]^d$, $l \in \{k-1, k\}$, delivered by the mixed dG approximation of Sect. 4.4 is not appropriate for our purposes, since generally $[\![\sigma_h]\!]\cdot n_F \neq 0$ across interfaces, while, owing to Lemma 1.24, we require $[\![\sigma_h^*]\!]\cdot n_F = 0$.

The reconstructed quantities $u_h^* \in H_0^1(\Omega)$ and $\sigma_h^* \in H(\mathrm{div};\Omega)$ are used in Sect. 5.6 in the context of *a posteriori* error estimates. The reconstructed diffusive flux can be relevant in other applications; it can be used, for instance, as advective velocity when modeling contaminant transport through porous media. An application to two-phase porous media flows has been investigated by Ern, Mozolevski, and Schuh [146].

In what follows, we abbreviate as $a \lesssim b$ the inequality $a \leq Cb$ with positive C that can depend on the space dimension d, the polynomial degree k, and the mesh regularity parameters ϱ, but not on the meshsize h (and the dG penalty parameter η). For any subset $\mathcal{T} \subset \mathcal{T}_h$, we also use the notation $\|\cdot\|_{L^2(\mathcal{T})} := \left(\sum_{T \in \mathcal{T}} \|\cdot\|_{L^2(T)}^2\right)^{1/2}$.

5.5.1 Local Residuals

We introduce the local residuals

$$\forall T \in \mathcal{T}_h, \qquad \mathcal{R}_{\mathrm{pde},T} := h_T \|f + \triangle u_h\|_{L^2(T)}, \qquad (5.32\mathrm{a})$$

$$\forall F \in \mathcal{F}_h^i, \qquad \mathcal{R}_{\mathrm{jdf},F} := h_F^{1/2} \|[\![\nabla_h u_h]\!]\cdot n_F\|_{L^2(F)}, \qquad (5.32\mathrm{b})$$

$$\forall F \in \mathcal{F}_h, \qquad \mathcal{R}_{\mathrm{jpt},F} := h_F^{-1/2} \|[\![u_h]\!]\|_{L^2(F)}. \qquad (5.32\mathrm{c})$$

The residual $\mathcal{R}_{\mathrm{pde},T}$ is associated with the elementwise PDE residual. The residual $\mathcal{R}_{\mathrm{jdf},F}$ contains the jump of the normal component of the broken gradient across interfaces, and results from the nonconformity of the broken gradient in $H(\mathrm{div};\Omega)$. Finally, the residual $\mathcal{R}_{\mathrm{jpt},F}$, which contains the interface and boundary jumps of the discrete potential, results from the nonconformity of the discrete potential in $H_0^1(\Omega)$.

For all $T \in \mathcal{T}_h$, we also define the local data oscillation residual

$$\mathcal{R}_{\mathrm{osc},T} := h_T \|f - \pi_h f\|_{L^2(T)}, \tag{5.33}$$

where π_h is the L^2-orthogonal projection onto $\mathbb{P}_d^k(\mathcal{T}_h)$. For any subset $\mathcal{T} \subset \mathcal{T}_h$, we set

$$\mathcal{R}_{\mathrm{osc},\mathcal{T}} := \left(\sum_{T \in \mathcal{T}} \mathcal{R}_{\mathrm{osc},T}^2 \right)^{1/2}. \tag{5.34}$$

If, locally in some $T \in \mathcal{T}_h$, $f|_T \in \mathbb{P}_d^k(T)$, then $\mathcal{R}_{\mathrm{osc},T} = 0$, and if, globally, $f \in \mathbb{P}_d^k(\mathcal{T}_h)$, then $\mathcal{R}_{\mathrm{osc},\mathcal{T}_h} = 0$.

The following bound on the local residuals $\mathcal{R}_{\mathrm{pde},T}$ and $\mathcal{R}_{\mathrm{jdf},F}$ in terms of the approximation error $(u - u_h)$ and the data oscillation plays an important role in Sects. 5.5 and 5.6. We observe that the proof of this result does not use the fact that u_h solves a specific discrete problem, but just that $u_h \in V_h$ and that $u \in H_0^1(\Omega)$.

Lemma 5.27 (Bound on local residuals). *There holds*

$$\forall T \in \mathcal{T}_h, \quad \mathcal{R}_{\mathrm{pde},T} \lesssim \|\nabla(u - u_h)\|_{[L^2(T)]^d} + \mathcal{R}_{\mathrm{osc},T}, \tag{5.35a}$$

$$\forall F \in \mathcal{F}_h^i, \quad \mathcal{R}_{\mathrm{jdf},F} \lesssim \|\nabla_h(u - u_h)\|_{[L^2(\mathcal{T}_F)]^d} + \mathcal{R}_{\mathrm{osc},\mathcal{T}_F}, \tag{5.35b}$$

where $\mathcal{T}_F$ collects the two mesh elements sharing F.

Proof. The proof uses the element and face bubble functions considered by Verfürth [298] in the context of a posteriori error estimates for conforming finite element methods. We present the proof on matching simplicial meshes. The bounds on general meshes are proven by combining the corresponding bounds on the matching simplicial submesh and using mesh regularity.
(i) Bound on $\mathcal{R}_{\mathrm{pde},T}$. Let $T \in \mathcal{T}_h$ and let b_T be the element bubble function in $H_0^1(T)$ equal on F to the product of barycentric coordinates in T rescaled so as to take the value 1 at the center of gravity of T. Set $r_T := (\pi_h f + \triangle_h u_h)|_T$, where $\triangle_h$ denotes the broken Laplacian acting elementwise, and $\psi_T := b_T r_T$. Owing to the properties of bubble functions and mesh regularity,

$$\|r_T\|_{L^2(T)}^2 \lesssim \int_T r_T b_T r_T, \qquad \|\psi_T\|_{L^2(T)} \lesssim \|r_T\|_{L^2(T)}.$$

Since the bubble function b_T vanishes on ∂T, integration by parts yields

$$\|r_T\|_{L^2(T)}^2 \lesssim \int_T r_T b_T r_T = \int_T r_T \psi_T = \int_T (\pi_h f + \triangle u_h)\psi_T$$

$$= \int_T \nabla(u - u_h)\cdot\nabla\psi_T + \int_T (\pi_h f - f)\psi_T.$$

5.5. Local Postprocessing

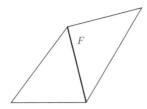

Fig. 5.2: Set ω_F associated with an interface $F \in \mathcal{F}_h^i$

Since ψ_T is a polynomial, the inverse inequality (1.36) implies that
$$\|\nabla \psi_T\|_{[L^2(T)]^d} \lesssim h_T^{-1} \|\psi_T\|_{L^2(T)} \lesssim h_T^{-1} \|r_T\|_{L^2(T)}.$$
Therefore, owing to the Cauchy–Schwarz inequality and the definition of $\mathcal{R}_{\mathrm{osc},T}$,
$$\|r_T\|_{L^2(T)}^2 \lesssim h_T^{-1}(\|\nabla(u-u_h)\|_{[L^2(T)]^d} + \mathcal{R}_{\mathrm{osc},T})\|r_T\|_{L^2(T)}.$$
Hence, $\|r_T\|_{L^2(T)} \lesssim h_T^{-1}(\|\nabla(u-u_h)\|_{[L^2(T)]^d} + \mathcal{R}_{\mathrm{osc},T})$, and using the triangle inequality yields (5.35a).

(ii) Bound on $\mathcal{R}_{\mathrm{jdf},F}$. Let $F \in \mathcal{F}_h^i$, let (cf. Fig. 5.2)
$$\omega_F := \mathrm{int}\left(\cup_{T \in \mathcal{T}_F} \overline{T}\right),$$
and let b_F be the face bubble function in $H_0^1(\omega_F)$ equal on F to the product of barycentric coordinates in F rescaled so as to take the value 1 at the center of gravity of F. Set $r_F := [\![\nabla_h u_h]\!] \cdot \mathbf{n}_F$, extend r_F to ω_F by constant values along $\mathbf{n}_F$, and set $\psi_F := b_F r_F$. Owing to the properties of bubble functions and mesh regularity,
$$\|r_F\|_{L^2(F)}^2 \lesssim \int_F r_F b_F r_F, \qquad \|\psi_F\|_{L^2(\omega_F)} \lesssim h_F^{1/2} \|r_F\|_{L^2(F)}.$$
Since the bubble function b_F vanishes on $\partial \omega_F$, integration by parts yields
$$\|r_F\|_{L^2(F)}^2 \lesssim \int_F r_F b_F r_F = \int_F r_F \psi_F = \int_F [\![\nabla_h(u_h - u)]\!] \cdot \mathbf{n}_F \psi_F$$
$$= \int_{\omega_F} (f + \triangle_h u_h) \psi_F - \int_{\omega_F} \nabla_h(u - u_h) \cdot \nabla \psi_F.$$
Using the inverse inequality (1.36) to bound $\|\nabla \psi_F\|_{[L^2(\omega_F)]^d}$ leads to
$$\|r_F\|_{L^2(F)}^2 \lesssim (\|f + \triangle_h u_h\|_{L^2(\omega_F)} + h_F^{-1}\|\nabla_h(u-u_h)\|_{[L^2(\mathcal{T}_F)]^d})\|\psi_F\|_{L^2(\omega_F)}$$
$$\lesssim (h_F^{1/2}\|f + \triangle_h u_h\|_{L^2(\omega_F)} + h_F^{-1/2}\|\nabla_h(u-u_h)\|_{[L^2(\mathcal{T}_F)]^d})\|r_F\|_{L^2(F)}.$$
Therefore,
$$\|r_F\|_{L^2(F)} \lesssim h_F^{1/2}\|f + \triangle_h u_h\|_{L^2(\omega_F)} + h_F^{-1/2}\|\nabla_h(u-u_h)\|_{[L^2(\mathcal{T}_F)]^d}.$$
Using (5.35a) for the two mesh elements in $\mathcal{T}_F$ together with mesh regularity leads to (5.35b). $\square$

5.5.2 Potential Reconstruction

We first assume that $\mathcal{T}_h$ is a *matching simplicial mesh*. In this case, we can consider the *averaging operator* (sometimes called Oswald interpolation operator)

$$\mathcal{I}_{\mathrm{av}}^{\mathcal{T}_h} : \mathbb{P}_d^k(\mathcal{T}_h) \to \mathbb{P}_d^k(\mathcal{T}_h) \cap H_0^1(\Omega)$$

such that, for all $v_h \in \mathbb{P}_d^k(\mathcal{T}_h)$, the value of $\mathcal{I}_{\mathrm{av}}^{\mathcal{T}_h}(v_h)$ is prescribed at the Lagrange interpolation nodes of the conforming finite element space $\mathbb{P}_d^k(\mathcal{T}_h) \cap H_0^1(\Omega)$ by setting at each interpolation node V located inside Ω,

$$\mathcal{I}_{\mathrm{av}}^{\mathcal{T}_h}(v_h)(V) := \frac{1}{\mathrm{card}(\mathcal{T}_V)} \sum_{T \in \mathcal{T}_V} v_h|_T(V),$$

where $\mathcal{T}_V \subset \mathcal{T}_h$ collects the simplices to which V belongs (see Fig. 5.3), while $\mathcal{I}_{\mathrm{av}}^{\mathcal{T}_h}(v_h)$ is set to zero at the interpolation nodes located on $\partial \Omega$. The averaging operator $\mathcal{I}_{\mathrm{av}}^{\mathcal{T}_h}$ has been considered in the context of a posteriori error estimates for nonconforming finite element methods by Achdou, Bernardi, and Coquel [3] and Ern, Nicaise, and Vohralík [140] and for dG methods by Karakashian and Pascal [207]. The following result is proven in [207] (see also Burman and Ern [64] for further discussion on the polynomial degree): For all $v_h \in \mathbb{P}_d^k(\mathcal{T}_h)$ and all $T \in \mathcal{T}_h$,

$$\|v_h - \mathcal{I}_{\mathrm{av}}^{\mathcal{T}_h}(v_h)\|_{L^2(T)} \lesssim \left(\sum_{F \in \tilde{\mathcal{F}}_T} h_F \|[\![v_h]\!]\|_{L^2(F)}^2 \right)^{1/2}, \tag{5.36}$$

where $\tilde{\mathcal{F}}_T$ collects the mesh faces having a nonempty intersection with ∂T; see Fig. 5.4. In what follows, we define, for any subset $\mathcal{F} \subset \mathcal{F}_h$ and for all $v \in H^1(\mathcal{T}_h)$,

$$|v|_{\mathrm{J},\mathcal{F}} := \left(\sum_{F \in \mathcal{F}} \frac{1}{h_F} \|[\![v]\!]\|_{L^2(F)}^2 \right)^{1/2}, \tag{5.37}$$

so that $|v|_{\mathrm{J},\mathcal{F}_h}$ coincides with the jump seminorm $|v|_{\mathrm{J}}$ defined by (4.18).

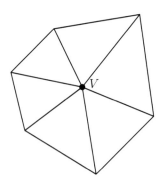

Fig. 5.3: Set $\mathcal{T}_V$ used for potential reconstruction

5.5. Local Postprocessing

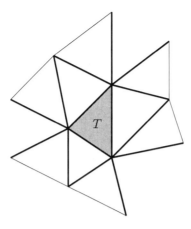

Fig. 5.4: Set $\tilde{\mathcal{F}}_T$ (thick lines)

Lemma 5.28 (Potential reconstruction, matching simplicial meshes). *Assume that $\mathcal{T}_h$ is a matching simplicial mesh. Then, the reconstructed potential*

$$u_h^* := \mathcal{I}_{\mathrm{av}}^{\mathcal{T}_h}(u_h) \in H_0^1(\Omega),$$

is such that, for all $T \in \mathcal{T}_h$,

$$\|\nabla(u - u_h^*)\|_{[L^2(T)]^d} \lesssim \|\nabla(u - u_h)\|_{[L^2(T)]^d} + |u - u_h|_{\mathrm{J},\tilde{\mathcal{F}}_T}, \tag{5.38}$$

where $\tilde{\mathcal{F}}_T$ collects the mesh faces having a nonempty intersection with ∂T. As a result,

$$\|\nabla_h(u - u_h^*)\|_{[L^2(\Omega)]^d} \lesssim \|u - u_h\|_{\mathrm{sip}}, \tag{5.39}$$

where the $\|\cdot\|_{\mathrm{sip}}$-norm is defined by (4.17).

Proof. Combining (5.36) with an inverse inequality and mesh regularity, we infer that, for all $v_h \in \mathbb{P}_d^k(\mathcal{T}_h)$ and all $T \in \mathcal{T}_h$,

$$\|\nabla(v_h - \mathcal{I}_{\mathrm{av}}^{\mathcal{T}_h}(v_h))\|_{[L^2(T)]^d} \lesssim |v_h|_{\mathrm{J},\tilde{\mathcal{F}}_T},$$

so that $\|\nabla(u_h - u_h^*)\|_{[L^2(T)]^d} \lesssim |u_h|_{\mathrm{J},\tilde{\mathcal{F}}_T}$. We then infer the bound (5.38) using the triangle inequality and the fact that for the exact solution, $[\![u]\!] = 0$ for all $F \in \mathcal{F}_h$. Finally, (5.39) is obtained by summing (5.38) over mesh elements and using the fact that the number of faces collected in each set $\tilde{\mathcal{F}}_T$ is uniformly bounded owing to mesh regularity. □

On *general meshes*, we can consider the matching simplicial submesh $\mathfrak{S}_h$ (cf. Definition 1.37) together with the averaging operator $\mathcal{I}_{\mathrm{av}}^{\mathfrak{S}_h}$ mapping $\mathbb{P}_d^k(\mathfrak{S}_h)$ onto the conforming finite element space $\mathbb{P}_d^k(\mathfrak{S}_h) \cap H_0^1(\Omega)$. On each element $T \in \mathcal{T}_h$, a discrete function $v_h \in \mathbb{P}_d^k(\mathcal{T}_h)$ can jump only across those faces of the submesh located on the boundary ∂T, while v_h is continuous at the remaining faces located inside T. Applying (5.36) for all $T' \in \mathfrak{S}_T$, regrouping the contributions of the submesh elements, and observing that $\text{card}(\mathfrak{F}_F)$ is uniformly bounded for all $F \in \mathcal{F}_h$, we again obtain the following bound: For all $v_h \in \mathbb{P}_d^k(\mathcal{T}_h)$ and all $T \in \mathcal{T}_h$,

$$\|v_h - \mathcal{I}_{\text{av}}^{\mathfrak{S}_h}(v_h)\|_{L^2(T)} \lesssim \left(\sum_{F \in \tilde{\mathcal{F}}_T} h_F \|[\![v_h]\!]\|_{L^2(F)}^2 \right)^{1/2}.$$

Proceeding as in the proof of Lemma 5.28, we then arrive at the following result.

Lemma 5.29 (Potential reconstruction, general meshes). *The reconstructed potential $u_h^* := \mathcal{I}_{\text{av}}^{\mathfrak{S}_h}(u_h) \in H_0^1(\Omega)$ satisfies the bounds (5.38) and (5.39).*

An interesting application of potential reconstruction is to prove that the jump seminorm $|u_h|_{\text{J}} = |u - u_h|_{\text{J}}$ defined by (4.18) is controlled globally in Ω by the approximation error plus a data oscillation term, provided the penalty parameter in the SIP method is large enough. This result has been established by Karakashian and Pascal [208], yet using a different proof as that presented below. An extension to nonmatching simplicial meshes has been derived by Bonito and Nochetto [46]. A sharper result for a first-order dG approximation on matching triangular meshes is derived by Ainsworth [7], showing that, in this case, the threshold on the penalty coefficient is the same as that required for discrete coercivity; see also Ainsworth and Rankin [9] for triangular meshes with hanging nodes.

Lemma 5.30 (Bound on the jumps). *Let $\mathcal{T}_h$ be a matching simplicial mesh. Let $u_h \in V_h = \mathbb{P}_d^k(\mathcal{T}_h)$ solve the discrete Poisson problem (5.8). Then, there is $\eta_{\text{av}} > 0$ such that for $\eta > \eta_{\text{av}}$,*

$$(\eta - \eta_{\text{av}})|u_h|_{\text{J}} \lesssim \|\nabla_h(u - u_h)\|_{[L^2(\Omega)]^d} + \mathcal{R}_{\text{osc},\Omega}. \tag{5.40}$$

The threshold η_{av}, which is independent of h, but can depend on the space dimension d, the polynomial degree k, and the mesh regularity parameters ϱ, is such that, for all $v_h \in \mathbb{P}_d^k(\mathcal{T}_h)$,

$$\left(\sum_{F \in \mathcal{F}_h} h_F \|\{\!\!\{\nabla_h(v_h - \mathcal{I}_{\text{av}}^{\mathcal{T}_h}(v_h))\}\!\!\}\|_{L^2(F)}^2 \right)^{1/2} \leq \eta_{\text{av}} |v_h|_{\text{J}}, \tag{5.41}$$

resulting from the bound (5.36), the inverse inequality (1.36), the discrete trace inequality (1.37), and mesh regularity.

Proof. Since u_h solves (5.8), using the expression (4.13) for the SIP bilinear form yields, for all $v_h \in \mathbb{P}_d^k(\mathcal{T}_h)$,

$$\sum_{F \in \mathcal{F}_h} \frac{\eta}{h_F} \int_F [\![u_h]\!][\![v_h]\!] = \sum_{T \in \mathcal{T}_h} \int_T (f + \Delta u_h)v_h - \sum_{F \in \mathcal{F}_h^i} \int_F [\![\nabla_h u_h]\!] \cdot \mathbf{n}_F \{\!\!\{v_h\}\!\!\}$$

$$+ \sum_{F \in \mathcal{F}_h} \int_F [\![u_h]\!]\{\!\!\{\nabla_h v_h\}\!\!\} \cdot \mathbf{n}_F.$$

5.5. Local Postprocessing

Using the definitions (5.32a) and (5.32b) of the local residuals $\mathcal{R}_{\mathrm{pde},T}$ and $\mathcal{R}_{\mathrm{jdf},F}$, together with the Cauchy–Schwarz inequality, leads to

$$\sum_{F \in \mathcal{F}_h} \frac{\eta}{h_F} \int_F [\![u_h]\!][\![v_h]\!] \leq \sum_{T \in \mathcal{T}_h} \mathcal{R}_{\mathrm{pde},T} h_T^{-1} \|v_h\|_{L^2(T)} + \sum_{F \in \mathcal{F}_h^i} \mathcal{R}_{\mathrm{jdf},F} h_F^{-1/2} \|\{\!\{v_h\}\!\}\|_{L^2(F)}$$
$$+ \sum_{F \in \mathcal{F}_h} \int_F [\![u_h]\!] \{\!\{\nabla_h v_h\}\!\} \cdot n_F.$$

Taking $v_h = u_h - u_h^*$ with the potential reconstruction $u_h^* = \mathcal{I}_{\mathrm{av}}^{\mathcal{T}_h}(u_h) \in \mathbb{P}_d^k(\mathcal{T}_h)$ yields, since $[\![u_h^*]\!] = 0$,

$$\eta |u_h|_{\mathrm{J}}^2 \leq \sum_{T \in \mathcal{T}_h} \mathcal{R}_{\mathrm{pde},T} h_T^{-1} \|u_h - u_h^*\|_{L^2(T)} + \sum_{F \in \mathcal{F}_h^i} \mathcal{R}_{\mathrm{jdf},F} h_F^{-1/2} \|\{\!\{u_h - u_h^*\}\!\}\|_{L^2(F)}$$
$$+ \sum_{F \in \mathcal{F}_h} \int_F \{\!\{\nabla_h(u_h - u_h^*)\}\!\} \cdot n_F [\![u_h]\!].$$

Denote by $\mathfrak{T}_1$, $\mathfrak{T}_2$, and $\mathfrak{T}_3$ the three terms on the right-hand side. Using (5.36), together with the discrete trace inequality (1.37) to estimate $(u_h - u_h^*)$, and Lemma 5.27 to estimate $\mathcal{R}_{\mathrm{pde},T}$ and $\mathcal{R}_{\mathrm{jdf},F}$, we infer

$$|\mathfrak{T}_1| + |\mathfrak{T}_2| \lesssim (\|\nabla_h(u - u_h)\|_{[L^2(\Omega)]^d} + \mathcal{R}_{\mathrm{osc},\Omega})|u_h|_{\mathrm{J}}.$$

Moreover, the Cauchy–Schwarz inequality combined with (5.41) yields $|\mathfrak{T}_3| \leq \eta_{\mathrm{av}}|u_h|_{\mathrm{J}}^2$. As a result, $(\eta - \eta_{\mathrm{av}})|u_h|_{\mathrm{J}}^2 \lesssim (\|\nabla_h(u - u_h)\|_{[L^2(\Omega)]^d} + \mathcal{R}_{\mathrm{osc},\Omega})|u_h|_{\mathrm{J}}$, yielding (5.40). □

Remark 5.31 (General meshes). The present proof cannot be extended to general meshes since $\mathcal{I}_{\mathrm{av}}^{\mathfrak{S}_h}(u_h) \notin \mathbb{P}_d^k(\mathcal{T}_h)$, so that it is no longer possible to use $v_h = u_h - \mathcal{I}_{\mathrm{av}}^{\mathfrak{S}_h}(u_h)$ as a test function in the discrete problem.

5.5.3 Diffusive Flux Reconstruction by Prescription

Our aim is to reconstruct a diffusive flux $\sigma_h^* \in H(\mathrm{div};\Omega)$ such that σ_h^* accurately approximates the exact diffusive flux σ in $H(\mathrm{div};\Omega)$. In particular, we want that, locally on each mesh element, the divergence of σ_h^* be close to that of σ. In the spirit of FV schemes, a minimal requirement is

$$\int_T \nabla \cdot \sigma_h^* = \int_T \nabla \cdot \sigma = \int_T f \qquad \forall T \in \mathcal{T}_h. \tag{5.42}$$

In the context of high-order dG methods with polynomial degree $k \geq 1$, we show that a tighter relationship can be achieved between $\nabla \cdot \sigma_h^*$ and f than just having the same mean value on each mesh element.

In this section, we focus on *matching simplicial meshes*, where the construction of σ_h^* can be achieved simply by prescribing degrees of freedom in

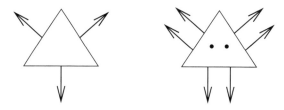

Fig. 5.5: Local degrees of freedom in $\mathbb{RTN}_d^l(\mathcal{T}_h)$ for $l=0$ (*left*) and $l=1$ (*right*); a single mesh element is represented

Raviart–Thomas–Nédélec finite element spaces. General meshes are considered in Sect. 5.5.4. Let $l \geq 0$ be an integer. We define

$$\mathbb{RTN}_d^l(\mathcal{T}_h) := \left\{ \tau_h \in H(\mathrm{div};\Omega) \mid \forall T \in \mathcal{T}_h,\ \tau_h|_T \in \mathbb{RTN}_d^l(T) \right\},$$

where

$$\mathbb{RTN}_d^l(T) := [\mathbb{P}_d^l(T)]^d + x\mathbb{P}_d^l(T).$$

For a detailed presentation of Raviart–Thomas–Nédélec finite element spaces, we refer the reader to Brezzi and Fortin [57, p. 116], Ern and Guermond [141, Sect. 1.4.7] (see also Nédélec [242], Raviart and Thomas [267], and Roberts and Thomas [273, p. 550]). In what follows, we need two important facts. Firstly, a function τ_h in $\mathbb{RTN}_d^l(\mathcal{T}_h)$ satisfies:

(a) For all $T \in \mathcal{T}_h$, $\nabla\cdot\tau_h \in \mathbb{P}_d^l(T)$.

(b) For all $F \in \mathcal{F}_h$, $\tau_h \cdot n_F \in \mathbb{P}_{d-1}^l(F)$, and $\tau_h \cdot n_F$ is single-valued at all $F \in \mathcal{F}_h^i$.

The last property in (b) ensures the $H(\mathrm{div};\Omega)$-conformity of the discrete space $\mathbb{RTN}_d^l(\mathcal{T}_h)$; cf. Lemma 1.24. Secondly, the degrees of freedom that allow one to uniquely define a function $\tau_h \in \mathbb{RTN}_d^l(\mathcal{T}_h)$ are:

(a) For all $F \in \mathcal{F}_h$, the moments $\int_F (\tau_h \cdot n_F) q$ of the normal component of τ_h against scalar-valued polynomials $q \in \mathbb{P}_{d-1}^l(F)$.

(b) For all $T \in \mathcal{T}_h$, the moments $\int_T \tau_h \cdot r$ of the function τ_h against vector-valued polynomials $r \in [\mathbb{P}_d^{l-1}(T)]^d$ (by convention, nothing is prescribed if $l=0$).

A symbolic representation of these degrees of freedom on a single mesh element is presented in Fig. 5.5 for $l \in \{0,1\}$.

5.5.3.1 Prescription of the Degrees of Freedom

Following Kim [212] and Ern, Nicaise, and Vohralík [147], we construct

$$\sigma_h^* \in \mathbb{RTN}_d^l(\mathcal{T}_h), \qquad l \in \{k-1, k\}.$$

5.5. Local Postprocessing

The choice $l = k-1$ is somewhat more natural since $u_h \in \mathbb{P}_d^k(\mathcal{T}_h)$ and σ_h^* approximates the gradient of u_h. However, the choice $l = k$ is also interesting since the divergence of σ_h^* is more accurate; cf. Theorem 5.34. Our starting point is the local formulation (4.49) derived in Sect. 4.3.4 for the discrete problem: For all $T \in \mathcal{T}_h$ and for all $\xi \in \mathbb{P}_d^k(T)$,

$$\int_T G_h^l(u_h)\cdot\nabla\xi + \sum_{F\in\mathcal{F}_T}\epsilon_{T,F}\int_F \phi_F(u_h)\xi = \int_T f\xi, \qquad (5.43)$$

with $l \in \{k-1, k\}$, $\epsilon_{T,F} = n_T \cdot n_F$, and the numerical flux

$$\phi_F(u_h) := -\{\!\!\{\nabla_h u_h\}\!\!\}\cdot n_F + \frac{\eta}{h_F}[\![u_h]\!]. \qquad (5.44)$$

Definition 5.32 (Diffusive flux reconstruction by prescription). The degrees of freedom of $\sigma_h^* \in \mathbb{RTN}_d^l(\mathcal{T}_h)$, $l \in \{k-1, k\}$, are prescribed as follows:

(i) For all $F \in \mathcal{F}_h$ and all $q \in \mathbb{P}_{d-1}^l(F)$,

$$\int_F (\sigma_h^* \cdot n_F) q := \int_F \phi_F(u_h) q. \qquad (5.45)$$

(If $l = k$, $\phi_F(u_h) \in \mathbb{P}_{d-1}^l(F)$ so that (5.45) yields $\sigma_h^* \cdot n_F = \phi_F(u_h)$.)

(ii) For all $T \in \mathcal{T}_h$ and all $r \in [\mathbb{P}_d^{l-1}(T)]^d$,

$$\int_T \sigma_h^* \cdot r := -\int_T G_h^l(u_h)\cdot r. \qquad (5.46)$$

Remark 5.33 (Other construction). A different construction has been proposed by Bastian and Rivière [37] using Brezzi–Douglas–Marini finite element spaces. The advantage of the present construction is that the degrees of freedom of σ_h^* are more directly linked to the dG scheme, thereby achieving an optimal approximation result on the divergence of σ_h^* (cf. Theorem 5.34 below). In addition, the error $(\sigma - \sigma_h^*)$ in the L^2-norm can be bounded locally by the approximation error $(u - u_h)$ measured in a suitable norm plus a so-called data oscillation term related to the subgrid fluctuations of the source term (cf. Theorem 5.36 below).

5.5.3.2 Evaluation of $\nabla\cdot\sigma_h^*$

The key motivation for the prescription of σ_h^* using (5.45) and (5.46) is the following result. Notice that $\nabla\cdot\sigma_h^* \in L^2(\Omega)$ since, by construction, $\sigma_h^* \in H(\text{div};\Omega)$.

Theorem 5.34 (Evaluation of $\nabla\cdot\sigma_h^*$). *Let $\sigma_h^* \in \mathbb{RTN}_d^l(\mathcal{T}_h)$, $l \in \{k-1, k\}$, be prescribed using (5.45) and (5.46). Then,*

$$\nabla\cdot\sigma_h^* = \pi_h^l f, \qquad (5.47)$$

where π_h^l denotes the L^2-orthogonal projector onto $\mathbb{P}_d^l(\mathcal{T}_h)$. In other words (compare with (5.42)),
$$\int_T (\nabla \cdot \sigma_h^*)\xi = \int_T f\xi \qquad \forall T \in \mathcal{T}_h,\ \forall \xi \in \mathbb{P}_d^l(\mathcal{T}_h).$$

Proof. Let $T \in \mathcal{T}_h$ and let $\xi \in \mathbb{P}_d^l(T)$. Integration by parts yields
$$\int_T (\nabla \cdot \sigma_h^*)\xi = -\int_T \sigma_h^* \cdot \nabla \xi + \int_{\partial T} (\sigma_h^* \cdot n_T)\xi$$
$$= -\int_T \sigma_h^* \cdot \nabla \xi + \sum_{F \in \mathcal{F}_T} \epsilon_{T,F} \int_F (\sigma_h^* \cdot n_F)\xi$$
$$= \int_T G_h^l(u_h) \cdot \nabla \xi + \sum_{F \in \mathcal{F}_T} \epsilon_{T,F} \int_F \phi_F(u_h)\xi = \int_T f\xi,$$

where we have used the fact that $\nabla \xi \in [\mathbb{P}_d^{l-1}(T)]^d$ and $\xi|_F \in \mathbb{P}_{d-1}^l(F)$ to exploit (5.45) and (5.46) and the local formulation (5.43). □

Remark 5.35 (Optimality of Theorem 5.34). Theorem 5.34 delivers an optimal approximation result in $L^2(\Omega)$ for the divergence of the diffusive flux. Indeed, since the exact diffusive flux satisfies $\nabla \cdot \sigma = f$, estimate (5.47) yields $\nabla \cdot \sigma_h^* = \pi_h^l(\nabla \cdot \sigma)$. Observing that $\nabla \cdot \sigma_h^* \in \mathbb{P}_d^l(\mathcal{T}_h)$, this result is optimal.

5.5.3.3 L^2-Norm Estimate

We now bound the error $(\sigma_h^* - \sigma)$ in the L^2-norm. To simplify the notation, we set $\eta_1 := \max(1, \eta)$.

Theorem 5.36 (L^2-estimate). *Let $\sigma_h^* \in \mathbb{RTN}_d^l(\mathcal{T}_h)$, $l \in \{k-1, k\}$, be prescribed using (5.45) and (5.46). Let $\sigma = -\nabla u$ be the exact diffusive flux. Then, for all $T \in \mathcal{T}_h$,*
$$\|\sigma_h^* - \sigma\|_{[L^2(T)]^d} \lesssim \|\nabla_h(u - u_h)\|_{[L^2(\mathcal{N}_T)]^d} + \mathcal{R}_{\mathrm{osc},\mathcal{N}_T} + \eta_1 |u_h|_{\mathrm{J},\mathcal{F}_T}, \qquad (5.48)$$

where $\mathcal{N}_T \subset \mathcal{T}_h$ collects T and its neighbors in the sense of faces, while the data oscillation term $\mathcal{R}_{\mathrm{osc},\mathcal{N}_T}$ is defined by (5.34). Moreover,
$$\|\sigma_h^* - \sigma\|_{[L^2(\Omega)]^d} \lesssim \eta_1 \|u - u_h\|_{\mathrm{sip}} + \mathcal{R}_{\mathrm{osc},\mathcal{T}_h}. \qquad (5.49)$$

Proof. Let σ_h^* be prescribed using (5.45) and (5.46) and set $\sigma_h' := \sigma_h^* + \nabla_h u_h$. Let $T \in \mathcal{T}_h$. We first observe that $\sigma_h'|_T \in \mathbb{RTN}_d^l(T)$ since $\sigma_h^*|_T \in \mathbb{RTN}_d^l(T)$ by construction and $\nabla u_h|_T \in \mathbb{RTN}_d^l(T)$ since $l \geq k-1$. Then, we use the fact that, for all $\tau \in \mathbb{RTN}_d^l(T)$,
$$\|\tau\|_{[L^2(T)]^d} \lesssim h_T^{1/2} \|\tau \cdot n_F\|_{L^2(\partial T)} + \sup_{r \in [\mathbb{P}_d^{l-1}(T)]^d \setminus \{0\}} \frac{\int_T \tau \cdot r}{\|r\|_{[L^2(T)]^d}}.$$

5.5. Local Postprocessing

This estimate is classically proven using the Piola transformation onto the reference simplex, norm equivalence in finite dimension, and transforming back onto the current simplex T. For all $F \in \mathcal{F}_T$ and all $q \in \mathbb{P}^l_{d-1}(F)$, (5.44) and (5.45) yield

$$\int_F (\sigma'_h \cdot n_F) q = \int_F (\sigma^*_h \cdot n_F + (\nabla u_h)|_T \cdot n_F) q$$
$$= \int_F (\phi_F(u_h) + (\nabla u_h)|_T \cdot n_F) q$$
$$= \int_F (\varpi_F [\![\nabla_h u_h]\!] \cdot n_T + \eta h_F^{-1} [\![u_h]\!]) q,$$

where $\varpi_F := 1/2$ if $F \in \mathcal{F}_h^i$ and $\varpi_F := 0$ if $F \in \mathcal{F}_h^b$. Since $\sigma'_h \cdot n_F \in \mathbb{P}^l_{d-1}(F)$, we obtain $\sigma'_h \cdot n_F|_F = \varpi_F [\![\nabla_h u_h]\!] \cdot n_T + \eta h_F^{-1} \Pi_F^l [\![u_h]\!]$, where Π_F^l denotes the $L^2(F)$-orthogonal projection onto $\mathbb{P}^l_{d-1}(F)$. Using the triangle inequality, mesh regularity, and the fact that $\|\Pi_F^l [\![u_h]\!]\|_{L^2(F)} \leq \|[\![u_h]\!]\|_{L^2(F)}$, we infer

$$h_T^{1/2} \|\sigma'_h \cdot n_T\|_{L^2(\partial T)} \lesssim \left(\sum_{F \in \mathcal{F}_T \cap \mathcal{F}_h^i} \mathcal{R}^2_{\mathrm{jdf},F} \right)^{1/2} + \eta \left(\sum_{F \in \mathcal{F}_T} \mathcal{R}^2_{\mathrm{jpt},F} \right)^{1/2},$$

recalling the local residuals (cf. (5.32b) and (5.32c)) $\mathcal{R}_{\mathrm{jdf},F} = h_F^{1/2} \|[\![\nabla_h u_h]\!] \cdot n_F\|_{L^2(F)}$ and $\mathcal{R}_{\mathrm{jpt},F} = h_F^{-1/2} \|[\![u_h]\!]\|_{L^2(F)}$. Moreover, since $l \leq k$, it is easily seen that, for all $r \in [\mathbb{P}^{l-1}_d(T)]^d$, (5.46) yields

$$\int_T \sigma'_h \cdot r = \int_T \mathrm{R}_h^{k-1}([\![u_h]\!]) \cdot r = \sum_{F \in \mathcal{F}_T} \int_F (1 - \varpi_F)(r \cdot n_F) [\![u_h]\!],$$

so that, using the Cauchy–Schwarz inequality, the fact that $h_F \leq h_T$ for all $F \in \mathcal{F}_T$, the discrete trace inequality (1.40), and the uniform bound on N_∂, we obtain

$$\int_T \sigma'_h \cdot r \lesssim \left(\sum_{F \in \mathcal{F}_T} \mathcal{R}^2_{\mathrm{jpt},F} \right)^{1/2} \|r\|_{[L^2(T)]^d}.$$

Combining the two above inequalities yields

$$\|\sigma^*_h + \nabla u_h\|_{[L^2(T)]^d} \lesssim \left(\sum_{F \in \mathcal{F}_T \cap \mathcal{F}_h^i} \mathcal{R}^2_{\mathrm{jdf},F} \right)^{1/2} + \eta_1 \left(\sum_{F \in \mathcal{F}_T} \mathcal{R}^2_{\mathrm{jpt},F} \right)^{1/2}.$$

Using the bound (5.35b) on $\mathcal{R}_{\mathrm{jdf},F}$ and the definition (5.37) for the jump seminorm, we obtain

$$\|\sigma^*_h + \nabla u_h\|_{[L^2(T)]^d} \lesssim \|\nabla_h(u - u_h)\|_{[L^2(\mathcal{N}_T)]^d} + \mathcal{R}_{\mathrm{osc},\mathcal{N}_T} + \eta_1 |u_h|_{\mathrm{J},\mathcal{F}_T}.$$

The local bound (5.48) then follows from the triangle inequality, while the global bound (5.49) is obtained by summing (5.48) over mesh elements. $\square$

Remark 5.37 (Optimality). If $l = k$, estimate (5.49) is suboptimal since the upper bound converges as h^k, while the Raviart–Thomas–Nédélec interpolate in $\mathbb{RTN}_d^l(\mathcal{T}_h)$ of the exact diffusive flux converges as h^{l+1} in the $[L^2(\Omega)]^d$-norm if the exact solution is smooth enough. Optimality is recovered if $l = k - 1$.

5.5.4 Diffusive Flux Reconstruction by Solving Local Problems

A simple and cost-effective approach recently proposed by Ern and Vohralík [151] to reconstruct the diffusive flux on general meshes consists in solving, on each mesh element, a local Neumann problem approximated by mixed finite elements. The present methodology can also be applied within matching simplicial meshes; cf. Remark 5.42.

5.5.4.1 The Local Neumann Problem

Let $\mathcal{T}_h$ be a general mesh and let $T \in \mathcal{T}_h$. We define the local PDE residual $\rho_{\text{pde},T} \in L^2(T)$ such that, for all $T \in \mathcal{T}_h$,

$$\rho_{\text{pde},T} := (f + \triangle u_h)|_T.$$

We also define the local jump residual $\rho_{\text{nc},T} \in L^2(\partial T)$ such that, for all $T \in \mathcal{T}_h$,

$$\rho_{\text{nc},T}|_F := \varpi_F [\![\nabla_h u_h]\!] \cdot \mathbf{n}_T + \eta h_F^{-1} [\![u_h]\!] \quad \forall F \in \mathcal{F}_T,$$

where, as before, $\varpi_F := 1/2$ if $F \in \mathcal{F}_h^i$ and $\varpi_F := 0$ if $F \in \mathcal{F}_h^b$. We observe that

$$\rho_{\text{nc},T}|_F = \phi_F(u_h) + (\nabla u_h)|_T \cdot \mathbf{n}_F \quad \forall F \in \mathcal{F}_T, \qquad (5.50)$$

where $\phi_F(u_h)$ is the numerical flux defined by (5.44). We recall that (cf. (4.51))

$$\sum_{F \in \mathcal{F}_T} \epsilon_{T,F} \int_F \phi_F(u_h) = \int_T f, \qquad (5.51)$$

with $\epsilon_{T,F} = \mathbf{n}_T \cdot \mathbf{n}_F$.

We consider the following local Neumann problem: Find $p : T \to \mathbb{R}$ such that

$$\begin{cases} -\triangle p = \rho_{\text{pde},T} & \text{in } T, \\ -\nabla p \cdot \mathbf{n}_T = \rho_{\text{nc},T} & \text{on } \partial T, \\ \langle p \rangle_T = 0, \end{cases}$$

where $\langle \cdot \rangle_T$ denotes the mean value over T. Letting $H_*^1(T) = \{q \in H^1(T) \mid \langle q \rangle_T = 0\}$, this problem amounts to finding $p \in H_*^1(T)$ such that

$$\int_T \nabla p \cdot \nabla q = \int_T q \rho_{\text{pde},T} - \int_{\partial T} q \rho_{\text{nc},T} \text{ for all } q \in H_*^1(T). \qquad (5.52)$$

Lemma 5.38 (Well-posedness). *The local Neumann problem (5.52) is well-posed.*

5.5. Local Postprocessing

Proof. It is well-known that (5.52) is well-posed if and only if the data satisfy the compatibility condition

$$\int_T \rho_{\text{pde},T} = \int_{\partial T} \rho_{\text{nc},T}.$$

Using (5.51) and integrating by parts $\triangle u_h$ in T yields

$$\int_T \rho_{\text{pde},T} = \int_T (f + \triangle u_h) = \sum_{F \in \mathcal{F}_T} \epsilon_{T,F} \int_F \phi_F(u_h) + \sum_{F \in \mathcal{F}_T} \int_F (\nabla u_h)|_T \cdot n_T.$$

We conclude using (5.50). □

5.5.4.2 Approximation of Local Neumann Problem by Mixed Finite Elements

Let $\mathfrak{S}_h$ be the matching simplicial submesh of $\mathcal{T}_h$. For all $T \in \mathcal{T}_h$, we approximate the local Neumann problem (5.52) using mixed finite elements in Raviart–Thomas–Nédélec finite element spaces based on the simplicial submesh $\mathfrak{S}_T$. In what follows, the subfaces in $\mathfrak{F}_h$ included in $\overline{T}$ are split into the set $\mathfrak{F}_T^i$ collecting the subfaces located inside T and the set $\mathfrak{F}_T^b$ collecting the subfaces located on ∂T; cf. Fig. 5.6.

Let $l \in \{k-1, k\}$. We consider the following spaces for the potential

$$\mathbb{P}_d^l(\mathfrak{S}_T) := \{q_h \in L^2(T) \mid \forall T' \in \mathfrak{S}_T,\ q_h|_{T'} \in \mathbb{P}_d^l(T')\},$$
$$\mathbb{P}_{d,*}^l(\mathfrak{S}_T) := \{q_h \in \mathbb{P}_d^l(\mathfrak{S}_T) \mid \langle q_h \rangle_T = 0\},$$

together with the following spaces for the diffusive flux

$$\mathbb{RTN}_d^l(\mathfrak{S}_T) := \{\tau_h \in H(\text{div};T) \mid \forall T' \in \mathfrak{S}_T,\ \tau_h|_{T'} \in \mathbb{RTN}_d^l(T')\},$$
$$\mathbb{RTN}_{d,\rho}^l(\mathfrak{S}_T) := \{\tau_h \in \mathbb{RTN}_d^l(\mathfrak{S}_T) \mid \forall F' \in \mathfrak{F}_T^b,\ \forall \xi \in \mathbb{P}_{d-1}^l(F'),$$
$$\int_{F'} (\tau_h \cdot n_T - \rho_{\text{nc},T})\xi = 0\},$$

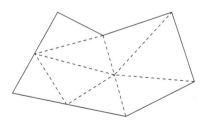

Fig. 5.6: Example of matching simplicial submesh $\mathfrak{S}_T$ of a mesh element T consisting of eight subelements; there are eight faces belonging to $\mathfrak{F}_T^i$ (*dashed lines*) and eight faces belonging to $\mathfrak{F}_T^b$ (*solid lines*)

$$\mathbb{RTN}_{d,0}^l(\mathfrak{S}_T) := \{\tau_h \in \mathbb{RTN}_d^l(\mathfrak{S}_T) \mid \forall F' \in \mathfrak{F}_T^b, \forall \xi \in \mathbb{P}_{d-1}^l(F'),$$
$$\int_{F'} (\tau_h \cdot n_T)\xi = 0\}.$$

The local discrete problem (with nonhomogeneous Neumann boundary conditions) consists in finding $(\sigma_h', p_h) \in \mathbb{RTN}_{d,\rho}^l(\mathfrak{S}_T) \times \mathbb{P}_{d,*}^l(\mathfrak{S}_T)$ such that

$$\int_T \sigma_h' \cdot \tau_h + \int_T p_h(\nabla \cdot \tau_h) = 0 \quad \forall \tau_h \in \mathbb{RTN}_{d,0}^l(\mathfrak{S}_T), \qquad (5.53\text{a})$$

$$\int_T (\nabla \cdot \sigma_h') q_h = \int_T \rho_{\text{pde},T} q_h \quad \forall q_h \in \mathbb{P}_{d,*}^l(\mathfrak{S}_T). \qquad (5.53\text{b})$$

Solving the local Neumann problem (5.53) for all $T \in \mathcal{T}_h$ yields a vector-valued field $\sigma_h' \in [L^2(\Omega)]^d$ defined by its restrictions to mesh elements. This field is used as a correction to the broken gradient $\nabla_h u_h$.

Definition 5.39 (Diffusive flux reconstruction by solving local problems). We set
$$\sigma_h^* := -\nabla_h u_h + \sigma_h'. \qquad (5.54)$$

A crucial point to verify is that $\sigma_h^* \in H(\text{div}; \Omega)$.

Lemma 5.40 ($H(\text{div}; \Omega)$-conformity of σ_h^*). *There holds*
$$\sigma_h^* \in \mathbb{RTN}_d^l(\mathfrak{S}_h).$$

Proof. It is clear that, for all $T' \in \mathfrak{S}_h$, $\sigma_h^*|_{T'} \in \mathbb{RTN}_d^l(T')$ since $l \geq k-1$. It remains to verify that $\sigma_h^* \in H(\text{div}; \Omega)$. By construction, for all $T \in \mathcal{T}_h$,
$$\sigma_h^*|_T \in H(\text{div}; T),$$
and, therefore, owing to Lemma 1.24, it suffices to check that the normal component of σ_h^* is single-valued on the boundaries of the mesh elements in $\mathcal{T}_h$. Let $T \in \mathcal{T}_h$, let $F' \in \mathfrak{F}_T^b$, and let $\xi \in \mathbb{P}_{d-1}^l(F')$. Then, using (5.50),
$$\int_{F'} (\sigma_h^* \cdot n_T)\xi = \int_{F'} (\rho_{\text{nc},T} - (\nabla u_h)|_T \cdot n_T)\xi = \int_{F'} \epsilon_{T,F} \phi_F(u_h)\xi, \qquad (5.55)$$
where F is the mesh face of $\mathcal{F}_h$ to which the subface F' belongs. This shows that the moments on F' of the normal component of σ_h^* against polynomials in $\mathbb{P}_{d-1}^l(F')$ are single-valued, and since $\sigma_h^* \cdot n_T|_{F'} \in \mathbb{P}_{d-1}^l(F')$, this concludes the proof. $\square$

5.5.4.3 Evaluation of $\nabla \cdot \sigma_h^*$

Let $\mathbb{P}_d^l(\mathfrak{S}_h) := \{q_h \in L^2(\Omega) \mid \forall T' \in \mathfrak{S}_h, q_h|_{T'} \in \mathbb{P}_d^l(T')\}$. We observe that
$$\nabla \cdot \sigma_h^* \in \mathbb{P}_d^l(\mathfrak{S}_h).$$

5.5. Local Postprocessing

Theorem 5.41 (Evaluation of $\nabla\cdot\sigma_h^*$). *Let $\sigma_h^* \in \mathrm{RTN}_d^l(\mathfrak{S}_h)$, $l \in \{k-1, k\}$, be defined by (5.54) where $\sigma_h' \in \mathrm{RTN}_d^l(\mathfrak{S}_h)$ is obtained on each mesh element by solving the local problem (5.53). Then,*

$$\nabla\cdot\sigma_h^* = \widehat{\pi}_h^l f,$$

where $\widehat{\pi}_h^l$ denotes the L^2-orthogonal projection onto $\mathbb{P}_d^l(\mathfrak{S}_h)$. In other words,

$$\int_{T'} (\nabla\cdot\sigma_h^*)\xi = \int_{T'} f\xi, \qquad \forall T' \in \mathfrak{S}_h,\ \forall \xi \in \mathbb{P}_d^l(T'). \tag{5.56}$$

Proof. Let $T \in \mathcal{T}_h$ and let $T' \in \mathfrak{S}_T$. Let $\psi \in \mathbb{P}_d^l(T')$ and extend ψ to T by zero outside T'. We first observe that

$$\int_{T'} f\psi = \int_T f\psi = \int_T f(\psi - \langle\psi\rangle_T) + \int_T f\langle\psi\rangle_T.$$

Since $\psi - \langle\psi\rangle_T \in \mathbb{P}_{d,*}^l(\mathfrak{S}_T)$, (5.53b) implies

$$\int_T (f + \triangle u_h)(\psi - \langle\psi\rangle_T) = \int_T \rho_{\mathrm{pde},T}(\psi - \langle\psi\rangle_T)$$
$$= \int_T (\nabla\cdot\sigma_h')(\psi - \langle\psi\rangle_T)$$
$$= \int_T (\nabla\cdot\sigma_h^* + \triangle u_h)(\psi - \langle\psi\rangle_T),$$

so that

$$\int_T f(\psi - \langle\psi\rangle_T) = \int_T (\nabla\cdot\sigma_h^*)(\psi - \langle\psi\rangle_T).$$

Moreover, using (5.55) with $\xi \equiv 1$, summing over all subfaces in $\mathfrak{F}_T^b$, and using (5.51) yields

$$\int_T \nabla\cdot\sigma_h^* = \sum_{F'\in\mathfrak{F}_T^b}\int_{F'}(\sigma_h^*\cdot n_T) = \sum_{F\in\mathcal{F}_T}\epsilon_{T,F}\int_F \phi_F(u_h) = \int_T f.$$

As a result, since $\langle\psi\rangle_T$ is constant in T, $\int_T f\langle\psi\rangle_T = \int_T (\nabla\cdot\sigma_h^*)\langle\psi\rangle_T$. Collecting the above identities yields, for all $\psi \in \mathbb{P}_d^l(T')$,

$$\int_{T'} f\psi = \int_{T'} (\nabla\cdot\sigma_h^*)\psi,$$

whence the assertion. $\square$

Remark 5.42 (Link with the reconstruction by prescription). Whenever $\mathcal{T}_h$ is a matching simplicial mesh, the same result is obtained for $l \in \{0, 1\}$ when using the reconstruction by the prescription (5.45) and (5.46) and the reconstruction of this section based on solving the local Neumann problems (5.53). The two reconstructions differ for $l \geq 2$. Furthermore, on general meshes, the reconstruction by prescription can still be performed on the matching simplicial submesh $\mathfrak{S}_h$, yielding, say, $\sigma_h^{**} \in \mathbb{RTN}_d^l(\mathfrak{S}_h)$. Then, proceeding as in the proof of Theorem 5.34 leads to

$$\int_T (\nabla \cdot \sigma_h^{**}) \xi = \int_T f\xi, \qquad \forall T \in \mathcal{T}_h, \ \forall \xi \in \mathbb{P}_d^l(T).$$

This is a less stringent conservation property than (5.56).

5.5.4.4 L^2-Norm Estimate

We now bound the error $(\sigma_h^* - \sigma)$ in the L^2-norm.

Theorem 5.43 (L^2-estimate). *Let $\sigma_h^* \in \mathbb{RTN}_d^l(\mathfrak{S}_h)$, $l \in \{k-1, k\}$, be defined by (5.54) where $\sigma_h' \in \mathbb{RTN}_d^l(\mathfrak{S}_h)$ is obtained on each mesh element by solving the local problem (5.53). Let $\sigma = -\nabla u$ be the exact diffusive flux. Then, the estimates (5.48) and (5.49) hold true.*

Proof. We focus on the local estimate, since the global estimate is a direct consequence of the local one. Let $\sigma_h^* \in \mathbb{RTN}_d^l(\mathfrak{S}_h)$ be defined by (5.54). Let $T \in \mathcal{T}_h$. Similarly to the proof of Theorem 5.36, it is sufficient to prove that

$$\|\sigma_h^* + \nabla u_h\|_{[L^2(T)]^d} \lesssim \mathcal{R}_{\mathrm{pde},T} + \left(\sum_{F \in \mathcal{F}_T \cap \mathcal{F}_h^i} \mathcal{R}_{\mathrm{jdf},F}^2 \right)^{1/2} + \eta \left(\sum_{F \in \mathcal{F}_T} \mathcal{R}_{\mathrm{jpt},F}^2 \right)^{1/2}.$$

Recall that $\sigma_h' = \sigma_h^* + \nabla_h u_h$. The idea is to postprocess the mixed finite element solution (p_h, σ_h') of (5.53) following the ideas of Arnold and Brezzi [15], Chen [88], Arbogast and Chen [13], and, more recently, Vohralík [301, 302]. In particular, there is a space $W(\mathfrak{S}_T)$ of piecewise polynomials on $\mathfrak{S}_T$ supplemented with bubble functions and a function $\tilde{p}_h \in W(\mathfrak{S}_T)$ such that, for all $T' \in \mathfrak{S}_T$,

$$\int_{T'} (-\nabla \tilde{p}_h) \cdot \tau_h = \int_{T'} \sigma_h' \cdot \tau_h \qquad \forall \tau_h \in \mathbb{RTN}_d^l(T'), \tag{5.57a}$$

$$\int_{T'} \tilde{p}_h q_h = \int_{T'} p_h q_h \qquad \forall q_h \in \mathbb{P}_d^l(T'). \tag{5.57b}$$

Moreover, for all $F' \in \mathfrak{F}_T^i$ and all $\xi \in \mathbb{P}_{d-1}^l(F')$,

$$\int_{F'} [\![\tilde{p}_h]\!] \xi = 0, \tag{5.58}$$

where $[\![\cdot]\!]$ denotes the jump across F' and $\tilde{p}_h$ satisfies the stability estimate

$$\left(\sum_{T' \in \mathfrak{S}_T} \|\nabla \tilde{p}_h\|_{[L^2(T')]^d}^2 \right)^{1/2} \lesssim \|\sigma_h'\|_{[L^2(T)]^d}. \tag{5.59}$$

Equations (5.57) mean that $(-\nabla \tilde{p}_h)$ and $\tilde{p}_h$ are close approximations of σ_h' and p_h, while (5.58) means that $\tilde{p}_h$ is close to being conforming in $H^1(T)$. Moreover, since $\langle p_h \rangle_T = 0$, (5.57b) implies

$$\langle \tilde{p}_h \rangle_T = 0.$$

Now, we observe using (5.57a) and integration by parts that

$$\|\sigma_h'\|_{[L^2(T)]^d}^2 = \sum_{T' \in \mathfrak{S}_T} \int_{T'} (-\nabla \tilde{p}_h) \cdot \sigma_h' = \sum_{T' \in \mathfrak{S}_T} \int_{T'} \tilde{p}_h (\nabla \cdot \sigma_h') - \sum_{F' \in \mathfrak{F}_T^b} \int_{F'} \tilde{p}_h (\sigma_h' \cdot n_T).$$

The contribution of the subfaces $F' \in \mathfrak{F}_T^i$ has been discarded in the integration by parts owing to (5.58) and the fact that the normal component of σ_h' on each $F' \in \mathfrak{F}_T^i$ is continuous and in $\mathbb{P}_{d-1}^l(F')$. Let $\widehat{\pi}_{\mathfrak{S}_T}^l$ denote the L^2-orthogonal projection onto $\mathbb{P}_d^l(\mathfrak{S}_T)$ and let $\widehat{\pi}_{\mathfrak{F}_T^b}^l$ denote the L^2-orthogonal projection onto $\prod_{F' \in \mathfrak{F}_T^b} \mathbb{P}_{d-1}^l(F')$. Since $\nabla \cdot \sigma_h' = \widehat{\pi}_{\mathfrak{S}_T}^l \rho_{\text{pde},T}$ owing to (5.53b) and since $\sigma_h' \cdot n_T|_{\partial T} = \widehat{\pi}_{\mathfrak{F}_T^b}^l \rho_{\text{nc},T}$ because $\sigma_h' \in \text{RTN}_{d,\rho}^l(\mathfrak{S}_T)$, this yields, using the Cauchy–Schwarz inequality,

$$\|\sigma_h'\|_{[L^2(T)]^d}^2 \leq \|\widehat{\pi}_{\mathfrak{S}_T}^l \rho_{\text{pde},T}\|_{L^2(T)} \|\tilde{p}_h\|_{L^2(T)} + \|\widehat{\pi}_{\mathfrak{F}_T^b}^l \rho_{\text{nc},T}\|_{L^2(\partial T)} \|\tilde{p}_h\|_{L^2(\partial T)}$$

$$\leq \|\rho_{\text{pde},T}\|_{L^2(T)} \|\tilde{p}_h\|_{L^2(T)} + \|\rho_{\text{nc},T}\|_{L^2(\partial T)} \|\tilde{p}_h\|_{L^2(\partial T)}.$$

Owing to the discrete trace inequality (1.40) and the uniform bound on N_∂, $h_T^{1/2} \|\tilde{p}_h\|_{L^2(\partial T)} \lesssim \|\tilde{p}_h\|_{L^2(T)}$, so that

$$\|\sigma_h'\|_{[L^2(T)]^d}^2 \lesssim \left(\|\rho_{\text{pde},T}\|_{L^2(T)} + h_T^{-1/2} \|\rho_{\text{nc},T}\|_{L^2(\partial T)} \right) \|\tilde{p}_h\|_{L^2(T)}.$$

Owing to (5.58) and the fact that $\langle \tilde{p}_h \rangle_T = 0$, we infer

$$\|\tilde{p}_h\|_{L^2(T)} \lesssim h_T \left(\sum_{T' \in \mathfrak{S}_T} \|\nabla \tilde{p}_h\|_{[L^2(T')]^d}^2 \right)^{1/2}.$$

This estimate is proven by Vohralík [300] on polyhedral domains with simplicial meshes (here, the mesh element T plays the role of the domain and $\mathfrak{S}_T$ that of the simplicial mesh). Combining the last two bounds with (5.59) yields

$$\|\sigma_h'\|_{[L^2(T)]^d} \lesssim h_T \|\rho_{\text{pde},T}\|_{L^2(T)} + h_T^{1/2} \|\rho_{\text{nc},T}\|_{L^2(\partial T)}.$$

Recalling that $\rho_{\text{pde},T} = (f + \triangle u_h)|_T$ and $\rho_{\text{nc},T}|_F = \varpi_F [\![\nabla_h u_h]\!] \cdot n_T + \eta h_F^{-1} [\![u_h]\!]$ for all $F \in \mathcal{F}_T$, using the triangle inequality, and mesh regularity yields the assertion. □

5.6 A Posteriori Error Estimates

A posteriori error estimates constitute a useful tool in practical computations for error control and enhanced computational efficiency. Herein, we focus on energy-norm upper and lower error bounds. In particular, we exploit the ideas of Sect. 5.5 to formulate a posteriori error estimates using a potential and a diffusive flux reconstruction.

5.6.1 Overview

Let u be the exact solution of the Poisson problem and let u_h be an approximate solution. A posteriori error estimates deliver upper and lower bounds on the approximation error $(u - u_h)$ measured in some global norm over Ω, say $\|\cdot\|_\Omega$. The upper bound takes the form

$$\|u - u_h\|_\Omega \leq \mathcal{E}(f, u_h), \tag{5.60}$$

where $\mathcal{E}(f, u_h)$ is called an estimator. We require that $\mathcal{E}(f, u_h)$ be fully computable from the source term f and the approximate solution u_h without featuring undetermined multiplicative factors. As such, (5.60) can be used for guaranteed error control to determine whether a desired level of accuracy has been reached by the approximate solution. The computational costs for evaluating $\mathcal{E}(f, u_h)$ turn out to be significantly lower than those incurred to obtain u_h, so that this error control procedure adds little effort. It is worthwhile to mention that, in the literature, error upper bounds are often given in the form $\|u - u_h\|_\Omega \leq \tilde{C}\mathcal{E}(f, u_h)$ with the value of $\tilde{C}$ left undetermined; as such, these estimates cannot be used for guaranteed error control. Furthermore, to avoid overestimating the error, it is necessary to prove also a lower bound of the form

$$\mathcal{E}(f, u_h) \leq C\|u - u_h\|_\Omega,$$

with C independent of h. Thus, the so-called effectivity index $\mathcal{E}(f, u_h)/\|u-u_h\|_\Omega$ takes values in the interval $[1, C]$, and the a posteriori error estimate becomes sharper as the effectivity index approaches one.

Another important property is that the estimator $\mathcal{E}(f, u_h)$ can be localized elementwise in the form

$$\mathcal{E}(f, u_h) = \left(\sum_{T \in \mathcal{T}_h} \mathcal{E}_T(f, u_h)^2 \right)^{1/2}.$$

The so-called local estimators $\mathcal{E}_T(f, u_h)$ can be used to drive adaptive mesh procedures whereby mesh elements are possibly marked for refinement (or coarsening) depending on the relative size of their associated local estimator. It is in general possible to localize elementwise the $\|\cdot\|_\Omega$-norm in the form

$$\|\cdot\|_\mathcal{T} = \left(\sum_{T \in \mathcal{T}} \|\cdot\|_T^2 \right)^{1/2},$$

for any subset $\mathcal{T} \subset \mathcal{T}_h$, and to prove local lower bounds on the error of the form

$$\mathcal{E}_T(f, u_h) \leq C \left(\|u - u_h\|_{\tilde{\mathcal{N}}_T} + \mathcal{R}_{\mathrm{osc}, \mathcal{N}_T} \right) \qquad \forall T \in \mathcal{T}_h, \tag{5.61}$$

5.6. A Posteriori Error Estimates

with C independent of h, while the sets $\mathcal{N}_T$ and $\tilde{\mathcal{N}}_T$ collect the mesh element T and some of its neighbors (a more precise statement is given in Sect. 5.6.3). The local data oscillation term $\mathcal{R}_{\mathrm{osc},\mathcal{N}_T}$, defined by (5.33) and (5.34), results from the small-scale fluctuations of the source term f that cannot be captured by the discretization. It is usually not necessary to determine the value of C in actual computations. The local lower bounds (5.61) explain why refining mesh elements with large estimators usually tends to equidistribute approximation errors. A general overview of a posteriori error estimates including various discretization methods and applications can be found in the textbooks by Verfürth [298] and by Ainsworth and Oden [8]. We also refer the reader to Morin, Nochetto, and Siebert [240] for a convergence proof of adaptive conforming finite element methods and to Karakashian and Pascal [208], Hoppe, Kanschat, and Warburton [194], Nicaise and Cochez-Dhondt [246], and Bonito and Nochetto [46] for adaptive dG methods.

5.6.2 Energy-Norm Error Upper Bounds

We recall that the exact solution u of the Poisson problem solves the weak problem (cf. (4.2))

$$\text{Find } u \in H_0^1(\Omega) \text{ s.t. } a(u,v) = \int_\Omega fv \text{ for all } v \in H_0^1(\Omega), \qquad (5.62)$$

where $a(u,v) = \int_\Omega \nabla u \cdot \nabla v$. At this stage, we only need the fact that the approximate solution u_h belongs to some discrete space that is nonconforming in $H_0^1(\Omega)$, say the broken polynomial space $\mathbb{P}_d^k(\mathcal{T}_h)$ defined by (1.15) with polynomial degree $k \geq 1$ and $\mathcal{T}_h$ belonging to an admissible mesh sequence. Henceforth, we set

$$V_h := \mathbb{P}_d^k(\mathcal{T}_h).$$

Because of nonconformity, the approximation error $(u - u_h)$ belongs to the extended space

$$V_{\mathrm{E}} := H_0^1(\Omega) + V_h,$$

which we equip with the so-called energy (semi)norm such that, for all $v \in V_{\mathrm{E}}$,

$$\|v\|_{\mathrm{E}}^2 = \sum_{T \in \mathcal{T}_h} \|v\|_{\mathrm{E},T}^2, \qquad \|v\|_{\mathrm{E},T} := \|\nabla v\|_{[L^2(T)]^d}.$$

The bilinear form a, originally defined on $H_0^1(\Omega) \times H_0^1(\Omega)$, can be extended to $V_{\mathrm{E}} \times V_{\mathrm{E}}$ by setting $a(u,v) := \int_\Omega \nabla_h u \cdot \nabla_h v$. A useful fact is that

$$a(v,v) = \|v\|_{\mathrm{E}}^2 \qquad \forall v \in V_{\mathrm{E}}. \qquad (5.63)$$

We notice that $\|\cdot\|_{\mathrm{E}}$ is a norm on $H_0^1(\Omega)$, but not on V_{E} where it acts as a seminorm, since for a piecewise constant function $v \in V_{\mathrm{E}}$, $\|v\|_{\mathrm{E}} = 0$.

In what follows, we focus on energy-norm a posteriori error estimates. We first show that such estimates are associated with the dual norm of a residual.

Then, we examine two ways to bound the dual norm of the residual, taking inspiration from Ern, Stephansen, and Vohralík [148, 149] (where heterogeneous diffusion-advection-reaction problems are also treated). One approach relies on Galerkin orthogonality and the other on the diffusive flux reconstruction introduced in Sect. 5.5. Energy-norm a posteriori error estimates for dG methods have been first obtained by Becker, Hansbo, and Larson [40] and by Karakashian and Pascal [207] using Galerkin orthogonality and an interpolation operator to bound the dual norm of the residual (see also Houston, Schötzau, and Wihler for the hp-analysis [199]). Such estimates, however, are expressed in terms of undetermined constants which can be avoided by the present analysis. We also refer the reader to the work of Ainsworth [7] for a different approach, also avoiding undetermined constants. Finally, we mention that it is possible to consider a posteriori error estimates in other norms, e.g., the $L^2(\Omega)$-norm, as derived by Becker, Hansbo, and Stenberg [41] and Rivière and Wheeler [270].

5.6.2.1 An Estimate with the Dual Norm of the Residual

For all $v_h \in V_h$, the residual $\mathcal{R}(v_h) \in H^{-1}(\Omega)$ is defined such that, for all $\varphi \in H_0^1(\Omega)$,

$$\langle \mathcal{R}(v_h), \varphi \rangle_{H^{-1}, H_0^1} := a(u - v_h, \varphi) = \int_\Omega f\varphi - \int_\Omega \nabla_h v_h \cdot \nabla\varphi.$$

The dual norm of the residual is measured as

$$\|\mathcal{R}(v_h)\|_{H^{-1}(\Omega)} = \sup_{\varphi \in H_0^1(\Omega), \|\varphi\|_E = 1} \langle \mathcal{R}(v_h), \varphi \rangle_{H^{-1}, H_0^1}.$$

Lemma 5.44 (Abstract estimate). *Let $u \in H_0^1(\Omega)$ solve (5.62). Let $v_h \in V_h$ be arbitrary. Then,*

$$\|u - v_h\|_E^2 \leq \inf_{s \in H_0^1(\Omega)} \|v_h - s\|_E^2 + \|\mathcal{R}(v_h)\|_{H^{-1}(\Omega)}^2$$
$$\leq 2\|u - v_h\|_E^2. \tag{5.64}$$

Proof. Proceeding as Kim in [212, Lemma 4.4], let $\psi \in H_0^1(\Omega)$ be such that

$$a(\psi, y) = a(v_h, y) \qquad \forall y \in H_0^1(\Omega),$$

that is, $(\psi - v_h)$ is a-orthogonal to $H_0^1(\Omega)$. Since a is $H_0^1(\Omega)$-coercive, the Lax–Milgram Lemma implies that this problem is well-posed, so that ψ is well-defined. For all $s \in H_0^1(\Omega)$, we obtain using (5.63) together with the symmetry of a,

$$\|s - v_h\|_E^2 = a(s - v_h, s - v_h)$$
$$= a(s - \psi, s - \psi) + 2a(\psi - v_h, s - \psi) + a(\psi - v_h, \psi - v_h)$$
$$= a(s - \psi, s - \psi) + a(\psi - v_h, \psi - v_h),$$

since $(\psi - v_h)$ is a-orthogonal to $(s - \psi)$. Hence, using again (5.63) yields

$$\|s - v_h\|_E^2 = \|v_h - \psi\|_E^2 + \|s - \psi\|_E^2, \tag{5.65}$$

5.6. A Posteriori Error Estimates

and this implies
$$\|v_h - \psi\|_E^2 = \inf_{s \in H_0^1(\Omega)} \|v_h - s\|_E^2.$$

Moreover, taking $s = u$ in (5.65) yields $\|u - v_h\|_E^2 = \|v_h - \psi\|_E^2 + \|u - \psi\|_E^2$ and, provided $\psi \neq u$ which yields a trivial case, elementary algebra leads to

$$a\left(u - v_h, \frac{u - \psi}{\|u - \psi\|_E}\right) = \frac{1}{\|u - \psi\|_E} a(u - v_h, u - \psi)$$

$$= \frac{1}{\|u - \psi\|_E} \{a(u - \psi, u - \psi) + a(\psi - v_h, u - \psi)\}$$

$$= \frac{1}{\|u - \psi\|_E} a(u - \psi, u - \psi) = \|u - \psi\|_E,$$

where we have used the linearity of a, the fact that $(\psi - v_h)$ is a-orthogonal to $(u - \psi)$, and (5.63). Hence,

$$\|u - v_h\|_E^2 = \inf_{s \in H_0^1(\Omega)} \|v_h - s\|_E^2 + a\left(u - v_h, \frac{u - \psi}{\|u - \psi\|_E}\right)^2.$$

Observing that

$$a\left(u - v_h, \frac{u - \psi}{\|u - \psi\|_E}\right) \leq \sup_{\varphi \in H_0^1(\Omega), \|\varphi\|_E = 1} a(u - v_h, \varphi) = \|\mathcal{R}(v_h)\|_{H^{-1}(\Omega)},$$

we infer the first bound in (5.64). To prove the second bound, we choose $s = u$ in the infimum and observe that the second term is bounded by $\|u - v_h\|_E$ since, for all $\varphi \in H_0^1(\Omega)$ with $\|\varphi\|_E = 1$,

$$\langle \mathcal{R}(v_h), \varphi \rangle_{H^{-1}, H_0^1} \leq \|u - v_h\|_E \|\varphi\|_E = \|u - v_h\|_E,$$

so that $\|\mathcal{R}(v_h)\|_{H^{-1}(\Omega)} \leq \|u - v_h\|_E$. This concludes the proof. □

The abstract estimate (5.64) is valid for any function $v_h \in V_h$ (the estimate can even be extended to functions in V_E). As such, it can be applied to the discrete solution $u_h \in V_h$ yielding

$$\|u - u_h\|_E^2 \leq \inf_{s \in H_0^1(\Omega)} \|u_h - s\|_E^2 + \|\mathcal{R}(u_h)\|_{H^{-1}(\Omega)}^2. \tag{5.66}$$

To turn (5.66) into a computable upper bound on the error requires choosing a specific (discrete) $s_h \in H_0^1(\Omega)$ and bounding the dual norm of the residual $\mathcal{R}(u_h)$. In what follows, we consider the $H_0^1(\Omega)$-conforming potential reconstruction u_h^* presented in Sect. 5.5.2, that is, we set

$$s_h := u_h^* \in H_0^1(\Omega), \tag{5.67}$$

so that (5.66) becomes

$$\|u - u_h\|_E^2 \leq \|u_h - u_h^*\|_E^2 + \|\mathcal{R}(u_h)\|_{H^{-1}(\Omega)}^2. \tag{5.68}$$

We now aim at bounding the dual norm $\|\mathcal{R}(u_h)\|_{H^{-1}(\Omega)}$ using locally computable quantities. To this purpose, it is necessary to use some local information on the discrete solution u_h. There are two ways to proceed. The first one uses explicitly the discrete scheme by considering suitable test functions; this leads to residual-based estimates. The second way consists in exploiting the local conservation property of the scheme by introducing a reconstructed diffusive flux. Combined with the choice (5.67), both approaches lead to guaranteed error upper bounds. We establish in Sect. 5.6.3 the corresponding lower error bounds.

5.6.2.2 Residual-Based Estimates

For the sake of simplicity, we consider matching simplicial meshes; cf. Remark 5.46 for the general case. Let π_h^0 denote the L^2-orthogonal projection onto the discrete space $\mathbb{P}_d^0(\mathcal{T}_h)$; this space is spanned by piecewise constant functions on $\mathcal{T}_h$. The operator π_h^0 satisfies the following approximation properties: For all $T \in \mathcal{T}_h$ and for all $\varphi \in H^1(T)$,

$$\|\varphi - \pi_h^0 \varphi\|_{L^2(T)} \le C_{\mathrm{P},T} h_T \|\nabla \varphi\|_{[L^2(T)]^d} = C_{\mathrm{P},T} h_T \|\varphi\|_{\mathrm{E},T}, \tag{5.69}$$

$$\|\varphi - \pi_h^0 \varphi\|_{L^2(\partial T)} \le C_{\mathrm{F},T}^{1/2} h_T^{1/2} \|\nabla \varphi\|_{[L^2(T)]^d} = C_{\mathrm{F},T}^{1/2} h_T^{1/2} \|\varphi\|_{\mathrm{E},T}. \tag{5.70}$$

The real number $C_{\mathrm{P},T}$ in the Poincaré-type inequality (5.69) can be evaluated as $C_{\mathrm{P},T} = \pi^{-1}$ whenever T is convex (recall that T is assumed to be a simplex here); see Bebendorf [39] and Payne and Weinberger [255]. The real number $C_{\mathrm{F},T}$ in the Friedrichs-type inequality (5.70) can be evaluated on simplices as $C_{\mathrm{F},T} = (h_T|\partial T|_{d-1}|T|_d^{-1})(2d^{-1} + C_{\mathrm{P},T})C_{\mathrm{P},T}$; this can be proven by applying the bound (1.42) to the function $(\varphi - \pi_h^0 \varphi)$, using (5.69), and summing over $F \in \mathcal{F}_T$.

Let u_h solve the discrete problem:

$$\text{Find } u_h \in V_h \text{ s.t. } a_h^{\mathrm{sip}}(u_h, v_h) = \int_\Omega f v_h \text{ for all } v_h \in V_h, \tag{5.71}$$

where a_h^{sip} is the SIP bilinear form defined by (4.12); variations on symmetry and penalty can be easily accommodated. Then, for all $T \in \mathcal{T}_h$, we define the local *nonconformity* estimator $\mathcal{E}_{\mathrm{nc},T}$, the local *residual* estimator $\mathcal{E}_{\mathrm{res},T}$, and the local *diffusive flux* estimator $\mathcal{E}_{\mathrm{df},T}$ as

$$\mathcal{E}_{\mathrm{nc},T} := \|\nabla(u_h - u_h^*)\|_{[L^2(T)]^d}, \tag{5.72a}$$

$$\mathcal{E}_{\mathrm{res},T} := C_{\mathrm{P},T} h_T \|(f + \triangle u_h) - \pi_h^0 (f + \triangle u_h)\|_{L^2(T)}, \tag{5.72b}$$

$$\mathcal{E}_{\mathrm{df},T} := C_{\mathrm{F},T}^{1/2} h_T^{1/2} \|\varpi_F [\![\nabla_h u_h]\!] \cdot n_T + \frac{\eta}{h_F} [\![u_h]\!]\|_{L^2(\partial T)}, \tag{5.72c}$$

where for $F \in \mathcal{F}_T$, $\varpi_F = 1/2$ if $F \in \mathcal{F}_h^i$ and $\varpi_F = 0$ if $F \in \mathcal{F}_h^b$.

Theorem 5.45 (Error upper bound). *Let $u \in H_0^1(\Omega)$ solve (5.62). Let $u_h \in V_h$ solve (5.71). Let $\mathcal{E}_{\mathrm{nc},T}$, $\mathcal{E}_{\mathrm{df},T}$, and $\mathcal{E}_{\mathrm{res},T}$ be defined by (5.72). Then,*

$$\|u - u_h\|_{\mathrm{E}} \le \left(\sum_{T \in \mathcal{T}_h} \left\{ \mathcal{E}_{\mathrm{nc},T}^2 + (\mathcal{E}_{\mathrm{res},T} + \mathcal{E}_{\mathrm{df},T})^2 \right\} \right)^{1/2}. \tag{5.73}$$

5.6. A Posteriori Error Estimates

Proof. Let $\varphi \in H_0^1(\Omega)$ with $\|\varphi\|_E = 1$. Integration by parts yields

$$a(u - u_h, \varphi) = \sum_{T \in \mathcal{T}_h} \int_T (f + \triangle u_h)\varphi - \sum_{F \in \mathcal{F}_h^i} \int_F [\![\nabla_h u_h]\!] \cdot \mathbf{n}_F \varphi.$$

Moreover, since u_h solves (5.71), considering the test function $\pi_h^0 \varphi \in \mathbb{P}_d^0(\mathcal{T}_h) \subset V_h$, and using (4.9) to evaluate $a_h^{\text{sip}}(u_h, \pi_h^0 \varphi)$ yields

$$\sum_{T \in \mathcal{T}_h} \int_T (f + \triangle u_h) \pi_h^0 \varphi = \sum_{F \in \mathcal{F}_h^i} \int_F [\![\nabla_h u_h]\!] \cdot \mathbf{n}_F \{\!\{\pi_h^0 \varphi\}\!\} - \sum_{F \in \mathcal{F}_h} \int_F [\![u_h]\!] \{\!\{\nabla_h \pi_h^0 \varphi\}\!\} \cdot \mathbf{n}_F$$

$$+ \sum_{F \in \mathcal{F}_h} \frac{\eta}{h_F} \int_F [\![u_h]\!] [\![\pi_h^0 \varphi]\!].$$

Since $\pi_h^0 \varphi$ is piecewise constant, $\{\!\{\nabla_h \pi_h^0 \varphi\}\!\} = 0$ for all $F \in \mathcal{F}_h$, so that

$$\sum_{T \in \mathcal{T}_h} \int_T (f + \triangle u_h) \pi_h^0 \varphi = \sum_{F \in \mathcal{F}_h^i} \int_F [\![\nabla_h u_h]\!] \cdot \mathbf{n}_F \{\!\{\pi_h^0 \varphi\}\!\} + \sum_{F \in \mathcal{F}_h} \frac{\eta}{h_F} \int_F [\![u_h]\!] [\![\pi_h^0 \varphi]\!].$$

As a result,

$$a(u - u_h, \varphi) = \sum_{T \in \mathcal{T}_h} \int_T (f + \triangle u_h)(\varphi - \pi_h^0 \varphi) - \sum_{F \in \mathcal{F}_h^i} \int_F [\![\nabla_h u_h]\!] \cdot \mathbf{n}_F \{\!\{\varphi - \pi_h^0 \varphi\}\!\}$$

$$+ \sum_{F \in \mathcal{F}_h} \frac{\eta}{h_F} \int_F [\![u_h]\!] [\![\pi_h^0 \varphi - \varphi]\!],$$

where, in the last term, we have exploited the fact that $[\![\pi_h^0 \varphi - \varphi]\!] = [\![\pi_h^0 \varphi]\!]$ since $\varphi \in H_0^1(\Omega)$. Expressing the last two terms as sums over mesh elements while using the L^2-orthogonality of $\pi_h^0(f + \triangle u_h)$ and $(\varphi - \pi_h^0 \varphi)$ in the first term yields

$$a(u - u_h, \varphi) = \sum_{T \in \mathcal{T}_h} \int_T ((f + \triangle u_h) - \pi_h^0(f + \triangle u_h))(\varphi - \pi_h^0 \varphi)$$

$$+ \sum_{T \in \mathcal{T}_h} \sum_{F \in \mathcal{F}_T} \epsilon_{T,F} \int_F \left(\varpi_F [\![\nabla_h u_h]\!] \cdot \mathbf{n}_T + \frac{\eta}{h_F} [\![u_h]\!] \right) (\varphi - (\pi_h^0 \varphi)|_T).$$

Owing to the Cauchy–Schwarz inequality,

$$a(u - u_h, \varphi) \le \sum_{T \in \mathcal{T}_h} \|(f + \triangle u_h) - \pi_h^0(f + \triangle u_h)\|_{L^2(T)} \|\varphi - \pi_h^0 \varphi\|_{L^2(T)}$$

$$+ \sum_{T \in \mathcal{T}_h} \|\varpi_F [\![\nabla_h u_h]\!] \cdot \mathbf{n}_T + \tfrac{\eta}{h_F}[\![u_h]\!]\|_{L^2(\partial T)} \|\varphi - \pi_h^0 \varphi\|_{L^2(\partial T)},$$

so that using the definitions (5.72b) and (5.72c) together with the bounds (5.69) and (5.70) leads to

$$a(u - u_h, \varphi) \le \sum_{T \in \mathcal{T}_h} (\mathcal{E}_{\text{res},T} + \mathcal{E}_{\text{df},T}) \|\varphi\|_{E,T}.$$

As a result,
$$\|\mathcal{R}(u_h)\|_{H^{-1}(\Omega)} \leq \left(\sum_{T \in \mathcal{T}_h} (\mathcal{E}_{\text{res},T} + \mathcal{E}_{\text{df},T})^2 \right)^{1/2}.$$

Furthermore, $\|u_h - u_h^*\|_E^2 = \sum_{T \in \mathcal{T}_h} \mathcal{E}_{\text{nc},T}^2$. We conclude using (5.68). □

Estimate (5.73) has been derived by Ern and Stephansen [148]. The idea of subtracting a piecewise constant function in the residual estimator can be traced back to Carstensen [76].

Remark 5.46 (General meshes). The local estimators defined by (5.72) can be used without modification on nonmatching simplicial meshes. On general meshes with nonconvex elements, the evaluation of $C_{P,T}$ in the Poincaré-type inequality (5.69) is more complex; upper bounds have been derived, e.g., by Vohralík [300]. Concerning (5.70), the proof on a polyhedron in $\mathbb{R}^d$ can be found in the work of Eymard, Gallouët, and Herbin [157].

5.6.2.3 Estimates Based on Diffusive Flux Reconstruction

An alternative way to estimate the dual norm $\|\mathcal{R}(u_h)\|_{H^{-1}(\Omega)}$ is possible as soon as there is a diffusive flux $t_h \in H(\text{div}; \Omega)$ satisfying the local conservation property
$$\forall T \in \mathcal{T}_h, \quad \int_T \nabla \cdot t_h = \int_T f. \tag{5.74}$$

For all $T \in \mathcal{T}_h$, we (re)define the local *nonconformity* estimator $\mathcal{E}_{\text{nc},T}$, the local *residual* estimator $\mathcal{E}_{\text{res},T}$, and the local *diffusive flux* estimator $\mathcal{E}_{\text{df},T}$ as

$$\mathcal{E}_{\text{nc},T} := \|\nabla(u_h - u_h^*)\|_{[L^2(T)]^d}, \tag{5.75a}$$

$$\mathcal{E}_{\text{res},T} := C_{P,T} h_T \|f - \nabla \cdot t_h\|_{L^2(T)}, \tag{5.75b}$$

$$\mathcal{E}_{\text{df},T} := \|\nabla u_h + t_h\|_{[L^2(T)]^d}. \tag{5.75c}$$

Theorem 5.47 (Error upper bound). *Let $u \in H_0^1(\Omega)$ solve (5.62). Let $u_h \in V_h$ solve (5.71). Let $\mathcal{E}_{\text{nc},T}$, $\mathcal{E}_{\text{df},T}$, and $\mathcal{E}_{\text{res},T}$ be defined by (5.75). Then,*

$$\|u - u_h\|_E \leq \left(\sum_{T \in \mathcal{T}_h} \left\{ \mathcal{E}_{\text{nc},T}^2 + (\mathcal{E}_{\text{res},T} + \mathcal{E}_{\text{df},T})^2 \right\} \right)^{1/2}. \tag{5.76}$$

Proof. Let $\varphi \in H_0^1(\Omega)$ with $\|\varphi\|_E = 1$. Owing to the Green theorem, the local conservation property (5.74), the discrete Poincaré-type inequality (5.69), and the Cauchy–Schwarz inequality,

5.6. A Posteriori Error Estimates

$$\begin{aligned}\langle \mathcal{R}(u_h), \varphi\rangle_{H^{-1}, H_0^1} &= \int_\Omega f\varphi - \int_\Omega \nabla_h u_h \cdot \nabla\varphi \\ &= \int_\Omega (f - \nabla \cdot t_h)\varphi - \int_\Omega (\nabla_h u_h + t_h)\cdot \nabla\varphi \\ &= \int_\Omega (f - \nabla \cdot t_h)(\varphi - \pi_h^0\varphi) - \int_\Omega (\nabla_h u_h + t_h)\cdot \nabla\varphi \\ &\leq \sum_{T\in\mathcal{T}_h}(\mathcal{E}_{\text{res},T} + \mathcal{E}_{\text{df},T})\|\varphi\|_{E,T},\end{aligned}$$

so that

$$\|\mathcal{R}(u_h)\|_{H^{-1}(\Omega)} \leq \left(\sum_{T\in\mathcal{T}_h}(\mathcal{E}_{\text{res},T} + \mathcal{E}_{\text{df},T})^2\right)^{1/2}.$$

The conclusion is straightforward. □

Remark 5.48 (Extension of Theorem 5.47). Theorem 5.47 can be extended to bound the norm $\|u-v\|_E$ for any $v \in V_E$ by just replacing u_h by v in the definition of $\mathcal{E}_{\text{nc},T}$ and $\mathcal{E}_{\text{df},T}$.

The diffusive flux t_h can be constructed using the ideas of Sect. 5.5. On matching simplicial meshes, we set $t_h = \sigma_h^*$, where $\sigma_h^* \in \text{RTN}_d^l(\mathcal{T}_h)$, $l \in \{k-1, k\}$, is prescribed locally using (5.45) and (5.46). Theorem 5.34 implies that

$$\nabla \cdot t_h = \pi_h^l f,$$

so that the local conservation property (5.74) is satisfied since $l \geq 0$. The local residual estimator $\mathcal{E}_{\text{res},T}$ becomes

$$\mathcal{E}_{\text{res},T} = C_{P,T} h_T \|f - \pi_h^l f\|_{L^2(T)},$$

with $C_{P,T} = \pi^{-1}$ since the simplex T is convex.

Remark 5.49 (Superconvergence of $\mathcal{E}_{\text{res},T}$). Whenever the source term f is smooth enough, the contribution of $\mathcal{E}_{\text{res},T}$ is superconvergent. Indeed, the energy error $\|u - u_h\|_E$ converges as h^k if u is smooth enough, while $\{\sum_{T\in\mathcal{T}_h}\mathcal{E}_{\text{res},T}^2\}^{1/2}$ converges as h^{l+2}, that is, one order faster if $l = k-1$ and two orders faster if $l = k$.

On general meshes, we set $t_h = \sigma_h^* = -\nabla_h u_h + \sigma_h'$, where $\sigma_h' \in \text{RTN}_d^l(\mathfrak{S}_h)$, $l \in \{k-1, k\}$, is obtained by solving the local Neumann problems introduced in Sect. 5.5.4. Theorem 5.41 yields that

$$\nabla \cdot t_h = \widehat{\pi}_h^l f,$$

where $\widehat{\pi}_h^l$ denotes the L^2-orthogonal projection onto $\mathbb{P}_d^k(\mathfrak{S}_h)$. As a result, a more local version of the local conservation property (5.74) is obtained, namely

$$\forall T' \in \mathfrak{S}_h, \quad \int_{T'}\nabla \cdot t_h = \int_{T'} f. \tag{5.77}$$

Then, proceeding as in the proof of Theorem 5.47, but using $\widehat{\pi}_h^0 \varphi$ instead of $\pi_h^0 \varphi$, leads to the error upper bound (5.76) with the local residual estimators $\mathcal{E}_{\mathrm{res},T}$, $T \in \mathcal{T}_h$, modified as

$$\mathcal{E}_{\mathrm{res},T} = \left(\sum_{T' \in \mathfrak{S}_T} C_{\mathrm{P},T'}^2 h_{T'}^2 \| f - \widehat{\pi}_h^l f \|_{L^2(T')}^2 \right)^{1/2},$$

and, as before, exploiting that the submesh is simplicial, so that T' is convex, yields $C_{\mathrm{P},T'} = \pi^{-1}$.

Energy-norm a posteriori error estimates based on $H(\mathrm{div};\Omega)$-conforming diffusive flux reconstruction can be traced back to the seminal work of Prager and Synge [261]. In the context of dG methods, such estimates have been developed by Kim [212], Cochez-Dhondt and Nicaise [93], Ern, Stephansen, and Vohralík [149], and Lazarov, Repin, and Tomar [224]. A closely related approach is based on equilibrated fluxes. This approach can be traced back to the seminal works of Ladevèze [218] and Haslinger and Hlaváček [187], and has been further explored by Ladevèze and Leguillon [219], Hlaváček, Haslinger, Nečas, and Lovíšek [193], Destuynder and Métivet [126], Ainsworth and Oden [8], Luce and Wohlmuth [233], and Braess and Schöberl [50]. Energy-norm dG a posteriori error estimates based on equilibrated fluxes have been derived by Ainsworth [7].

5.6.3 Error Lower Bounds

The goal of this section is to bound the local estimators $\mathcal{E}_{\mathrm{nc},T}$, $\mathcal{E}_{\mathrm{res},T}$, and $\mathcal{E}_{\mathrm{df},T}$ defined by (5.72) or (5.75) by the approximation error evaluated in a neighborhood of T plus a data oscillation term also evaluated in a neighborhood of T. For any subset $\mathcal{T} \subset \mathcal{T}_h$, we define

$$\| u - u_h \|_{\mathrm{E},\mathcal{T}} := \left(\sum_{T \in \mathcal{T}} \| u - u_h \|_{\mathrm{E},T}^2 \right)^{1/2}.$$

In what follows, the notation $a \lesssim b$ means the inequality $a \leq Cb$ with positive C that can depend on the space dimension d, the polynomial degree k, and the mesh regularity parameters ϱ, but not on the meshsize h (and the dG penalty parameter η). To shorten the notation, we set $\eta_1 = \max(1, \eta)$.

Theorem 5.50 (Local lower bound). *Let $\mathcal{E}_{\mathrm{nc},T}$, $\mathcal{E}_{\mathrm{res},T}$, and $\mathcal{E}_{\mathrm{df},T}$ be defined by (5.72) or (5.75). Then, for all $T \in \mathcal{T}_h$,*

$$\mathcal{E}_{\mathrm{nc},T} + \mathcal{E}_{\mathrm{res},T} + \mathcal{E}_{\mathrm{df},T} \lesssim \| u - u_h \|_{\mathrm{E},\mathcal{N}_T} + \mathcal{R}_{\mathrm{osc},\mathcal{N}_T} + \eta_1 |u_h|_{\mathrm{J},\tilde{\mathcal{F}}_T}, \qquad (5.78)$$

where $\mathcal{N}_T$ collects T and its neighbors in the sense of faces, the data oscillation $\mathcal{R}_{\mathrm{osc},\mathcal{N}_T}$ is defined according to (5.34), the jump seminorm $|u_h|_{\mathrm{J},\tilde{\mathcal{F}}_T}$ according to (5.37), and $\tilde{\mathcal{F}}_T$ collects mesh faces having a nonempty intersection with ∂T.

5.6. A Posteriori Error Estimates

Proof. Recall the local residuals introduced in Sect. 5.5.1, namely

$$\forall T \in \mathcal{T}_h, \qquad \mathcal{R}_{\text{pde},T} := h_T \|f + \triangle u_h\|_{L^2(T)},$$

$$\forall F \in \mathcal{F}_h^i, \qquad \mathcal{R}_{\text{jdf},F} := h_F^{1/2} \|[\![\nabla_h u_h]\!] \cdot n_F\|_{L^2(F)},$$

$$\forall F \in \mathcal{F}_h, \qquad \mathcal{R}_{\text{jpt},F} := h_F^{-1/2} \|[\![u_h]\!]\|_{L^2(F)}.$$

Let $T \in \mathcal{T}_h$. To prove (5.78), we proceed in two steps. To shorten the notation, Hilbertian sums of the local residuals $\mathcal{R}_{\text{jdf},F}$ or $\mathcal{R}_{\text{jpt},F}$ over face subsets are denoted by the corresponding subscript. We also set $\mathcal{F}_T^i := \mathcal{F}_T \cap \mathcal{F}_h^i$.
(i) Let us first prove that

$$\mathcal{E}_{\text{nc},T} + \mathcal{E}_{\text{res},T} + \mathcal{E}_{\text{df},T} \lesssim \mathcal{R}_{\text{pde},T} + \mathcal{R}_{\text{jdf},\mathcal{F}_T^i} + \eta_1 \mathcal{R}_{\text{jpt},\tilde{\mathcal{F}}_T}. \qquad (5.79)$$

Whenever the local estimators are defined by (5.72), we obtain, owing to Lemma 5.28 or 5.29,

$$\mathcal{E}_{\text{nc},T} \lesssim \mathcal{R}_{\text{jpt},\tilde{\mathcal{F}}_T}.$$

Moreover, it is clear that

$$\mathcal{E}_{\text{res},T} \lesssim \mathcal{R}_{\text{pde},T},$$
$$\mathcal{E}_{\text{df},T} \lesssim \mathcal{R}_{\text{jdf},\mathcal{F}_T^i} + \eta \mathcal{R}_{\text{jpt},\mathcal{F}_T}.$$

Whenever the local estimators are defined by (5.75), we obtain the same bound on $\mathcal{E}_{\text{nc},T}$, while the proof of Theorems 5.36 or 5.43 shows that

$$\mathcal{E}_{\text{df},T} \lesssim \mathcal{R}_{\text{pde},T} + \mathcal{R}_{\text{jdf},\mathcal{F}_T^i} + \eta \mathcal{R}_{\text{jpt},\mathcal{F}_T}.$$

Finally, using the triangle inequality and the inverse inequality (1.36) yields

$$\mathcal{E}_{\text{res},T} \lesssim h_T \|f - \nabla \cdot t_h\|_{L^2(T)}$$
$$\leq h_T \|f + \triangle u_h\|_{L^2(T)} + h_T \|\triangle u_h + \nabla \cdot t_h\|_{L^2(T)}$$
$$\lesssim \mathcal{R}_{\text{pde},T} + \mathcal{E}_{\text{df},T} \lesssim \mathcal{R}_{\text{pde},T} + \mathcal{R}_{\text{jdf},\mathcal{F}_T^i} + \eta \mathcal{R}_{\text{jpt},\mathcal{F}_T}.$$

This proves (5.79).
(ii) Combining (5.79) with the upper bounds (5.35) on $\mathcal{R}_{\text{pde},T}$ and $\mathcal{R}_{\text{jdf},F}$ and using mesh regularity yields (5.78). $\square$

Finally, under the assumptions of Lemma 5.30, a global lower bound on the error without the jump seminorm can be inferred.

Corollary 5.51 (Global lower bound). *Let $\mathcal{T}_h$ be a matching simplicial mesh. Assume that $\eta > \eta_{\text{av}}$ with η_{av} defined in Lemma 5.30. Then,*

$$\left(\sum_{T \in \mathcal{T}_h} \left\{ \mathcal{E}_{\text{nc},T}^2 + (\mathcal{E}_{\text{res},T} + \mathcal{E}_{\text{df},T})^2 \right\} \right)^{1/2} \lesssim \|u - u_h\|_{\text{E}} + \mathcal{R}_{\text{osc},\Omega},$$

where the multiplicative factor depends on $\eta_1 (\eta - \eta_{\text{av}})^{-1}$.

Part III

Systems

Chapter 6

Incompressible Flows

The equations governing fluid motion are the Navier–Stokes equations, which express the fundamental laws of mass and momentum conservation. These equations were first derived using a molecular approach by Navier [241] and Poisson [260], while a more specific derivation was found by Saint-Venant [30] and Stokes [284] based on the assumption that the stresses are linear functions of the strain rates (or deformation velocities), that is, for Newtonian fluids. An exhaustive presentation of the Navier–Stokes equations can be found in many textbooks on fluid dynamics; see, e.g., Batchelor [38]. In this chapter, we are concerned with the special case of incompressible (that is, constant density) Newtonian flows, thereby leading to the so-called Incompressible Navier–Stokes (INS) equations. In these equations, the dependent variables are the velocity and the pressure; other formulations of the INS equations are possible (see, e.g., Quartapelle [263]). The mass conservation equation enforces zero divergence on the velocity field (because of incompressibility), while the momentum conservation equation expresses the balance between diffusion (due to viscosity), nonlinear convection, pressure gradient, and external forcings.

Focusing first on the steady case, the main difficulties in the discretization of the steady INS equations are (1) the zero-divergence constraint on the velocity and (2) the contribution of the nonlinear convection term to the kinetic energy balance. The first issue is addressed in Sect. 6.1 in the simpler context of the steady Stokes equations. In these equations, convection is neglected, thereby leading to a linear system of PDEs. At the discrete level, we consider discontinuous velocities in the usual broken polynomial spaces and use the SIP method presented in Sect. 4.2 to approximate the diffusion term. We also introduce a discrete divergence operator in the same spirit as the discrete gradient operator introduced in Sect. 4.3.2. Several possibilities can be considered to discretize the pressure. We discuss in some detail the choice of discontinuous pressures with the same polynomial order as for velocities, as originally analyzed by Cockburn, Kanschat, Schötzau, and Schwab [105]. Although this choice does not properly match the required polynomial approximation properties in the convergence analysis, computational practice often indicates more efficiency for the equal-order choice than the mixed-order counterpart. One important fact is that, in the equal-order case, penalizing pressure jumps is needed to achieve discrete stability. Moreover, we present convergence proofs to smooth solutions (using classical arguments leading to optimal energy- and L^2-norm estimates) and to minimal regularity solutions (following the recent work by the authors [131]). Finally, we briefly discuss other discretization approaches which, in particular, circumvent the need to stabilize the pressure. In all cases, the discrete velocity is only weakly divergence-free. Approaches using Brezzi–Douglas–Marini or Raviart–Thomas–Nédélec velocity spaces, so that the discrete velocity is exactly divergence-free, have been investigated by Cockburn, Kanschat, and Schötzau [103]. Alternative approaches are also discussed by Hansbo and Larson [185] and by Montlaur, Fernandez-Mendez, and Huerta [239].

In Sect. 6.2, we turn to the steady INS equations. At the continuous level, existence of a solution can always be asserted, but uniqueness requires a smallness assumption on the data. At the discrete level, the central issue is now the discretization of the nonlinear convection term. An important ingredient is to mimic the fact that, at the continuous level, this term does not contribute to the kinetic energy balance. Since the divergence-free nature of the exact velocity plays an instrumental role in asserting this property, and since the discrete velocity is not exactly divergence-free, we consider a technique proposed by Temam [291, 292] which consists in adding a consistent term to the discrete momentum equation so as to recover the correct kinetic energy balance at the discrete level. The existence of a discrete solution is then proven by means of a topological degree argument, while uniqueness is recovered by invoking a smallness condition on the data, as for the exact problem. Then, without any smallness assumption on the data, we prove the convergence, up to a subsequence, of the sequence of discrete solutions to a solution of the INS equations with minimal regularity; convergence of the whole sequence can be asserted if the exact solution is unique. Finally, returning to the discretization of the nonlinear convection term, we identify the abstract design conditions required in the convergence analysis and briefly discuss an alternative to Temam's stabilization leading to a locally conservative formulation, but with somewhat weaker stability properties.

The last section of this chapter deals with a pressure-correction algorithm for the unsteady INS equations. In order to reduce the coupling between the momentum and the mass conservation equations, the space discretization uses continuous pressures. The interest of this formulation is that the coupling between the two equations is less tight than in the fully discontinuous case, since the discrete bilinear form coupling velocity and pressure is inherently inf-sup stable. The resulting space semidiscrete problem lends itself to fractional step time-marching methods, such as the pressure-correction scheme. The main interest of pressure-correction methods is to circumvent the need of a monolithic solver for the unsteady INS equations, which leads to linear systems whose condition number behaves extremely poorly with the time-step. For simplicity, we focus on pressure-correction schemes with the time derivative discretized using backward Euler or BDF2 schemes.

6.1 Steady Stokes Flows

In this section, we consider the steady Stokes equations. These equations describe incompressible viscous flows under the assumption that the fluid motion is sufficiently slow so that diffusion dominates over convection in the transport of momentum. We first study the well-posedness of the steady Stokes equations at the continuous level. At the discrete level, we focus on equal-order approximations using both discontinuous velocities and pressures, whereby pressure jumps are penalized to achieve stability. We prove the well-posedness of the discrete problem and establish convergence to both smooth and minimal regularity solutions. Finally, we briefly discuss alternative discretizations avoiding the need to stabilize the pressure.

6.1.1 The Continuous Setting

Let $\Omega \subset \mathbb{R}^d$, $d \geq 2$, be a polyhedron. The steady Stokes equations can be expressed in the form

$$-\triangle u + \nabla p = f \quad \text{in } \Omega, \tag{6.1a}$$

$$\nabla \cdot u = 0 \quad \text{in } \Omega, \tag{6.1b}$$

$$u = 0 \quad \text{on } \partial\Omega, \tag{6.1c}$$

$$\langle p \rangle_\Omega = 0, \tag{6.1d}$$

where $u : \Omega \to \mathbb{R}^d$ with Cartesian components $(u_i)_{1 \leq i \leq d}$ is the velocity field, $p : \Omega \to \mathbb{R}$ the pressure, and $f : \Omega \to \mathbb{R}^d$ with Cartesian components $(f_i)_{1 \leq i \leq d}$ the forcing term. Equation (6.1a), which expresses the conservation of momentum, can be written in component form as

$$-\triangle u_i + \partial_i p = f_i \quad \forall i \in \{1, \ldots, d\}.$$

Equation (6.1b), which expresses the conservation of mass, enforces the abovementioned divergence-free constraint on the velocity. Equation (6.1c) enforces a homogeneous Dirichlet boundary condition on the velocity; other boundary conditions can be considered, as discussed, e.g., by Ern and Guermond [141, p. 179]. Finally, condition (6.1d), where $\langle \cdot \rangle_\Omega$ denotes the mean value over Ω, is added to avoid leaving the pressure undetermined up to an additive constant.

Remark 6.1 (Stress and strain tensors, viscosity). A more general form of the momentum conservation equation (6.1a) takes the form

$$-\nabla \cdot \sigma + \nabla p = f \quad \text{in } \Omega,$$

where $\sigma : \Omega \to \mathbb{R}^{d,d}$ is the *stress tensor*. In component form, this equation can be rewritten as

$$-\sum_{j=1}^{d} \partial_j \sigma_{ij} + \partial_i p = f_i \quad \forall i \in \{1, \ldots, d\}.$$

In Newtonian flows, stresses are proportional to strain rates. More specifically, introducing for a given velocity field u the (linearized) *strain tensor* $\varepsilon : \Omega \to \mathbb{R}^{d,d}$ such that
$$\varepsilon = \frac{1}{2}(\nabla u + \nabla u^t),$$
or, in component form, $\varepsilon_{ij} = 1/2(\partial_j u_i + \partial_i u_j)$, there holds
$$\sigma = 2\nu\varepsilon,$$
where $\nu > 0$ is the (kinematic) *viscosity*. Taking the viscosity to be constant for simplicity, we obtain
$$-\nu\nabla\cdot(\nabla u + \nabla u^t) + \nabla p = f, \qquad (6.2)$$
and up to rescaling of the pressure and the source term, we can assume that $\nu = 1$. Then, observing that $\nabla\cdot(\nabla u) = \triangle u$ and $\nabla\cdot(\nabla u)^t = \nabla(\nabla\cdot u) = 0$ because of incompressibility, we recover (6.1a). Considering the form (6.2) of the momentum conservation equation is useful when dealing with other boundary conditions than (6.1c), e.g., when weakly enforcing the Navier slip boundary condition $(\sigma\cdot n + \lambda u)\cdot t = 0$ where t is a tangent vector to the boundary $\partial\Omega$ and $\lambda \geq 0$ a given parameter. We also observe that using the form (6.2) requires a different form of coercivity for the diffusion term; cf. Remark 6.4 below.

Remark 6.2 (Nonzero divergence). In some situations, it is interesting to consider a nonzero right-hand side in the mass conservation equation (6.1b). This case can be treated with straightforward changes in the analysis presented below.

6.1.1.1 Weak Formulation

We assume that the forcing term f is in $[L^2(\Omega)]^d$. Owing to (6.1c), the natural space for the velocity is $[H_0^1(\Omega)]^d$, while owing to (6.1d), the natural space for the pressure is $L_0^2(\Omega) \subset L^2(\Omega)$ where
$$L_0^2(\Omega) := \{q \in L^2(\Omega) \mid \langle q \rangle_\Omega = 0\}.$$
We set
$$U := [H_0^1(\Omega)]^d, \qquad P := L_0^2(\Omega), \qquad X := U \times P. \qquad (6.3)$$
The spaces U, P, and X are Hilbert spaces when equipped with the inner products inducing the norms
$$\|v\|_U := \|v\|_{[H^1(\Omega)]^d} = \left(\sum_{i=1}^d \|v_i\|_{H^1(\Omega)}^2\right)^{1/2},$$
$$\|q\|_P := \|q\|_{L^2(\Omega)}, \qquad \|(v,q)\|_X := \left(\|v\|_U^2 + \|q\|_P^2\right)^{1/2}.$$

6.1. Steady Stokes Flows

We define, for all $u, v \in U$ and for all $q \in P$, the bilinear forms

$$a(u,v) := \int_\Omega \nabla u : \nabla v = \sum_{i,j=1}^d \int_\Omega \partial_j u_i \, \partial_j v_i = (\nabla u, \nabla v)_{[L^2(\Omega)]^{d,d}}, \qquad (6.4a)$$

$$b(v,q) := -\int_\Omega q \nabla \cdot v = -(\nabla \cdot v, q)_P. \qquad (6.4b)$$

The weak formulation of problem (6.1) reads: Find $(u,p) \in X$ such that

$$a(u,v) + b(v,p) = \int_\Omega f \cdot v \qquad \forall v \in U, \qquad (6.5a)$$

$$-b(u,q) = 0 \qquad \forall q \in P, \qquad (6.5b)$$

or, equivalently,

$$\text{Find } (u,p) \in X \text{ s.t. } c((u,p),(v,q)) = \int_\Omega f \cdot v \text{ for all } (v,q) \in X,$$

with

$$c((u,p),(v,q)) := a(u,v) + b(v,p) - b(u,q).$$

While the bilinear form c is clearly not coercive on X, we observe that the bilinear form a is coercive on U. Indeed, applying the continuous Poincaré inequality (4.4) to each velocity component, we infer that there exists $\alpha_\Omega > 0$, only depending on Ω, such that

$$\forall v \in U, \qquad a(v,v) = \|\nabla v\|^2_{[L^2(\Omega)]^{d,d}} \geq \alpha_\Omega \|v\|^2_U. \qquad (6.6)$$

This yields a so-called *partial coercivity* for the bilinear form c in the form

$$\forall (v,q) \in X, \qquad c((v,q),(v,q)) = a(v,v) \geq \alpha_\Omega \|v\|^2_U. \qquad (6.7)$$

Remark 6.3 (Saddle-point problem). A problem of the form (6.5) is said to have a *saddle-point* structure since $(u,p) \in X$ solves (6.5) if and only if (u,p) is a saddle-point of the Lagrangian $\mathcal{L} : X \to \mathbb{R}$ such that, for all $(v,q) \in X$,

$$\mathcal{L}(v,q) = \frac{1}{2} a(v,v) + b(v,q).$$

In this context, the pressure plays the role of the Lagrange multiplier associated with the incompressibility constraint.

Remark 6.4 (Coercivity with the strain formulation (6.2)). Working with the strain formulation (6.2) (with $\nu = 1$) leads to the bilinear form $a(u,v) = 2\int_\Omega \varepsilon(u):\varepsilon(v)$ where, for all $v \in U$, we have set $\varepsilon(v) = \frac{1}{2}(\nabla v + \nabla v^t)$. The U-coercivity of the bilinear form a then results from Korn's First Inequality (see, e.g., Ciarlet [92, p. 24]) stating that there exists $\kappa_\Omega > 0$, only depending on Ω, such that

$$\forall v \in U, \qquad \|\varepsilon(v)\|_{[L^2(\Omega)]^{d,d}} \geq \kappa_\Omega \|v\|_U.$$

Discrete Korn inequalities in the context of piecewise smooth vector-valued fields are derived by Brenner [52] and Duarte, do Carmo, and Rochinha [137].

6.1.1.2 The Divergence Operator

We introduce the divergence operator $B \in \mathcal{L}(U, P)$ such that

$$B : U \ni v \longmapsto Bv := -\nabla\cdot v \in P. \tag{6.8}$$

(The fact that Bv has zero mean is a consequence of the divergence theorem since $\int_\Omega Bv = -\int_\Omega \nabla\cdot v = -\int_{\partial\Omega}(v\cdot \mathrm{n}) = 0$.) The operator B is readily linked to the bilinear form b since there holds

$$(Bv, q)_P = b(v, q) \qquad \forall (v, q) \in X.$$

The well-posedness of the Stokes problem (6.5) hinges on the surjectivity of the operator B or, equivalently, on an inf-sup condition on the bilinear form b, so that we first focus on these properties.

Theorem 6.5 (Surjectivity of divergence operator, inf-sup condition on b). *Let $\Omega \in \mathbb{R}^d$, $d \geq 1$, be a connected domain. Then, the operator B is surjective. Equivalently, there exists a real number $\beta_\Omega > 0$, only depending on Ω, such that, for all $q \in P$, there is $v_q \in U$ satisfying*

$$q = -Bv_q = \nabla\cdot v_q \quad \text{and} \quad \beta_\Omega \|v_q\|_U \leq \|q\|_P. \tag{6.9}$$

Moreover, property (6.9) *is equivalent to the following* inf-sup condition *on the bilinear form b:*

$$\forall q \in P, \qquad \beta_\Omega \|q\|_P \leq \sup_{w \in U \setminus \{0\}} \frac{b(w, q)}{\|w\|_U}. \tag{6.10}$$

Definition 6.6 (Velocity lifting). *For all $q \in P$, a field $v_q \in U$ satisfying* (6.9) *is called a* velocity lifting *of q.*

Proof. The proof can be found, e.g., in the textbook by Girault and Raviart [170, Sect. 2.2]; see also Bogovskiĭ [45], Solonnikov [280], and Durán and Muschietti [139]. We briefly sketch the main arguments.
(i) The fact that the surjectivity of B is equivalent to (6.9) is a simple application of the Open Mapping Theorem (see, e.g., Ern and Guermond [141, Theorem A.35]).
(ii) We now verify the equivalence between (6.9) and (6.10). We introduce the dual space $U' := [H^{-1}(\Omega)]^d$ and we identify P with its dual space. Let $B^* \in \mathcal{L}(P, U')$ be the adjoint operator of B and observe that B^* coincides with the gradient operator. Moreover, (6.10) can be reformulated as

$$\forall q \in P, \qquad \beta_\Omega \|q\|_P \leq \|B^* q\|_{U'}. \tag{6.11}$$

Using classical results in Hilbert spaces, we infer

$$(6.11) \iff \mathrm{Im}(B^*) \text{ is closed and } \mathrm{Ker}(B^*) = \{0\}$$
$$\iff \mathrm{Im}(B) \text{ is closed and } \mathrm{Ker}(B^*) = \{0\}$$
$$\iff \mathrm{Im}(B) \text{ is closed and } \overline{\mathrm{Im}(B)} = P$$
$$\iff B \text{ is surjective}$$

6.1. Steady Stokes Flows

where the second equivalence results from the Closed Range Theorem (see, e.g., [141, Theorem A.34]).
(iii) Finally, we prove inequality (6.11). We start with the following inequality (see Nečas [244]): There exists $\beta'_\Omega > 0$ such that

$$\forall q \in L^2(\Omega), \qquad \beta'_\Omega \|q\|_{L^2(\Omega)} \leq \|q\|_{H^{-1}(\Omega)} + \|\nabla q\|_{[H^{-1}(\Omega)]^d}.$$

Then, a classical argument by contradiction using the compact injection of $L^2(\Omega)$ into $H^{-1}(\Omega)$ and the fact that Ω is connected shows that by restricting q to have zero mean-value, there exists $\beta_\Omega > 0$ such that

$$\forall q \in P, \qquad \beta_\Omega \|q\|_{L^2(\Omega)} \leq \|\nabla q\|_{[H^{-1}(\Omega)]^d},$$

which is exactly (6.11). □

Remark 6.7 (Parameter β_Ω in (6.9) and (6.11)). The fact that the same parameter $\beta_\Omega > 0$ appears in (6.9) and (6.11) is a classical result. We briefly sketch the argument (which provides a more constructive link between (6.9) and (6.11)). Assume first that (6.9) is satisfied with parameter β_Ω. We obtain, for all $q \in P$,

$$\|q\|_P^2 = -(Bv_q, q)_P = -\langle B^*q, v_q\rangle_{U',U}$$
$$\leq \|B^*q\|_{U'}\|v_q\|_U \leq \beta_\Omega^{-1}\|B^*q\|_{U'}\|q\|_P,$$

so that $\|B^*q\|_{U'} \geq \beta_\Omega \|q\|_P$. Conversely, assume that (6.11) is satisfied with parameter β_Ω. Let $q \in P$. The problem of finding $q_* \in P$ such that, for all $r \in P$, $\langle B^*q_*, RB^*r\rangle_{U',U} = (q, r)_P$, where $R \in \mathcal{L}(U', U)$ is the Riesz isomorphism between U and its dual space U', is well-posed. Indeed, the bilinear form on the left-hand side is P-coercive since, for all $r \in P$, owing to (6.11), $\langle B^*r, RB^*r\rangle_{U',U} = \|B^*r\|_{U'}^2 \geq \beta_\Omega^2 \|r\|_P^2$. Moreover, the unique solution $q_* \in P$ to the above problem satisfies the a priori estimate $\|RB^*q_*\|_U \leq \beta_\Omega^{-2}\|q\|_P$. Then, a velocity lifting of q satisfying (6.9) with parameter β_Ω is obtained by setting $v_q = -RB^*q_* \in U$.

6.1.1.3 Well-Posedness

We can now prove the well-posedness of the Stokes problem (6.5).

Theorem 6.8 (Well-posedness). *Problem (6.5) is well-posed.*

Proof. We consider the bilinear form c and check the two conditions of the BNB Theorem.
(i) *Proof of* (1.4). Let $(v, q) \in X$ and set $\mathbb{S} := \sup_{(w,r) \in X\setminus\{0\}} \frac{c((v,q),(w,r))}{\|(w,r)\|_X}$. The partial coercivity (6.7) of c yields

$$\alpha_\Omega \|v\|_U^2 \leq c((v,q),(v,q)) \leq \mathbb{S}\|(v,q)\|_X,$$

while the inf-sup condition (6.10) leads to

$$\beta_\Omega \|q\|_P \le \sup_{w \in U \setminus \{0\}} \frac{b(w,q)}{\|w\|_U}$$

$$\le \sup_{w \in U \setminus \{0\}} \frac{c((v,q),(w,0)) - a(v,w)}{\|w\|_U}$$

$$\le \sup_{w \in U \setminus \{0\}} \frac{c((v,q),(w,0))}{\|(w,0)\|_X} + \sup_{w \in U \setminus \{0\}} \frac{a(v,w)}{\|w\|_U} \le \mathsf{S} + \|v\|_U,$$

since $\|w\|_U = \|(w,0)\|_X$ and $|a(v,w)| \le \|v\|_U \|w\|_U$. As a result, up to positive factors depending on α_Ω and β_Ω,

$$\|(v,q)\|_X^2 = \|v\|_U^2 + \|q\|_P^2 \lesssim \mathsf{S}\|(v,q)\|_X + \mathsf{S}^2.$$

Using Young's inequality for the first term on the right-hand side, we conclude that there exists $\gamma_\Omega > 0$ (depending on α_Ω and β_Ω) such that $\gamma_\Omega \|(v,q)\|_X \le \mathsf{S}$.
(ii) *Proof of* (1.5). Let $(w,r) \in X$ be such that $c((v,q),(w,r)) = 0$ for all $(v,q) \in X$. Then, taking $(v,q) = (w,r)$ and using the partial coercivity (6.7), we infer

$$\alpha_\Omega \|w\|_U^2 \le a(w,w) = c((w,r),(w,r)) = 0,$$

so that $w = 0$. To prove that $r = 0$, we let $(v,q) = -(v_r, 0)$, with v_r the velocity lifting of r, to infer $\|r\|_P^2 = -b(v_r, r) = c((v_r, 0),(0,r)) = 0$. □

Remark 6.9 (Pressure elimination). An alternative proof for the well-posedness of (6.5) consists in first eliminating the pressure. Specifically, let

$$V := \mathrm{Ker}(B) = \{v \in U \mid \nabla \cdot v = 0\}.$$

If the couple $(u,p) \in X$ solves (6.5), then the velocity u solves the problem:

$$\text{Find } u \in V \text{ s.t. } a(u,v) = \int_\Omega f \cdot v \text{ for all } v \in V. \tag{6.12}$$

Problem (6.12) is well-posed owing to the Lax–Milgram Lemma combined with the U-coercivity of the bilinear form a on V; cf. (6.6). Moreover, the unique solution $u \in V$ of (6.12) is such that

$$f + \Delta u \in V^\perp = (\mathrm{Ker}(B))^\perp = \overline{\mathrm{Im}(B^*)} = \mathrm{Im}(B^*),$$

so that there is $p \in P$ such that $f + \Delta u = B^* p = \nabla p$. Since p has zero mean-value, p is unique.

6.1.2 Equal-Order Discontinuous Velocities and Pressures

In this section, we consider one possible dG discretization of the steady Stokes equations based on equal-order discontinuous velocities and pressures. Other approaches are discussed in Sect. 6.1.5. DG methods based on equal-order

6.1. Steady Stokes Flows

discontinuous velocities and pressures have been introduced by Cockburn, Kanschat, Schötzau, and Schwab [105] for the Stokes equations and extended to the Oseen equations in [101] and to the INS equations in [104]; see also Ern and Guermond [144].

Let $\mathcal{T}_h$ be a mesh of Ω belonging to an admissible mesh sequence with mesh regularity parameters denoted by ϱ. Recalling the broken polynomial space $\mathbb{P}_d^k(\mathcal{T}_h)$ defined by (1.15) with polynomial degree $k \geq 1$, we define the discrete spaces

$$U_h := [\mathbb{P}_d^k(\mathcal{T}_h)]^d, \qquad P_h := \mathbb{P}_{d,0}^k(\mathcal{T}_h), \qquad X_h := U_h \times P_h, \tag{6.13}$$

where $\mathbb{P}_{d,0}^k(\mathcal{T}_h)$ denotes the subspace of $\mathbb{P}_d^k(\mathcal{T}_h)$ spanned by functions having zero mean-value over Ω; cf. Remark 6.14 for discarding this constraint. The discrete solution is sought in the space X_h.

For further use, we denote by π_h the L^2-projector onto the broken polynomial space $\mathbb{P}_d^k(\mathcal{T}_h)$ and by Π_h the L^2-projector onto U_h. A useful remark is that π_h preserves the mean value since

$$\forall q \in L^2(\Omega), \qquad \langle q \rangle_\Omega = \frac{1}{|\Omega|_d} \int_\Omega q = \frac{1}{|\Omega|_d} \int_\Omega \pi_h q = \langle \pi_h q \rangle_\Omega. \tag{6.14}$$

Hence, for all $q \in P$, $\pi_h q \in P_h$.

6.1.2.1 Discretization of the Diffusion Term

To discretize the diffusion term, we use, for each velocity component, the SIP bilinear form (cf. Sect. 4.2). We define on $U_h \times U_h$ the bilinear form

$$a_h(v_h, w_h) := \sum_{i=1}^d a_h^{\text{sip}}(v_{h,i}, w_{h,i}), \tag{6.15}$$

where $(v_{h,i})_{1 \leq i \leq d}$ and $(w_{h,i})_{1 \leq i \leq d}$ denote the Cartesian components of v_h and w_h, respectively, and where a_h^{sip} is defined by (4.12), so that

$$a_h^{\text{sip}}(v_{h,i}, w_{h,i}) = \int_\Omega \nabla_h v_{h,i} \cdot \nabla_h w_{h,i} + \sum_{F \in \mathcal{F}_h} \frac{\eta}{h_F} \int_F [\![v_{h,i}]\!][\![w_{h,i}]\!]$$
$$- \sum_{F \in \mathcal{F}_h} \int_F \left(\{\!\{\nabla_h v_{h,i}\}\!\} \cdot n_F [\![w_{h,i}]\!] + [\![v_{h,i}]\!] \{\!\{\nabla_h w_{h,i}\}\!\} \cdot n_F \right).$$

As in Chap. 4 (cf. Definition 4.5), in dimension $d \geq 2$, h_F denotes the diameter of the face F (other local length scales can be chosen, cf. Remark 4.6), while in dimension 1, we set $h_F := \min(h_{T_1}, h_{T_2})$ if $F \in \mathcal{F}_h^i$ with $F = \partial T_1 \cap \partial T_2$ and $h_F := h_T$ if $F \in \mathcal{F}_h^b$ with $F = \partial T \cap \partial \Omega$.

It is natural to equip the discrete velocity space U_h with the $\|\cdot\|_{\text{sip}}$-norm defined by (4.17) for each Cartesian component, so that we set

$$\|v_h\|_{\text{vel}} := \left(\sum_{i=1}^d \|v_{h,i}\|_{\text{sip}}^2 \right)^{1/2} = \left(\|\nabla_h v_h\|_{[L^2(\Omega)]^{d,d}}^2 + |v_h|_{\text{J}}^2 \right)^{1/2}, \tag{6.16}$$

with the $|\cdot|_J$-seminorm acting now on vector-valued arguments as

$$|v_h|_J = \left(\sum_{F \in \mathcal{F}_h} h_F^{-1} \|[\![v_h]\!]\|_{[L^2(F)]^d}^2 \right)^{1/2}.$$

The discrete bilinear form a_h and the $\|\cdot\|_{\text{vel}}$-norm are extended to larger spaces in Sect. 6.1.3. In what follows, referring to Lemma 4.12, we assume that the penalty parameter η is such that $\eta > \underline{\eta}$ so that

$$\forall v_h \in U_h, \qquad a_h(v_h, v_h) \geq \alpha \|v_h\|_{\text{vel}}^2, \tag{6.17}$$

where $\alpha = C_\eta$.

6.1.2.2 Discretization of the Pressure-Velocity Coupling

To discretize the pressure-velocity coupling, we need a discrete counterpart of the bilinear form b defined on $U \times P$ by (6.4). We define on $U_h \times P_h$ the discrete bilinear form

$$b_h(v_h, q_h) = -\int_\Omega q_h \nabla_h \cdot v_h + \sum_{F \in \mathcal{F}_h} \int_F [\![v_h]\!] \cdot n_F \{\!\{q_h\}\!\}, \tag{6.18}$$

where the broken divergence operator $\nabla_h\cdot$ acts elementwise, like the broken gradient operator ∇_h defined by (1.20). We observe that elementwise integration by parts yields

$$b_h(v_h, q_h) = \int_\Omega v_h \cdot \nabla_h q_h - \sum_{F \in \mathcal{F}_h^i} \int_F \{\!\{v_h\}\!\} \cdot n_F [\![q_h]\!]. \tag{6.19}$$

One motivation for the above definition is that, extending the domain of b_h to the broken Sobolev spaces $[H^1(\mathcal{T}_h)]^d \times H^1(\mathcal{T}_h)$, we obtain the consistency properties

$$\forall (v, q_h) \in U \times P_h, \qquad b_h(v, q_h) = -\int_\Omega q_h \nabla \cdot v,$$

$$\forall (v_h, q) \in U_h \times H^1(\Omega), \qquad b_h(v_h, q) = \int_\Omega v_h \cdot \nabla q,$$

since, for all $v \in U$, $[\![v]\!] = 0$ for all $F \in \mathcal{F}_h$, while, for all $q \in H^1(\Omega)$, $[\![q]\!] = 0$ for all $F \in \mathcal{F}_h^i$. The discrete bilinear form b_h can also be derived by defining first a discrete divergence operator like the discrete gradient operator introduced in Sect. 4.3.2; cf. Sect. 6.1.4 for further discussion. Finally, as detailed later in this section, the discrete bilinear form b_h can be associated with the use of centered fluxes to discretize the gradient and divergence operators in the pressure-velocity coupling.

Similarly to the operator B at the continuous level, we introduce the discrete operator $B_h : U_h \to P_h$ such that, for all $(v_h, q_h) \in X_h$,

$$(B_h v_h, q_h)_P = b_h(v_h, q_h).$$

It turns out that, contrary to the exact operator B, the discrete operator B_h is not surjective. As a result, the L^2-norm of a function in P_h cannot be controlled uniquely in terms of b_h. To recover control, it is necessary to add the following pressure seminorm defined on $H^1(\mathcal{T}_h)$:

$$|q|_p := \left(\sum_{F \in \mathcal{F}_h^i} h_F \|[\![q]\!]\|_{L^2(F)}^2 \right)^{1/2}.$$

Lemma 6.10 (Stability for b_h). *There exists $\beta > 0$, independent of h, such that*

$$\forall q_h \in P_h, \quad \beta \|q_h\|_P \le \sup_{w_h \in U_h \setminus \{0\}} \frac{b_h(w_h, q_h)}{\|w_h\|_{\mathrm{vel}}} + |q_h|_p. \tag{6.20}$$

Proof. Let $q_h \in P_h$. Owing to Theorem 6.5, there is $v_{q_h} \in U$ such that $\nabla \cdot v_{q_h} = q_h$ and $\beta_\Omega \|v_{q_h}\|_U \le \|q_h\|_P$. Since $v_{q_h} \in U$, integration by parts yields

$$\|q_h\|_P^2 = \int_\Omega q_h (\nabla \cdot v_{q_h}) = -\int_\Omega \nabla_h q_h \cdot v_{q_h} + \sum_{F \in \mathcal{F}_h^i} \int_F [\![q_h]\!] v_{q_h} \cdot n_F.$$

Since $\nabla_h P_h \subset U_h$, we infer from the definition of the L^2-projector Π_h that $\int_\Omega \nabla_h q_h \cdot (v_{q_h} - \Pi_h v_{q_h}) = 0$ so that

$$\|q_h\|_P^2 = -\int_\Omega \nabla_h q_h \cdot \Pi_h v_{q_h} + \sum_{F \in \mathcal{F}_h^i} \int_F [\![q_h]\!] v_{q_h} \cdot n_F$$

$$= -b_h(\Pi_h v_{q_h}, q_h) + \sum_{F \in \mathcal{F}_h^i} \int_F [\![q_h]\!] \{\!\{v_{q_h} - \Pi_h v_{q_h}\}\!\} \cdot n_F = \mathfrak{T}_1 + \mathfrak{T}_2.$$

Owing to Lemma 6.11 (see below),

$$|\mathfrak{T}_1| = \frac{|b_h(\Pi_h v_{q_h}, q_h)|}{\|\Pi_h v_{q_h}\|_{\mathrm{vel}}} \|\Pi_h v_{q_h}\|_{\mathrm{vel}} \le \left(\sup_{w_h \in U_h \setminus \{0\}} \frac{b_h(w_h, q_h)}{\|w_h\|_{\mathrm{vel}}} \right) C_\Pi \|v_{q_h}\|_U$$

$$\le \beta_\Omega^{-1} C_\Pi \left(\sup_{w_h \in U_h \setminus \{0\}} \frac{b_h(w_h, q_h)}{\|w_h\|_{\mathrm{vel}}} \right) \|q_h\|_P,$$

where C_Π only depends on the mesh regularity parameters ϱ and the polynomial degree k. To treat the second term, we use the Cauchy–Schwarz inequality together with Lemmata 1.42 and 1.59 to infer that, up to positive factors only depending on ϱ and k,

$$|\mathfrak{T}_2| \le \left(\sum_{F \in \mathcal{F}_h^i} h_F |[\![q_h]\!]|^2 \right)^{1/2} \times \left(\sum_{F \in \mathcal{F}_h^i} \frac{1}{h_F} \int_F |\{\!\{v_{q_h} - \Pi_h v_{q_h}\}\!\}|^2 \right)^{1/2}$$

$$\lesssim |q_h|_p \left(\sum_{T \in \mathcal{T}_h} \sum_{F \in \mathcal{F}_T \cap \mathcal{F}_h^i} \frac{1}{h_T} \|v_{q_h} - \Pi_h v_{q_h}\|_{[L^2(F)]^d}^2 \right)^{1/2}$$

$$\lesssim |q_h|_p \|v_{q_h}\|_U \lesssim \beta_\Omega^{-1} |q_h|_p \|q_h\|_P.$$

This concludes the proof. □

Lemma 6.11 (Stability of Π_h). *There is C_Π, only depending on the mesh regularity parameters ϱ and the polynomial degree k, such that, for all $v \in U$,*

$$\|\Pi_h v\|_{\mathrm{vel}} \leq C_\Pi \|v\|_U.$$

Proof. Let $v \in U$. Owing to the H^1-stability of the L^2-projector onto the broken polynomial space $\mathbb{P}_d^k(\mathcal{T}_h)$, we infer, up to positive factors only depending on ϱ and k,

$$\|\nabla_h \Pi_h v\|_{[L^2(\Omega)]^{d,d}} \lesssim \|v\|_U.$$

Moreover,

$$\sum_{F \in \mathcal{F}_h} h_F^{-1} \|[\![\Pi_h v]\!]\|^2_{[L^2(F)]^d} = \sum_{F \in \mathcal{F}_h} h_F^{-1} \|[\![\Pi_h v - v]\!]\|^2_{[L^2(F)]^d} \lesssim \|v\|^2_U,$$

where we have used, as above, Lemmata 1.42 and 1.59. □

Remark 6.12 (Ladyzhenskaya–Babuška–Brezzi (LBB) condition). In the setting of conforming mixed finite element approximations, the stability of the discrete bilinear form coupling velocity and pressure takes the form of an inf-sup condition without stabilization term, the so-called Ladyzhenskaya–Babuška–Brezzi (LBB) condition (see Babuška [19] and Brezzi [56]). Condition (6.20) can be viewed as an extended LBB condition owing to the additional presence of the pressure seminorm on the right-hand side.

6.1.2.3 Discrete Problem and Well-Posedness

We consider the following discretization of problem (6.5): Find $(u_h, p_h) \in X_h$ such that

$$a_h(u_h, v_h) + b_h(v_h, p_h) = \int_\Omega f \cdot v_h \qquad \forall v_h \in U_h, \qquad (6.21\mathrm{a})$$

$$-b_h(u_h, q_h) + s_h(p_h, q_h) = 0 \qquad \forall q_h \in P_h, \qquad (6.21\mathrm{b})$$

where the discrete bilinear form a_h is defined by (6.15), the discrete bilinear form b_h by (6.18) (or, equivalently, by (6.19)), and where

$$s_h(q_h, r_h) := \sum_{F \in \mathcal{F}_h^i} h_F \int_F [\![q_h]\!][\![r_h]\!]. \qquad (6.22)$$

The stabilization bilinear form s_h is meant to control pressure jumps across interfaces, thereby allowing to control the L^2-norm of the discrete pressure by virtue of Lemma 6.10. The following formulation, equivalent to (6.21), is obtained by summing equations (6.21a) and (6.21b): Find $(u_h, p_h) \in X_h$ such that

$$c_h((u_h, p_h), (v_h, q_h)) = \int_\Omega f \cdot v_h \text{ for all } (v_h, q_h) \in X_h, \qquad (6.23)$$

6.1. Steady Stokes Flows

where
$$c_h((u_h,p_h),(v_h,q_h)) := a_h(u_h,v_h) + b_h(v_h,p_h) - b_h(u_h,q_h) + s_h(p_h,q_h). \quad (6.24)$$

Owing to (6.17), we infer partial coercivity for c_h in the form
$$\forall (v_h,q_h) \in X_h, \quad c_h((v_h,q_h),(v_h,q_h)) = a_h(v_h,v_h) + s_h(q_h,q_h)$$
$$\geq \alpha \|v_h\|_{\mathrm{vel}}^2 + |q_h|_p^2. \quad (6.25)$$

Discrete well-posedness hinges on the discrete inf-sup stability of the bilinear form c_h. We equip the discrete space X_h with the norm
$$\|(v_h,q_h)\|_{\mathrm{sto}} := \left(\|v_h\|_{\mathrm{vel}}^2 + \|q_h\|_P^2 + |q_h|_p^2 \right)^{1/2}. \quad (6.26)$$

Lemma 6.13 (Discrete inf-sup stability). *Assume that the penalty parameter η in the SIP method is such that $\eta > \underline{\eta}$ with $\underline{\eta}$ defined in Lemma 4.12. Then, there is $\gamma > 0$, independent of h, such that, for all $(v_h,q_h) \in X_h$,*
$$\gamma \|(v_h,q_h)\|_{\mathrm{sto}} \leq \sup_{(w_h,r_h)\in X_h \setminus \{0\}} \frac{c_h((v_h,q_h),(w_h,r_h))}{\|(w_h,r_h)\|_{\mathrm{sto}}}. \quad (6.27)$$

Proof. Let $(v_h,q_h) \in X_h$ and, for brevity of notation, let $\mathbb{S}$ denote the supremum on the right-hand side of (6.27). Owing to (6.25),
$$\alpha \|v_h\|_{\mathrm{vel}}^2 + |q_h|_p^2 \leq \mathbb{S}\|(v_h,q_h)\|_{\mathrm{sto}}.$$

It only remains to estimate $\|q_h\|_P$. To this end, we use Lemma 6.10 and the fact that, for all $w_h \in U_h$, $b_h(w_h,q_h) = c_h((v_h,q_h),(w_h,0)) - a_h(v_h,w_h)$ to infer
$$\beta_\Omega \|q_h\|_P \leq \sup_{w_h \in U_h \setminus \{0\}} \left(\frac{-a_h(v_h,w_h)}{\|w_h\|_{\mathrm{vel}}} + \frac{c_h((v_h,q_h),(w_h,0))}{\|(w_h,0)\|_{\mathrm{sto}}} \right) + |q_h|_p$$
$$\leq \sup_{w_h \in U_h \setminus \{0\}} \frac{a_h(v_h,w_h)}{\|w_h\|_{\mathrm{vel}}} + \mathbb{S} + |q_h|_p.$$

Owing to the boundedness of the bilinear form a_h^{sip} (cf. Lemma 4.16) and the discrete equivalence of the $\|\cdot\|_{\mathrm{sip}}$ and $\|\cdot\|_{\mathrm{sip},*}$-norms (cf. Lemma 4.20), we infer, up to positive factors independent of h,
$$\sup_{w_h \in U_h \setminus \{0\}} \frac{a_h(v_h,w_h)}{\|w_h\|_{\mathrm{vel}}} \lesssim \|v_h\|_{\mathrm{vel}}.$$

Gathering the above estimates yields
$$\|(v_h,q_h)\|_{\mathrm{sto}}^2 \lesssim \mathbb{S}\|(v_h,q_h)\|_{\mathrm{sto}} + \mathbb{S}^2,$$
whence the conclusion follows from Young's inequality. $\square$

As a consequence of Lemma 1.30, the discrete problem (6.21) or, equivalently, (6.23) is well-posed.

Remark 6.14 (Discarding the zero mean-value constraint). In practice, the discrete problem (6.21) can be formulated by discarding the zero mean-value constraint on the discrete pressures, that is, using the broken polynomial space $\mathbb{P}_d^k(\mathcal{T}_h)$ for the discrete pressures. With this choice, the problem matrix has a one-dimensional kernel. The discrete solution can be obtained using a direct solver with full pivoting or an iterative solver which only requires matrix-vector products. After a discrete solution has been obtained, the zero mean-value constraint on the pressure is enforced by postprocessing.

Remark 6.15 (Pressure penalty). An alternative choice for the local length scale in the pressure penalty consists in replacing h_F with the ratio $\frac{\{\!\{|T|_d\}\!\}}{|F|_{d-1}}$ in (6.22) and redefining the seminorm $|\cdot|_p$ accordingly (cf. Remark 4.6 for the corresponding choice in the penalty term for the SIP bilinear form). In the work of Bassi and coworkers [32, 33, 127], this pressure penalty term is related to the solution of a local Riemann problem at the interfaces.

6.1.2.4 Numerical Fluxes

To formulate the discrete problem (6.21) locally, we consider test functions having support localized to a single mesh element. We define the numerical fluxes

$$\phi_F^{\mathrm{grad}}(p_h) := \begin{cases} \{\!\{p_h\}\!\}\mathrm{n}_F & \text{if } F \in \mathcal{F}_h^i, \\ p_h \mathrm{n} & \text{if } F \in \mathcal{F}_h^b, \end{cases} \qquad (6.28)$$

$$\phi_F^{\mathrm{div}}(u_h, p_h) := \begin{cases} \{\!\{u_h\}\!\}\cdot\mathrm{n}_F + h_F[\![p_h]\!] & \text{if } F \in \mathcal{F}_h^i, \\ 0 & \text{if } F \in \mathcal{F}_h^b, \end{cases} \qquad (6.29)$$

and observe that $\phi_F^{\mathrm{grad}}(p_h)$ is vector-valued whereas $\phi_F^{\mathrm{div}}(u_h, p_h)$ is scalar-valued. Moreover, referring to Sect. 4.3.4 and, in particular, to (4.50) for the numerical fluxes associated with the SIP method, we consider here the vector-valued numerical fluxes

$$\phi_F^{\mathrm{diff}}(u_h) = -\{\!\{\nabla_h u_h\}\!\}\cdot\mathrm{n}_F + \frac{\eta}{h_F}[\![u_h]\!]. \qquad (6.30)$$

Let $T \in \mathcal{T}_h$ and let $\xi \in [\mathbb{P}_d^k(T)]^d$ with Cartesian components $(\xi_i)_{1 \le i \le d}$. Using $v_h = \xi \chi_T$ as a test function in the discrete momentum conservation equation (6.21a) (where χ_T denotes the characteristic function of T), we obtain

$$\int_T \sum_{i=1}^d G_h^l(u_{h,i})\cdot\nabla\xi_i - \int_T p_h \nabla\cdot\xi$$
$$+ \sum_{F \in \mathcal{F}_T} \epsilon_{T,F} \int_F \left[\phi_F^{\mathrm{diff}}(u_h) + \phi_F^{\mathrm{grad}}(p_h)\right]\cdot\xi = \int_T f\cdot\xi, \qquad (6.31)$$

where $l \in \{k-1, k\}$, G_h^l is the discrete gradient operator defined in Sect. 4.3.2, and $\epsilon_{T,F} = \mathrm{n}_T\cdot\mathrm{n}_F$.

6.1. Steady Stokes Flows

Similarly, let $\zeta \in \mathbb{P}_d^k(T)$. Using $q_h = \zeta \chi_T - \langle \zeta \chi_T \rangle_\Omega$ as a test function in the discrete mass conservation equation (6.21b) and using the expression (6.19) of the discrete bilinear form b_h, we obtain

$$-\int_T u_h \cdot \nabla \zeta + \sum_{F \in \mathcal{F}_T} \epsilon_{T,F} \int_F \phi_F^{\mathrm{div}}(u_h, p_h) \zeta = 0. \qquad (6.32)$$

Equations (6.31) and (6.32) express the local conservation properties satisfied by the dG approximation. We observe that, in the numerical fluxes $\phi_F^{\mathrm{grad}}(p_h)$ and $\phi_F^{\mathrm{div}}(u_h, p_h)$, the centered part results from the discrete bilinear form b_h, while the presence of the pressure jump in the flux $\phi_F^{\mathrm{div}}(u_h, p_h)$ stems from stabilizing the pressure jumps across interfaces.

6.1.3 Convergence to Smooth Solutions

The goal of this section is to analyze the convergence of the solution of the discrete Stokes problem (6.21) or, equivalently, (6.23) in the case of smooth exact solutions. We proceed in the spirit of Theorem 1.35 and derive an error estimate in the $\|\cdot\|_{\mathrm{sto}}$-norm. We also derive an L^2-error estimate on the velocity under an additional regularity assumption on the Stokes problem.

6.1.3.1 Consistency

As in the previous chapters, some additional regularity of the exact solution $(u, p) \in X$ is needed to assert consistency by plugging the pair (u, p) into the discrete bilinear form c_h. Concerning the velocity, we hinge for simplicity on Assumption 4.4 for all the velocity components. Concerning the pressure, we need traces on all interfaces and that the resulting jumps vanish; again for simplicity, this requirement is matched by assuming $H^1(\Omega)$-regularity for the pressure.

Assumption 6.16 (Regularity of the exact solution and space X_*). *We assume that the exact solution (u, p) is in $X_* := U_* \times P_*$ where*

$$U_* := U \cap [H^2(\Omega)]^d, \qquad P_* := P \cap H^1(\Omega).$$

In the spirit of Sect. 1.3, we set

$$U_{*h} := U_* + U_h, \qquad P_{*h} := P_* + P_h, \qquad X_{*h} := X_* + X_h.$$

We extend the discrete bilinear form a_h defined by (6.15) to $U_{*h} \times U_h$ and the $\|\cdot\|_{\mathrm{vel}}$-norm to U_{*h}. We recall that the discrete bilinear form b_h is already defined on $[H^1(\mathcal{T}_h)]^d \times H^1(\mathcal{T}_h)$. Finally, we extend the discrete bilinear form c_h defined by (6.24) to $X_{*h} \times X_h$ and we extend the $\|\cdot\|_{\mathrm{sto}}$-norm defined by (6.26) to X_{*h}.

Before asserting consistency, we study the jumps of ∇u and p across interfaces.

Lemma 6.17 (Jumps of ∇u and p across interfaces). *Assume $(u,p) \in X_*$. Then,*
$$[\![\nabla u]\!] \cdot n_F = 0 \quad \text{and} \quad [\![p]\!] = 0 \quad \forall F \in \mathcal{F}_h^i. \tag{6.33}$$

Proof. The proof of (6.34) is similar to that of Lemma 1.24. For all $\varphi \in [C_0^\infty(\Omega)]^d$, elementwise integration by parts combined with the regularity assumption on the exact solution yields
$$0 = a(u,\varphi) + b(\varphi,p) - \int_\Omega f \cdot \varphi = \sum_{F \in \mathcal{F}_h^i} \int_F \left([\![\nabla u]\!] \cdot n_F - [\![p]\!] n_F \right) \cdot \varphi.$$

Taking the support of φ intersecting a single interface and using a density argument, we infer
$$[\![\nabla u]\!] \cdot n_F - [\![p]\!] n_F = 0 \quad \forall F \in \mathcal{F}_h^i. \tag{6.34}$$

Moreover, since $p \in P_*$, Lemma 1.23 implies that, for all $F \in \mathcal{F}_h^i$, $[\![p]\!] = 0$. Using (6.34), we finally infer $[\![\nabla u]\!] \cdot n_F = 0$. $\square$

Lemma 6.18 (Consistency). *Assume that $(u,p) \in X_*$. Then,*
$$c_h((u,p),(v_h,q_h)) = \int_\Omega f \cdot v_h \quad \forall (v_h, q_h) \in X_h.$$

Proof. Let $v_h \in U_h$. Observing that $\triangle u = \nabla p - f \in [L^2(\Omega)]^d$ since $p \in P_*$ and $f \in [L^2(\Omega)]^d$ and that $[\![u]\!] = 0$ across all $F \in \mathcal{F}_h$ since $u \in U$, we infer using (4.13) that
$$a_h(u,v_h) = -\int_\Omega \triangle u \cdot v_h + \sum_{F \in \mathcal{F}_h^i} \int_F ([\![\nabla u]\!] \cdot n_F) \cdot \{\!\{v_h\}\!\}.$$

Moreover, it follows from (6.19) that
$$b_h(v_h, p) = \int_\Omega v_h \cdot \nabla p - \sum_{F \in \mathcal{F}_h^i} \int_F [\![p]\!] n_F \cdot \{\!\{v_h\}\!\}.$$

As a result,
$$c_h((u,p),(v_h,0)) = a_h(u,v_h) + b(v_h,p)$$
$$= \int_\Omega (-\triangle u + \nabla p) \cdot v_h + \sum_{F \in \mathcal{F}_h^i} \int_F ([\![\nabla u]\!] \cdot n_F - [\![p]\!] n_F) \cdot \{\!\{v_h\}\!\}$$
$$= \int_\Omega f \cdot v_h,$$

where we have concluded using (6.34). Let now $q_h \in P_h$. Owing to (6.18),
$$b_h(u,q_h) = -\int_\Omega (\nabla \cdot u) q_h = 0,$$

since $[\![u]\!] = 0$ for all $F \in \mathcal{F}_h$. Moreover, owing to (6.33), $s_h(p,q_h) = 0$. As a result, $c_h((u,p),(0,q_h)) = -b_h(u,q_h) + s_h(p_h,q_h) = 0$, completing the proof. $\square$

6.1. Steady Stokes Flows

Remark 6.19 (Role of assumption $p \in P_*$). Property (6.34) is sufficient to infer $c_h((u,p),(v_h,0)) = \int_\Omega f \cdot v_h$ for all $v_h \in U_h$. The fact that $[\![p_h]\!] = 0$ across all $F \in \mathcal{F}_h^i$ only serves to assert the consistency of the pressure stabilization in the discrete mass conservation equation.

6.1.3.2 Error Estimate in the $\|\cdot\|_{\text{sto}}$-Norm

Owing to Theorem 1.35 and recalling that discrete inf-sup stability holds true using the $\|\cdot\|_{\text{sto}}$-norm (cf. Lemma 6.13), it remains to investigate the boundedness of the discrete bilinear form c_h. To this purpose, we define on X_{*h} the norm

$$\|(v,q)\|_{\text{sto},*}^2 := \|(v,q)\|_{\text{sto}}^2 + \sum_{T \in \mathcal{T}_h} h_T \|\nabla v|_T \cdot n_T\|_{L^2(\partial T)}^2 + \sum_{T \in \mathcal{T}_h} h_T \|q\|_{L^2(\partial T)}^2.$$

Recalling the definition (4.22) of the $\|\cdot\|_{\text{sip},*}$-norm, we observe that

$$\|(v,q)\|_{\text{sto},*}^2 = \sum_{i=1}^d \|v_i\|_{\text{sip},*}^2 + \|q\|_P^2 + |q|_p^2 + \sum_{T \in \mathcal{T}_h} h_T \|q\|_{L^2(\partial T)}^2.$$

Lemma 6.20 (Boundedness). *There exists C_{bnd}, independent of h, such that, for all $(v,q) \in X_{*h}$ and all $(w_h, r_h) \in X_h$,*

$$c_h((v,q),(w_h,r_h)) \leq C_{\text{bnd}} \|(v,q)\|_{\text{sto},*} \|(w_h,r_h)\|_{\text{sto}}.$$

Proof. Let $(v,q) \in X_{*h}$ and let $(w_h, r_h) \in X_h$. We bound the four terms in the definition (6.24) of c_h. Working componentwise, it follows from Lemma 4.16 that, up to positive factors independent of h,

$$a_h(v, w_h) \lesssim \sum_{i=1}^d \|v_i\|_{\text{sip},*} \|w_{h,i}\|_{\text{sip}} \leq \|(v,0)\|_{\text{sto},*} \|(w_h, 0)\|_{\text{sto}}.$$

To treat $b_h(v, r_h)$, we consider the expression (6.18) for b_h. Letting $\mathfrak{T}_1 := -\int_\Omega r_h \nabla_h \cdot v$ and $\mathfrak{T}_2 := \sum_{F \in \mathcal{F}_h} \int_F [\![v]\!] \cdot n_F \{\!\{r_h\}\!\}$, we bound each term separately. Clearly,

$$|\mathfrak{T}_1| \leq \|\nabla_h v\|_{[L^2(\Omega)]^{d,d}} \|r_h\|_P \leq \|(v,0)\|_{\text{sto}} \|(0, r_h)\|_{\text{sto}}.$$

For the second term, we use the Cauchy–Schwarz inequality followed by the discrete trace inequality (1.37) to infer

$$|\mathfrak{T}_2| \leq \left(\sum_{F \in \mathcal{F}_h} h_F^{-1} \int_F |[\![v]\!]|^2\right)^{1/2} \times \left(\sum_{F \in \mathcal{F}_h} h_F \int_F |\{\!\{r_h\}\!\}|^2\right)^{1/2}$$

$$\lesssim \|(v,0)\|_{\text{sto}} \|r_h\|_{L^2(\Omega)} \leq \|(v,0)\|_{\text{sto}} \|(0, r_h)\|_{\text{sto}}.$$

To bound $b_h(w_h, q)$, we consider again the expression (6.18) for b_h, and letting $\mathfrak{T}_3 := -\int_\Omega q \nabla_h \cdot w_h$ and $\mathfrak{T}_4 := \sum_{F \in \mathcal{F}_h} \int_F \{\!\{q\}\!\} [\![w_h]\!] \cdot n_F$, we obtain as before

$|\mathfrak{T}_3| \leq \|q\|_{L^2(\Omega)} \|\nabla_h w_h\|_{[L^2(\Omega)]^{d,d}} \leq \|(0,q)\|_{\text{sto}} \|(w_h, 0)\|_{\text{sto}}$, while, owing to the Cauchy–Schwarz inequality,

$$|\mathfrak{T}_4| \leq \left(\sum_{F \in \mathcal{F}_h} h_F \int_F |\{\!\{q\}\!\}|^2 \right)^{1/2} \times \left(\sum_{F \in \mathcal{F}_h} h_F^{-1} \int_F |[\![w_h]\!]|^2 \right)^{1/2}$$
$$\leq \|(0,q)\|_{\text{sto},*} \|(w_h, 0)\|_{\text{sto}}.$$

Finally, we observe that $s_h(q, r_h) \leq |q|_p |r_h|_p \leq \|(0,q)\|_{\text{sto}} \|(0, r_h)\|_{\text{sto}}$. □

A straightforward consequence of the above results together with Theorem 1.35 is the following error estimate.

Theorem 6.21 ($\|\cdot\|_{\text{sto}}$-norm error estimate). *Let $(u, p) \in X_*$ denote the unique solution of problem (6.5). Let $(u_h, p_h) \in X_h$ solve (6.23) with c_h defined by (6.24). Then, there is C, independent of h, such that*

$$\|(u - u_h, p - p_h)\|_{\text{sto}} \leq C \inf_{(v_h, q_h) \in X_h} \|(u - v_h, p - q_h)\|_{\text{sto},*}. \qquad (6.35)$$

To infer a convergence result from (6.35), we take $(v_h, q_h) = (\Pi_h u, \pi_h p)$ and observe that $\pi_h p \in P_h$ owing to (6.14). Then, assuming that the exact solution is smooth enough, we use Lemmata 1.58 and 1.59. The resulting estimate is optimal both for the velocity and the pressure.

Corollary 6.22 (Convergence rate in the $\|\cdot\|_{\text{sto}}$-norm). *Besides the hypotheses of Theorem 6.21, assume $(u, p) \in [H^{k+1}(\Omega)]^d \times H^k(\Omega)$. Then,*

$$\|(u - u_h, p - p_h)\|_{\text{sto}} \leq C_{u,p} h^k,$$

with $C_{u,p} = C \left(\|u\|_{[H^{k+1}(\Omega)]^d} + \|p\|_{H^k(\Omega)} \right)$ and C independent of h.

Remark 6.23 (Regularity assumption on the pressure). The regularity assumption $p \in H^k(\Omega)$ is just what is needed to achieve the overall convergence rate in the $\|\cdot\|_{\text{sto}}$-norm of order h^k. Since polynomials of degree $\leq k$ are used for the pressure, the contribution of the pressure terms to the upper bound $\|(u - \Pi_h u, p - \pi_h p)\|_{\text{sto},*}$ would be of order h^{k+1} if $p \in H^{k+1}(\Omega)$. In this case, the error would be dominated by the velocity error which is still of order h^k.

6.1.3.3 L^2-Norm Error Estimate on the Velocity

An L^2-error estimate on the velocity can be obtained using a duality argument in the same spirit as in Sect. 4.2.4 for the Poisson problem. To apply the Aubin–Nitsche argument [17], we need additional regularity for the solution of the Stokes problem (see Cattabriga [82] and Amrouche and Girault [11]).

Assumption 6.24 (Cattabriga's regularity). *We assume that there is C_{Cat}, only depending on Ω, such that, for all $f \in [L^2(\Omega)]^d$, the solution (u, p) of the steady Stokes problem (6.1) satisfies*

$$\|u\|_{[H^2(\Omega)]^d} + \|p\|_{H^1(\Omega)} \leq C_{\text{Cat}} \|f\|_{[L^2(\Omega)]^d}.$$

6.1. Steady Stokes Flows

To proceed, we extend the discrete bilinear form a_h to $U_{*h} \times U_{*h}$ and the discrete bilinear form c_h to $X_{*h} \times X_{*h}$. We can now establish the L^2-norm velocity error estimate.

Theorem 6.25 (L^2-norm velocity error estimate). *Besides the hypotheses of Theorem 6.21, assume Cattabriga's regularity. Then, there is C, independent of h, such that*
$$\|u - u_h\|_{[L^2(\Omega)]^d} \leq Ch \|(u - u_h, p - p_h)\|_{\mathrm{sto},*}. \tag{6.36}$$

Proof. Let $(\zeta, \xi) \in X$ denote the solution of the steady Stokes problem (6.1) with forcing term $(u - u_h) \in [L^2(\Omega)]^d$. Then, owing to Cattabriga's regularity,
$$\|\zeta\|_{[H^2(\Omega)]^d} + \|\xi\|_{H^1(\Omega)} \leq C_{\mathrm{Cat}} \|u - u_h\|_{[L^2(\Omega)]^d}. \tag{6.37}$$

Moreover, since $\zeta \in [H^2(\Omega) \cap H_0^1(\Omega)]^d$, $[\![\nabla \zeta]\!] \cdot n_F = 0$ across all $F \in \mathcal{F}_h^i$ and $[\![\zeta]\!] = 0$ across all $F \in \mathcal{F}_h$, so that using the symmetry of a_h together with the expression (4.13) of a_h^{sip} yields
$$a_h(u - u_h, \zeta) = a_h(\zeta, u - u_h) = \int_\Omega (-\triangle \zeta) \cdot (u - u_h).$$

Since $\xi \in H^1(\Omega)$, $[\![\xi]\!] = 0$ across all $F \in \mathcal{F}_h^i$, so that using the expression (6.19) for b_h yields
$$b_h(u - u_h, \xi) = \int_\Omega (u - u_h) \cdot \nabla \xi.$$

Hence,
$$\|u - u_h\|_{[L^2(\Omega)]^d}^2 = \int_\Omega (-\triangle \zeta + \nabla \xi) \cdot (u - u_h) = a_h(u - u_h, \zeta) + b_h(u - u_h, \xi).$$

Additionally, since $\nabla \cdot \zeta = 0$ and $[\![\zeta]\!] = 0$ across all $F \in \mathcal{F}_h$, we obtain using the expression (6.18) for b_h that
$$b_h(\zeta, p - p_h) = 0,$$
while it is clear that $s_h(p - p_h, \zeta) = 0$ since $\zeta \in H^1(\Omega)$. As a result,
$$\|u - u_h\|_{[L^2(\Omega)]^d}^2 = c_h((u - u_h, p - p_h), (\zeta, \xi)).$$

Using consistency (cf. Lemma 6.18), the boundedness of c_h on $X_{*h} \times X_{*h}$ with respect to the $\|\cdot\|_{\mathrm{sto},*}$-norm (this result can be derived by proceeding as in the proof of Lemma 6.20), the approximation properties of Π_h and π_h, and the regularity estimate (6.37), we infer, up to positive factors independent of h,
$$\begin{aligned}\|u - u_h\|_{[L^2(\Omega)]^d}^2 &= c_h((u - u_h, p - p_h), (\zeta - \Pi_h \zeta, \xi - \pi_h \xi)) \\ &\lesssim \|(u - u_h, p - p_h)\|_{\mathrm{sto},*} \|(\zeta - \Pi_h \zeta, \xi - \pi_h \xi)\|_{\mathrm{sto},*} \\ &\lesssim h \|(u - u_h, p - p_h)\|_{\mathrm{sto},*} \left(\|\zeta\|_{[H^2(\Omega)]^d} + \|\xi\|_{H^1(\Omega)} \right), \\ &\lesssim h \|(u - u_h, p - p_h)\|_{\mathrm{sto},*} \|u - u_h\|_{[L^2(\Omega)]^d}.\end{aligned}$$

Simplifying by $\|u - u_h\|_{[L^2(\Omega)]^d}$ yields the assertion. $\square$

We can now infer a convergence rate for the velocity error in the L^2-norm.

Corollary 6.26 (Convergence rate for velocity in L^2-norm). *Besides the hypotheses of Theorem 6.25, assume $(u,p) \in [H^{k+1}(\Omega)]^d \times H^k(\Omega)$. Then, there holds*
$$\|u - u_h\|_{[L^2(\Omega)]^d} \leq C_{u,p} h^{k+1},$$
with $C_{u,p} = C \left(\|u\|_{[H^{k+1}(\Omega)]^d} + \|p\|_{H^k(\Omega)} \right)$ and C independent of h.

Proof. The assertion results from (6.36) and the $\|\cdot\|_{\text{sto},*}$-norm error estimate
$$\|(u - u_h, p - p_h)\|_{\text{sto},*} \leq C_{u,p} h^k. \tag{6.38}$$
To prove this fact, we proceed similarly to the Poisson problem. The $\|\cdot\|_{\text{sto}}$- and $\|\cdot\|_{\text{sto},*}$-norms are uniformly equivalent on X_h (cf. Lemma 4.20). Hence, discrete inf-sup stability holds for the bilinear form c_h on X_h using the $\|\cdot\|_{\text{sto},*}$-norm. Since the bilinear form c_h is also bounded on $X_{*h} \times X_{*h}$ with respect to the $\|\cdot\|_{\text{sto},*}$-norm, we infer the error estimate
$$\|(u - u_h, p - p_h)\|_{\text{sto},*} \leq C \inf_{(v_h, q_h) \in X_h} \|(u - v_h, p - q_h)\|_{\text{sto},*},$$
with C independent of h, whence (6.38) follows owing to the approximation properties of Π_h and π_h. $\square$

6.1.4 Convergence to Minimal Regularity Solutions

In this section, we study the convergence of the sequence
$$(u_{\mathcal{H}}, p_{\mathcal{H}}) := ((u_h, p_h))_{h \in \mathcal{H}},$$
where, for all $h \in \mathcal{H}$, (u_h, p_h) solves the discrete problem (6.23), to the unique solution (u, p) of the steady Stokes problem (6.5) using minimal regularity on (u, p), that is to say, $(u, p) \in X$. This result is an important building block in the convergence study of the dG discretization of the INS equations undertaken in Sect. 6.2.3. For conciseness of notation, subsequences are not renumbered in what follows.

6.1.4.1 Reformulation of Diffusion Term Using Discrete Gradients

To analyze the convergence of the diffusion term, we rely on the convergence analysis to minimal regularity solutions for the Poisson problem; cf. Sect. 5.2. In particular, we formulate the discrete bilinear form a_h using the discrete gradients introduced in Sect. 4.3.2, namely, for all $v_h, w_h \in U_h$,
$$a_h(v_h, w_h) = \int_\Omega \sum_{i=1}^d G_h^l(v_{h,i}) \cdot G_h^l(w_{h,i}) + \hat{s}_h(v_h, w_h), \tag{6.39}$$

6.1. Steady Stokes Flows

with $l \in \{k-1, k\}$ and

$$\hat{s}_h(v_h, w_h) := \sum_{F \in \mathcal{F}_h} \frac{\eta}{h_F} \int_F [\![v_h]\!] \cdot [\![w_h]\!] - \int_\Omega \sum_{i=1}^d \mathrm{R}_h^l([\![v_{h,i}]\!]) \cdot \mathrm{R}_h^l([\![w_{h,i}]\!]).$$

The expression (6.39) is equivalent to (6.15) on $U_h \times U_h$.

Sequences of vector-valued functions in $U_{\mathcal{H}} := (U_h)_{h \in \mathcal{H}}$ bounded in the $\|\cdot\|_{\mathrm{vel}}$-norm are such that all their components are bounded in the $\|\cdot\|_{\mathrm{dG}}$-norm. Hence, Theorem 5.7, and in particular the weak asymptotic consistency result (5.7), can be applied to such sequences componentwise. Additionally, the strong asymptotic consistency result (5.10) can be extended to vector-valued smooth functions. For convenience, we collect these results in the following proposition.

Proposition 6.27 (Asymptotic consistency of discrete gradient). *Let $v_{\mathcal{H}}$ be a sequence in $U_{\mathcal{H}}$ bounded in the $\|\cdot\|_{\mathrm{vel}}$-norm. Then, there exists a function $v \in U$ such that as $h \to 0$, up to a subsequence, $v_h \to v$ strongly in $[L^2(\Omega)]^d$ and, for all $l \geq 0$,*

$$G_h^l(v_{h,i}) \rightharpoonup \nabla v_i \qquad \text{weakly in } [L^2(\Omega)]^d \text{ for all } i \in \{1, \ldots, d\}.$$

Moreover, for all $\Phi \in [C_0^\infty(\Omega)]^d$, letting $\Phi_h := \Pi_h \Phi$, there holds, for all $l \geq 0$,

$$G_h^l(\Phi_{h,i}) \to \nabla \Phi_i \qquad \text{in } [L^2(\Omega)]^d \text{ for all } i \in \{1, \ldots, d\}.$$

6.1.4.2 Reformulation of Pressure-Velocity Coupling Using Discrete Divergence

For any integer $l \geq 0$, we define the discrete divergence operator

$$D_h^l : [H^1(\mathcal{T}_h)]^d \to L^2(\Omega)$$

such that, for all $v \in [H^1(\mathcal{T}_h)]^d$ with Cartesian components $(v_i)_{1 \leq i \leq d}$,

$$D_h^l(v) := \sum_{i=1}^d G_h^l(v_i) \cdot e_i,$$

where e_i denotes the ith vector of the Cartesian basis of $\mathbb{R}^d$. Then, using the expression (6.18) for b_h, we observe that, for all $(v_h, q_h) \in X_h$,

$$b_h(v_h, q_h) = -\int_\Omega q_h D_h^k(v_h). \tag{6.40}$$

We also introduce a new discrete gradient operator

$$\mathcal{G}_h^l : H^1(\mathcal{T}_h) \to L^2(\Omega)$$

such that, for all $q \in H^1(\mathcal{T}_h)$,

$$\mathcal{G}_h^l(q) := \nabla_h q - \sum_{F \in \mathcal{F}_h^i} \mathrm{r}_F^l([\![q]\!]). \tag{6.41}$$

The only difference with respect to the discrete gradient operator G_h^l defined by (4.44) is that boundary faces are not included in the summation on the right-hand side of (6.41). A motivation for this modification is that there holds

$$\forall (v_h, q_h) \in X_h, \qquad \int_\Omega v_h \cdot \mathcal{G}_h^k(q_h) = -\int_\Omega q_h D_h^k(v_h),$$

so that an alternative expression for b_h on X_h is

$$b_h(v_h, q_h) = \int_\Omega v_h \cdot \mathcal{G}_h^k(q_h).$$

The discrete divergence operator enjoys asymptotic consistency properties similar to those satisfied by the discrete gradient operator.

Proposition 6.28 (Asymptotic consistency of discrete divergence). *Let $v_\mathcal{H}$ be a sequence in $U_\mathcal{H}$ bounded in the $\|\cdot\|_\mathrm{vel}$-norm and let $v \in U$ be given by Proposition 6.27. Then, as $h \to 0$, for all $l \geq 0$, up to a subsequence,*

$$D_h^l(v_h) \rightharpoonup \nabla \cdot v \qquad \text{weakly in } L^2(\Omega).$$

Moreover, for all $\Phi \in [C_0^\infty(\Omega)]^d$, letting $\Phi_h := \Pi_h \Phi$, there holds, for all $l \geq 0$,

$$D_h^l(\Phi_h) \to \nabla \cdot \Phi \qquad \text{in } L^2(\Omega).$$

Finally, the modified discrete gradient operator $\mathcal{G}_h^l$ enjoys the same asymptotic consistency properties as G_h^l. Indeed, discarding the liftings on boundary faces does not affect weak and strong asymptotic consistency since these properties are established by considering compactly supported functions in Ω. Note, however, that the operator $\mathcal{G}_h^l$ cannot be used in the proof of Theorem 5.7 where extensions to $\mathbb{R}^d$ are considered.

6.1.4.3 Asymptotic Consistency for c_h

We now prove an asymptotic consistency result for the discrete bilinear form c_h similar to the result established in Sect. 5.2.1 for the Poisson problem. Recalling the definition (6.26) of the $\|\cdot\|_\mathrm{sto}$-norm, we preliminarily observe that if the sequence $(v_\mathcal{H}, q_\mathcal{H})$ in $X_\mathcal{H}$ is bounded in the $\|\cdot\|_\mathrm{sto}$-norm, the sequence $v_\mathcal{H}$ is bounded in the $\|\cdot\|_\mathrm{vel}$-norm, and the sequence $q_\mathcal{H}$ is bounded in the L^2-norm. Hence, there is $(v, q) \in X$ such that v is given by Proposition 6.27 while $q_h \rightharpoonup q$ weakly in $L^2(\Omega)$.

Lemma 6.29 (Asymptotic consistency for c_h). *Let $(v_\mathcal{H}, q_\mathcal{H})$ be a sequence in $X_\mathcal{H}$ bounded in the $\|\cdot\|_\mathrm{sto}$-norm. Then, for all $\Phi \in [C_0^\infty(\Omega)]^d$ and for all $\varphi \in C_0^\infty(\Omega)$, there holds, up to a subsequence,*

$$\lim_{h \to 0} c_h((v_h, q_h), (\Pi_h \Phi, \pi_h \varphi)) = c((v, q), (\Phi, \varphi)),$$

where Π_h denotes the L^2-projector onto U_h and π_h the L^2-projector onto $\mathbb{P}_d^k(\mathcal{T}_h)$.

6.1. Steady Stokes Flows

Proof. Let $\Phi \in [C_0^\infty(\Omega)]^d$ and set $\Phi_h := \Pi_h \Phi$. In particular, we obtain

$$c_h((v_h, q_h), (\Phi_h, 0)) = a_h(v_h, \Phi_h) + b_h(\Phi_h, q_h).$$

Working componentwise for the diffusion term and using Lemma 5.10, we infer

$$\lim_{h \to 0} a_h(v_h, \Phi_h) = a(v, \Phi).$$

Consider now the second term and observe that, using (6.40), $b_h(\Phi_h, q_h) = -\int_\Omega q_h D_h^k(\Phi_h)$. Owing to the weak convergence of $q_\mathcal{H}$ to q in $L^2(\Omega)$ and to the strong convergence of $(D_h^k(\Phi_h))_{h \in \mathcal{H}}$ to $\nabla \cdot \Phi$ in $L^2(\Omega)$, $b_h(\Phi_h, q_h)$ tends to $-\int_\Omega q \nabla \cdot \Phi = b(\Phi, q)$ as $h \to 0$. As a result,

$$\lim_{h \to 0} c_h((v_h, q_h), (\Phi_h, 0)) = a(v, \Phi) + b(\Phi, q).$$

Let now $\varphi \in C_0^\infty(\Omega)$ and set $\varphi_h := \pi_h \varphi$. We obtain

$$c_h((v_h, q_h), (0, \varphi_h)) = -b_h(v_h, \varphi_h) + s_h(q_h, \varphi_h).$$

Clearly, $-b_h(v_h, \varphi_h) = \int_\Omega \varphi_h D_h^k(v_h)$ tends to $\int_\Omega \varphi(\nabla \cdot v) = -b(v, \varphi)$ as $h \to 0$ since $(D_h^k(v_h))_{h \in \mathcal{H}}$ weakly converges to $\nabla \cdot v$ in $L^2(\Omega)$ and $\varphi_\mathcal{H}$ strongly converges to φ in $L^2(\Omega)$. Finally, concerning pressure stabilization, $|s_h(p_h, \varphi_h)| \leq |p_h|_p |\varphi_h|_p = |p_h|_p |\varphi - \varphi_h|_p$, and this upper bound tends to zero since the first factor is bounded by assumption and the second factor tends to zero. As a result,

$$\lim_{h \to 0} c_h((v_h, q_h), (0, \varphi_h)) = -b(v, \varphi),$$

which completes the proof. □

6.1.4.4 Main Result

We can now state and prove the main result of this section.

Theorem 6.30 (Convergence to minimal regularity solutions). *Let $k \geq 1$. Let $(u_\mathcal{H}, p_\mathcal{H})$ be the sequence of approximate solutions generated by solving the discrete problems (6.23) on the admissible mesh sequence $\mathcal{T}_\mathcal{H}$. Then, as $h \to 0$,*

$$\begin{aligned} u_h &\to u & &\text{in } [L^2(\Omega)]^d, \\ \nabla_h u_h &\to \nabla u & &\text{in } [L^2(\Omega)]^{d,d}, \\ |u_h|_J &\to 0, \\ p_h &\to p & &\text{in } L^2(\Omega), \\ |p_h|_p &\to 0, \end{aligned}$$

where $(u, p) \in X$ denotes the unique solution to (6.5).

Proof. (i) *A priori estimate.* Owing to the inf-sup condition (6.27), the regularity assumptions on f, and the discrete Poincaré inequality (5.6), there holds

$$\gamma \|(u_h, p_h)\|_{\text{sto}} \leq \sup_{(w_h, r_h) \in X_h \setminus \{0\}} \frac{c_h((u_h, p_h), (w_h, r_h))}{\|(w_h, r_h)\|_{\text{sto}}}$$

$$= \sup_{(w_h, r_h) \in X_h \setminus \{0\}} \frac{\int_\Omega f \cdot w_h}{\|(w_h, r_h)\|_{\text{sto}}}$$

$$\leq \|f\|_{[L^2(\Omega)]^d} \sup_{(w_h, r_h) \in X_h \setminus \{0\}} \frac{\|w_h\|_{[L^2(\Omega)]^d}}{\|(w_h, r_h)\|_{\text{sto}}} \leq \sigma_2 \|f\|_{[L^2(\Omega)]^d},$$

so that the sequence $(u_\mathcal{H}, p_\mathcal{H})$ is bounded in the $\|\cdot\|_{\text{sto}}$-norm. Hence, there is $(u, p) \in X$ such that, up to a subsequence, as $h \to 0$, $u_h \to u$ strongly in $[L^2(\Omega)]^d$, $G_h^l(u_{h,i}) \rightharpoonup \nabla u_i$ weakly in $[L^2(\Omega)]^d$ for all $l \geq 0$ and all $i \in \{1, \ldots, d\}$, and $p_h \rightharpoonup p$ weakly in $L^2(\Omega)$.

(ii) *Identification of the limit and convergence of the whole sequence.* Let $\Phi \in [C_0^\infty(\Omega)]^d$ and let $\varphi \in C_0^\infty(\Omega)$. Observing that

$$\int_\Omega f \cdot \Pi_h \Phi = c_h((u_h, p_h), (\Pi_h \Phi, \pi_h \varphi - \langle \pi_h \varphi \rangle_\Omega)) = c_h((u_h, p_h), (\Pi_h \Phi, \pi_h \varphi)),$$

and using asymptotic consistency (cf. Lemma 6.29), we infer by passing to the limit $h \to 0$ that

$$\int_\Omega f \cdot \Phi = c((u, p), (\Phi, \varphi)).$$

By density of $[C_0^\infty(\Omega)]^d \times C_0^\infty(\Omega)$ in $U \times L^2(\Omega)$, this shows that (u, p) solves the Stokes equations (6.5). Since the solution to this problem is unique, the whole sequence $(u_\mathcal{H}, p_\mathcal{H})$ converges to (u, p).

(iii) *Strong convergence of the velocity gradient and convergence of velocity and pressure jumps.* Using the partial coercivity (6.25) and the stability property (5.11) for the SIP bilinear form, we infer, for $l \in \{k-1, k\}$,

$$\int_\Omega f \cdot u_h = c_h((u_h, p_h), (u_h, p_h)) = a_h(u_h, u_h) + s_h(p_h, p_h)$$

$$\geq a_h(u_h, u_h) \geq \sum_{i=1}^d \|G_h^l(u_{h,i})\|_{[L^2(\Omega)]^d}^2.$$

Thus,

$$\limsup_{h \to 0} \sum_{i=1}^d \|G_h^l(u_{h,i})\|_{[L^2(\Omega)]^{d,d}}^2 \leq \limsup_{h \to 0} \int_\Omega f \cdot u_h = \int_\Omega f \cdot u = \|\nabla u\|_{[L^2(\Omega)]^{d,d}}^2.$$

Proceeding as in point (iv) of Theorem 5.12, we infer $G_h^l(u_{h,i}) \to \nabla u_i$ in $[L^2(\Omega)]^d$ for all $i \in \{1, \ldots, d\}$ and $|u_{h,i}|_J \to 0$ for all $i \in \{1, \ldots, d\}$. Finally, since

$$|p_h|_p^2 = b_h(u_h, p_h) = \int_\Omega f \cdot u_h - a_h(u_h, u_h),$$

and the right-hand side converges to $\int_\Omega f \cdot u - a(u,u) = 0$ since $b(u,p) = 0$, we conclude that $|p_h|_p \to 0$.

(iv) *Strong convergence of the pressure.* Using Theorem 6.5, let $v_{p_h} \in U$ be such that $\nabla \cdot v_{p_h} = p_h$ with $\beta_\Omega \|v_{p_h}\|_U \le \|p_h\|_P$ and set $v_h := \Pi_h v_{p_h}$. Then, proceeding as in the proof of Lemma 6.10 yields

$$\|p_h\|_P^2 = -b_h(v_h, p_h) + \sum_{F \in \mathcal{F}_h^i} \int_F [\![p_h]\!]\{\!\{v_{p_h} - v_h\}\!\} \cdot n_F$$

$$\le -b_h(v_h, p_h) + C|p_h|_p \|p_h\|_P,$$

with C independent of h. Since $|p_h|_p$ tends to zero and $\|p_h\|_P$ is bounded, the second term on the right-hand side converges to zero. Let us now prove that

$$\lim_{h \to 0} -b_h(v_h, p_h) = \|p\|_P^2.$$

We first observe that

$$-b_h(v_h, p_h) = a_h(u_h, v_h) - \int_\Omega f \cdot v_h = \mathfrak{T}_1 + \mathfrak{T}_2. \tag{6.42}$$

The sequence $v_\mathcal{H}$ is bounded in the $\|\cdot\|_{\text{vel}}$-norm since owing to Lemma 6.11,

$$\|v_h\|_{\text{vel}} = \|\Pi_h v_{p_h}\|_{\text{vel}} \le C_\Pi \|v_{p_h}\|_U \le \beta_\Omega^{-1} C_\Pi \|p_h\|_P,$$

and $\|p_h\|_P$ is uniformly bounded in h. Hence, there is $v \in U$ such that, up to a subsequence, $v_h \to v$ strongly in $[L^2(\Omega)]^d$ and, for all $l \ge 0$, $G_h^l(v_{h,i}) \rightharpoonup \nabla v_i$ weakly in $[L^2(\Omega)]^d$ for all $i \in \{1,\ldots,d\}$. Moreover, $(v_h - v_{p_h}) \to 0$ strongly in $[L^2(\Omega)]^d$. Hence, for all $\varphi \in C_0^\infty(\Omega)$, observing that

$$\int_\Omega v_h \cdot \nabla \varphi = -\int_\Omega p_h \varphi + \int_\Omega (v_h - v_{p_h}) \cdot \nabla \varphi,$$

and letting $h \to 0$, the left-hand side converges to $\int_\Omega v \cdot \nabla \varphi$ and the right-hand side to $-\int_\Omega p\varphi$. These two quantities are then equal, so that $\nabla \cdot v = p$. Consider now the terms $\mathfrak{T}_1$ and $\mathfrak{T}_2$ in (6.42). Using expression (6.39), we obtain, for $l \in \{k-1, k\}$,

$$\mathfrak{T}_1 = a_h(u_h, v_h) = \int_\Omega \sum_{i=1}^d G_h^l(u_{h,i}) \cdot G_h^l(v_{h,i}) + \hat{s}_h(u_h, v_h) = \mathfrak{T}_{1,1} + \mathfrak{T}_{1,2}.$$

Owing to the strong convergence of $(G_h^l(u_{h,i}))_{h \in \mathcal{H}}$ to ∇u_i in $[L^2(\Omega)]^d$ and to the weak convergence of $(G_h^l(v_{h,i}))_{h \in \mathcal{H}}$ to ∇v_i in $[L^2(\Omega)]^d$, we infer

$$\mathfrak{T}_{1,1} \to \int_\Omega \sum_{i=1}^d \nabla u_i \cdot \nabla v_i = \int_\Omega \nabla u : \nabla v.$$

Moreover, up to positive factors independent of h, $|\mathfrak{T}_{1,2}| \lesssim |u_h|_J |v_h|_J$, which converges to zero since the first factor converges to zero while the second factor is bounded. Finally, it is clear that $\mathfrak{T}_2 \to \int_\Omega f \cdot v$. Collecting the above limits leads to

$$\limsup_{h \to 0} \|p_h\|_P^2 \leq \lim_{h \to 0} -b_h(v_h, p_h)$$

$$= \int_\Omega \nabla u{:}\nabla v - \int_\Omega f \cdot v = \int_\Omega p \nabla \cdot v = \|p\|_P^2,$$

classically yielding the strong convergence of the pressure in $L^2(\Omega)$. □

6.1.5 Formulations Without Pressure Stabilization

Fully discontinuous formulations, such as the one presented in Sect. 6.1.2, are appealing in problems where corner singularities are present (e.g., the well-known lid-driven cavity problem; see Ghia, Ghia, and Shin [168]), since, in this context, discontinuous pressures are generally less prone to spurious oscillations. Using equal-order velocity and pressure spaces, however, requires penalizing pressure jumps across interfaces to achieve discrete stability (cf. Lemma 6.10). Such a term introduces a tighter coupling between the discrete momentum and mass conservation equations, since the pressure is also explicitly present in the mass conservation equation. In practice, this can be a drawback when using classical solution methods (such as the Uzawa method) for saddle-point problems in the steady case or projection methods in the unsteady case (cf. Sect. 6.3).

It turns out that the pressure penalty term can be omitted in various cases which, however, do not accommodate the same level of mesh generality as in Sect. 6.1.2. On matching affine quadrilateral or hexahedral meshes, formulations without pressure stabilization have been analyzed by Toselli [296] for different couples of polynomial degrees for velocity and pressure. On matching simplicial meshes with polynomials for the pressure one degree less than for the velocity, inf-sup stability has been proven by Hansbo and Larson [183] in the incompressible limit of two-dimensional linear elasticity and by Girault, Rivière, and Wheeler [171] for the two- and three-dimensional Stokes equations in the context of domain decomposition methods (with polynomial degree for the velocity between 1 and 3). Still on matching simplicial meshes for $d \in \{2, 3\}$, a parameter-free dG approximation using piecewise affine discrete velocities supplemented by element bubble functions coupled with continuous piecewise affine and/or piecewise constant discrete pressures has been analyzed by Burman and Stamm [72].

A means to achieve discrete inf-sup stability on matching simplicial meshes is to consider a discontinuous approximation of the velocity together with a continuous approximation of the pressure. This approach constitutes the basis for the projection method discussed in Sect. 6.3 in the unsteady case. More precisely, let $\mathcal{T}_\mathcal{H}$ be an admissible sequence of matching simplicial meshes and define

$$\mathcal{P}_{d,0}^k(\mathcal{T}_h) := \left\{ v \in C^0(\overline{\Omega}),\ \langle v \rangle_\Omega = 0 \mid \forall T \in \mathcal{T}_h,\ v|_T \in \mathbb{P}_d^k(T) \right\}. \qquad (6.43)$$

We consider the approximation of the Stokes problem obtained using the discrete space $U_h = [\mathbb{P}_d^k(\mathcal{T}_h)]^d$ for the velocity and replacing the discrete space P_h by $\mathcal{P}_h := \mathcal{P}_{d,0}^k(\mathcal{T}_h)$ for the pressure. In this case, all terms involving the jumps of functions in $\mathcal{P}_h$ vanish, so that $s_h(q_h, r_h) = 0$ for all $q_h, r_h \in \mathcal{P}_h$. Moreover, for all $(v_h, q_h) \in \mathcal{X}_h := U_h \times \mathcal{P}_h$,

$$b_h(v_h, q_h) = \int_\Omega v_h \cdot \nabla q_h = -\int_\Omega q_h \nabla_h \cdot v_h + \sum_{F \in \mathcal{F}_h} \int_F [\![v_h]\!] \cdot n_F q_h.$$

The discrete problem then reads: Find $(u_h, p_h) \in \mathcal{X}_h$ such that

$$c_h((u_h, p_h), (v_h, q_h)) = \int_\Omega f \cdot v_h \text{ for all } (v_h, q_h) \in \mathcal{X}_h, \qquad (6.44)$$

where

$$c_h((u_h, p_h), (v_h, q_h)) = a_h(u_h, v_h) + b_h(v_h, p_h) - b_h(u_h, q_h).$$

A straightforward consequence of Lemma 6.10 is the LBB condition

$$\forall q_h \in \mathcal{P}_h, \qquad \beta \|q_h\|_P \leq \sup_{w_h \in U_h \setminus \{0\}} \frac{b_h(w_h, q_h)}{\|w_h\|_{\text{vel}}}.$$

Therefore, the discrete problem (6.44) is well-posed. Moreover, the error estimates (6.35) and (6.36) hold true, along with the convergence rates derived in Corollaries 6.22 and 6.26 for smooth solutions.

6.2 Steady Navier–Stokes Flows

In this section, we consider steady Navier–Stokes flows. The main difference with respect to Sect. 6.1 is the inclusion of a nonlinear term modeling the convective transport of momentum. The discretization with dG methods of this nonlinear term is the main focus of this section. We also account for the viscosity ν in the momentum conservation equation. We refer to Remark 6.1 for why the viscosity can be omitted in the context of steady Stokes flows, up to rescaling of the pressure and the forcing term. For steady Navier–Stokes flows, the viscosity ν is important since it quantifies the relative importance of convective and diffusive momentum transport.

6.2.1 The Continuous Setting

Let $\Omega \subset \mathbb{R}^d$, $d \in \{2, 3, 4\}$, be a polyhedron, let $f \in [L^2(\Omega)]^d$ be the forcing term, and let $\nu > 0$ be the viscosity. The discussion of this section is confined to space dimensions up to 4 since the nonlinear term requires embeddings of functional spaces valid for $d \leq 4$. The *steady INS problem* reads

$$-\nu \triangle u + (u \cdot \nabla) u + \nabla p = f \quad \text{in } \Omega, \qquad (6.45a)$$
$$\nabla \cdot u = 0 \quad \text{in } \Omega, \qquad (6.45b)$$
$$u = 0 \quad \text{on } \partial\Omega, \qquad (6.45c)$$
$$\langle p \rangle_\Omega = 0. \qquad (6.45d)$$

In component form, the momentum conservation equation becomes

$$-\nu\triangle u_i + \sum_{j=1}^{d} u_j \partial_j u_i + \partial_i p = f_i \qquad \forall i \in \{1,\ldots,d\}.$$

Remark 6.31 (Conservative formulation). Since $(u\cdot\nabla)u = \nabla\cdot(u\otimes u)$ because $\nabla\cdot u = 0$, the momentum conservation equation (6.45a) can be rewritten in the *conservative form*

$$-\nu\triangle u + \nabla\cdot(u\otimes u) + \nabla p = f,$$

or, in component form,

$$-\nu\triangle u_i + \sum_{j=1}^{d} \partial_j(u_i u_j) + \partial_i p = f_i \qquad \forall i \in \{1,\ldots,d\}.$$

In contrast, (6.45a) is said to be in *nonconservative form*. The conservative formulation is further discussed in Sect. 6.2.4 in the context of dG discretizations.

6.2.1.1 Weak Formulation and Trilinear Form

The weak formulation of system (6.45) reads: Find $(u,p) \in X$ such that

$$c((u,p),(v,q)) + t(u,u,v) = \int_\Omega f\cdot v \text{ for all } (v,q) \in X, \qquad (6.46)$$

where $X = U \times P$ is defined by (6.3), the bilinear form $c \in \mathcal{L}(X \times X, \mathbb{R})$ now accounts for the viscosity and is given by

$$c((u,p),(v,q)) = \nu a(u,v) + b(v,p) - b(u,q),$$

with a and b still defined by (6.4), and the trilinear form $t \in \mathcal{L}(U \times U \times U, \mathbb{R})$ is such that

$$t(w,u,v) := \int_\Omega (w\cdot\nabla u)\cdot v = \int_\Omega \sum_{i,j=1}^{d} w_j(\partial_j u_i) v_i. \qquad (6.47)$$

The trilinear form is indeed bounded on $U \times U \times U$.

Lemma 6.32 (Boundedness of trilinear form). *There is τ_Ω, only depending on Ω, such that, for all $w,u,v \in U$,*

$$t(w,u,v) \leq \tau_\Omega \|w\|_U \|u\|_U \|v\|_U. \qquad (6.48)$$

Proof. We use the Sobolev embedding of $H_0^1(\Omega)$ into $L^4(\Omega)$ (valid in space dimension up to 4) stating that there is $C_{(2,4)}$ such that, for all $\psi \in H_0^1(\Omega)$, $\|\psi\|_{L^4(\Omega)} \leq C_{(2,4)} \|\nabla\psi\|_{[L^2(\Omega)]^d}$ so that, for all $v \in U$, $\|v\|_{[L^4(\Omega)]^d} = (\int_\Omega |v|_{\ell^4}^4)^{1/4} \leq C_{(2,4)} \|\nabla v\|_{[L^2(\Omega)]^{d,d}}$. Using Hölder's inequality, we then infer, for all $w,u,v \in U$,

$$t(w,u,v) \leq d^{1/2} \|w\|_{[L^4(\Omega)]^d} \|\nabla u\|_{[L^2(\Omega)]^{d,d}} \|v\|_{[L^4(\Omega)]^d}$$
$$\leq d^{1/2} C_{(2,4)}^2 \|\nabla w\|_{[L^2(\Omega)]^{d,d}} \|\nabla u\|_{[L^2(\Omega)]^{d,d}} \|\nabla v\|_{[L^2(\Omega)]^{d,d}},$$

whence the conclusion with $\tau_\Omega := d^{1/2} C_{(2,4)}^2$. □

6.2. Steady Navier–Stokes Flows

An important property of the trilinear form t defined by (6.47) is skew-symmetry with respect to the last two arguments whenever the first argument is divergence-free and has zero normal component on the boundary. For simplicity, we consider that the three arguments of the trilinear form are in U.

Lemma 6.33 (Skew-symmetry of trilinear form). *For all $w \in U$, there holds*

$$\forall v \in U, \qquad t(w, v, v) = -\frac{1}{2} \int_\Omega (\nabla \cdot w) |v|^2. \qquad (6.49)$$

Moreover, if $w \in V := \{v \in U \mid \nabla \cdot v = 0\}$,

$$\forall v \in U, \qquad t(w, v, v) = 0. \qquad (6.50)$$

Proof. Let $w \in U$. We observe that, for all $v \in U$,

$$t(w, v, v) + \frac{1}{2} \int_\Omega (\nabla \cdot w)|v|^2 = \int_\Omega \frac{1}{2} w \cdot \nabla |v|^2 + \frac{1}{2} \int_\Omega (\nabla \cdot w)|v|^2 = \int_\Omega \frac{1}{2} \nabla \cdot (w|v|^2),$$

so that the divergence theorem yields

$$t(w, v, v) + \frac{1}{2} \int_\Omega (\nabla \cdot w)|v|^2 = \frac{1}{2} \int_{\partial\Omega} (w \cdot n)|v|^2 = 0,$$

since $(w \cdot n)$ vanishes on $\partial\Omega$. This proves (6.49), and (6.50) is a direct consequence of (6.49). $\square$

A crucial consequence of Lemma 6.33 is that, using $(v, q) = (u, p)$ as a test function in (6.46) and since u is divergence-free, we obtain, up to the viscosity scaling, the same *energy balance* as for steady Stokes flows, namely

$$\nu \|\nabla u\|^2_{[L^2(\Omega)]^{d,d}} = \int_\Omega f \cdot u.$$

In other words, convection does not influence energy balance. More generally, combining the U-coercivity (6.6) of the bilinear form a with the skew-symmetry (6.50) of the trilinear form t leads to the following result.

Lemma 6.34 (Dissipativity). *There holds, for all $w \in V$,*

$$\forall v \in U, \qquad \nu a(v, v) + t(w, v, v) \geq \nu \alpha_\Omega \|v\|^2_U, \qquad (6.51)$$

where α_Ω denotes the coercivity parameter of the bilinear form a.

The last ingredient in the analysis of the continuous setting is the following weak sequential continuity of the trilinear form t (see Girault and Raviart [170, p. 115]).

Lemma 6.35 (Weak sequential continuity). *Let $(v_m)_{m \in \mathbb{N}}$ be a sequence in V bounded in the $\|\cdot\|_U$-norm, so that there is $v \in V$ such that, up to a subsequence, as $m \to \infty$, $\nabla v_m \rightharpoonup \nabla v$ weakly in $[L^2(\Omega)]^{d,d}$ and $v_m \to v$ in $[L^2(\Omega)]^d$. Then, for all $\Phi \in [C_0^\infty(\Omega)]^d$, there holds*

$$\lim_{m \to \infty} t(v_m, v_m, \Phi) = t(v, v, \Phi).$$

6.2.1.2 Existence and Uniqueness

We now address the existence and uniqueness of the solution to (6.46).

Theorem 6.36 (Existence and uniqueness). *There exists at least one $(u, p) \in X$ solving (6.46). Moreover, under the smallness condition on the data*

$$\tau_\Omega \|f\|_{U'} < (\nu \alpha_\Omega)^2, \tag{6.52}$$

the solution is unique.

Proof. For an existence proof, we refer the reader to Girault and Raviart [170, Theorem 2.4] (cf. also Remark 6.48); the proof uses Lemmata 6.32, 6.33, and 6.35. We sketch the uniqueness proof which is based on a fixed point argument; see [170, Theorem 2.5]. As for the linear Stokes equations (cf. Remark 6.9), the pressure can be eliminated by incorporating the zero-divergence constraint in an essential way, that is, using the space $V = \{v \in U \mid \nabla \cdot v = 0\}$ for the velocity. Indeed, if the couple $(u, p) \in X$ solves (6.46), then the velocity u solves the problem

$$\text{Find } u \in V \text{ s.t. } \nu a(u, v) + t(u, u, v) = \int_\Omega f \cdot v \text{ for all } v \in V. \tag{6.53}$$

For all $w \in V$, we consider the bounded linear operator $A(w) \in \mathcal{L}(V, V')$ such that, for all $u, v \in V$,

$$\langle A(w)u, v \rangle_{V', V} = \nu a(u, v) + t(w, u, v).$$

Owing to (6.51),

$$\forall v \in V, \quad \langle A(w)v, v \rangle_{V', V} \geq \nu \alpha_\Omega \|v\|_U^2.$$

Hence, $A(w)$ is an isomorphism mapping V onto V' for all $w \in V$. Its inverse, say $S(w)$, belongs to $\mathcal{L}(V', V)$ and satisfies

$$\forall w \in V, \quad \|S(w)\|_{\mathcal{L}(V', V)} \leq (\nu \alpha_\Omega)^{-1}. \tag{6.54}$$

Moreover, problem (6.53) can be reformulated as

$$\text{Find } u \in V \text{ such that } u = S(u)f \text{ in } V.$$

Let us prove that, under the smallness condition (6.52), the map $V \ni w \mapsto S(w)f \in V$ is a strict contraction. Let $w_1, w_2 \in V$. Observing that $S(w_1) - S(w_2) = S(w_1)[A(w_2) - A(w_1)]S(w_2)$ and using the bound (6.54), we obtain

$$\|S(w_1) - S(w_2)\|_{\mathcal{L}(V', V)} \leq (\nu \alpha_\Omega)^{-2} \|A(w_2) - A(w_1)\|_{\mathcal{L}(V, V')} \|f\|_{U'}$$
$$\leq \tau_\Omega \|f\|_{U'} (\nu \alpha_\Omega)^{-2} \|w_1 - w_2\|_U$$
$$< \|w_1 - w_2\|_U,$$

6.2. Steady Navier–Stokes Flows

since owing to (6.48) and the trilinearity of t, we infer

$$\|A(w_2) - A(w_1)\|_{\mathcal{L}(V,V')} = \sup_{u,v \in V \setminus \{0\}} \frac{\langle (A(w_2) - A(w_1))u, v \rangle_{V',V}}{\|u\|_U \|v\|_U}$$

$$= \sup_{u,v \in V \setminus \{0\}} \frac{t(w_2 - w_1, u, v)}{\|u\|_U \|v\|_U} \leq \tau_\Omega \|w_1 - w_2\|_U.$$

This proves the strict contraction property. Hence, owing to Banach's Fixed Point Theorem, there is a unique $u \in V$ such that $u = S(u)f$, and, hence, there is a unique $u \in V$ solving (6.53). Finally, the existence and uniqueness of the pressure $p \in P$ is obtained as for the linear Stokes equations. □

Remark 6.37 (Interpretation of condition (6.52)). At fixed viscosity ν, condition (6.52) means that the forcing term f must be small enough. Alternatively, at fixed f, condition (6.52) means that the viscosity ν must be large enough (so that sufficient energy is dissipated by the flow).

6.2.2 The Discrete Setting

In this section, we derive a dG discretization of the INS equations (6.46). For the Stokes part (resulting from the bilinear form c), we follow the approach of Sect. 6.1.2 and consider equal-order discontinuous velocities and pressures. Alternative dG methods to approximate the INS equations have been explored by Karakashian and Jureidini [206], Girault, Rivière, and Wheeler [171], and Cockburn, Kanschat, and Schötzau [102–104].

Let $\mathcal{T}_\mathcal{H}$ denote an admissible mesh sequence and let $k \geq 1$ be an integer. We consider the discrete spaces (cf. (6.13))

$$U_h := [\mathbb{P}^k_d(\mathcal{T}_h)]^d, \qquad P_h := \mathbb{P}^k_{d,0}(\mathcal{T}_h), \qquad X_h := U_h \times P_h.$$

Our first goal is to derive a discrete trilinear form t_h that satisfies suitable discrete counterparts of Lemmata 6.32, 6.33, and 6.35. Then, we prove the existence of a solution to the discrete INS equations, as well as uniqueness under a smallness assumption on the data. The material in this section is restricted to $d \leq 3$; cf. Remark 6.46.

6.2.2.1 Temam's Modification of the Trilinear Form t

When working with dG approximations, the convective velocity is generally not divergence-free (but only weakly divergence-free), so that the important property (6.50) is generally not satisfied. Following Temam [291, 292], a possible way to circumvent this difficulty is to modify the trilinear form t and to consider instead, for all $w, u, v \in U$,

$$t'(w,u,v) = t(w,u,v) + \frac{1}{2}\int_\Omega (\nabla \cdot w) u \cdot v = \int_\Omega (w \cdot \nabla u) \cdot v + \frac{1}{2}\int_\Omega (\nabla \cdot w) u \cdot v. \quad (6.55)$$

The following result is then a straightforward consequence of (6.49).

Lemma 6.38 (Skew-symmetry of modified trilinear form). *For all $w \in U$, there holds*
$$\forall v \in U, \qquad t'(w,v,v) = 0. \tag{6.56}$$

Moreover, $(u,p) \in X$ solves (6.46) if and only if $(u,p) \in X$ is such that
$$c((u,p),(v,q)) + t'(u,u,v) = \int_\Omega f \cdot v \text{ for all } (v,q) \in X.$$

6.2.2.2 Discrete Trilinear Form

We start with Temam's modification of the trilinear form t. Specifically, we consider broken differential operators in the trilinear form t' defined by (6.55) and set, for all $w_h, u_h, v_h \in U_h$,
$$t_h^{(0)}(w_h, u_h, v_h) := \int_\Omega (w_h \cdot \nabla_h u_h) \cdot v_h + \frac{1}{2} \int_\Omega (\nabla_h \cdot w_h) u_h \cdot v_h.$$

Our first goal is to derive a discrete counterpart of (6.56). For all $w_h, v_h \in U_h$, integrating by parts elementwise and proceeding as usual, we obtain
$$t_h^{(0)}(w_h, v_h, v_h) = \frac{1}{2} \sum_{F \in \mathcal{F}_h} \int_F [\![w_h]\!] \cdot \mathbf{n}_F \{\!\{v_h \cdot v_h\}\!\} + \sum_{F \in \mathcal{F}_h^i} \int_F \{\!\{w_h\}\!\} \cdot \mathbf{n}_F [\![v_h]\!] \cdot \{\!\{v_h\}\!\}.$$

Since the right-hand side of the above equation is nonzero, we modify $t_h^{(0)}$ as
$$t_h(w_h, u_h, v_h) := \int_\Omega (w_h \cdot \nabla_h u_h) \cdot v_h - \sum_{F \in \mathcal{F}_h^i} \int_F \{\!\{w_h\}\!\} \cdot \mathbf{n}_F [\![u_h]\!] \cdot \{\!\{v_h\}\!\}$$
$$+ \frac{1}{2} \int_\Omega (\nabla_h \cdot w_h)(u_h \cdot v_h) - \frac{1}{2} \sum_{F \in \mathcal{F}_h} \int_F [\![w_h]\!] \cdot \mathbf{n}_F \{\!\{u_h \cdot v_h\}\!\}. \tag{6.57}$$

This choice, which incorporates Temam's modification *at the discrete level*, possesses the following important property which is the discrete counterpart of Lemma 6.33.

Lemma 6.39 (Skew-symmetry of discrete trilinear form). *For all $w_h \in U_h$, there holds*
$$\forall v_h \in U_h, \qquad t_h(w_h, v_h, v_h) = 0. \tag{6.58}$$

We now address the boundedness of the discrete trilinear form t_h on $U_h \times U_h \times U_h$. Recall that the discrete velocity space U_h is equipped with the $\|\cdot\|_{\text{vel}}$-norm defined by (6.16). The following result is the discrete counterpart of Lemma 6.32.

Lemma 6.40 (Boundedness of discrete trilinear form). *There is τ, independent of h, such that, for all $w_h, u_h, v_h \in U_h$, there holds*
$$t_h(w_h, u_h, v_h) \le \tau \|w_h\|_{\text{vel}} \|u_h\|_{\text{vel}} \|v_h\|_{\text{vel}}.$$

6.2. Steady Navier–Stokes Flows

Proof. Let $\mathfrak{T}_i$, $i \in \{1,\ldots,4\}$ denote the terms on the right-hand side of (6.57). For the first term, Hölder's inequality yields

$$|\mathfrak{T}_1| \leq d^{1/2} \|w_h\|_{[L^4(\Omega)]^d} \|\nabla_h u_h\|_{[L^2(\Omega)]^{d,d}} \|v_h\|_{[L^4(\Omega)]^d}.$$

Moreover, applying componentwise the discrete Sobolev embedding of Theorem 5.3 for $p=2$ and $q=4$ (valid for $d \leq 4$), we infer

$$\forall v_h \in U_h, \qquad \|v_h\|_{[L^4(\Omega)]^d} \leq \sigma_{2,4} \|v_h\|_{\text{vel}}. \qquad (6.59)$$

As a result,

$$|\mathfrak{T}_1| \leq d^{1/2} \sigma_{2,4}^2 \|w_h\|_{\text{vel}} \|u_h\|_{\text{vel}} \|v_h\|_{\text{vel}}.$$

The third term can be bounded in a similar way. To estimate the second term, we use again Hölder's inequality to infer

$$|\mathfrak{T}_2| \leq d \left(\sum_{F \in \mathcal{F}_h^i} h_F \int_F |\{\!\!\{w_h\}\!\!\}|^4 \right)^{1/4} |u_h|_{\text{J}} \left(\sum_{F \in \mathcal{F}_h^i} h_F \int_F |\{\!\!\{v_h\}\!\!\}|^4 \right)^{1/4}.$$

Using the bound $(a+b)^4 \leq 8(a^4+b^4)$ for real numbers a and b, together with the discrete trace inequality (1.44) with $p=4$, we infer

$$\sum_{F \in \mathcal{F}_h^i} h_F \int_F |\{\!\!\{w_h\}\!\!\}|^4 \leq \sum_{F \in \mathcal{F}_h^i} \frac{1}{2} h_F \left(\|w_h|_{T_1}\|_{[L^4(F)]^d}^4 + \|w_h|_{T_2}\|_{[L^4(F)]^d}^4 \right)$$

$$\leq \frac{1}{2} N_\partial C_{\text{tr},4}^4 \|w_h\|_{[L^4(\Omega)]^d}^4,$$

where, for all $F \in \mathcal{F}_h^i$, we employed the usual notation $F = \partial T_1 \cap \partial T_2$. Proceeding similarly for the factor involving v_h, we arrive at the bound

$$|\mathfrak{T}_2| \leq 2^{-1/2} d N_\partial^{1/2} C_{\text{tr},4}^2 \|w_h\|_{[L^4(\Omega)]^d} |u_h|_{\text{J}} \|v_h\|_{[L^4(\Omega)]^d},$$

so that, using again the discrete Sobolev embedding (6.59), we infer

$$|\mathfrak{T}_2| \leq 2^{-1/2} d N_\partial^{1/2} C_{\text{tr},4}^2 \sigma_{2,4}^2 \|w_h\|_{\text{vel}} \|u_h\|_{\text{vel}} \|v_h\|_{\text{vel}}.$$

Finally, the fourth term can be bounded in a similar way. $\square$

6.2.2.3 Discrete Problem

Let a_h and b_h be the discrete bilinear forms considered for the linear Stokes equations, cf. (6.15) for a_h and (6.18) or, equivalently, (6.19) for b_h. Let t_h be the discrete trilinear form defined by (6.57). The discrete INS problem reads: Find $(u_h, p_h) \in X_h$ such that

$$\nu a_h(u_h, v_h) + t_h(u_h, u_h, v_h) + b_h(v_h, p_h) = \int_\Omega f \cdot v_h \qquad \forall v_h \in U_h, \qquad (6.60a)$$

$$-b_h(u_h, q_h) + \nu^{-1} s_h(p_h, q_h) = 0 \qquad \forall q_h \in P_h, \qquad (6.60b)$$

or, equivalently, such that

$$c_h((u_h, p_h), (v_h, q_h)) + t_h(u_h, u_h, v_h) = \int_\Omega f \cdot v_h \qquad \forall (v_h, q_h) \in X_h, \qquad (6.61)$$

with

$$c_h((u_h, p_h), (v_h, q_h)) := \nu a_h(u_h, v_h) + b_h(v_h, p_h) - b_h(u_h, q_h) + \nu^{-1} s_h(p_h, q_h).$$

We observe that both the diffusion and pressure stabilization terms differ from the case of the linear Stokes equations, cf. (6.24), since the former is scaled by the viscosity and the latter by the reciprocal of the viscosity.

Recalling (6.17), let $\alpha > 0$ denote the coercivity parameter of the discrete bilinear form a_h such that

$$\forall v_h \in U_h, \qquad a_h(v_h, v_h) \geq \alpha \|v_h\|_{\text{vel}}^2.$$

This leads to partial coercivity for the discrete bilinear form c_h in the form

$$\forall (v_h, q_h) \in X_h, \qquad c_h((v_h, q_h), (v_h, q_h)) \geq \nu \alpha \|v_h\|_{\text{vel}}^2 + \nu^{-1} |q_h|_p^2. \qquad (6.62)$$

Moreover, we define the $\|\cdot\|_{\text{ns}}$-norm as

$$\|(v_h, q_h)\|_{\text{ns}} := \left(\nu \|v_h\|_{\text{vel}}^2 + \|q_h\|_P^2 + \nu^{-1} |q_h|_p^2 \right)^{1/2}.$$

It is straightforward to verify, as in the proof of Lemma 6.13, the following discrete inf-sup condition: There is $\gamma > 0$, independent of h and of the viscosity ν, such that, for all $(v_h, q_h) \in X_h$,

$$\gamma \|(v_h, q_h)\|_{\text{ns}} \leq \sup_{(w_h, r_h) \in X_h \setminus \{0\}} \frac{c_h((v_h, q_h), (w_h, r_h))}{\|(w_h, r_h)\|_{\text{ns}}}. \qquad (6.63)$$

We observe that the fact that γ is independent of ν results from the scaling used in the pressure stabilization. Finally, owing to Lemma 6.39, the discrete counterpart of Lemma 6.34 is the following: For all $w_h \in U_h$,

$$\forall v_h \in U_h, \qquad \nu a_h(v_h, v_h) + t_h(w_h, v_h, v_h) \geq \nu \alpha \|v_h\|_{\text{vel}}^2.$$

6.2.2.4 Existence and Uniqueness

Our first step is to derive an a priori estimate on the solution.

Lemma 6.41 (A priori estimate). *If the couple* $(u_h, p_h) \in X_h$ *solves* (6.61), *then*

$$\gamma \|(u_h, p_h)\|_{\text{ns}} \leq \sigma_2 \|f\|_{[L^2(\Omega)]^d} + \tau(\nu \alpha)^{-2} \left(\sigma_2 \|f\|_{[L^2(\Omega)]^d} \right)^2, \qquad (6.64)$$

where σ_2 *results from the discrete Poincaré inequality* (5.6).

6.2. Steady Navier–Stokes Flows

Proof. Using (u_h, p_h) as test function in (6.61), we infer from (6.62) and the discrete Poincaré inequality (5.6) applied componentwise that

$$\nu\alpha\|u_h\|_{\text{vel}}^2 + \nu^{-1}|p_h|_p^2 \leq \int_\Omega f \cdot u_h \leq \|f\|_{[L^2(\Omega)]^d}\|u_h\|_{[L^2(\Omega)]^d}$$
$$\leq \sigma_2\|f\|_{[L^2(\Omega)]^d}\|u_h\|_{\text{vel}}. \tag{6.65}$$

Hence, owing to (6.63) and Lemma 6.40,

$$\gamma\|(u_h,p_h)\|_{\text{ns}} \leq \sup_{(w_h,r_h)\in X_h\setminus\{0\}} \frac{c_h((u_h,p_h),(w_h,r_h))}{\|(w_h,r_h)\|_{\text{ns}}}$$
$$= \sup_{(w_h,r_h)\in X_h\setminus\{0\}} \frac{\int_\Omega f\cdot w_h - t_h(u_h,u_h,w_h)}{\|(w_h,r_h)\|_{\text{ns}}}$$
$$\leq \sigma_2\|f\|_{[L^2(\Omega)]^d} + \tau\|u_h\|_{\text{vel}}^2.$$

The a priori estimate then results from (6.65). □

To prove the existence of a solution to the discrete problem (6.61), we use the a priori estimate of Lemma 6.41 together with a topological degree argument. We refer the reader to Eymard, Gallouët, Ghilani, and Herbin [155] or Eymard, Herbin, and Latché [160] for the use of this argument in the convergence analysis of finite volume schemes and to Deimling [122] for a general presentation.

Lemma 6.42 (Topological degree argument). *Let V be a finite-dimensional space equipped with a norm $\|\cdot\|_V$. Let $M > 0$ and let $\Psi : V \times [0,1] \to V$ satisfy the following assumptions:*

(i) *Ψ is continuous.*

(ii) *$\Psi(\cdot, 0)$ is an affine function and the equation $\Psi(v, 0) = 0$ has a solution $v \in V$ such that $\|v\|_V < M$.*

(iii) *For any $(v, \rho) \in V \times [0,1]$, $\Psi(v, \rho) = 0$ implies $\|v\|_V < M$.*

Then, there exists $v \in V$ such that $\Psi(v, 1) = 0$ and $\|v\|_V < M$.

Theorem 6.43 (Existence and uniqueness). *There exists at least one $(u_h, p_h) \in X_h$ solving (6.61). Moreover, under the smallness condition*

$$\tau\|f\|_{[L^2(\Omega)]^d} < (\nu\alpha)^2, \tag{6.66}$$

the solution is unique.

Proof. To prove existence, we verify conditions (i), (ii), and (iii) in Lemma 6.42. Let $V = X_h$ be equipped with the $\|\cdot\|_{\text{ns}}$-norm and define the mapping

$$\Psi : X_h \times [0,1] \to X_h$$

such that, for (u_h, p_h) given in X_h and ρ given in $[0,1]$, $(\xi_h, \zeta_h) := \Psi((u_h, p_h), \rho) \in X_h$ is such that, for all $(v_h, q_h) \in X_h$,

$$(\xi_h, v_h)_{[L^2(\Omega)]^d} = c_h((u_h, p_h), (v_h, 0)) + \rho t_h(u_h, u_h, v_h) - \int_\Omega f \cdot v_h,$$
$$(\zeta_h, q_h)_{L^2(\Omega)} = c_h((u_h, p_h), (0, q_h)).$$

Observing that c_h is bounded on $X_h \times X_h$ for the $\|\cdot\|_{\text{ns}}$-norm, using the boundedness of t_h (cf. Lemma 6.40) and the equivalence of norms in finite dimension, we infer that Ψ is continuous (condition (i)). Moreover, it is clear that $\Psi(\cdot, 0)$ is an affine function since the nonlinear term disappears for $\rho = 0$. Moreover, $\Psi(x_h, 0) = 0$ means that $x_h = (v_h, q_h) \in X_h$ solves the discrete Stokes equations. Hence, the equation $\Psi(x_h, 0) = 0$ has a solution in X_h which satisfies the a priori estimate

$$\gamma \|x_h\|_{\text{ns}} \leq \sigma_2 \|f\|_{[L^2(\Omega)]^d}.$$

Thus, condition (ii) is satisfied for any real number M such that

$$\gamma M > \sigma_2 \|f\|_{[L^2(\Omega)]^d}.$$

In addition, condition (iii) is satisfied, owing to Lemma 6.41, if we choose M such that
$$\gamma M > \sigma_2 \|f\|_{[L^2(\Omega)]^d} + \tau(\nu\alpha)^{-2} \left(\sigma_2 \|f\|_{[L^2(\Omega)]^d}\right)^2.$$

As a result, there exists $x_h \in X_h$ such that $\Psi(x_h, 1) = 0$, which means that x_h solves (6.61). Finally, the proof of uniqueness under the smallness condition (6.66) is similar to the proof of uniqueness in the continuous case (cf. Theorem 6.36). $\square$

6.2.3 Convergence Analysis

In this section, we investigate the convergence of the sequence $(u_\mathcal{H}, p_\mathcal{H})$ of solutions to the discrete problem (6.61) on the admissible mesh sequence $\mathcal{T}_\mathcal{H}$ to a solution (u, p) of the INS equations (6.46).

6.2.3.1 Asymptotic Consistency of Discrete Trilinear Form

As for the Poisson problem (cf. Sect. 5.2) and the linear Stokes equations (cf. Sect. 6.1.4), an instrumental ingredient in the convergence analysis is the asymptotic consistency of the discrete trilinear form t_h. To this purpose, we record the following equivalent expression of t_h in terms of discrete gradients and discrete divergence. We observe that the polynomial degree used for the discrete gradients and divergence is $2k$ owing to the nonlinearities.

Lemma 6.44 (Reformulation of discrete trilinear form). *There holds, for all* $w_h, u_h, v_h \in U_h$,

6.2. Steady Navier–Stokes Flows

$$t_h(w_h, u_h, v_h) = \int_\Omega \sum_{i=1}^d w_h \cdot \mathcal{G}_h^{2k}(u_{h,i}) v_{h,i} + \frac{1}{2} \int_\Omega D_h^{2k}(w_h) u_h \cdot v_h$$
$$+ \frac{1}{4} \sum_{F \in \mathcal{F}_h^i} \int_F [\![w_h]\!] \cdot n_F [\![u_h]\!] \cdot [\![v_h]\!]. \tag{6.67}$$

Proof. We start with the expression (6.57) and observe that

$$\int_\Omega (\nabla_h \cdot w_h)(u_h \cdot v_h) - \sum_{F \in \mathcal{F}_h^i} \int_F [\![w_h]\!] \cdot n_F \{\!\{u_h \cdot v_h\}\!\} = \int_\Omega D_h^{2k}(w_h) u_h \cdot v_h,$$

and that

$$\int_\Omega (w_h \cdot \nabla_h u_h) \cdot v_h - \sum_{F \in \mathcal{F}_h^i} \int_F \{\!\{w_h\}\!\} \cdot n_F [\![u_h]\!] \cdot \{\!\{v_h\}\!\} = \int_\Omega \sum_{i=1}^d w_h \cdot \mathcal{G}_h^{2k}(u_{h,i}) v_{h,i}$$
$$- \sum_{F \in \mathcal{F}_h^i} \int_F \sum_{i=1}^d [\![u_{h,i}]\!] \left(\{\!\{v_{h,i}\}\!\} \{\!\{w_h\}\!\} \cdot n_F - \{\!\{v_{h,i} w_h\}\!\} \cdot n_F \right).$$

The conclusion follows from $\{\!\{v_{h,i}\}\!\} \{\!\{w_h\}\!\} - \{\!\{v_{h,i} w_h\}\!\} = -\frac{1}{4}[\![v_{h,i}]\!][\![w_h]\!]$. □

We now turn to the asymptotic consistency of the discrete trilinear form t_h. This property is the discrete counterpart of Lemma 6.35.

Lemma 6.45 (Asymptotic consistency for t_h). *Let $v_\mathcal{H}$ be a sequence in $U_\mathcal{H}$ bounded in the $\|\cdot\|_\mathrm{vel}$-norm and let $v \in U$ be given by Proposition 6.27. Then, for all $\Phi \in [C_0^\infty(\Omega)]^d$, there holds, up to a subsequence,*

$$\lim_{h \to 0} t_h(v_h, v_h, \Pi_h \Phi) = t'(v, v, \Phi),$$

where Π_h denotes the L^2-projector onto U_h.

Proof. Let $\Phi \in [C_0^\infty(\Omega)]^d$ and set $\Phi_h = \Pi_h \Phi$. Owing to (6.67),

$$t_h(v_h, v_h, \Phi_h) = \int_\Omega \sum_{i=1}^d v_h \cdot \mathcal{G}_h^{2k}(v_{h,i}) \Phi_{h,i} + \frac{1}{2} \int_\Omega D_h^{2k}(v_h) v_h \cdot \Phi_h$$
$$+ \frac{1}{4} \sum_{F \in \mathcal{F}_h^i} \int_F [\![v_h]\!] \cdot n_F [\![v_h]\!] \cdot [\![\Phi_h]\!] := \mathfrak{T}_1 + \mathfrak{T}_2 + \mathfrak{T}_3.$$

Owing to Theorem 5.6, the sequences $v_\mathcal{H}$ and $\Phi_\mathcal{H}$ are relatively compact in $[L^4(\Omega)]^d$ (since $d \le 3$), and these sequences converge (up to a subsequence) in $[L^4(\Omega)]^d$ to v and Φ respectively. As a result, for all $i \in \{1, \dots, d\}$, the sequence

$(v_h\Phi_{h,i})_{h\in\mathcal{H}}$ converges in $[L^2(\Omega)]^d$ to $v\Phi_i$. In addition, $(\mathcal{G}_h^{2k}(v_{h,i}))_{h\in\mathcal{H}}$ weakly converges to ∇v_i in $[L^2(\Omega)]^d$. As a result,

$$\lim_{h\to 0}\mathfrak{T}_1 = \int_\Omega (v\cdot\nabla v)\cdot\Phi.$$

Similarly,

$$\lim_{h\to 0}\mathfrak{T}_2 = \frac{1}{2}\int_\Omega (\nabla\cdot v)(v\cdot\Phi).$$

Finally, using Hölder's inequality, together with the inverse inequality (1.43) to control the $\|\cdot\|_{L^4(F)}$-norm by the $\|\cdot\|_{L^2(F)}$-norm, we infer, up to a positive factor independent of h,

$$|\mathfrak{T}_3| \lesssim h^{(4-d)/2}|v_h|_{\mathrm{J}}^2|\Phi_h|_{\mathrm{J}} = h^{(4-d)/2}|v_h|_{\mathrm{J}}^2|\Phi_h - \Phi|_{\mathrm{J}},$$

so that $\mathfrak{T}_3 \to 0$ since $|v_h|_{\mathrm{J}}$ is bounded, while the first and last factors on the right-hand side converge to zero. This concludes the proof. □

Remark 6.46 (Assumption $d \le 3$). The assumption $d \le 3$ is used in the above proof to assert the relative compactness of the sequences $v_\mathcal{H}$ and $\Phi_\mathcal{H}$ in $[L^4(\Omega)]^d$. Since the modified discrete gradients only weakly converge, asserting the boundedness of these sequences is not sufficient to pass to the limit in the nonlinear terms.

6.2.3.2 Main Result

We can now state and prove our main convergence result.

Theorem 6.47 (Convergence). *Let $(u_\mathcal{H}, p_\mathcal{H})$ be a sequence of approximate solutions generated by solving the discrete problems (6.61) on the admissible mesh sequence $\mathcal{T}_\mathcal{H}$. Then, as $h \to 0$, up to a subsequence,*

$$\begin{aligned}
u_h &\to u & &\text{in } [L^2(\Omega)]^d,\\
\nabla_h u_h &\to \nabla u & &\text{in } [L^2(\Omega)]^{d,d},\\
|u_h|_{\mathrm{J}} &\to 0,\\
p_h &\to p & &\text{in } L^2(\Omega),\\
|p_h|_p &\to 0,
\end{aligned}$$

where $(u,p) \in X$ is a solution of (6.46). Moreover, under the smallness condition (6.66), the whole sequence converges to the unique solution of (6.46).

Remark 6.48 (Alternative existence proof). We incidentally observe that the existence result stated in Theorem 6.36 in the continuous setting can also be inferred from the convergence result stated in Theorem 6.47 in the context of dG approximations (at least for $d \le 3$).

6.2. Steady Navier–Stokes Flows

Proof. (i) *Existence of a limit.* Owing to the a priori estimate (6.64), we infer that there is $(u, p) \in X$ such that, up to a subsequence, $u_h \to u$ strongly in $[L^2(\Omega)]^d$, $G_h^l(u_{h,i}) \rightharpoonup \nabla u_i$ weakly in $[L^2(\Omega)]^d$ for all $l \geq 0$ and all $i \in \{1, \ldots, d\}$, and $p_h \rightharpoonup p$ weakly in $L^2(\Omega)$.

(ii) *Identification of the limit.* Using Lemma 6.45 and treating the linear part as for the Stokes equations, we infer that, for all $\Phi \in [C_0^\infty(\Omega)]^d$,

$$\nu a(u, \Phi) + t'(u, u, \Phi) - b(\Phi, p) = \int_\Omega f \cdot \Phi,$$

and that, for all $\varphi \in C_0^\infty(\Omega)$, $\int_\Omega \varphi \nabla \cdot u = 0$. Hence, (u, p) solves (6.46).

(iii) *Strong convergence of the velocity gradient and of the velocity and pressure jumps.* Proceeding as for the linear Stokes equations and using the skew-symmetry of t_h (cf. Lemma 6.39) yields the strong convergence of the broken gradient of each velocity component in $[L^2(\Omega)]^d$ as well as the convergence to zero of the jump seminorms $|u_h|_J$ and $|p_h|_P$.

(v) *Strong convergence of the pressure.* To prove the strong convergence of the pressure, we use, as for the linear Stokes equations, the velocity lifting of p_h provided by Theorem 6.5, say $v_{p_h} \in U$ such that $\nabla \cdot v_{p_h} = p_h$ and set $v_h := \Pi_h v_{p_h}$. We recall that the sequence $v_\mathcal{H}$ is bounded in the $\|\cdot\|_{\text{vel}}$-norm so that there is $v \in U$ given by Proposition 6.27. Then, to assert the strong convergence of the pressure, it suffices to show that $\lim_{h \to 0} b_h(v_h, p_h) = -\|p\|_P^2$. We obtain

$$b_h(v_h, p_h) = \nu a_h(u_h, v_h) + t_h(u_h, u_h, v_h) - \int_\Omega f \cdot v_h = \mathfrak{T}_1 + \mathfrak{T}_2 + \mathfrak{T}_3.$$

Clearly,

$$\lim_{h \to 0} (\mathfrak{T}_1 + \mathfrak{T}_3) = \nu a(u, v) - \int_\Omega f \cdot v.$$

Moreover, owing to Lemma 6.49 below,

$$\lim_{h \to 0} t_h(u_h, u_h, v_h) = t'(u, u, v).$$

As a result,

$$\lim_{h \to 0} -b_h(v_h, p_h) = \nu a(u, v) + t'(u, u, v) - \int_\Omega f \cdot v = -b(v, p) = \int_\Omega p \nabla \cdot v = \|p\|_P^2.$$

(v) Finally, under the smallness condition (6.66), the exact solution is unique owing to Theorem 6.43, so that the whole sequence $(u_\mathcal{H}, p_\mathcal{H})$ converges. □

Lemma 6.49 (Asymptotic consistency for t_h). *Let $u_\mathcal{H}$ be a sequence in $U_\mathcal{H}$ such that, for all $l \geq 0$ and all $i \in \{1, \ldots, d\}$, $G_h^l(u_{h,i}) \to \nabla u_i$ in $[L^2(\Omega)]^d$ and $|u_h|_J \to 0$. Let $v_\mathcal{H}$ be another sequence in U_h, bounded in the $\|\cdot\|_{\text{vel}}$-norm. Then, as $h \to 0$, up to a subsequence,*

$$\lim_{h \to 0} t_h(u_h, u_h, v_h) \to t'(u, u, v),$$

where $v \in U$ is given by Proposition 6.27.

Proof. The proof is similar to that of Lemma 6.45. Owing to (6.67),

$$t_h(u_h, u_h, v_h) = \int_\Omega \sum_{i=1}^d u_h \cdot \mathcal{G}_h^{2k}(u_{h,i}) v_{h,i} + \frac{1}{2} \int_\Omega D_h^{2k}(u_h) u_h \cdot v_h$$
$$+ \frac{1}{4} \sum_{F \in \mathcal{F}_h^i} \int_F [\![u_h]\!] \cdot n_F [\![u_h]\!] \cdot [\![v_h]\!] := \mathfrak{T}_1 + \mathfrak{T}_2 + \mathfrak{T}_3.$$

Proceeding as above yields

$$\lim_{h \to 0} (\mathfrak{T}_1 + \mathfrak{T}_2) = t'(u, u, v).$$

Moreover, proceeding as in the proof of Lemma 6.45, we infer

$$|\mathfrak{T}_3| \lesssim h^{(4-d)/2} |u_h|_J^2 |v_h|_J,$$

so that $\mathfrak{T}_3 \to 0$ since the first two factors on the right-hand side tend to zero, while the last factor is bounded. □

To sum up, we observe that the general design conditions to be fulfilled by the discrete trilinear form t_h are Lemma 6.39 (skew-symmetry), Lemma 6.40 (boundedness), and Lemmata 6.45 and 6.49 (asymptotic consistency). Moreover, the skew-symmetry property (6.58) can be generalized to the requirement that t_h be non-dissipative, namely that, for all $w_h \in U_h$, there holds

$$\forall v_h \in U_h, \qquad t_h(w_h, v_h, v_h) \geq 0. \tag{6.68}$$

Such a property is indeed sufficient to derive all the necessary a priori estimates and prove the convergence result of Theorem 6.47. An example of modification of the discrete trilinear form t_h leading to (6.68) is the use of *upwinding* terms by adding to t_h a term of the form

$$\sum_{F \in \mathcal{F}_h^i} \frac{1}{2} \int_F |\{w_h\} \cdot n_F| [\![u_h]\!] \cdot [\![v_h]\!].$$

6.2.4 A Conservative Formulation

Recalling Remark 6.31, the momentum conservation equation (6.45a) can be rewritten in the *conservative form*

$$-\nu \Delta u + \nabla \cdot (u \otimes u) + \nabla p = f.$$

This section briefly discusses this approach both in the continuous and discrete settings.

6.2.4.1 The Continuous Setting

In conservative form, the weak formulation (6.46) can be equivalently reformulated as follows: Find $(u, p) \in X$ such that

$$c((u,p),(v,q)) + \tilde{t}(u,u,v) = \int_\Omega f \cdot v \text{ for all } (v,q) \in X,$$

with the trilinear form $\tilde{t} \in \mathcal{L}(U \times U \times U, \mathbb{R})$ such that

$$\tilde{t}(w,u,v) := -\int_\Omega (w \otimes u) : \nabla v = -\int_\Omega \sum_{i,j=1}^d w_i u_j \partial_j v_i.$$

Indeed, testing with $(0, q)$, both formulations yield that u is divergence-free, while testing with $(v, 0)$, we obtain

$$\nu a(u,v) + b(v,p) + \tilde{t}(u,u,v) = \nu a(u,v) + b(v,p) + t(u,u,v),$$

since, for divergence-free u,

$$\tilde{t}(u,u,v) = -\int_\Omega (u \otimes u) : \nabla v = \int_\Omega (\nabla \cdot (u \otimes u)) \cdot v = \int_\Omega (u \cdot \nabla u) \cdot v = t(u,u,v).$$

For the modified trilinear form $\tilde{t}$, the counterpart of the skew-symmetry property (6.50) is less general since it requires to use the three same arguments in $\tilde{t}$, namely

$$\forall v \in U, \qquad \tilde{t}(v,v,v) = \frac{1}{2}\int_\Omega (\nabla \cdot v)|v|^2,$$

so that

$$\forall v \in V, \qquad \tilde{t}(v,v,v) = 0. \tag{6.69}$$

When working with convective velocities that are not divergence-free, a possible modification of the trilinear form $\tilde{t}$ is to consider, for all $w, u, v \in U$,

$$\tilde{t}'(w,u,v) = \tilde{t}(w,u,v) - \frac{1}{2}\int_\Omega (w \cdot u)\nabla \cdot v$$

$$= -\int_\Omega (w \otimes u):\nabla v - \frac{1}{2}\int_\Omega (w \cdot u)\nabla \cdot v,$$

so that

$$\forall v \in U, \qquad \tilde{t}'(v,v,v) = 0.$$

Finally, $(u,p) \in X$ solves (6.46) if and only if $(u, \tilde{p} := p - \frac{1}{2}|u|^2)$ is such that

$$c((u,\tilde{p}),(v,q)) + \tilde{t}'(u,u,v) = \int_\Omega f \cdot v \text{ for all } (v,q) \in X.$$

Note that the kinetic energy $\frac{1}{2}|u|^2$ is subtracted from, and not added to, the pressure p, so that the modified pressure $\tilde{p}$ differs from the Bernoulli pressure. Testing with $(0,q)$, both formulations yield that u is divergence-free, while testing with $(v,0)$, we obtain

$$\nu a(u,v) + b(v,\tilde{p}) + \tilde{t}'(u,u,v) = \nu a(u,v) + b(v,p) + \frac{1}{2}\int_\Omega |u|^2 \nabla\cdot v + \tilde{t}'(u,u,v)$$
$$= \nu a(u,v) + b(v,p) + \tilde{t}(u,u,v).$$

This approach based on the modified trilinear form $\tilde{t}'$ and the corresponding pressure modification has been hinted to by Cockburn, Kanschat, and Schötzau [102] and has been further investigated by the authors [131].

6.2.4.2 The Discrete Setting

In the context of dG methods, the main advantage of the above approach is to yield a locally conservative formulation, while ensuring the discrete counterpart of property (6.69). The discrete counterpart of the modified triliner form $\tilde{t}'$ can be designed by setting for all $w_h, u_h, v_h \in U_h$ (see [131]),

$$\tilde{t}_h(w_h, u_h, v_h) = -\int_\Omega (w_h \otimes u_h) : \nabla_h v_h + \sum_{F\in\mathcal{F}_h^i} \int_F \{u_h\}\cdot n_F \{w_h\}\cdot [\![v_h]\!]$$
$$- \frac{1}{2}\int_\Omega (w_h \cdot u_h)\nabla_h\cdot v_h + \frac{1}{2}\sum_{F\in\mathcal{F}_h^i}\int_F \{w_h\cdot u_h\}[\![v_h]\!]\cdot n_F,$$

or, equivalently,

$$\tilde{t}_h(w_h, u_h, v_h) = -\int_\Omega \sum_{i=1}^d w_{h,i} u_h \cdot \mathcal{G}_h^{2k}(v_{h,i}) - \frac{1}{2}\int_\Omega (w_h\cdot u_h) D_h^{2k}(v_h)$$
$$- \frac{1}{4}\sum_{F\in\mathcal{F}_h^i}\int_F [\![u_h]\!]\cdot n_F [\![w_h]\!]\cdot [\![v_h]\!].$$

It is readily seen that the discrete trilinear form satisfies Lemma 6.40 (boundedness), and Lemmata 6.45 and 6.49 (asymptotic consistency). As in the continuous case, skew-symmetry requires to use the three same arguments, namely,

$$\forall v_h \in U_h, \qquad \tilde{t}_h(v_h, v_h, v_h) = 0. \tag{6.70}$$

This property is sufficient to derive all the necessary a priori estimates and prove the convergence result of Theorem 6.47, but this property is not sufficient to prove uniqueness for the continuous problem. Moreover, the fact that the three same arguments must be used in (6.70) indicates (as reflected by numerical experiments) weaker stability than with the use of Temam's modification. Another potential drawback is that, loosely speaking, the modification of the pressure with the kinetic energy adds high frequencies to the pressure field.

Yet, using $\tilde{t}_h$ offers one attractive feature, namely that the discrete problem is locally conservative. Let $T \in \mathcal{T}_h$ and let $\xi \in [\mathbb{P}_d^k(T)]^d$ with Cartesian components $(\xi_i)_{1 \le i \le d}$. Then, the local formulation of the discrete momentum conservation equation takes the form

$$\int_T \sum_{i=1}^d \nu G_h^l(u_{h,i}) \cdot \nabla \xi_i - \int_T (u_h \otimes u_h) : \nabla \xi - \int_T \left(\tilde{p}_h + \tfrac{1}{2}|u_h|^2\right) \nabla \cdot \xi$$
$$+ \sum_{F \in \mathcal{F}_T} \epsilon_{T,F} \int_F \left[\nu \phi_F^{\mathrm{diff}}(u_h) + \phi_F^{\mathrm{conv}}(u_h) + \phi_F^{\mathrm{grad}}(p_h)\right] \cdot \xi = \int_T f \cdot \xi,$$

where the fluxes $\phi_F^{\mathrm{grad}}(p_h)$ and $\phi_F^{\mathrm{diff}}(u_h)$ are defined by (6.28) and (6.30), respectively, $\epsilon_{T,F} = n_T \cdot n_F$, and where the convective flux $\phi_F^{\mathrm{conv}}(u_h)$ is defined as

$$\phi_F^{\mathrm{conv}}(u_h) := \begin{cases} \{\!\{u_h\}\!\} \cdot n_F \{\!\{u_h\}\!\} + \tfrac{1}{2}\{\!\{|u_h|^2\}\!\} n_F & \text{if } F \in \mathcal{F}_h^i, \\ \tfrac{1}{2}|u_h|^2 n & \text{if } F \in \mathcal{F}_h^b. \end{cases}$$

Similarly, letting $\zeta \in \mathbb{P}_d^k(T)$, the local formulation of the discrete mass conservation equation takes the form

$$-\int_T u_h \cdot \nabla \zeta + \sum_{F \in \mathcal{F}_T} \epsilon_{T,F} \int_F \phi_F^{\mathrm{div}}(u_h, p_h) \zeta = 0,$$

with the fluxes

$$\phi_F^{\mathrm{div}}(u_h, p_h) := \begin{cases} \{\!\{u_h\}\!\} \cdot n_F + \nu^{-1} h_F [\![p_h]\!] & \text{if } F \in \mathcal{F}_h^i, \\ 0 & \text{if } F \in \mathcal{F}_h^b. \end{cases}$$

6.3 The Unsteady Case

In this section, we present a pressure-correction algorithm for the unsteady INS equations discretized using discontinuous velocities and continuous pressures. Fully implicit time discretizations of the INS equations coupling velocity and pressure yield linear systems that are often difficult to solve efficiently. Pressure-correction methods, inspired by the pioneering works of Chorin [90, 91] and Temam [291], circumvent this problem by dealing with convective-diffusive momentum transport and incompressibility in two separate steps. This approach leads to solving, at each time step, first a system of convection-diffusion equations and then a projection problem which takes the form of a Poisson problem with homogeneous Neumann boundary conditions for the pressure time increment. The method discussed herein has been proposed by Botti and Di Pietro [47] in the context of dG methods, and is closely inspired by the work of Guermond and Quartapelle [180] for conforming finite element space discretizations.

6.3.1 The Continuous Setting

For a given finite time $t_F > 0$ and an initial velocity field $u_0 \in U$, we consider the *unsteady INS equations* in the form

$$\partial_t u - \nu \triangle u + (u \cdot \nabla) u + \nabla p = f \quad \text{in } \Omega \times (0, t_F), \tag{6.71a}$$
$$\nabla \cdot u = 0 \quad \text{in } \Omega \times (0, t_F), \tag{6.71b}$$
$$u = 0 \quad \text{on } \partial\Omega \times (0, t_F), \tag{6.71c}$$
$$u(\cdot, t = 0) = u_0 \quad \text{in } \Omega, \tag{6.71d}$$
$$\langle p \rangle_\Omega = 0. \tag{6.71e}$$

We recall (cf. Sect. 3.1.1) that, for a function ψ defined on the space-time cylinder $\Omega \times (0, t_F)$, we can consider ψ as a function of the time variable with values in a Hilbert space V of functions of the space variable, in such a way that

$$\psi : (0, t_F) \ni t \longmapsto \psi(t) \equiv \psi(\cdot, t) \in V.$$

We also recall that, for an integer $l \geq 0$, $C^l(V)$ denotes the space of V-valued functions that are l times continuously differentiable in the interval $[0, t_F]$. We are interested here in smooth solutions of the evolutive problem (6.71). For the source term, we assume $f \in C^0([L^2(\Omega)]^d)$, and for the exact solution,

$$u \in C^0(U) \cap C^1([L^2(\Omega)]^d), \qquad p \in C^0(P),$$

with spaces U and P defined by (6.3). Recalling the operator $B \in \mathcal{L}(U, P)$ defined by (6.8) and its adjoint $B^* \in \mathcal{L}(P, U')$, we also consider, for all $w \in U$, the bounded linear operator $A(w) \in \mathcal{L}(U, U')$ such that, for all $y, z \in U$,

$$\langle A(w) y, z \rangle_{U', U} = \nu a(y, z) + t'(w, y, z)$$
$$= \int_\Omega \nu \nabla y {:} \nabla z + \int_\Omega (w \cdot \nabla y) \cdot z + \frac{1}{2} \int_\Omega (\nabla \cdot w) y \cdot z,$$

noticing that Temam's modification of the trilinear form has been used. Then, problem (6.71) can be recast in the equivalent form

$$d_t u(t) + A(u(t)) u(t) + B^* p(t) = f(t), \tag{6.72a}$$
$$B u(t) = 0, \tag{6.72b}$$
$$u(0) = u_0. \tag{6.72c}$$

6.3.2 The Projection Method

In this section, we first present the projection method using the backward Euler scheme for time discretization and $H^1(\Omega)$-conforming discrete pressures. Then, we discuss the reformulation of the projection step as a Poisson problem, the modifications leading to BDF2 time discretization, and the use of discontinuous approximations for the pressure which introduces additional difficulties related to the stabilization of the projection step.

6.3.2.1 Backward Euler and $H^1(\Omega)$-Conforming Discrete Pressures

Let δt be the time step, taken constant for simplicity and such that $t_F = N\delta t$ where N is an integer. For all $n \in \{0, \ldots, N\}$, we define the discrete times $t^n := n\delta t$. A superscript n indicates the value of a function at the discrete time t^n, so that, e.g., $u^n = u(t^n)$ and $f^n = f(t^n)$. Let $\mathcal{T}_h$ belong to an admissible sequence of matching simplicial meshes and let $k \geq 1$. We asume that the meshes are kept fixed in time. We consider the discrete spaces

$$U_h := [\mathbb{P}_d^k(\mathcal{T}_h)]^d, \qquad \mathcal{P}_h := \mathcal{P}_{d,0}^k(\mathcal{T}_h), \tag{6.73}$$

with $\mathcal{P}_{d,0}^k(\mathcal{T}_h)$ defined by (6.43). For $k \geq 2$, the choice $\mathcal{P}_h = \mathcal{P}_{d,0}^{k-1}(\mathcal{T}_h)$ can also be accommodated with minor modifications. We define the operators $A_h : U_h \to U_h$ and $B_h : U_h \to \mathcal{P}_h$ such that, for all $w_h, u_h, v_h \in U_h$ and all $q_h \in \mathcal{P}_h$,

$$(A_h(w_h)u_h, v_h)_{L^2(\Omega)} = \nu a_h(u_h, v_h) + t_h(w_h, u_h, v_h), \tag{6.74}$$

$$(B_h w_h, q_h)_{L^2(\Omega)} = b_h(w_h, q_h), \tag{6.75}$$

with bilinear forms a_h and b_h defined by (6.15) and (6.18), respectively, and trilinear form t_h defined by (6.57). We observe that the transpose operator B_h^t coincides with the usual gradient operator ∇ since $\mathcal{P}_h$ is conforming in $H^1(\Omega)$. Indeed, for all $(w_h, q_h) \in \mathcal{X}_h = U_h \times \mathcal{P}_h$,

$$(B_h w_h, q_h)_{L^2(\Omega)} = \int_\Omega w_h \cdot \nabla q_h = (w_h, B_h^t q_h)_{L^2(\Omega)}.$$

The projection method produces, at each discrete time t^n, $n \in \{1, \ldots, N\}$, a triplet

$$(\widehat{u}_h^n, \widetilde{u}_h^n, p_h^n) \in U_h \times U_h \times \mathcal{P}_h$$

containing two approximations for the velocity, $\widehat{u}_h^n$ and $\widetilde{u}_h^n$, and one for the pressure, p_h^n. For all $n \in \{0, \ldots, N-1\}$, two steps are performed sequentially.

(a) Convection-diffusion step. The approximate velocity $\widehat{u}_h^{n+1}$ is computed from the values of $\widetilde{u}_h^n$ and p_h^n by solving the nonlinear convection-diffusion problem

$$\frac{\widehat{u}_h^{n+1} - \widetilde{u}_h^n}{\delta t} + A_h(\widehat{u}_h^{n+1})\widehat{u}_h^{n+1} + \nabla p_h^n = f_h^{n+1}, \tag{6.76}$$

where $f_h^{n+1} := \Pi_h f^{n+1}$ and Π_h denotes the L^2-orthogonal projection onto U_h. In problem (6.76), the boundary condition (6.71c) is (weakly) enforced on $\widehat{u}_h^{n+1}$, but the divergence-free constraint (6.71b) is not included.

(b) Projection step. The approximate velocity $\widetilde{u}_h^{n+1}$ and the pressure p_h^{n+1} are then obtained by projecting $\widehat{u}_h^{n+1}$ onto the space of discrete functions w_h that satisfy the weak divergence-free constraint $B_h w_h = 0$. Specifically, $\widetilde{u}_h^{n+1}$ and p_h^{n+1} are computed from $\widehat{u}_h^{n+1}$ and p_h^n by solving

$$\frac{\widetilde{u}_h^{n+1} - \widehat{u}_h^{n+1}}{\delta t} + \nabla(p_h^{n+1} - p_h^n) = 0, \tag{6.77a}$$

$$B_h \widetilde{u}_h^{n+1} = 0. \tag{6.77b}$$

The condition $B_h \widetilde{u}_h^{n+1} = 0$ implies that $\widetilde{u}_h^{n+1}$ is weakly divergence-free.

The algorithm is initialized by prescribing values for $\widetilde{u}_h^0$ and p_h^0 inferred from (a projection of) the initial condition (6.71d). Whenever an initial guess p^0 for the pressure is available, one possibility is to use the Stokes projection by solving the problem: Find $(w_h, r_h) \in \mathcal{X}_h$ such that

$$a_h(w_h, v_h) + b_h(v_h, r_h) = a_h(u^0, v_h) + b_h(v_h, p^0) \quad \forall v_h \in U_h,$$
$$b_h(w_h, q_h) = b_h(u^0, q_h) \quad \forall q_h \in \mathcal{P}_h.$$

The algorithm is then initialized by setting $\widetilde{u}_h^0 = w_h$ and $p_h^0 = r_h$.

6.3.2.2 The Projection Step As a Poisson Problem

The projection step (6.77) can be recast as a Poisson problem with homogeneous Neumann boundary conditions for the pressure increment $\delta p_h^{n+1} := p_h^{n+1} - p_h^n$ plus an explicit formula for the velocity approximation $\widetilde{u}_h^{n+1}$. Indeed, applying the operator B_h to (6.77a) and using (6.77b), we obtain

$$B_h \nabla \delta p_h^{n+1} = \delta t^{-1} B_h \widehat{u}_h^{n+1}.$$

Using the expression (6.19) for the discrete bilinear form b_h and observing that, for all $q_h \in \mathcal{P}_h$, $[\![q_h]\!] = 0$ across interfaces, we obtain the following problem for the pressure increment: Find $\delta p_h^{n+1} \in \mathcal{P}_h$ such that, for all $q_h \in \mathcal{P}_h$,

$$(B_h \nabla (\delta p_h^{n+1}), q_h)_{L^2(\Omega)} = \int_\Omega \nabla(\delta p_h^{n+1}) \cdot \nabla q_h = \delta t^{-1} (B_h \widehat{u}_h^{n+1}, q_h)_{L^2(\Omega)}. \quad (6.78)$$

The expression on the right-hand side can be computed using the value of $\widehat{u}_h^{n+1}$ resulting from (6.76). The velocity approximation $\widetilde{u}_h^{n+1}$ then results from

$$\widetilde{u}_h^{n+1} = \widehat{u}_h^{n+1} - \delta t \nabla \delta p_h^{n+1}. \quad (6.79)$$

We observe that (6.78) is a Poisson problem with homogeneous Neumann boundary conditions. This problem is well-posed owing to the zero-mean condition (6.71e).

6.3.2.3 Reformulation of the Convection-Diffusion Step

The convection-diffusion step (6.76) can be reformulated by eliminating the velocity $\widetilde{u}_h^n$ and introducing an exptrapolated pressure $\widehat{p}_h^n$. More specifically, plugging (6.79) at step n into (6.76), we obtain

$$\delta_t^{(1)} \widehat{u}_h^{n+1} + A_h(\widehat{u}_h^{n+1}) \widehat{u}_h^{n+1} = f_h^{n+1} - \nabla \widehat{p}_h^n,$$

where, for all $n \in \{1, \ldots, N\}$, we have set

$$\delta_t^{(1)} \widehat{u}_h^n = \frac{\widehat{u}_h^n - \widehat{u}_h^{n-1}}{\delta t}, \qquad \widehat{p}_h^n = 2p_h^n - p_h^{n-1}.$$

These expressions make it clear that the backward Euler method is used to discretize the time derivative of the velocity, while a first-order extrapolation is used to reconstruct the pressure at the discrete time t^{n+1}.

6.3. The Unsteady Case

6.3.2.4 BDF2 Time Discretization

To enhance time accuracy, a possible choice, considered by Guermond, Minev, and Shen [179] in the conforming case and by Botti and Di Pietro [47] in the dG case, is to use the BDF2 method (cf. Sect. 4.7.4), which amounts to taking, for all $n \in \{2,\ldots,N\}$,

$$\delta_t^{(2)} \widehat{u}_h^n = \delta t^{-1} \left(\frac{3}{2} \widehat{u}_h^n - 2\widehat{u}_h^{n-1} + \frac{1}{2} \widehat{u}_h^{n-2} \right),$$

$$\widehat{p}_h^n = \frac{7}{3} p_h^n - \frac{5}{3} p_h^{n-1} + \frac{1}{3} p_h^{n-2}.$$

This choice allows one, in particular, to enhance the accuracy in time. When using the BDF2 formula, the values at discrete times t^0 and t^1 are needed to initialize the algorithm. One possibility is to initialize $\widetilde{u}_h^0$ and p_h^0 as above, and perform a first time step using the Crank–Nicolson method (cf. Sect. 4.7.4).

6.3.2.5 Discontinuous Pressures

We briefly pinpoint the modifications of the projection step (6.77) required to deal with discontinuous pressures. The discrete velocity space U_h is still defined by (6.73), and we consider the pressure discrete space

$$P_h := \mathbb{P}_{d,0}^k(\mathcal{T}_h).$$

Let A_h be defined as in (6.74) and let $\tilde{B}_h$ be the straightforward extension of B_h to $U_h \times P_h$ such that, for all $w_h \in U_h$ and all $q_h \in P_h$,

$$(\tilde{B}_h w_h, q_h)_{L^2(\Omega)} = b_h(w_h, q_h).$$

Since P_h is not $H^1(\Omega)$-conforming, the transpose operator $\tilde{B}_h^t$ no longer coincides with the usual gradient operator. Moreover, stabilization is required for the well-posedness of the projection step. More specifically, for a given time step $n \ge 0$, the projection step reads: Find $\delta p_h^{n+1} \in P_h$ such that, for all $q_h \in P_h$,

$$(\tilde{B}_h \tilde{B}_h^t \delta p_h^{n+1}, q_h)_{L^2(\Omega)} + s_h(\delta p_h^{n+1}, q_h) = \delta t^{-1} (\tilde{B}_h \widehat{u}_h^{n+1}, q_h)_{L^2(\Omega)},$$

where $\widehat{u}_h^{n+1}$ has a known value resulting from the convection-diffusion step (6.76) and the stabilization bilinear form s_h has to be specified. Using the definition of $\tilde{B}_h$, it is readily inferred that, for all $r_h, q_h \in P_h$ and $l \in \{k-1, k\}$,

$$(\tilde{B}_h \tilde{B}_h^t r_h, q_h)_{L^2(\Omega)} = \int_\Omega \mathcal{G}_h^l(r_h) \cdot \mathcal{G}_h^l(q_h),$$

where we recall that, for all $q_h \in P_h$, the modified discrete gradient is such that

$$\mathcal{G}_h^l(q_h) = \nabla_h q_h - \mathcal{R}_h^l(q_h), \qquad \mathcal{R}_h^l(q_h) := \sum_{F \in \mathcal{F}_h^i} r_F^l(\llbracket q_h \rrbracket).$$

According to the discussion at the end of Sect. 4.2.1, the bilinear form s_h can be chosen so as to obtain the SIP method as follows:

$$s_h(r_h, q_h) := \sum_{F \in \mathcal{F}_h^i} \frac{\eta}{h_F} \int_F [\![r_h]\!][\![q_h]\!] - \int_\Omega \mathcal{R}_h^l(r_h) \cdot \mathcal{R}_h^l(q_h).$$

Indeed, using the definition of lifting operators, we obtain, for all $r_h, q_h \in P_h$,

$$\int_\Omega \mathcal{G}_h^l(r_h) \cdot \mathcal{G}_h^l(q_h) + s_h(r_h, q_h) = \int_\Omega \nabla_h r_h \cdot \nabla_h q_h + \sum_{F \in \mathcal{F}_h^i} \frac{\eta}{h_F} \int_F [\![r_h]\!][\![q_h]\!]$$

$$- \sum_{F \in \mathcal{F}_h^i} \int_\Omega (\{\!\{\nabla_h r_h\}\!\} \cdot \mathbf{n}_F [\![q_h]\!] + [\![r_h]\!] \{\!\{\nabla_h q_h\}\!\} \cdot \mathbf{n}_F),$$

which corresponds to the SIP bilinear form when a Neumann boundary condition is prescribed; cf. (4.16) with $\gamma = 0$.

6.3.2.6 Numerical Examples

In practice, the use of discontinuous pressures can significantly increase the effort required to solve the projection step. To illustrate this point, we consider the three-dimensional solution proposed by Ethier and Steinman [152]

$$u_1 = -a \left[e^{ax_1} \sin(ax_2 + bx_3) + e^{ax_3} \cos(ax_1 + bx_2) \right] e^{-\nu b^2 t},$$

$$u_2 = -a \left[e^{ax_2} \sin(ax_3 + bx_1) + e^{ax_1} \cos(ax_2 + bx_3) \right] e^{-\nu b^2 t},$$

$$u_3 = -a \left[e^{ax_3} \sin(ax_1 + bx_2) + e^{ax_2} \cos(ax_3 + bx_1) \right] e^{-\nu b^2 t},$$

$$p = -\frac{a^2}{2} \Big[e^{2ax_1} + e^{2ax_2} + e^{2ax_3}$$

$$+ 2 \sin(ax_1 + bx_2) \cos(ax_3 + bx_1) e^{a(x_2 + x_3)}$$

$$+ 2 \sin(ax_2 + bx_3) \cos(ax_1 + bx_2) e^{a(x_3 + x_1)}$$

$$+ 2 \sin(ax_3 + bx_1) \cos(ax_2 + bx_3) e^{a(x_1 + x_2)} \Big] e^{-2\nu b^2 t} - \overline{p},$$

with $a = \pi/4$, $b = \pi/2$, and $\overline{p}$ is chosen so as to satisfy the zero-mean constraint (6.71e). An inspection of the fifth and seventh columns of Table 6.1 displaying percentages of CPU times reveals that the computational effort required to solve the projection step can become dominant when discontinuous pressures are used.

We present now some numerical examples to assess convergence rates in space and time for various choices of the discrete spaces. To evaluate the convergence rate in time, we consider the Taylor vortex test case. The unsteady INS equations are solved in the time-space cylinder $\Omega \times (0.1, 6.1)$ with $\Omega = (-\pi/2, \pi/2)^2$. Both the Dirichlet boundary condition and the initial condition are deduced from the

6.3. The Unsteady Case

Table 6.1: Comparison of the use of discontinuous and continuous pressures on the Ethier–Steinman test case. Linear systems are solved using the GMRes algorithm. Legend: `ksp` dim. of Krilov space, `nit`$_1$ average number of GMRes iterations per Newton iteration, `nit`$_2$ average number of GMRes iterations

δt	ksp	Avg time/step (s)	Convection-diffusion		Projection	
			nit$_1$	% time	nit$_2$	% time
		Continuous pressures, $k=2$				
0.1	150	206	110+70	49	4,300	51
	200	113		79	700	21
0.05	150	193	70+40	39	4,200	61
	200	100		76	600	24
0.025	150	135	50+20	47	2,200	53
	200	80		76	490	24
0.0125	150	111	30+10	48	1,800	52
	200	77		75	490	25
		Discontinuous pressures, $k=2$				
0.1	200	589	110+70	16	>8,000	84
	400	201		48	1,000	52
0.05	200	543	70+40	13	>8,000	87
	400	178		41	900	59
0.025	200	612	50+20	10	>8,000	90
	400	176		37	900	63
0.0125	200	667	30+10	9	>8,000	81
	400	183		30	1,100	70

exact solution
$$u_1 = -\cos(\pi x_1)\sin(\pi x_2)e^{-2\pi\nu t},$$
$$u_2 = \sin(\pi x_1)\cos(\pi x_2)e^{-2\pi\nu t},$$
$$p = -\cos(2\pi x_1)\cos(2\pi x_2)e^{-4\pi\nu t} - \overline{p},$$

with constant $\overline{p}$ adjusted so as to satisfy the zero-mean constraint (6.71e). To ensure that the time error dominates over the space error, we solve the discrete problem on a very fine 300×300 quadrilateral mesh. The time steps are in $\{0.2 \times 2^{-i}\}_{0 \le i \le 4}$, while $\nu \in \{10^{-2}, 10^{-3}, 10^{-4}\}$. Figure 6.1 collects the results for the BDF2 time discretization. We observe, in particular, that second-order convergence in time is achieved for the velocity and the pressure, except for $dt = 0.2 \times 2^{-4}$ and $\nu = 10^{-4}$, where the space error is no longer negligible with respect to the time error.

Finally, the convergence in space is assessed using the steady test case proposed by Kovasznay [214]. The space domain is $\Omega = (-0.5, 1.5) \times (0, 2)$. The pressure-correction method is used to integrate over a sufficiently long time to reach a steady-state solution. The exact solution is

$$u_1 = 1 - e^{\lambda x_1}\cos(2\pi x_2), \quad u_2 = \frac{\lambda}{2\pi}e^{\lambda x_1}\sin(2\pi x_2), \quad p = -\frac{1}{2}e^{2\lambda x_1} - \overline{p},$$

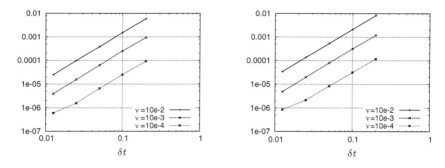

Fig. 6.1: Convergence rates in time, Taylor test case (*left*) $\|u(t_\mathrm{F}) - u_h(t_\mathrm{F})\|_{[L^2(\Omega)]^d}$ (*right*) $\|p(t_\mathrm{F}) - p_h(t_\mathrm{F})\|_{L^2(\Omega)}$

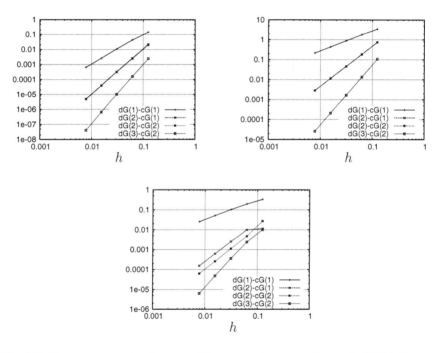

Fig. 6.2: Convergence rates in space, Kovasznay test case. Legend: dG(k) discontinuous polynomial spaces of degree $\leq k$, cG(k) continuous polynomial spaces of degree $\leq k$ (*top left*) $\|u - u_h\|_{[L^2(\Omega)]^d}$ (*top right*) $\|\nabla u - \nabla_h u_h\|_{[L^2(\Omega)]^{d,d}}$ (*bottom*) $\|p - p_h\|_{L^2(\Omega)}$

6.3. The Unsteady Case

with constant $\bar{p}$ adjusted so as to satisfy the zero-mean constraint (6.71e), $\lambda = 1/2\nu - \left(1/4\nu^2 + 4\pi^2\right)^{1/2}$, and $\nu = 0.025$. Optimal convergence rates are observed when using polynomial order k for the velocity and k or $(k-1)$ for the pressure. We notice, however, that equal-order approximations are beneficial in terms of accuracy for the pressure, as reflected by the dG(2)-cG(1) and dG(2)-cG(2) errors in Fig. 6.2.

Chapter 7
Friedrichs' Systems

Symmetric positive systems of first-order PDEs were introduced by Friedrichs in 1958 [163] as a means to handle within a single functional framework transonic flow problems which are partly elliptic and partly hyperbolic in different parts of the computational domain. Friedrichs introduced a very elegant technique to characterize admissible boundary conditions for such systems based on a (nonuniquely defined) nonnegative matrix-valued boundary field with certain algebraic properties. The formalism introduced by Friedrichs turns out to be sufficiently large to encompass a wide range of model problems. Those considered herein are, in the steady case, advection-reaction equations and various elliptic PDEs written in mixed form, e.g., the div-grad problem related to diffusion, the linear elasticity equations in the stress-displacement formulation, and the curl-curl problem related to Maxwell's equations in the diffusive regime.

In the steady case, the symmetric positive systems of first-order PDEs introduced by Friedrichs lend themselves quite nicely to space discretization by dG methods. The analysis of dG methods in the context of Friedrichs' systems was started in the 1970s by Lesaint [227,228] and Lesaint and Raviart [229], and the analysis was further refined a decade later by Johnson and Pitkäranta [204] for the advection-reaction equation; see also Johnson, Nävert, and Pitkäranta [203]. A systematic treatment can be found in a recent series of papers by Ern and Guermond [142–145], from which most of the material in this chapter is inspired. An alternative presentation can be found in the PhD thesis of Jensen [201].

The first goal of this chapter is to derive and analyze various dG methods to approximate in space steady Friedrichs' systems. In doing so, our aim is also to revisit in a unified setting various ideas presented in the previous chapters. Although the presentation is to a large extent self-contained, we refer the reader to [142–145] for various aspects that are not covered herein for the sake of brevity. We first describe the basic ingredients to formulate Friedrichs' systems and present various examples to illustrate the generality of the formalism. Then, using the ideas of [145], we present a mathematical setting so as to formulate Friedrichs' systems as well-posed problems with weakly enforced boundary conditions. The next step deals with the derivation and analysis of dG methods for such problems. A dG method with minimal stability is derived following the approach based on centered fluxes considered in Sect. 2.2 for the advection-reaction equation. The stability of the approximation can be tightened by means of suitable interface and boundary operators, similarly to the approach based on upwinding considered in Sect. 2.3. Then, we examine in more detail the situation where the Friedrichs' system possesses a two-field structure related to the mixed formulation of an elliptic PDE. Finally, we address the approximation of unsteady Friedrichs' systems using explicit time-marching schemes for time discretization and dG methods for space discretization.

7.1 Basic Ingredients and Examples

The purpose of this section is to present the basic ingredients to formulate Friedrichs' systems. The precise mathematical meaning of the PDE system and its boundary conditions are specified in Sect. 7.2. We also present various examples of Friedrichs' systems to emphasize the generality of the formalism.

7.1.1 Basic Ingredients

Let $m \geq 1$ be an integer which corresponds to the number of scalar-valued PDEs in the system. A Friedrichs' system is formulated using $(d+1)$ $\mathbb{R}^{m,m}$-valued fields defined in the domain Ω, say $\mathcal{A}^0, \mathcal{A}^1, \ldots, \mathcal{A}^d$. For convenience, we set

$$\mathcal{X} := \sum_{k=1}^{d} \partial_k \mathcal{A}^k, \qquad (7.1)$$

where ∂_k denotes the distributional derivative in the kth spatial direction; the field $\mathcal{X}$ can be loosely interpreted as the divergence associated with the fields $\mathcal{A}^1, \ldots, \mathcal{A}^d$.

The assumptions on the fields $\mathcal{A}^0, \mathcal{A}^1, \ldots, \mathcal{A}^d$ are

$$\forall k \in \{0, \ldots, d\}, \quad \mathcal{A}^k \in [L^\infty(\Omega)]^{m,m} \quad \text{and} \quad \mathcal{X} \in [L^\infty(\Omega)]^{m,m}, \qquad (7.2\text{a})$$

$$\forall k \in \{1, \ldots, d\}, \quad \mathcal{A}^k = (\mathcal{A}^k)^t \quad \text{a.e. in } \Omega, \qquad (7.2\text{b})$$

$$\exists \mu_0 > 0, \quad \mathcal{A}^0 + (\mathcal{A}^0)^t - \mathcal{X} \geq 2\mu_0 \mathbb{1}_m \quad \text{a.e. in } \Omega, \qquad (7.2\text{c})$$

where $\mathbb{1}_m$ denotes the identity matrix in $\mathbb{R}^{m,m}$ and the superscript ^{t} indicates the transpose; the same superscript is used for the scalar product of vectors in $\mathbb{R}^m$. Henceforth, inequalities between (symmetric) matrices are understood on the associated quadratic forms, i.e., for $\mathcal{G}_1$ and $\mathcal{G}_2$ in $\mathbb{R}^{m,m}$, the inequality $\mathcal{G}_1 \geq \mathcal{G}_2$ means that, for all $\xi \in \mathbb{R}^m$, $\xi^t \mathcal{G}_1 \xi \geq \xi^t \mathcal{G}_2 \xi$. Assumptions (7.2b) and (7.2c) are, respectively, the symmetry and positivity properties mentioned in the introduction.

The material of the previous chapters should have made it clear that dG methods are L^2-based methods. Therefore, it is natural to work in a Hilbertian

7.1. Basic Ingredients and Examples

setting. Let $L := [L^2(\Omega)]^m$ be equipped with its natural scalar product

$$(f,g)_L := \int_\Omega f^t g,$$

and associated norm $\|\cdot\|_L$. We are interested in the differential operator

$$A : [C^1(\overline{\Omega})]^m \ni z \longmapsto Az := A_{(0)}z + A_{(1)}z \in L, \qquad (7.3)$$

where

$$A_{(0)}z := \mathcal{A}^0 z, \qquad A_{(1)}z := \sum_{k=1}^d \mathcal{A}^k \partial_k z. \qquad (7.4)$$

The domain of A is extended to more general functions z in Sect. 7.2.1. We observe that the field $\mathcal{A}^0$ specifies the zero-order term in the differential operator A, while the fields $\mathcal{A}^1, \ldots, \mathcal{A}^d$ specify the first-order terms. We also consider the differential operator

$$\tilde{A} : [C^1(\overline{\Omega})]^m \ni z \longmapsto \tilde{A}z := \tilde{A}_{(0)}z - A_{(1)}z \in L, \qquad (7.5)$$

where $\tilde{A}_{(0)}z := ((\mathcal{A}^0)^t - \mathcal{X})z$. Using integration by parts and the symmetry property (7.2b), we infer that, for all $\varphi, \psi \in [C_0^\infty(\Omega)]^m$,

$$(\varphi, \mathcal{X}\psi)_L + (A_{(1)}\varphi, \psi)_L + (\varphi, A_{(1)}\psi)_L = 0,$$

and since $(\mathcal{A}^0 \varphi, \psi)_L = (\varphi, (\mathcal{A}^0)^t \psi)_L$, we obtain

$$(A\varphi, \psi)_L = (\varphi, \tilde{A}\psi)_L.$$

This identity shows that $\tilde{A}$ is the formal adjoint of A.

In practice, the fields $\mathcal{A}^1, \ldots, \mathcal{A}^d$ are smooth enough to be defined at the boundary $\partial\Omega$. It is then possible to introduce the boundary field $\mathcal{D} : \partial\Omega \to \mathbb{R}^m$ such that, for a.e. $x \in \partial\Omega$,

$$\mathcal{D} := \sum_{k=1}^d n_k \mathcal{A}^k, \qquad (7.6)$$

where $(n_1, \ldots, n_d)$ are the Cartesian components of the outward unit normal n to Ω. We observe that, by construction, $\mathcal{D}$ takes symmetric values. The idea of Friedrichs is to specify boundary conditions using a second boundary field, namely $\mathcal{M} : \partial\Omega \to \mathbb{R}^m$, such that, for a.e. $x \in \partial\Omega$,

$$\mathcal{M} \text{ is nonnegative, i.e., for all } \xi \in \mathbb{R}^m,\ \xi^t \mathcal{M} \xi \geq 0, \qquad (7.7a)$$
$$\mathbb{R}^m = \mathrm{Ker}(\mathcal{D} - \mathcal{M}) + \mathrm{Ker}(\mathcal{D} + \mathcal{M}). \qquad (7.7b)$$

Let $f \in L$. Friedrichs proved in [163] the uniqueness of the strong solution $z \in [C^1(\overline{\Omega})]^m$ such that

$$Az = f \qquad \text{in } \Omega, \qquad (7.8a)$$
$$(\mathcal{D} - \mathcal{M})z = 0 \quad \text{on } \partial\Omega, \qquad (7.8b)$$

and the existence of a so-called ultraweak solution $z \in L$ such that $(z, \tilde{A}y)_L = (f, y)_L$ for all $y \in [C^1(\overline{\Omega})]^m$ such that $(\mathcal{D} + \mathcal{M}^t)y = 0$ on $\partial\Omega$. In the above formalism, boundary conditions are expressed explicitly, and this makes it difficult to obtain existence and uniqueness simultaneously. Furthermore, with an eye toward the design of a dG approximation method, it is desirable to formulate the boundary conditions weakly. These observations are further pursued in Sect. 7.2.

Remark 7.1 (Nonlocal zero-order term). The theory of Friedrichs' systems can also be developed for more general zero-order terms. Indeed, it is possible to consider an operator $A_{(0)} \in \mathcal{L}(L, L)$ more general than $\mathcal{A}^0 z$ in (7.4), e.g., associated with an integral kernel. In this case, the zero-order term is said to be *nonlocal*. Such terms are useful, for instance, when modeling radiative heat transfer.

7.1.2 Examples

This section collects various examples of Friedrichs' systems to highlight the generality of the formalism.

7.1.2.1 Advection-Reaction

The advection-reaction equation has already been studied thoroughly in Chap. 2. Let $f \in L^2(\Omega)$. We consider the PDE

$$\mu z + \beta \cdot \nabla z = f,$$

with $\mu \in L^\infty(\Omega)$ and $\beta \in [L^\infty(\Omega)]^d$ such that $\nabla \cdot \beta \in L^\infty(\Omega)$. Then, the differential operator

$$Az = \mu z + \beta \cdot \nabla z$$

fits the above framework by setting $m = 1$, $\mathcal{A}^0 = \mu$, and, for all $k \in \{1, \ldots, d\}$, $\mathcal{A}^k = \beta_k$, the kth component of the field β, so that $\mathcal{X} = \nabla \cdot \beta$ (cf. (7.1)). Assumptions (7.2a) and (7.2b) clearly hold true. To enforce (7.2c), we assume that there is $\mu_0 > 0$ such that a.e. in Ω,

$$\mu - \frac{1}{2}\nabla \cdot \beta \geq \mu_0 > 0,$$

thereby recovering assumption (2.6).

The boundary field $\mathcal{D}$ is given by

$$\mathcal{D} = \beta \cdot \mathrm{n}.$$

An admissible choice for the boundary field $\mathcal{M}$ is

$$\mathcal{M} = |\beta \cdot \mathrm{n}|.$$

Let us verify (7.7). First, $\mathcal{M}$ is obviously nonnegative, so that (7.7a) holds true. Moreover, for all $x \in \partial\Omega$,

- If $\beta(x)\cdot n(x) > 0$, $\mathrm{Ker}(\mathcal{D} - \mathcal{M}) = \mathbb{R}$ and $\mathrm{Ker}(\mathcal{D} + \mathcal{M}) = \{0\}$.
- If $\beta(x)\cdot n(x) < 0$, $\mathrm{Ker}(\mathcal{D} - \mathcal{M}) = \{0\}$ and $\mathrm{Ker}(\mathcal{D} + \mathcal{M}) = \mathbb{R}$.
- Finally, if $\beta(x)\cdot n(x) = 0$, $\mathrm{Ker}(\mathcal{D} - \mathcal{M}) = \mathrm{Ker}(\mathcal{D} + \mathcal{M}) = \mathbb{R}$.

Hence, in all cases, (7.7b) holds true. The boundary condition $(\mathcal{D}-\mathcal{M})z|_{\partial\Omega} = 0$ amounts to the homogeneous Dirichlet boundary condition $z = 0$ at the inflow boundary $\partial\Omega^-$ defined by (2.2) (recall that $\partial\Omega^- = \{x \in \partial\Omega \mid \beta(x)\cdot n(x) < 0\}$).

7.1.2.2 The Div-Grad Problem and Diffusion-Advection-Reaction

Let $f \in L^2(\Omega)$. The PDE
$$-\triangle u + u = f$$
can be rewritten in the mixed form (cf. Sect. 4.4)
$$\begin{pmatrix} \sigma + \nabla u \\ \nabla\cdot\sigma + u \end{pmatrix} = \begin{pmatrix} 0 \\ f \end{pmatrix}. \tag{7.9}$$

In the mixed formulation, we recall that u is termed the potential and σ the diffusive flux (cf. Definition 4.1). To recover the setting of Friedrichs' systems, we set $m = d+1$ and
$$\mathcal{A}^0 = \begin{bmatrix} \mathbb{1}_d & \mathbb{0}_{d,1} \\ \hline \mathbb{0}_{1,d} & 1 \end{bmatrix}, \qquad \mathcal{A}^k = \begin{bmatrix} \mathbb{0}_{d,d} & e_k \\ \hline (e_k)^t & 0 \end{bmatrix}, \quad k \in \{1,\ldots,d\},$$
where e_k is the kth vector in the Cartesian basis of $\mathbb{R}^d$ and $\mathbb{0}_{s,t}$ denotes the null matrix in $\mathbb{R}^{s,t}$. This yields $\mathcal{X} = \mathbb{0}_{m,m}$. Clearly, assumptions (7.2) hold true. At this stage, it is not possible to discard the zero-order term and consider the problem $-\triangle u = f$ without violating assumption (7.2c). A more general framework circumventing the need for full positivity in $\mathbb{R}^{m,m}$ is presented in Sect. 7.4.3.

The boundary field $\mathcal{D}$ is given by
$$\mathcal{D} = \begin{bmatrix} \mathbb{0}_{d,d} & n \\ \hline n^t & 0 \end{bmatrix}.$$

To enforce the homogeneous Dirichlet boundary condition $u = 0$ on $\partial\Omega$, we can take
$$\mathcal{M} = \begin{bmatrix} \mathbb{0}_{d,d} & -n \\ \hline n^t & 0 \end{bmatrix}. \tag{7.10}$$

Clearly, (7.7a) holds true since $\mathcal{M}$ is skew-symmetric and, hence, nonnegative; moreover, (7.7b) results from the fact that
$$\mathrm{Ker}(\mathcal{D} - \mathcal{M}) = \{(\sigma, 0),\, \sigma \in \mathbb{R}^d\},$$
$$\mathrm{Ker}(\mathcal{D} + \mathcal{M}) = \{(0, u),\, u \in \mathbb{R}\}.$$

The formalism easily incorporates other types of boundary conditions. For instance, the Robin/Neumann boundary condition $\sigma\cdot n = \gamma u$ with $\gamma \in L^\infty(\partial\Omega)$ and $\gamma \geq 0$ a.e. on $\partial\Omega$ is enforced by taking

$$\mathcal{M} = \begin{bmatrix} \mathbb{0}_{d,d} & n \\ -n^t & 2\gamma \end{bmatrix}.$$

Assumption (7.7a) results from the fact that γ is nonnegative, while assumption (7.7b) can be proven by observing that, for all $(\tau, v) \in \mathbb{R}^{d+1}$, $(\tau, v) = (\gamma v n, v) + (\tau - \gamma v n, 0)$ with $(\gamma v n, v) \in \mathrm{Ker}(\mathcal{D}-\mathcal{M})$ and $(\tau - \gamma v n, 0) \in \mathrm{Ker}(\mathcal{D}+\mathcal{M})$.

The extension to the diffusion-advection-reaction equation

$$-\nabla\cdot(\kappa\nabla u) + \beta\cdot\nabla u + \mu u = f,$$

is also straightforward. Here, the diffusion tensor $\kappa \in [L^\infty(\Omega)]^{d,d}$ is assumed to be symmetric and uniformly positive definite, i.e., $\kappa \geq \kappa_0 \mathbb{1}_d$ a.e. in Ω with a real number $\kappa_0 > 0$, while the advective velocity β and the reaction coefficient μ satisfy the assumptions stated above for the advection-reaction problem. The framework of Friedrichs' systems is recovered by setting

$$\mathcal{A}^0 = \begin{bmatrix} \kappa^{-1} & \mathbb{0}_{d,1} \\ \mathbb{0}_{1,d} & \mu \end{bmatrix}, \qquad \mathcal{A}^k = \begin{bmatrix} \mathbb{0}_{d,d} & e_k \\ (e_k)^t & \beta_k \end{bmatrix},$$

so that the boundary field $\mathcal{D}$ becomes

$$\mathcal{D} = \begin{bmatrix} \mathbb{0}_{d,d} & n \\ n^t & \beta\cdot n \end{bmatrix}.$$

The homogeneous Dirichlet boundary condition $u = 0$ on $\partial\Omega$ can be enforced by taking the boundary field $\mathcal{M}$ defined by (7.10). The Robin/Neumann boundary condition $\sigma\cdot n = \gamma u$ with $\gamma \in L^\infty(\partial\Omega)$ and $2\gamma + \beta\cdot n \geq 0$ can be enforced by taking

$$\mathcal{M} = \begin{bmatrix} \mathbb{0}_{d,d} & n \\ -n^t & 2\gamma + \beta\cdot n \end{bmatrix}.$$

In practice, a natural choice is $\gamma = (\beta\cdot n)^\ominus$, the negative part of $\beta\cdot n$ (cf. (2.12)) so that $2\gamma + \beta\cdot n = |\beta\cdot n|$. Then, at the inflow boundary (where $\beta\cdot n < 0$), the Robin boundary condition means that the normal component of the diffusive-advective flux $(\sigma + \beta u)$ is prescribed to zero, while a homogeneous Neumann boundary condition is enforced in the remaining part of the boundary.

7.1.2.3 Linear Elasticity

We consider the equations of compressible linear elasticity in the stress-displacement formulation. Let σ be the stress tensor with values in $\mathbb{R}^{d,d}$,

7.1. Basic Ingredients and Examples

let u be the displacement vector with values in $\mathbb{R}^d$, and let $f \in [L^2(\Omega)]^d$ be the body force. The governing equations can be written as

$$\sigma - \tfrac{1}{d+\lambda}\operatorname{tr}(\sigma)\mathbb{1}_d - \tfrac{1}{2}(\nabla u + (\nabla u)^t) = 0, \qquad (7.11a)$$

$$-\tfrac{1}{2}\nabla\cdot(\sigma + \sigma^t) + \alpha u = f, \qquad (7.11b)$$

where the coefficients α and λ are both in $L^\infty(\Omega)$ and are assumed to be uniformly positive in Ω. The coefficient λ is related to the compressibility of the material; the incompressible limit $\lambda \to 0$ is further discussed in Sect. 7.4.3. The coefficient α is added at this stage to fulfill the positivity assumption (7.2c). It can be removed, as discussed in Sect. 7.4.3. Taking the trace of (7.11a) yields $\tfrac{\lambda}{d+\lambda}\operatorname{tr}(\sigma) = \nabla\cdot u$ so that

$$\sigma = \tfrac{1}{2}(\nabla u + (\nabla u)^t) + \lambda^{-1}(\nabla\cdot u)\mathbb{1}_d,$$

that is, the usual relation between stress and strain tensors is recovered. Moreover, (7.11b) expresses, up to the zero-order term, the equilibrium of forces. We also observe that the symmetry of the stress tensor is not enforced a priori, but is a consequence of (7.11a).

To recover the framework of Friedrichs' systems, the tensor field σ with values in $\mathbb{R}^{d,d}$ is identified with the vector field $\overline{\sigma}$ with values in $\mathbb{R}^{d^2}$ by setting $\overline{\sigma}_{[ij]} = \sigma_{ij}$ with $1 \leq i,j \leq d$ and $[ij] := d(j-1) + i$. To alleviate the notation, we use the same symbol for both fields. Then, problem (7.11) fits the above framework by setting $m = d^2 + d$ and

$$\mathcal{A}^0 = \begin{bmatrix} \mathbb{1}_{d^2} - \tfrac{1}{d+\lambda}\mathcal{Z} & \mathbb{0}_{d^2,d} \\ \hline \mathbb{0}_{d,d^2} & \alpha\mathbb{1}_d \end{bmatrix}, \qquad \mathcal{A}^k = \begin{bmatrix} \mathbb{0}_{d^2,d^2} & \mathcal{E}^k \\ \hline (\mathcal{E}^k)^t & \mathbb{0}_{d,d} \end{bmatrix}, \quad k \in \{1,\ldots,d\},$$

where $\mathcal{Z} \in \mathbb{R}^{d^2,d^2}$ is such that $\mathcal{Z}_{[ij][kl]} = \delta_{ij}\delta_{kl}$ with $1 \leq i,j,k,l \leq d$, and, for all $k \in \{1,\ldots,d\}$, $\mathcal{E}^k \in \mathbb{R}^{d^2,d}$ is such that $\mathcal{E}^k_{[ij],l} = -\tfrac{1}{2}(\delta_{ik}\delta_{jl} + \delta_{il}\delta_{jk})$ with $1 \leq i,j,l \leq d$; here, the δ's are Kronecker symbols. This yields $\mathcal{X} = \mathbb{0}_{m,m}$. Clearly, hypotheses (7.2a) and (7.2b) hold true. To verify (7.2c), we denote by $(\mathcal{A}^0)^{\sigma\sigma} \in \mathbb{R}^{d^2,d^2}$ the upper left block of the matrix $\mathcal{A}^0$ and observe that, for any $\sigma \in \mathbb{R}^{d^2}$,

$$\sigma^t (\mathcal{A}^0)^{\sigma\sigma} \sigma = \sigma^t(\sigma - \tfrac{1}{d+\lambda}\operatorname{tr}(\sigma)\mathbb{1}_d) = \tfrac{\lambda}{d+\lambda}|\sigma|^2_{\ell^2} + \tfrac{d}{d+\lambda}|\sigma - \tfrac{1}{d}\operatorname{tr}(\sigma)\mathbb{1}_d|^2_{\ell^2}, \quad (7.12)$$

where for all $\tau \in \mathbb{R}^{d^2}$, $|\tau|_{\ell^2} := (\tau^t\tau)^{1/2} = (\sum_{1\leq i,j\leq d}\tau^2_{[ij]})^{1/2}$ denotes its Frobenius norm. Then, evaluating $(\sigma,u)^t \mathcal{A}^0 (\sigma,u)$, we obtain positivity on the σ-component (resp., the u-component) since λ (resp., α) is positive.

The boundary field $\mathcal{D}$ and a boundary field $\mathcal{M}$ enforcing homogeneous Dirichlet boundary conditions on the displacement are

$$\mathcal{D} = \begin{bmatrix} \mathbb{0}_{d^2,d^2} & \mathcal{N} \\ \hline \mathcal{N}^t & \mathbb{0}_{d,d} \end{bmatrix}, \qquad \mathcal{M} = \begin{bmatrix} \mathbb{0}_{d^2,d^2} & -\mathcal{N} \\ \hline \mathcal{N}^t & \mathbb{0}_{d,d} \end{bmatrix},$$

where $\mathcal{N} \in \mathbb{R}^{d^2,d}$ is such that, for all $\xi \in \mathbb{R}^d$,

$$\mathcal{N}\xi := -\frac{1}{2}(n\otimes\xi + \xi\otimes n). \tag{7.13}$$

Robin/Neumann boundary conditions can also be accommodated by proceeding as for the div-grad problem.

DG methods for linear elasticity, with a focus on the incompressible limit, have been developed by Hansbo and Larson [183, 184], Wihler [307], and Cockburn, Schötzau, and Wang [109].

7.1.2.4 The Curl-Curl Problem

We close this series of examples with a problem related to the (three-dimensional) Maxwell equations in the *diffusive regime*, namely the curl-curl problem. Let H be the magnetic field with values in $\mathbb{R}^3$, let E be the electric field with values in $\mathbb{R}^3$, and let $f, g \in [L^2(\Omega)]^3$. The governing equations are

$$\begin{cases} \mu H + \nabla\times E = f, \\ \sigma E - \nabla\times H = g, \end{cases}$$

where the coefficients μ and σ are both in $L^\infty(\Omega)$ and are assumed to be uniformly positive in Ω. The above problem fits the framework of Friedrichs' systems by setting $m = 6$ and

$$\mathcal{A}^0 = \begin{bmatrix} \mu\mathbb{1}_3 & \mathbb{0}_{3,3} \\ \mathbb{0}_{3,3} & \sigma\mathbb{1}_3 \end{bmatrix}, \quad \mathcal{A}^k = \begin{bmatrix} \mathbb{0}_{3,3} & \mathcal{R}^k \\ (\mathcal{R}^k)^t & \mathbb{0}_{3,3} \end{bmatrix}, \quad k \in \{1,\ldots,d\},$$

where the entries of the matrices $\mathcal{R}^k \in \mathbb{R}^{3,3}$ are those of the Levi–Civita permutation tensor, i.e., $\mathcal{R}^k_{ij} = \epsilon_{ikj}$ for $1 \leq i,j,k \leq 3$, that is,

$$\mathcal{R}^1 = \begin{pmatrix} 0 & 0 & 0 \\ 0 & 0 & -1 \\ 0 & 1 & 0 \end{pmatrix}, \quad \mathcal{R}^2 = \begin{pmatrix} 0 & 0 & 1 \\ 0 & 0 & 0 \\ -1 & 0 & 0 \end{pmatrix}, \quad \mathcal{R}^3 = \begin{pmatrix} 0 & -1 & 0 \\ 1 & 0 & 0 \\ 0 & 0 & 0 \end{pmatrix}.$$

This yields $\mathcal{X} = \mathbb{0}_{m,m}$. Clearly, assumptions (7.2) hold true.

The boundary field $\mathcal{D}$ and a boundary field $\mathcal{M}$ enforcing homogeneous Dirichlet boundary conditions on the tangential component of the electric field are

$$\mathcal{D} = \begin{bmatrix} \mathbb{0}_{3,3} & \mathcal{T} \\ \mathcal{T}^t & \mathbb{0}_{3,3} \end{bmatrix}, \quad \mathcal{M} = \begin{bmatrix} \mathbb{0}_{3,3} & -\mathcal{T} \\ \mathcal{T}^t & \mathbb{0}_{3,3} \end{bmatrix},$$

where $\mathcal{T} \in \mathbb{R}^{3,3}$ is such that, for all $\xi \in \mathbb{R}^3$,

$$\mathcal{T}\xi = n\times\xi. \tag{7.14}$$

Homogeneous Dirichlet boundary conditions on the tangential component of the magnetic field can be enforced by using $-\mathcal{M}$.

DG methods have been extensively developed for the Maxwell equations in various forms, and we mention some recent work on the subject. DG methods for the mixed form of the Maxwell equations in the time domain have been investigated by Cockburn, Li, and Shu [107] and Cohen, Ferrieres, and Pernet [115]. DG methods for the indefinite time-harmonic Maxwell equations in the frequency-domain have been studied by Houston, Perugia, Schneebeli, and Schötzau in both mixed [198] and non-mixed [197] forms. The analysis of the SIP method for the unsteady Maxwell problem in second-order form has been performed by Grote, Schneebeli, and Schötzau [178]. Finally, dG approximations of the Maxwell eigenvalue problem have been investigated by Buffa, Perugia, and Warburton [62, 63].

7.2 The Continuous Setting

In this section, we give a different mathematical meaning to the solution of problem (7.8) so as to arrive at a problem in weak form that is well-posed. To this purpose, we extend the domain of the differential operator A defined by (7.3) to a larger space, the so-called graph space V with dual space V', and we give a different meaning to the boundary conditions using boundary operators D and M in $\mathcal{L}(V, V')$ instead of the pointwise representation by the boundary fields $\mathcal{D}$ and $\mathcal{M}$. In particular, the boundary condition (7.8b) is replaced by

$$z \in V_0 := \mathrm{Ker}(D - M) \subset V.$$

We consider the problem:

$$\text{Find } z \in V_0 \text{ s.t. } Az = f \text{ in } L. \tag{7.15}$$

To formulate this problem in weak form, we introduce the bilinear form such that, for all $z, y \in V$,

$$a(z, y) := (Az, y)_L + \frac{1}{2} \langle (M - D)z, y \rangle_{V', V}. \tag{7.16}$$

The weak formulation of (7.15) is

$$\text{Find } z \in V \text{ s.t. } a(z, y) = (f, y)_L \text{ for all } y \in V. \tag{7.17}$$

Observe that the boundary condition $z \in V_0$ is weakly enforced in (7.17). Theorem 7.14 below states that under suitable assumptions, problem (7.17) is well-posed.

The main goal of this section is to provide some mathematical insight into the derivation of the weak formulation (7.17) and why it leads to a well-posed problem. The presentation hinges on the ideas of [145]. Assumptions (7.2) are reformulated in an abstract setting leading to the notion of Friedrichs' operators $(A, \tilde{A})$, and a so-called cone formalism is introduced to identify subspaces of the graph space where the Friedrichs' operators are isomorphisms onto L. Then, it is shown that such subspaces can be constructed using a boundary operator with suitable algebraic properties. We observe that we do not work with a single operator A, but instead with an operator pair $(A, \tilde{A})$ where $\tilde{A}$ is the formal adjoint of A. Indeed, an important point is that the bijectivity of the operator A goes hand in hand with that of $\tilde{A}$. As a matter of fact, we prove simultaneously the well-posedness of (7.17) together with that of the adjoint problem:

$$\text{Find } z^* \in V \text{ s.t. } a^*(z^*, y) = (f, y)_L \text{ for all } y \in V, \qquad (7.18)$$

with the bilinear form a^* such that, for all $z, y \in V$,

$$a^*(z, y) := (\tilde{A}z, y)_L + \frac{1}{2}\langle (M^* + D)z, y \rangle_{V', V}, \qquad (7.19)$$

where $M^* \in \mathcal{L}(V, V')$ denotes the adjoint operator of M.

The reader mainly interested in dG approximations can directly jump to Sect. 7.3. The starting point in the design of the dG bilinear form is the expression (7.16) for the bilinear form a and the assumption that, for z and y smooth enough,

$$a(z, y) = (Az, y)_L + \frac{1}{2}\int_{\partial\Omega} y^t (\mathcal{M} - \mathcal{D})z.$$

7.2.1 Friedrichs' Operators

Let L be a Hilbert space equipped with scalar product $(\cdot, \cdot)_L$ and associated norm $\|\cdot\|_L$. Let Υ be a dense subspace of L. We assume that we have at hand two linear operators $A : \Upsilon \to L$ and $\tilde{A} : \Upsilon \to L$ such that

$$\forall (\varphi, \psi) \in \Upsilon \times \Upsilon, \quad (A\varphi, \psi)_L = (\varphi, \tilde{A}\psi)_L, \qquad (7.20\text{a})$$

$$\exists C \text{ s.t. } \forall \varphi \in \Upsilon, \quad \|(A + \tilde{A})\varphi\|_L \le C\|\varphi\|_L. \qquad (7.20\text{b})$$

Owing to (7.20a), the operator $\tilde{A}$ can be viewed as the formal adjoint of A.

In the context of first-order PDEs,

$$\Upsilon = [C_0^\infty(\Omega)]^m, \qquad L = [L^2(\Omega)]^m,$$

and the operator A is defined by (7.3) and (7.4). Integration by parts in (7.20a) yields, for all $\psi \in [C_0^\infty(\Omega)]^m$,

$$\tilde{A}\psi = ((\mathcal{A}^0)^t - \mathcal{X})\psi - \sum_{k=1}^d (\mathcal{A}^k)^t \partial_k \psi. \qquad (7.21)$$

For assumption (7.20b) to hold, the first-order terms in A and $\tilde{A}$ must compensate each other, and this requires the symmetry of the fields $\mathcal{A}^1, \ldots, \mathcal{A}^d$, that is, assumption (7.2b). Then, the operator $\tilde{A}$ defined by (7.21) coincides on $[C_0^\infty(\Omega)]^m$ with that defined by (7.5).

Let V_Υ be the completion of Υ with respect to the scalar product $(\cdot, \cdot)_L + (A\cdot, A\cdot)_L$. Identifying L and its dual, we obtain

$$\Upsilon \hookrightarrow V_\Upsilon \hookrightarrow L \equiv L' \hookrightarrow V_\Upsilon' \hookrightarrow \Upsilon',$$

7.2. The Continuous Setting

with dense and continuous injections, where Υ' is the algebraic dual of Υ, while L' and V'_Υ are topological duals. Owing to (7.20b), the completion of Υ with respect to the scalar product $(\cdot,\cdot)_L + (\tilde{A}\cdot,\tilde{A}\cdot)_L$ is also V_Υ. By a slight abuse of notation, the unique extensions of A and $\tilde{A}$ to V_Υ are still denoted by A and $\tilde{A}$ respectively, yielding $A : V_\Upsilon \to L$ and $\tilde{A} : V_\Upsilon \to L$. Following the terminology of Aubin [18, Sect. 5.5], we say that V_Υ is the minimal domain of A and $\tilde{A}$. Moreover, the adjoint operator of $\tilde{A}$, say $(\tilde{A})^* : L \to V'_\Upsilon$, is the unique extension of $A : V_\Upsilon \to L$ to L. We abuse again the notation by setting $A = (\tilde{A})^* \in \mathcal{L}(L, V'_\Upsilon)$. Similarly, we set $\tilde{A} = T^* \in \mathcal{L}(L, V'_\Upsilon)$. Since $L \subset V'_\Upsilon$, it makes sense to define the *graph space* as

$$V := \{z \in L \mid Az \in L\}, \tag{7.22}$$

Clearly, $V_\Upsilon \subset V$. The graph space is the maximal domain of A. The graph space is also the maximal domain of $\tilde{A}$ since $V = \{z \in L \mid \tilde{A}z \in L\}$ owing to (7.20b).

Lemma 7.2 (Hilbert space). *The graph space V defined by (7.22) is a Hilbert space when equipped with the scalar product $(\cdot,\cdot)_V := (\cdot,\cdot)_L + (A\cdot, A\cdot)_L$.*

Proof. Let $(z_n)_{n\in\mathbb{N}}$ be a Cauchy sequence in V. Then, $(z_n)_{n\in\mathbb{N}}$ and $(Az_n)_{n\in\mathbb{N}}$ are Cauchy sequences in L. Denote by z and y their respective limits in L. Owing to (7.20a) and to the way A and $\tilde{A}$ were extended, $(z_n, \tilde{A}\psi)_L = (Az_n, \psi)_L$ for all $\psi \in V_\Upsilon$. As a result,

$$\langle Az, \psi\rangle_{V'_\Upsilon, V_\Upsilon} = (z, \tilde{A}\psi)_L \leftarrow (z_n, \tilde{A}\psi)_L = (Az_n, \psi)_L \to (y, \psi)_L,$$

proving that Az is in L with $Az = y$. $\square$

Lemma 7.3 (Selfadjointness of $(A + \tilde{A})$). *The operator $(A + \tilde{A})$ is in $\mathcal{L}(L, L)$ and is selfadjoint.*

Proof. The fact that $(A+\tilde{A})$ is in $\mathcal{L}(L, L)$ results from (7.20b) and our extending A and $\tilde{A}$ to $\mathcal{L}(L, V'_\Upsilon)$. To prove that $(A + \tilde{A})$ is selfadjoint, let $z, y \in L$. Since Υ is dense in L, there exist sequences $(z_n)_{n\in\mathbb{N}}$ and $(y_n)_{n\in\mathbb{N}}$ in Υ converging to z and y in L, respectively. Property (7.20a) implies that

$$((A + \tilde{A})z_n, y_n)_L = ((A + \tilde{A})y_n, z_n)_L.$$

Letting $n \to \infty$ and since $(A + \tilde{A})$ is in $\mathcal{L}(L, L)$ yields the assertion. $\square$

We focus on the class of operator pairs $(A, \tilde{A})$ satisfying the following positivity property: There exists a real number $\mu_0 > 0$ such that

$$\forall z \in L, \qquad ((A + \tilde{A})z, z)_L \geq 2\mu_0 \|z\|_L^2, \tag{7.23}$$

that is, $(A + \tilde{A})$ is L-coercive.

Definition 7.4 (Friedrichs' operators). We say that A and $\tilde{A}$ are *Friedrichs' operators* if the pair $(A, \tilde{A})$ satisfies assumptions (7.20) and (7.23).

We introduce the operator $D \in \mathcal{L}(V, V')$ such that, for all $z, y \in V$,

$$\langle Dz, y \rangle_{V', V} := (Az, y)_L - (z, \tilde{A}y)_L. \qquad (7.24)$$

This definition makes sense since both A and $\tilde{A}$ are in $\mathcal{L}(V, L)$.

Lemma 7.5 (Selfadjointness of D). *The operator D defined by (7.24) is selfadjoint.*

Proof. For all $z, y \in V$, we observe that

$$\begin{aligned}\langle Dz, y \rangle_{V', V} - \langle Dy, z \rangle_{V', V} &= (Az, y)_L - (z, \tilde{A}y)_L - (Ay, z)_L + (y, \tilde{A}z)_L \\ &= ((A + \tilde{A})z, y)_L - ((A + \tilde{A})y, z)_L = 0,\end{aligned}$$

owing to Lemma 7.3. □

In the context of symmetric first-order PDEs with $Az = \mathcal{A}_{(0)}z + \mathcal{A}_{(1)}z$ and $\tilde{A}z = ((\mathcal{A}^0)^t - \mathcal{X})z - \mathcal{A}_{(1)}z$, we obtain, for all $z \in L$,

$$((A + \tilde{A})z, z)_L = ((\mathcal{A}^0 + (\mathcal{A}^0)^t - \mathcal{X})z, z)_L.$$

Hence, assumption (7.23) amounts to assumption (7.2c). Moreover, for z, y smooth enough, e.g., $z, y \in C^1(\overline{\Omega})$,

$$\begin{aligned}\langle Dz, y \rangle_{V', V} &= (\mathcal{A}^0 z, y)_L + (\mathcal{A}_{(1)} z, y)_L - (z, (\mathcal{A}^0)^t y)_L + (z, \mathcal{X}y)_L + (z, \mathcal{A}_{(1)} y)_L \\ &= (\mathcal{A}_{(1)} z, y)_L + (z, \mathcal{X}y)_L + (z, \mathcal{A}_{(1)} y)_L \\ &= \int_\Omega \left(\sum_{k=1}^d \partial_k(y^t \mathcal{A}^k z) \right) = \int_{\partial\Omega} y^t \left(\sum_{k=1}^d n_k \mathcal{A}^k \right) z = \int_{\partial\Omega} y^t \mathcal{D} z,\end{aligned}$$

where $\mathcal{D}$ is defined by (7.6).

Remark 7.6 (Kernel and range of D). It is proven in [145] that

$$\text{Ker}(D) = V_\Upsilon, \qquad \text{Im}(D) = V_\Upsilon^\perp,$$

where $V_\Upsilon^\perp = \{y \in V' \mid \forall z \in V_\Upsilon, \langle y, z \rangle_{V', V} = 0\}$ is the so-called annihilator of the space V_Υ. The fact that $\text{Ker}(D) = V_\Upsilon$ means that D is a boundary operator. Since $\text{Im}(D) = V_\Upsilon^\perp$, the range of D is closed. The proof in [145] uses the Riesz–Fréchet representation theorem; cf. Lemma 2.11 for a related proof in the case of the advection-reaction equation.

7.2.2 The Cone Formalism

The goal of this section is to identify a pair of subspaces of V, say (V_0, V_0^*), such that the restricted operators $A : V_0 \to L$ and $\tilde{A} : V_0^* \to L$ are isomorphisms. In the context of PDEs, the spaces V_0 and V_0^* are usually identified by enforcing boundary conditions. However, boundary conditions can be difficult to express since they require a notion of trace for functions in the graph space. Traces

7.2. The Continuous Setting

in the graph space of Friedrichs' operators constitute a subtle issue; we refer the reader to Rauch [265, 266] and Jensen [201] for results in this direction. Here, following [145], we adopt a more intrinsic approach, the so-called cone formalism, whereby sufficient conditions for well-posedness are identified on the pair of spaces (V_0, V_0^*) without relying explicitly on boundary conditions. The links between the cone formalism and the use of boundary operators with suitable algebraic properties is investigated in Sect. 7.2.3 below.

For a subset $Y \subset V'$, identifying the bidual space V'' with V, we let $Y^\perp \subset V$ be the annihilator of the space Y, namely

$$Y^\perp := \{z \in V \mid \forall y \in Y, \ \langle y, z \rangle_{V', V} = 0\}.$$

We also define the cones $C^+ \subset V$ and $C^- \subset V$ such that

$$C^\pm = \{w \in V \mid \pm \langle Dw, w \rangle_{V', V} \geq 0\}.$$

The two key assumptions on which the cone formalism is based are

$$V_0 \subset C^+ \quad \text{and} \quad V_0^* \subset C^-, \tag{7.25a}$$
$$V_0 = D(V_0^*)^\perp \quad \text{and} \quad V_0^* = D(V_0)^\perp. \tag{7.25b}$$

It is clear that, owing to (7.25b), V_0 and V_0^* are closed in V and that $V_\Upsilon = \text{Ker}(D)$ is a subspace of both V_0 and V_0^*.

Lemma 7.7 (*L-coercivity of A and $\tilde{A}$*). *Let $(A, \tilde{A})$ be Friedrichs' operators. Let (V_0, V_0^*) be a pair of spaces satisfying (7.25). Then, A is L-coercive on V_0 and $\tilde{A}$ is L-coercive on V_0^*.*

Proof. For all $z \in V$, we observe that

$$(Az, z)_L = \frac{1}{2}\Big((Az, z)_L + (\tilde{A}z, z)_L\Big) + \frac{1}{2}\Big((Az, z)_L - (\tilde{A}z, z)_L\Big)$$
$$= \frac{1}{2}((A + \tilde{A})z, z)_L + \frac{1}{2}\langle Dz, z \rangle_{V', V}.$$

Let now $z \in V_0$. Using (7.23) and since $V_0 \subset C^+$ owing to (7.25a), we infer

$$(Az, z)_L \geq \mu_0 \|z\|_L^2.$$

We proceed similarly to prove the L-coercivity of $\tilde{A}$ on V_0^*. □

We can now state and prove the main result of this section [145].

Theorem 7.8 (*Well-posedness*). *Let $(A, \tilde{A})$ be Friedrichs' operators. Let (V_0, V_0^*) be a pair of spaces satisfying (7.25). Then, the restricted operators $A : V_0 \to L$ and $\tilde{A} : V_0^* \to L$ are isomorphisms.*

Proof. We only prove that $A : V_0 \to L$ is an isomorphism, the proof for $\tilde{A}$ being similar. To this purpose, we verify the conditions (1.6) and (1.7) of the BNB Theorem (cf. Theorem 1.1) with spaces $X = V_0$ and $Y = L$. Observe that in the present setting, $L \equiv L'$.

(i) Proof of (1.6). Let $z \in V_0$. The L-coercivity of A on V_0 proven in Lemma 7.7 implies
$$\mu_0 \|z\|_L^2 \le (Az,z)_L \le \sup_{y \in L \setminus \{0\}} \frac{(y, Az)_L}{\|y\|_L}.$$
Hence,
$$\|z\|_V \le (\|z\|_L^2 + \|Az\|_L^2)^{1/2} \le (\mu_0^{-1} + 1)^{1/2} \sup_{y \in L \setminus \{0\}} \frac{(y, Az)_L}{\|y\|_L},$$
yielding (1.6).

(ii) Proof of (1.7). Let $y \in L$ be such that, for all $z \in V_0$, $(Az, y)_L = 0$. Since $V_\Upsilon \subset V_0$, we infer, for all $w \in V_\Upsilon$,
$$\langle w, \tilde{A}y \rangle_{V_\Upsilon, V_\Upsilon'} = (Aw, y)_L = 0.$$
As a result, $\tilde{A}y = 0$ in V_Υ', and hence in L since V_Υ is dense in L. This implies that $y \in V$. Owing to the definition (7.24) of D, we infer, for all $z \in V_0$,
$$\langle Dz, y \rangle_{V', V} = (Az, y)_L - (z, \tilde{A}y)_L = 0 - 0 = 0,$$
which means that $y \in D(V_0)^\perp$ and hence $y \in V_0^*$ by (7.25b). Since $\tilde{A}$ is L-coercive on V_0^* owing to Lemma 7.7 and $\tilde{A}y = 0$, this yields $y = 0$. □

Remark 7.9 (Maximality of V_0 and V_0^*). It is proven in [145] that if the pair of spaces (V_0, V_0^*) satisfies (7.25), then V_0 is maximal in the cone C^+, meaning that there is no $x \in V$ such that $V_x = V_0 + \text{span}(x) \subset C^+$ and V_0 is a proper subspace of V_x, and, similarly, V_0^* is maximal in the cone C^-. The converse statement also holds true, as proven recently by Antonić and Burazin [12] using the formalism of Kreĭn spaces. The idea of maximal boundary conditions for positive symmetric PDE systems has been introduced by Lax; we refer the reader to Lax and Phillips [223] and Friedrichs [163] for further discussion.

7.2.3 The Boundary Operator M

We now present a practical way to construct a pair of spaces (V_0, V_0^*) satisfying (7.25). We assume that we have at hand an operator $M \in \mathcal{L}(V, V')$ such that

$$M \text{ is monotone, i.e., for all } z \in V, \langle Mz, z \rangle_{V', V} \ge 0, \qquad (7.26a)$$
$$V = \text{Ker}(D - M) + \text{Ker}(D + M), \qquad (7.26b)$$

and we denote by $M^* \in \mathcal{L}(V, V')$ the adjoint operator of M.

Lemma 7.10 (Identification of V_0 and V_0^*). *Let $M \in \mathcal{L}(V, V')$ satisfy (7.26). Then, setting*
$$V_0 := \text{Ker}(D - M),$$
$$V_0^* := \text{Ker}(D + M^*),$$
the pair of spaces (V_0, V_0^*) satisfies (7.25).

7.2. The Continuous Setting

Proof. Let $z \in V_0$; then, $Dz = Mz$ so that

$$\langle Dz, z\rangle_{V',V} = \langle Mz, z\rangle_{V',V} \geq 0,$$

owing to (7.26a). Hence, $V_0 \subset C^+$. Similarly, we prove that $V_0^* \subset C^-$, completing the proof of (7.25a). Let us now prove that $D(V_0)^\perp \subset V_0^*$. Let $z \in D(V_0)^\perp$. Then, for all $y \in V$, using (7.26b) to write $y = y_+ + y_-$ with $y_\pm \in \text{Ker}(D \pm M)$, we infer

$$\langle (D+M^*)z, y\rangle_{V',V} = \langle (D+M^*)z, y_+\rangle_{V',V} + \langle (D+M^*)z, y_-\rangle_{V',V}$$
$$= \langle (D+M)y_-, z\rangle_{V',V} = 2\langle Dy_-, z\rangle_{V',V} = 0,$$

where we have used the selfadjointness of D, $(D+M)y_+ = 0$, $Dy_- = My_-$, $y_- \in V_0$, and $z \in D(V_0)^\perp$. This proves that $z \in \text{Ker}(D+M^*)$, that is, $z \in V_0^*$. Let us now prove the converse inclusion $V_0^* \subset D(V_0)^\perp$. Let $z \in V_0^*$. Then, for all $y \in V_0$, since $Dy = My$, we infer

$$\langle Dy, z\rangle_{V',V} = \frac{1}{2}\langle (D+M)y, z\rangle_{V',V} = \frac{1}{2}\langle (D+M^*)z, y\rangle_{V',V} = 0,$$

so that $z \in D(V_0)^\perp$. Hence, $V_0^* = D(V_0)^\perp$. The proof of $V_0 = D(V_0^*)^\perp$ is similar. □

Remark 7.11 (Kernel and range of M). It is proven in [145] that, under the assumptions (7.26), there holds

$$\text{Ker}(D) = \text{Ker}(M) = \text{Ker}(M^*),$$
$$\text{Im}(D) = \text{Im}(M) = \text{Im}(M^*).$$

In particular, M is, as D, a boundary operator.

Remark 7.12 (Converse of Lemma 7.10). The converse statement of Lemma 7.10 has been established recently by Antonić and Burazin [12], namely if there is a pair of spaces (V_0, V_0^*) satisfying (7.25), there is $M \in \mathcal{L}(V, V')$ satisfying (7.26).

It is convenient to reformulate Lemma 7.7 in terms of the boundary operator M and the bilinear forms a and a^*.

Lemma 7.13 (L-coercivity of a and a^*). *Let $(A, \tilde{A})$ be Friedrichs' operators, let $M \in \mathcal{L}(V, V')$ satisfy (7.26), and let the bilinear forms a and a^* be defined by (7.16) and (7.19) respectively. Then, a and a^* are L-coercive on V. More precisely, defining, for all $y \in V$, the seminorm $|y|_M := \langle My, y\rangle_{V',V}^{1/2}$, there holds*

$$\forall y \in V, \quad a(y,y) \geq \mu_0 \|y\|_L^2 + \frac{1}{2}|y|_M^2,$$
$$\forall y \in V, \quad a^*(y,y) \geq \mu_0 \|y\|_L^2 + \frac{1}{2}|y|_M^2.$$

Proof. We only prove the result for a since the proof for a^* is similar. Proceeding as in the proof of Lemma 7.7, we observe that, for all $y \in V$,

$$a(y,y) = (Ay,y)_L + \frac{1}{2}\langle (M-D)y, y\rangle_{V',V}$$
$$= \frac{1}{2}((A+\tilde{A})y, y)_L + \frac{1}{2}\langle Dy, y\rangle_{V',V} + \frac{1}{2}\langle (M-D)y, y\rangle_{V',V}$$
$$= \frac{1}{2}((A+\tilde{A})y, y)_L + \frac{1}{2}\langle My, y\rangle_{V',V},$$

whence the assertion. □

7.2.4 The Well-Posedness Result

We have now at our disposal all the necessary tools to establish the well-posedness of problems (7.17) and (7.18).

Theorem 7.14 (Well-posedness). *Let $(A, \tilde{A})$ be Friedrichs' operators, let $M \in \mathcal{L}(V, V')$ satisfy (7.26), and let the bilinear forms a and a^* be defined by (7.16) and (7.19) respectively. Then, problems (7.17) and (7.18) are well-posed. Moreover, the unique solution z of (7.17) satisfies*

$$Az = f \quad \text{in } L, \tag{7.27a}$$
$$(M-D)z = 0 \quad \text{in } V', \tag{7.27b}$$

while the unique solution z^* of (7.18) satisfies

$$\tilde{A}z^* = f \quad \text{in } L,$$
$$(M^* + D)z^* = 0 \quad \text{in } V'.$$

Proof. We only consider problem (7.17) since the proof for problem (7.18) is similar. Owing to Lemma 7.10 and Theorem 7.8, letting $V_0 = \text{Ker}(D - M)$, we infer that the restricted operator $A : V_0 \to L$ is an isomorphism. Let $f \in L$ and let $z \in V_0$ be such that $Az = f$. Then, z solves (7.17) since, for all $y \in V$,

$$a(z,y) = (Az, y)_L + \frac{1}{2}\langle (M-D)z, y\rangle_{V',V} = (f,y)_L + 0 = (f,y)_L.$$

This proves the existence of the solution to (7.17). Uniqueness results from the L-coercivity of the bilinear form a on V established in Lemma 7.13. Finally, the exact solution z satisfies (7.27a) since, for all $\varphi \in \Upsilon$, $(Az - f, \varphi)_L = 0$, while, for all $y \in V$, we obtain

$$\frac{1}{2}\langle (M-D)z, y\rangle_{V',V} = \frac{1}{2}\langle (M-D)z, y\rangle_{V',V} + (Az, y)_L - (f,y)_L$$
$$= a(z,y) - (f,y)_L = 0,$$

i.e., (7.27b) also holds true. □

7.2. The Continuous Setting

7.2.5 Examples

In this section, we review the examples introduced in Sect. 7.1.2. Assumptions (7.2) have already been verified. Thus, setting

$$Az = \mathcal{A}^0 z + \sum_{k=1}^{d} \mathcal{A}^k \partial_k z,$$

$$\tilde{A}z = ((\mathcal{A}^0)^t - \mathcal{X})z - \sum_{k=1}^{d} \mathcal{A}^k \partial_k z,$$

A and $\tilde{A}$ are Friedrichs' operators. It remains to specify the graph space V, the boundary operator D, and a suitable boundary operator M satisfying (7.26) to enforce boundary conditions.

7.2.5.1 Advection-Reaction

The graph space associated with the advection-reaction problem has been identified in (2.8), namely

$$V = \left\{ z \in L^2(\Omega) \mid \beta \cdot \nabla z \in L^2(\Omega) \right\}.$$

In Sect. 2.1.3, it was shown that under the assumptions that $\beta \in [\text{Lip}(\Omega)]^d$ and that the inflow and outflow boundaries are well-separated, functions in the graph space V admit traces in the space $L^2(|\beta \cdot n|; \partial\Omega)$ defined by (2.10), and that the following integration by parts formula holds true (cf. (2.11)): For all $z, y \in V$,

$$\int_\Omega [(\beta \cdot \nabla y)z + (\beta \cdot \nabla z)y + (\nabla \cdot \beta)yz] = \int_{\partial\Omega} (\beta \cdot n)yz.$$

As a result, the boundary operator D admits the following representation: For all $z, y \in V$,

$$\langle Dz, y \rangle_{V',V} = \int_{\partial\Omega} (\beta \cdot n)yz.$$

Furthermore, we can set, for all $z, y \in V$,

$$\langle Mz, y \rangle_{V',V} = \int_{\partial\Omega} |\beta \cdot n|yz.$$

Assumption (7.26a) is evident. To verify assumption (7.26b), we recall that there are two functions ψ^- and ψ^+ in $C^1(\overline{\Omega})$ such that

$$\psi^- + \psi^+ \equiv 1 \text{ in } \Omega, \quad \psi^-|_{\partial\Omega^+} = 0, \quad \psi^+|_{\partial\Omega^-} = 0.$$

Then, for all $z \in V$, letting $z_\pm = \psi^\pm z$, we obtain $z = z_- + z_+$ and $z_\pm \in \text{Ker}(D \mp M)$, thereby proving (7.26b). Finally, owing to the surjectivity of traces (cf. Lemma 2.11),

$$V_0 = \text{Ker}(D - M) = \{z \in V \mid z|_{\partial\Omega^-} = 0\}.$$

7.2.5.2 The Div-Grad Problem

The graph space associated with the div-grad problem is

$$V = H(\text{div}; \Omega) \times H^1(\Omega)$$
$$= \left\{ (z^\sigma, z^u) \in [L^2(\Omega)]^d \times L^2(\Omega) \mid \nabla \cdot z^\sigma \in L^2(\Omega),\ \nabla z^u \in [L^2(\Omega)]^d \right\}.$$

The boundary operator D takes the following form: For all $(\sigma, u), (\tau, v) \in V$,

$$\langle D(\sigma, u), (\tau, v) \rangle_{V', V} = \langle \sigma \cdot n, v \rangle_{-\frac{1}{2}, \frac{1}{2}} + \langle \tau \cdot n, u \rangle_{-\frac{1}{2}, \frac{1}{2}},$$

where $\langle \cdot, \cdot \rangle_{-\frac{1}{2}, \frac{1}{2}}$ denotes the duality pairing between the spaces $H^{-\frac{1}{2}}(\partial \Omega)$ and $H^{\frac{1}{2}}(\partial \Omega)$. For Dirichlet boundary conditions, the boundary operator M can be taken such that, for all $(\sigma, u), (\tau, v) \in V$,

$$\langle M(\sigma, u), (\tau, v) \rangle_{V', V} = \langle \sigma \cdot n, v \rangle_{-\frac{1}{2}, \frac{1}{2}} - \langle \tau \cdot n, u \rangle_{-\frac{1}{2}, \frac{1}{2}}.$$

Assumption (7.26a) clearly holds true since M is skew-symmetric. To verify assumption (7.26b), we observe that, for all $(\sigma, u) \in V$, $(\sigma, 0) \in \text{Ker}(D - M)$ and $(0, u) \in \text{Ker}(D + M)$. Furthermore, since $(\sigma, u) \in \text{Ker}(D - M)$ amounts to $\langle \tau \cdot n, u \rangle_{-\frac{1}{2}, \frac{1}{2}} = 0$ and since normal traces in $H(\text{div}; \Omega)$ are surjective onto $H^{-\frac{1}{2}}(\partial \Omega)$, we infer

$$V_0 = \text{Ker}(D - M) = \{ (\sigma, u) \in V \mid u|_{\partial \Omega} = 0 \}.$$

Whenever σ and τ are smooth enough so that their normal components are in $L^2(\partial \Omega)$, the boundary operators D and M can be represented by the boundary fields $\mathcal{D}$ and $\mathcal{M}$ introduced in Sect. 7.1.2, namely

$$\langle D(\sigma, u), (\tau, v) \rangle_{V', V} = \int_{\partial \Omega} (\tau, v)^t \begin{bmatrix} \mathbb{0}_{d,d} & n \\ \hline n^t & 0 \end{bmatrix} (\sigma, u),$$

and

$$\langle M(\sigma, u), (\tau, v) \rangle_{V', V} = \int_{\partial \Omega} (\tau, v)^t \begin{bmatrix} \mathbb{0}_{d,d} & -n \\ \hline n^t & 0 \end{bmatrix} (\sigma, u).$$

For Robin/Neumann boundary conditions, we can take, for all $(\sigma, u), (\tau, v) \in V$,

$$\langle M(\sigma, u), (\tau, v) \rangle_{V', V} = -\langle \sigma \cdot n, v \rangle_{-\frac{1}{2}, \frac{1}{2}} + \langle \tau \cdot n, u \rangle_{-\frac{1}{2}, \frac{1}{2}} + \int_{\partial \Omega} 2\gamma uv.$$

Finally, the diffusion-advection-reaction equation is accommodated by adding the term $\int_{\partial \Omega} (\beta \cdot n) uv$ to the expression of the boundary operators D and M.

7.2.5.3 Linear Elasticity

The formalism for treating (compressible) linear elasticity problems is quite close to that for the div-grad problem. The graph space is given by

$$V = H_\sigma \times [H^1(\Omega)]^d, \qquad H_\sigma = \left\{ \sigma \in [L^2(\Omega)]^{d,d} \mid \nabla \cdot (\sigma + \sigma^t) \in [L^2(\Omega)]^d \right\},$$

7.2. The Continuous Setting

and the boundary operator D is such that, for all $(\sigma, u), (\tau, v) \in V$,

$$\langle Dz, y\rangle_{V',V} = -\langle \tfrac{1}{2}(\tau + \tau^t)\cdot n, u\rangle_{-\frac{1}{2},\frac{1}{2}} - \langle \tfrac{1}{2}(\sigma + \sigma^t)\cdot n, v\rangle_{-\frac{1}{2},\frac{1}{2}},$$

where $\langle \cdot, \cdot \rangle_{-\frac{1}{2},\frac{1}{2}}$ now denotes the duality pairing between the spaces $[H^{-\frac{1}{2}}(\partial\Omega)]^d$ and $[H^{\frac{1}{2}}(\partial\Omega)]^d$. For Dirichlet boundary conditions, the boundary operator M can be taken such that, for all $(\sigma, u), (\tau, v) \in V$,

$$\langle Mz, y\rangle_{V',V} = \langle \tfrac{1}{2}(\tau + \tau^t)\cdot n, u\rangle_{-\frac{1}{2},\frac{1}{2}} - \langle \tfrac{1}{2}(\sigma + \sigma^t)\cdot n, v\rangle_{-\frac{1}{2},\frac{1}{2}},$$

and the verification of assumptions (7.26) proceeds similarly to the div-grad problem. Using the surjectivity of traces yields, as for the div-grad problem,

$$V_0 = \operatorname{Ker}(D - M) = \{(\sigma, u) \in V \mid u|_{\partial\Omega} = 0\}.$$

Whenever $(\sigma + \sigma^t)$ and $(\tau + \tau^t)$ are smooth enough so that their normal components are in $[L^2(\partial\Omega)]^d$, the boundary operators D and M can be represented by the boundary fields $\mathcal{D}$ and $\mathcal{M}$ introduced in Sect. 7.1.2, namely

$$\langle \mathcal{D}(\sigma, u), (\tau, v)\rangle_{V',V} = \int_{\partial\Omega} (\tau, v)^t \left[\begin{array}{c|c} \mathbb{O}_{d^2,d^2} & \mathcal{N} \\ \hline \mathcal{N}^t & \mathbb{O}_{d,d} \end{array}\right] (\sigma, u),$$

with $\mathcal{N} \in \mathbb{R}^{d^2,d}$ defined by (7.13), and

$$\langle \mathcal{M}(\sigma, u), (\tau, v)\rangle_{V',V} = \int_{\partial\Omega} (\tau, v)^t \left[\begin{array}{c|c} \mathbb{O}_{d^2,d^2} & -\mathcal{N} \\ \hline \mathcal{N}^t & \mathbb{O}_{d,d} \end{array}\right] (\sigma, u).$$

Robin/Neumann boundary conditions can be treated similarly.

7.2.5.4 The Curl-Curl Problem

For the curl-curl problem, the graph space is

$$V = H(\operatorname{curl}; \Omega) \times H(\operatorname{curl}; \Omega),$$

where

$$H(\operatorname{curl}; \Omega) := \{v \in [L^2(\Omega)]^3 \mid \nabla \times v \in [L^2(\Omega)]^3\}.$$

The boundary operator D takes the following form: For all $(H, E), (h, e) \in V$,

$$\langle D(H, E), (h, e)\rangle_{V',V} = (\nabla \times E, h)_{[L^2(\Omega)]^3} - (E, \nabla \times h)_{[L^2(\Omega)]^3}$$
$$+ (H, \nabla \times e)_{[L^2(\Omega)]^3} - (\nabla \times H, e)_{[L^2(\Omega)]^3}.$$

For Dirichlet boundary conditions on the tangential component of the electric field, the boundary operator M can be taken such that, for all $(H, E), (h, e) \in V$,

$$\langle M(H, E), (h, e)\rangle_{V',V} = -(\nabla \times E, h)_{[L^2(\Omega)]^3} + (E, \nabla \times h)_{[L^2(\Omega)]^3}$$
$$+ (H, \nabla \times e)_{[L^2(\Omega)]^3} - (\nabla \times H, e)_{[L^2(\Omega)]^3}.$$

Assumption (7.26a) clearly holds true since M is skew-symmetric. To verify assumption (7.26b), we observe that, for all $(H, E) \in V$, $(H, 0) \in \text{Ker}(D - M)$ and $(0, E) \in \text{Ker}(D + M)$. Assuming that Ω is smooth enough so that $[H^1(\Omega)]^3$ is dense in $H(\text{curl}; \Omega)$, it can be shown (see, e.g., [145, Sect. 5.4]) that

$$V_0 = \text{Ker}(D - M) = \{(H, E) \in V \mid n \times E|_{\partial\Omega} = 0\}.$$

Whenever all the fields are smooth enough so that their tangential components are in $[L^2(\partial\Omega)]^3$, the boundary operators D and M can be represented by the boundary fields $\mathcal{D}$ and $\mathcal{M}$ introduced in Sect. 7.1.2, namely

$$\langle D(H, E), (h, e)\rangle_{V', V} = \int_{\partial\Omega} (h, e)^t \begin{bmatrix} \mathbb{O}_{3,3} & \mathcal{T} \\ \hline \mathcal{T}^t & \mathbb{O}_{3,3} \end{bmatrix} (H, E),$$

with $\mathcal{T} \in \mathbb{R}^{3,3}$ defined by (7.14), and

$$\langle M(H, E), (h, e)\rangle_{V', V} = \int_{\partial\Omega} (h, e)^t \begin{bmatrix} \mathbb{O}_{3,3} & -\mathcal{T} \\ \hline \mathcal{T}^t & \mathbb{O}_{3,3} \end{bmatrix} (H, E).$$

A Dirichlet boundary condition on the tangential component of the magnetic field can be enforced by changing the sign of M.

7.3 Discretization

The goal of this section is to approximate by a dG method the unique weak solution $z \in V$ of the model problem (7.17). We follow the same path of ideas as that developed in Chap. 2 for the advection-reaction equation.

The approximate solution z_h is sought in the discrete space

$$V_h := [\mathbb{P}_d^k(\mathcal{T}_h)]^m,$$

where $\mathbb{P}_d^k(\mathcal{T}_h)$ is the broken polynomial space defined by (1.15) with polynomial degree $k \geq 0$ and $\mathcal{T}_h$ belonging to an admissible mesh sequence. We consider different polynomial approximation orders for the various components of z in Sect. 7.4.

For any subset $\omega \subset \Omega$, where ω is a mesh element, a face or a collection thereof, we denote by $(\cdot, \cdot)_{L,\omega}$ the usual scalar product in $[L^2(\omega)]^m$. The subscript ω is dropped whenever $\omega = \Omega$.

7.3.1 Assumptions on the Data and the Exact Solution

We assume that the fields $\mathcal{A}^1, \ldots, \mathcal{A}^d$ are smooth enough so that

$$\forall T \in \mathcal{T}_h, \quad \forall k \in \{1, \ldots, d\}, \quad \mathcal{A}^k|_T \in [C^{0,\frac{1}{2}}(\overline{T})]^{m,m}, \tag{7.28}$$

with norm bounded uniformly in Ω. In practice, we can proceed as for the heterogeneous diffusion problem discussed in Sect. 4.5. We assume that we are given

7.3. Discretization

a partition $P_\Omega := \{\Omega_i\}_{1 \le i \le N_\Omega}$ of Ω (cf. Assumption 4.43) such that the fields $\mathcal{A}^1|_{\Omega_i}, \ldots, \mathcal{A}^d|_{\Omega_i}$ are in $[C^{0,\frac{1}{2}}(\overline{\Omega}_i)]^{m,m}$ for all $1 \le i \le N_\Omega$, and we require that the mesh sequence $\mathcal{T}_\mathcal{H}$ is compatible with the partition P_Ω (cf. Assumption 4.45). This ensures that (7.28) holds true for all $h \in \mathcal{H}$.

We extend the definition (7.6) of the boundary field $\mathcal{D}$ to all interfaces in $\mathcal{F}_h^i$. Specifically, the field $\mathcal{D}$ is now two-valued at each interface $F \in \mathcal{F}_h^i$ with $F = \partial T_1 \cap \partial T_2$ and is defined such that

$$\mathcal{D}|_{T_i} := \sum_{k=1}^d \mathrm{n}_{T_i,k} \mathcal{A}^k|_{T_i}, \quad i \in \{1,2\},$$

where $(\mathrm{n}_{T_i,1}, \ldots, \mathrm{n}_{T_i,d})$ are the Cartesian components of the outward unit normal n_{T_i} to T_i. Owing to assumption (7.2a), $\sum_{k=1}^d \partial_k \mathcal{A}^k \in [L^\infty(\Omega)]^{m,m}$. A distributional argument (cf., e.g., the proof of Lemma 1.24) then yields, for all $F \in \mathcal{F}_h^i$,

$$\sum_{k=1}^d \mathrm{n}_{F,k} [\![\mathcal{A}^k]\!] = 0,$$

where $(\mathrm{n}_{F,1}, \ldots, \mathrm{n}_{F,d})$ are the Cartesian components of the fixed unit normal n_F to F. Hence,

$$\{\!\{\mathcal{D}\}\!\} = 0 \quad \forall F \in \mathcal{F}_h^i. \tag{7.29}$$

As a result, we can consider the single-valued field $\mathcal{D}_F$ such that, for all $F \in \mathcal{F}_h$,

$$\mathcal{D}_F = \sum_{k=1}^d \mathrm{n}_{F,k} \mathcal{A}^k.$$

As in Sect. 2.2 (cf. Assumption 2.13), we make a slightly more stringent regularity assumption on the exact solution z rather than just belonging to the graph space V. The objective is on the one hand to formulate the boundary conditions using the boundary fields $\mathcal{D}$ and $\mathcal{M}$ rather than the boundary operators D and M and on the other hand to verify the consistency of the approximation by plugging the exact solution into the dG bilinear form.

Assumption 7.15 (Regularity of exact solution and space V_*). *We assume that there is a partition $P_\Omega = \{\Omega_i\}_{1 \le i \le N_\Omega}$ of Ω into disjoint polyhedra such that, for the exact solution z,*

$$z \in V_* := V \cap [H^1(P_\Omega)]^m.$$

In the spirit of Sect. 1.3, we set $V_{*h} := V_* + V_h$.

Owing to the trace inequality (1.18) with $p = 2$, Assumption 7.15 implies that, for all $T \in \mathcal{T}_h$, the restriction $z|_T$ has traces a.e. on each face $F \in \mathcal{F}_T$, and these traces belong to $[L^2(F)]^m$. Moreover, the exact solution satisfies the boundary condition $(\mathcal{M} - \mathcal{D})z = 0$ for all $F \in \mathcal{F}_h^b$. Finally, proceeding as in the proof of Lemma 2.14, it is shown that

$$\mathcal{D}_F [\![z]\!] = 0 \quad \forall F \in \mathcal{F}_h^i. \tag{7.30}$$

7.3.2 Design of the Discrete Bilinear Form

The design of the discrete bilinear form proceeds in two steps, as in Chap. 2 for the advection-reaction equation, namely:

(a) We first formulate a consistent discrete bilinear form enjoying discrete L-coercivity on V_h; this property is the counterpart of Lemma 7.13 for the exact bilinear form a.

(b) Then, we tighten the stability of the discrete bilinear form by adding a least-squares penalty on (certain) interface jumps and boundary values.

In terms of numerical fluxes, the first step can be interpreted as the use of centered fluxes, while the second step leads to upwind-type fluxes. Numerical fluxes are further discussed in Sect. 7.3.4.

7.3.2.1 Step 1: Discrete L-Coercivity with Consistency

Similarly to the advection-reaction equation (cf. Sect. 2.2), we consider the following discrete bilinear form: For all $(z, y_h) \in V_{*h} \times V_h$,

$$a_h^{\mathrm{cf}}(z, y_h) := \sum_{T \in \mathcal{T}_h} (Az, y_h)_{L,T} + \frac{1}{2} \sum_{F \in \mathcal{F}_h^b} ((\mathcal{M} - \mathcal{D})z, y_h)_{L,F}$$
$$- \sum_{F \in \mathcal{F}_h^i} (\mathcal{D}_F[\![z]\!], \{\!\{y_h\}\!\})_{L,F}. \tag{7.31}$$

The superscript indicates the use of centered fluxes as shown in Sect. 7.3.4. We define on V_{*h} the seminorm

$$|y|_M^2 := \int_{\partial\Omega} y^t \mathcal{M} y.$$

Lemma 7.16 (Consistency and discrete L-coercivity). *The bilinear form a_h^{cf} defined by (7.31) satisfies the following two properties:*

(i) *Consistency: For the exact solution $z \in V_*$,*

$$a_h^{\mathrm{cf}}(z, y_h) = a(z, y_h) = (f, y_h)_L \qquad \forall y_h \in V_h, \tag{7.32}$$

where the exact bilinear form a is defined by (7.16).

(ii) *Discrete L-coercivity (compare with the result of Lemma 7.13): For all $y_h \in V_h$,*

$$a_h^{\mathrm{cf}}(y_h, y_h) \geq \mu_0 \|y_h\|_L^2 + \frac{1}{2}|y_h|_M^2, \tag{7.33}$$

where μ_0 results from assumption (7.2c).

7.3. Discretization

Proof. Consistency (7.32) results from Assumption 7.15 and property (7.30). To verify discrete L-coercivity, we use integration by parts to infer, for all $y_h \in V_h$,

$$\sum_{T \in \mathcal{T}_h} (Ay_h, y_h)_{L,T} = \sum_{T \in \mathcal{T}_h} [(A_{(0)}y_h, y_h)_{L,T} + (A_{(1)}y_h, y_h)_{L,T}]$$

$$= \sum_{T \in \mathcal{T}_h} ((\mathcal{A}^0 - \tfrac{1}{2}\mathcal{X})y_h, y_h)_{L,T} + \sum_{T \in \mathcal{T}_h} \tfrac{1}{2}(\mathcal{D}y_h, y_h)_{L,\partial T}$$

$$\geq \mu_0 \|y_h\|_L^2 + \sum_{T \in \mathcal{T}_h} \tfrac{1}{2}(\mathcal{D}y_h, y_h)_{L,\partial T},$$

To rearrange the second term, we observe that, for all $F \in \mathcal{F}_h^i$ with $F = \partial T_1 \cap \partial T_2$, $y_i = y_h|_{T_i}$, $\mathcal{D}_i = \mathcal{D}|_{T_i}$, $i \in \{1,2\}$, with n_F pointing from T_1 to T_2,

$$\{y_h^t \mathcal{D} y_h\} = \frac{1}{2}(y_1^t \mathcal{D}_1 y_1 + y_2 \mathcal{D}_2 y_2)$$

$$= \frac{1}{2}(y_1^t \mathcal{D}_1 y_1 - y_2 \mathcal{D}_1 y_2)$$

$$= \frac{1}{2}(y_1 + y_2)^t \mathcal{D}_1 (y_1 - y_2)$$

$$= \{y_h\}^t \mathcal{D}_F [\![y_h]\!],$$

where we have used (7.29) on the second line to infer $\mathcal{D}_2 = -\mathcal{D}_1$ and the symmetry of $\mathcal{D}_1$ on the third line. As a result,

$$\sum_{T \in \mathcal{T}_h} \tfrac{1}{2}(\mathcal{D}y_h, y_h)_{L,\partial T} = \sum_{F \in \mathcal{F}_h^b} \tfrac{1}{2}(\mathcal{D}y_h, y_h)_{L,F} + \sum_{F \in \mathcal{F}_h^i} (\mathcal{D}_F [\![y_h]\!], \{y_h\})_{L,F}.$$

The proof of (7.33) is now straightforward. $\square$

The discrete bilinear form a_h^{cf} can be used to formulate the discrete problem:

Find $z_h \in V_h$ s.t. $a_h^{\mathrm{cf}}(z_h, y_h) = (f, y_h)_L$ for all $y_h \in V_h$.

Since a_h^{cf} is L-coercive, this problem is well-posed. However, as for the advection-reaction equation, the stability of the bilinear form a_h^{cf} is not sufficient to derive quasi-optimal error estimates for smooth solutions. Indeed, for polynomial degrees $k \geq 1$, proceeding as in Sect. 2.2.2 (the arguments are not repeated here for brevity), we infer an error bound converging at the rate h^k for the error measured in the L-norm augmented by the $|\cdot|_M$-seminorm. The purpose of the next step is to inject additional stability into the discrete bilinear form using least-squares penalties of (certain) interface jumps and boundary values.

7.3.2.2 Step 2: Tightened Discrete Stability Using Penalties

In this second step, we consider the bilinear form, for all $(z, y_h) \in V_{*h} \times V_h$,

$$a_h(z, y_h) := a_h^{\mathrm{cf}}(z, y_h) + s_h(z, y_h), \tag{7.34}$$

with the stabilization bilinear form

$$s_h(z, y_h) := \sum_{F \in \mathcal{F}_h^b} (S_F^b z, y_h)_{L,F} + \sum_{F \in \mathcal{F}_h^i} (S_F^i [\![z]\!], [\![y_h]\!])_{L,F}. \qquad (7.35)$$

The $\mathbb{R}^{m,m}$-valued fields S_F^b and S_F^i involve some user-dependent parameters, and must satisfy some basic design conditions. First of all, they must not perturb the consistency and discrete L-coercivity satisfied by the bilinear form a_h^{cf}. To this purpose, we assume

$$\text{For all } F \in \mathcal{F}_h^b,\ S_F^b z = 0 \text{ and for all } F \in \mathcal{F}_h^i,\ S_F^i [\![z]\!] = 0, \qquad (7.36a)$$

$$S_F^b \text{ and } S_F^i \text{ are symmetric and nonnegative}, \qquad (7.36b)$$

where z is the exact solution. Since s_h is nonnegative, a coercivity argument shows that the discrete problem (7.39) is well-posed. Moreover, still owing to (7.36b), we can define the following seminorm: For all $y \in V_{*h}$,

$$|y|_S^2 := s_h(y, y),$$

and by construction, the bilinear form a_h satisfies the following strengthened discrete coercivity (compare with (7.33)): For all $y_h \in V_h$,

$$a_h(y_h, y_h) \geq \mu_0 \|y_h\|_L^2 + \tfrac{1}{2} |y_h|_M^2 + |y_h|_S^2. \qquad (7.37)$$

The fields S_F^b and S_F^i are assumed to satisfy the following additional design conditions: For all $y, z \in [L^2(F)]^m$,

$$S_F^b \leq \alpha_1 \mathbb{1}_m, \qquad \alpha_2 |\mathcal{D}_F| \leq S_F^i \leq \alpha_3 \mathbb{1}_m, \qquad (7.38a)$$

$$|((\mathcal{M} - \mathcal{D})y, z)_{L,F}| \leq \alpha_4 ((S_F^b + \mathcal{M})y, y)_{L,F}^{1/2} \|z\|_{L,F}, \qquad (7.38b)$$

$$|((\mathcal{M} + \mathcal{D})y, z)_{L,F}| \leq \alpha_5 \|y\|_{L,F} ((S_F^b + \mathcal{M})z, z)_{L,F}^{1/2}. \qquad (7.38c)$$

Here, $\alpha_1, \ldots, \alpha_5$ are user-dependent positive parameters, and the inequalities in (7.38a) are meant on the associated quadratic forms. We observe that the absolute value $|\mathcal{D}_F|$ is well-defined since $\mathcal{D}_F$ takes, by construction, symmetric values.

The discrete problem now becomes:

$$\text{Find } z_h \in V_h \text{ s.t. } a_h(z_h, y_h) = (f, y_h)_L \text{ for all } y_h \in V_h. \qquad (7.39)$$

Owing to the discrete coercivity (7.37), this problem is well-posed. However, the key point is that owing to the design conditions (7.38), the bilinear form a_h satisfies a discrete inf-sup stability condition in a stronger norm than the one used in (7.37). For all $y \in V_{*h}$, we define

$$\|y\|^2 := \|y\|_L^2 + |y|_M^2 + |y|_S^2 + \sum_{T \in \mathcal{T}_h} h_T \|A_{(1)} y\|_{L,T}^2. \qquad (7.40)$$

In what follows, the notation $a \lesssim b$ means the inequality $a \leq Cb$ with positive C independent of h; the real number C can depend on the mesh regularity parameters, the coefficients of the Friedrichs' system, and the parameters $\alpha_1, \ldots, \alpha_5$ introduced above. Without loss of generality, we assume $h \leq 1$.

7.3. Discretization

Lemma 7.17 (Discrete inf-sup stability). *There holds, for all $z_h \in V_h$,*
$$\|z_h\| \lesssim \sup_{y_h \in V_h \setminus \{0\}} \frac{a_h(z_h, y_h)}{\|y_h\|}.$$

Proof. The proof is similar to that for the advection-reaction equation in the case of upwinding; cf. Lemma 2.35. Let $z_h \in V_h$ and define
$$\mathbb{S} = \sup_{y_h \in V_h \setminus \{0\}} \frac{a_h(z_h, y_h)}{\|y_h\|}.$$

Our goal is to prove that $\|z_h\| \lesssim \mathbb{S}$. A direct consequence of (7.37) is that
$$\|z_h\|_L^2 + |z_h|_M^2 + |z_h|_S^2 \lesssim a_h(z_h, z_h) \leq \mathbb{S}\|z_h\|. \tag{7.41}$$

To bound the fourth term in the definition (7.40) of $\|z_h\|$, we consider the function $y_h \in V_h$ such that, for all $T \in \mathcal{T}_h$, $y_h|_T := h_T \langle \mathcal{A}_{(1)} \rangle_T z_h$ where $\langle \mathcal{A}_{(1)} \rangle_T := \sum_{k=1}^d \langle \mathcal{A}^k \rangle_T \partial_k$ and $\langle \mathcal{A}^k \rangle_T$ denotes the mean value of $\mathcal{A}^k$ on the mesh element T. First of all, we observe using the triangle inequality, an inverse inequality, the regularity assumption (7.28) on the fields $\mathcal{A}^1, \ldots, \mathcal{A}^d$, and $h \leq 1$ that
$$h_T \|\langle \mathcal{A}_{(1)} \rangle_T z_h\|_{L,T}^2 \lesssim h_T \|\mathcal{A}_{(1)} z_h\|_{L,T}^2 + h_T \|(\mathcal{A}_{(1)} - \langle \mathcal{A}_{(1)} \rangle_T) z_h\|_{L,T}^2$$
$$\lesssim h_T \|\mathcal{A}_{(1)} z_h\|_{L,T}^2 + \|z_h\|_{L,T}^2,$$

whence
$$\sum_{T \in \mathcal{T}_h} h_T^{-1} \|y_h\|_{L,T}^2 \lesssim \|z_h\|^2. \tag{7.42}$$

Moreover, using the boundedness of $\mathcal{M}$, the upper bounds in (7.38a), the discrete trace inequality (1.37), and the above bound, we infer
$$|y_h|_M^2 + |y_h|_S^2 \lesssim \|z_h\|^2.$$

Since $\|y_h\|_L \lesssim \|z_h\|_L$ and
$$\sum_{T \in \mathcal{T}_h} h_T \|\mathcal{A}_{(1)} y_h\|_{L,T}^2 \lesssim \|z_h\|^2,$$

using the inverse inequality (1.36), the boundedness of the fields $\mathcal{A}^1, \ldots, \mathcal{A}^d$, and (7.42), we finally arrive at
$$\|y_h\| \lesssim \|z_h\|. \tag{7.43}$$

Now, we observe that
$$\sum_{T \in \mathcal{T}_h} h_T \|\mathcal{A}_{(1)} z_h\|_{L,T}^2 = a_h(z_h, y_h) - (\mathcal{A}_{(0)} z_h, y_h)_L$$
$$+ \sum_{T \in \mathcal{T}_h} h_T (\mathcal{A}_{(1)} z_h, (\mathcal{A}_{(1)} - \langle \mathcal{A}_{(1)} \rangle_T) z_h)_{L,T}$$
$$- \sum_{F \in \mathcal{F}_h^b} \tfrac{1}{2}((\mathcal{M} - \mathcal{D}) z_h, y_h)_{L,F} + \sum_{F \in \mathcal{F}_h^i} (\mathcal{D}_F [\![z_h]\!], \{\!\{y_h\}\!\})_{L,F}$$
$$- s_h(z_h, y_h) = \mathfrak{T}_1 + \ldots + \mathfrak{T}_6,$$

and we estimate the terms $\mathfrak{T}_1, \ldots, \mathfrak{T}_6$ on the right-hand side. Clearly, $|\mathfrak{T}_1| \lesssim \mathbb{S}\|y_h\| \lesssim \mathbb{S}\|z_h\|$ owing to (7.43). Furthermore, $|\mathfrak{T}_2| \lesssim \|z_h\|_L \|y_h\|_L \lesssim \mathbb{S}^{1/2}\|z_h\|^{3/2}$ using (7.41) and (7.43). In addition,

$$|\mathfrak{T}_4| + |\mathfrak{T}_5| + |\mathfrak{T}_6| \lesssim (|z_h|_M + |z_h|_S) \left(\sum_{T \in \mathcal{T}_h} \|y_h\|_{L,\partial T}^2 \right)^{1/2},$$

where we have used (7.38b) to bound $\mathfrak{T}_4$, the lower bound for S_F^i in (7.38a) to bound $\mathfrak{T}_5$, and the upper bounds in (7.38a) to bound $\mathfrak{T}_6$. Using the discrete trace inequality (1.37) and the bounds (7.41) and (7.42), we infer

$$|\mathfrak{T}_4| + |\mathfrak{T}_5| + |\mathfrak{T}_6| \lesssim \mathbb{S}^{1/2}\|z_h\|^{3/2}.$$

Finally, to bound $\mathfrak{T}_3$, we use the Cauchy–Schwarz inequality, Young's inequality, and the regularity assumption (7.28) on the fields $\mathcal{A}^1, \ldots, \mathcal{A}^d$ to obtain

$$|\mathfrak{T}_3| \lesssim \|z_h\|_L^2 + \gamma \sum_{T \in \mathcal{T}_h} h_T \|A_{(1)} z_h\|_{L,T}^2,$$

where γ can be chosen as small as needed. Hence,

$$\sum_{T \in \mathcal{T}_h} h_T \|A_{(1)} z_h\|_{L,T}^2 \lesssim \mathbb{S}\|z_h\| + \mathbb{S}^{1/2}\|z_h\|^{3/2},$$

which combined with (7.41) leads to

$$\|z_h\|^2 \lesssim \mathbb{S}\|z_h\| + \mathbb{S}^{1/2}\|z_h\|^{3/2}.$$

Using Young's inequality yields $\|z_h\| \lesssim \mathbb{S}$. □

7.3.3 Convergence Analysis

To derive an error estimate on $\|z - z_h\|$, where z is the exact solution of the weak problem (7.17), we proceed in the spirit of Theorem 1.35. Since discrete stability and consistency have already been proven, we only need to address the boundedness of the discrete bilinear form a_h.

Integrating by parts in the definition (7.31), that is, using the formal adjoint $\tilde{A}$ of A, we infer, for all $z, y_h \in V_{*h} \times V_h$,

$$a_h^{\text{cf}}(z, y_h) = \sum_{T \in \mathcal{T}_h} (z, \tilde{A} y_h)_{L,T} + \frac{1}{2} \sum_{F \in \mathcal{F}_h^b} ((\mathcal{M} + \mathcal{D})z, y_h)_{L,F}$$
$$+ \sum_{F \in \mathcal{F}_h^i} (\mathcal{D}_F \{z\}, [\![y_h]\!])_{L,F}. \qquad (7.44)$$

This form is more convenient to examine the boundedness of the bilinear form $a_h = a_h^{\text{cf}} + s_h$. We define the following norm: For all $y \in V_{*h}$,

$$\|y\|_*^2 = \|y\|^2 + \sum_{T \in \mathcal{T}_h} \left(h_T^{-1} \|y\|_{L,T}^2 + \|y\|_{L,\partial T}^2 \right).$$

7.3. Discretization

Lemma 7.18 (Boundedness). *For all* $(w, y_h) \in V_{*h} \times V_h$, *there holds*

$$a_h(w, y_h) \lesssim \|w\|_* \|y_h\|.$$

Proof. Let $(w, y_h) \in V_{*h} \times V_h$. Denote by $\mathfrak{T}_1$, $\mathfrak{T}_2$, and $\mathfrak{T}_3$ the three terms on the right-hand side of (7.44). We first observe that

$$(w, \tilde{A} y_h)_{L,T} = (w, \tilde{A}_{(0)} y_h)_{L,T} - (w, A_{(1)} y_h)_{L,T},$$

and since $(w, A_{(1)} y_h)_{L,T} \leq h_T^{-1/2} \|w\|_{L,T} h_T^{1/2} \|A_{(1)} y_h\|_{L,T}$, we infer that

$$\mathfrak{T}_1 = \sum_{T \in \mathcal{T}_h} (w, \tilde{A} y_h)_{L,T} \lesssim \|w\|_* \|y_h\|.$$

Furthermore, using (7.38c) and the lower bound for S_F^i in (7.38a) yields

$$|\mathfrak{T}_2| + |\mathfrak{T}_3| \lesssim \left(\sum_{F \in \mathcal{F}_h} \|w\|_{L,F}^2 \right)^{1/2} (|y_h|_M + |y_h|_S) \leq \|w\|_* \|y_h\|.$$

Hence, $a_h^{\text{cf}}(w, y_h) \lesssim \|w\|_* \|y_h\|$. Finally, since $a_h = a_h^{\text{cf}} + s_h$ and

$$s_h(w, y_h) \leq |w|_S |y_h|_S \leq \|w\| \|y_h\|,$$

the desired bound on $a_h(w, y_h)$ is obtained. $\square$

We can now state the main result of this section.

Theorem 7.19 (Error estimate and convergence rate for smooth solutions). *Let* $z \in V_*$ *solve* (7.17). *Let* z_h *solve* (7.39) *with* a_h *defined by* (7.34). *Then, there holds*

$$\|z - z_h\| \lesssim \inf_{y_h \in V_h} \|z - y_h\|_*. \tag{7.45}$$

Moreover, if $z \in [H^{k+1}(\Omega)]^m$, *there holds*

$$\|z - z_h\| \leq C_z h^{k+1/2}, \tag{7.46}$$

where $C_z = C \|z\|_{[H^{k+1}(\Omega)]^m}$ *with* C *independent of* h.

Proof. The bound (7.45) follows from Theorem 1.35, while the bound (7.46) is obtained by taking $y_h = \pi_h z$, where π_h denotes the L-orthogonal projection onto V_h, and using Lemmata 1.58 and 1.59. $\square$

As for the advection-reaction problem, the above error estimate is optimal for the broken graph norm and the seminorms $|\cdot|_M$ and $|\cdot|_S$, while it is $1/2$-suboptimal for the L-norm.

Remark 7.20 (Convergence based on discrete coercivity). Proceeding as in Sect. 2.3.2, it is possible to derive a weaker convergence result solely based on the discrete coercivity (7.37) without requiring the discrete inf-sup condition established in Lemma 7.17. The analysis relies on the following boundedness on orthogonal subscales for the discrete bilinear form a_h: For all $(w, y_h) \in [H^1(\mathcal{T}_h)]^m \times V_h$, there holds

$$a_h(w - \pi_h w, y_h) \lesssim \|w - \pi_h w\|_* (\|y_h\|_L + |y_h|_M + |y_h|_S).$$

This result can be proven by proceeding similarly to the proof of Lemma 2.30.

7.3.4 Numerical Fluxes

The discrete problem (7.39) can be equivalently formulated by means of local (elementwise) problems and numerical fluxes defined on element faces. To this purpose, we use the formal adjoint $\tilde{A}$ after integration by parts (cf. (7.44)), similarly to the approach pursued in finite volume methods.

By localizing the test function y_h in (7.39) to mesh elements, the discrete problem (7.39) amounts to finding $z_h \in V_h$ such that, for all $T \in \mathcal{T}_h$ and all $y \in [\mathbb{P}_d^k(T)]^m$,

$$(z_h, \tilde{A} y)_{L,T} + \sum_{F \in \mathcal{F}_T} \epsilon_{T,F} (\phi_F(z_h), y)_{L,F} = (f, y)_{L,T},$$

where, as usual, $\epsilon_{T,F} := n_T \cdot n_F$, while the *numerical fluxes* are given by

$$\phi_F(z_h) = \begin{cases} \mathcal{D}_F \{\!\!\{z_h\}\!\!\} + S_F^i [\![z_h]\!] & \text{if } F \in \mathcal{F}_h^i, \\ \frac{1}{2}(\mathcal{M} + \mathcal{D}) z_h + S_F^b z_h & \text{if } F \in \mathcal{F}_h^b. \end{cases}$$

Centered fluxes are recovered when $S_F^i = 0$ (i.e., without interface penalties).

7.3.5 Examples

7.3.5.1 Advection-Reaction

For the advection-reaction problem, the fields S_F^b and S_F^i are scalar-valued and can be taken such that

$$S_F^b = 0, \qquad S_F^i = \eta,$$

where $\eta > 0$ is a user-dependent parameter (this parameter can vary from face to face). This yields the upwind dG method analyzed in Sect. 2.3. A simple choice is $\eta = 1$.

7.3.5.2 The Div-Grad Problem

For the div-grad problem with Dirichlet boundary conditions, the fields S_F^b and S_F^i can be taken such that

$$S_F^b = \begin{bmatrix} \mathbb{0}_{d,d} & \mathbb{0}_{d,1} \\ \hline \mathbb{0}_{1,d} & \eta_1 \end{bmatrix}, \qquad S_F^i = \begin{bmatrix} \eta_2 n_F \otimes n_F & \mathbb{0}_{d,1} \\ \hline \mathbb{0}_{1,d} & \eta_3 \end{bmatrix},$$

7.3. Discretization

where η_1, η_2, and η_3 are positive user-dependent parameters (these parameters can vary from face to face). The above choice amounts to penalizing the boundary values of the potential together with the jumps across interfaces of the potential and of the normal component of the diffusive flux. Furthermore, for Robin/Neumann boundary conditions, the field S_F^b can be taken such that

$$S_F^b = \eta_1 \left[\begin{array}{c|c} \mathbf{n} \otimes \mathbf{n} & -\gamma \mathbf{n} \\ \hline -\gamma \mathbf{n}^t & \gamma^2 \end{array}\right],$$

with positive user-dependent parameter η_1.

For $z = (\sigma, u)$, the numerical flux $\phi_F(z)$ is on an interface $F \in \mathcal{F}_h^i$,

$$\phi_F(z) = \begin{pmatrix} \mathbf{n}_F(\{\!\{u\}\!\} + \eta_2 \mathbf{n}_F \cdot [\![\sigma]\!]) \\ \mathbf{n}_F \cdot \{\!\{\sigma\}\!\} + \eta_3 [\![u]\!] \end{pmatrix},$$

while on a boundary face $F \in \mathcal{F}_h^b$, the numerical flux is

$$\phi_F(z) = \begin{pmatrix} 0 \\ \mathbf{n} \cdot \sigma + \eta_1 u \end{pmatrix}, \qquad \phi_F(z) = \begin{pmatrix} u\mathbf{n} + \eta_1 \mathbf{n} \cdot (\mathbf{n} \cdot \sigma - \gamma u) \\ \gamma u - \eta_1 \gamma (\mathbf{n} \cdot \sigma - \gamma u) \end{pmatrix},$$

for Dirichlet and Robin/Neumann boundary conditions, respectively.

The resulting dG methods differ from those considered in Chap. 4 for the Poisson problem (they were briefly mentioned at the end of Sect. 4.4.2). In particular, the present methods are more computationally demanding since they require solving simultaneously for the potential and the diffusive flux, instead of just solving for the potential and reconstructing the diffusive flux by post-processing the potential (cf. Sect. 5.5). The payoff of the present approach is to yield a more accurate approximation of the diffusive flux, provided the exact solution is smooth enough. Indeed, if the potential and the diffusive flux have $H^{k+1}(\Omega)$-regularity, the error estimate (7.46) yields an $h^{k+1/2}$ error estimate on the diffusive flux in the $[L^2(\Omega)]^d$-norm and an h^k error estimate on the broken divergence of the diffusive flux. Another interesting aspect of the present approach is that the choice $k = 0$ is possible; this choice entails the same number of degrees of freedom as the choice $k = 1$ when approximating only the potential.

7.3.5.3 Linear Elasticity

For the linear elasticity problem with Dirichlet boundary conditions on the displacement, the fields S_F^b and S_F^i can be taken such that

$$S_F^b = \left[\begin{array}{c|c} \mathbb{O}_{d^2,d^2} & \mathbb{O}_{d^2,d} \\ \hline \mathbb{O}_{d,d^2} & \eta_1 \mathbb{1}_d \end{array}\right], \qquad S_F^i = \left[\begin{array}{c|c} \eta_2 \mathcal{N}_F \mathcal{N}_F^t & \mathbb{O}_{d^2,d} \\ \hline \mathbb{O}_{d,d^2} & \eta_3 \mathbb{1}_d \end{array}\right],$$

where $\mathcal{N}_F$ is defined as $\mathcal{N}$ in (7.13) by just replacing $\mathbf{n}$ by $\mathbf{n}_F$, so that, for all $\tau \in \mathbb{R}^{d^2}$, $\mathcal{N}_F \tau = -1/2(\tau + \tau^t) \cdot \mathbf{n}_F \in \mathbb{R}^d$. Moreover, η_1, η_2, and η_3 are positive user-dependent parameters (these parameters can vary from face to face). The above choice amounts to penalizing the boundary values of the displacement together with the jumps across interfaces of the displacement and of the normal stress.

For $z = (\sigma, u)$ and with the notation $\sigma_S := \frac{1}{2}(\sigma + \sigma^t)$, the numerical flux $\phi_F(z)$ is

$$\phi_F(z) = \begin{pmatrix} \mathcal{N}_F(\{\!\{u\}\!\} + \eta_2 \mathrm{n}_F \cdot [\![\sigma_S]\!]) \\ \mathrm{n}_F \cdot \{\!\{\sigma_S\}\!\} + \eta_3 [\![u]\!] \end{pmatrix}, \qquad \phi_F(z) = \begin{pmatrix} 0 \\ \mathrm{n} \cdot \sigma_S + \eta_1 u \end{pmatrix},$$

on an interface $F \in \mathcal{F}_h^i$ and on a boundary face $F \in \mathcal{F}_h^b$, respectively.

7.3.5.4 The Curl-Curl Problem

For the curl-curl problem with Dirichlet boundary condition on the tangential component of the electric field, the fields S_F^b and S_F^i can be taken such that

$$S_F^b = \begin{bmatrix} 0_{3,3} & 0_{3,3} \\ \hline 0_{3,3} & \eta_1 \mathcal{T}^t \mathcal{T} \end{bmatrix}, \qquad S_F^i = \begin{bmatrix} \eta_2 \mathcal{T}_F^t \mathcal{T}_F & 0_{3,3} \\ \hline 0_{3,3} & \eta_3 \mathcal{T}_F^t \mathcal{T}_F \end{bmatrix},$$

where $\mathcal{T}_F$ is defined as $\mathcal{T}$ in (7.14) by just replacing n by n_F, so that, for all $\xi \in \mathbb{R}^3$, $\mathcal{T}_F \xi = \mathrm{n}_F \times \xi \in \mathbb{R}^3$. Moreover, η_1, η_2, and η_3 are positive user-dependent parameters (these parameters can vary from face to face). The above choice amounts to penalizing the boundary values of the tangential component of the electric field together with the jumps across interfaces of the tangential component of both the electric and magnetic fields.

For $z = (H, E)$, the numerical flux $\phi_F(z)$ is

$$\phi_F(z) = \begin{pmatrix} \mathrm{n}_F \times \{\!\{E\}\!\} + \eta_2 \mathrm{n}_F \times [\![H]\!] \\ -\mathrm{n}_F \times \{\!\{H\}\!\} + \eta_3 \mathrm{n}_F \times [\![E]\!] \end{pmatrix}, \qquad \phi_F(z) = \begin{pmatrix} 0 \\ -\mathrm{n} \times H + \eta_1 \mathrm{n} \times E \end{pmatrix},$$

on an interface $F \in \mathcal{F}_h^i$ and on a boundary face $F \in \mathcal{F}_h^b$, respectively.

7.4 Two-Field Systems

In this section, we particularize the continuous setting and the dG approximation to a special class of Friedrichs' systems featuring a two-field structure. Two-field Friedrichs' systems are obtained from the mixed formulation of elliptic-type second-order PDEs, so that the Friedrichs' system incorporates two components, that is, the $\mathbb{R}^m$-valued function z is decomposed as $z = (z^\sigma, z^u)$ where z^σ is $\mathbb{R}^{m_\sigma}$-valued and z^u is $\mathbb{R}^{m_u}$-valued with $m = m_\sigma + m_u$. Applications include the div-grad problem (together with diffusion-advection-reaction problems), linear elasticity, and the curl-curl problem (cf. Table 7.1).

The main difference with the dG approximation designed in the previous section is that we modify the penalty strategy in order to eliminate the σ-component

7.4. Two-Field Systems

Table 7.1: Examples of two-field Friedrichs' systems

Problem	z^σ	z^u	m_σ	m_u
Div-grad	$-\nabla u$	u	d	1
Linear elasticity	$1/2(\nabla u + \nabla u^t) + \lambda^{-1}(\nabla\cdot u)\mathbb{1}_d$	u	d^2	d
Curl-curl	H	E	3	3

locally by solving elementwise problems. This entails penalizing the jumps of the u-component with a weight that scales as the reciprocal of the local mesh-size, while the jumps of the σ-component are not penalized. The resulting dG method is more cost-effective than that presented in the previous section (since it only solves for the u-component), but the counterpart is that the approximation on the σ-component is (slightly) less accurate. For the div-grad problem, after elimination of the σ-component, we recover the SIP method or some of its variants discussed in Sect. 4.4 in the context of mixed dG methods.

For the sake of simplicity, we focus on Dirichlet boundary conditions on the u-component, and do not consider first-order terms in the PDEs. Robin/Neumann boundary conditions and first-order terms can be incorporated into the formalism as described in [143].

7.4.1 The Continuous Setting

It is convenient to define

$$L_\sigma := [L^2(\Omega)]^{m_\sigma} \quad \text{and} \quad L_u := [L^2(\Omega)]^{m_u}.$$

We assume that the fields $\mathcal{A}^0, \mathcal{A}^1, \ldots, \mathcal{A}^d$ admit the block-structure

$$\mathcal{A}^0 = \begin{bmatrix} \mathcal{K}^{\sigma\sigma} & \mathbb{0}_{m_\sigma,m_u} \\ \hline \mathbb{0}_{m_u,m_\sigma} & \mathcal{K}^{uu} \end{bmatrix}, \quad \mathcal{A}^k = \begin{bmatrix} \mathbb{0}_{m_\sigma,m_\sigma} & \mathcal{B}^k \\ \hline (\mathcal{B}^k)^t & \mathbb{0}_{m_u,m_u} \end{bmatrix}, \quad k \in \{1,\ldots,d\},$$
(7.47)

where $\mathcal{K}^{\sigma\sigma}$ is $\mathbb{R}^{m_\sigma,m_\sigma}$-valued, $\mathcal{K}^{uu}$ is $\mathbb{R}^{m_u,m_u}$-valued, and $\mathcal{B}^k$ is $\mathbb{R}^{m_\sigma,m_u}$-valued. We suppose that the fields $\mathcal{B}^k$ are constant in Ω for all $k \in \{1,\ldots,d\}$. The assumptions (7.2a) and (7.2b) stated in Sect. 7.1.1 hold true; indeed, (7.2a) is trivial since $\mathcal{X} = \mathbb{0}_{m,m}$ because the fields $\mathcal{B}^k$ are constant, while (7.2b) holds true by construction. Moreover, we assume (7.2c), meaning that the fields $\mathcal{K}^{\sigma\sigma}$ and $\mathcal{K}^{uu}$ are uniformly positive definite in Ω; a more general framework circumventing the need for full positivity is presented in Sect. 7.4.3.

In terms of PDEs, the two-field Friedrichs' system amounts to

$$\mathcal{K}^{\sigma\sigma} z^\sigma + Bz^u = f^\sigma,$$
$$\hat{B} z^\sigma + \mathcal{K}^{uu} z^u = f^u,$$

with data $f^\sigma \in L_\sigma$ and $f^u \in L_u$, and with the differential operators

$$B = \sum_{k=1}^d \mathcal{B}^k \partial_k \quad \text{and} \quad \hat{B} = \sum_{k=1}^d (\mathcal{B}^k)^t \partial_k,$$

so that $\hat{B}$ is the opposite of the formal adjoint of B. For instance, for the div-grad problem, B represents the gradient operator and $\hat{B}$ the divergence operator. Eliminating z^σ using the first equation (this is possible since $\mathcal{K}^{\sigma\sigma}$ is uniformly positive definite) yields

$$z^\sigma = -(\mathcal{K}^{\sigma\sigma})^{-1}(Bz^u - f^\sigma), \tag{7.48}$$

and substituting this expression into the second equation leads to a second-order PDE for z^u in the form

$$-\hat{B}(\mathcal{K}^{\sigma\sigma})^{-1}Bz^u + \mathcal{K}^{uu}z^u = f^u - \hat{B}(\mathcal{K}^{\sigma\sigma})^{-1}f^\sigma.$$

The boundary field $\mathcal{D}$ admits the block-structure

$$\mathcal{D} = \left[\begin{array}{c|c} \mathbb{0}_{m_\sigma,m_\sigma} & \mathcal{D}^{\sigma u} \\ \hline \mathcal{D}^{u\sigma} & \mathbb{0}_{m_u,m_u} \end{array}\right],$$

where

$$\mathcal{D}^{\sigma u} := \sum_{k=1}^d \mathcal{B}^k n_k$$

is $\mathbb{R}^{m_\sigma,m_u}$-valued and $\mathcal{D}^{u\sigma} = (\mathcal{D}^{\sigma u})^t$ is $\mathbb{R}^{m_u,m_\sigma}$-valued. We assume that $\mathcal{D}^{\sigma u}$ is injective, so that a Dirichlet boundary condition on z^u can be enforced by setting

$$\mathcal{M} = \left[\begin{array}{c|c} \mathbb{0}_{m_\sigma,m_\sigma} & -\mathcal{D}^{\sigma u} \\ \hline \mathcal{D}^{u\sigma} & \mathbb{0}_{m_u,m_u} \end{array}\right].$$

Indeed, $z \in \text{Ker}(\mathcal{M} - \mathcal{D})$ means that $\mathcal{D}^{\sigma u} z^u = 0$, and hence $z^u = 0$ since $\mathcal{D}^{\sigma u}$ is injective. Moreover, assumption (7.7) on $\mathcal{M}$ can also be verified.

7.4.2 Discretization

The two components of z can be approximated using polynomials of different degree, say l for the σ-component and k for the u-component. Correspondingly, we set

$$\Sigma_h := [\mathbb{P}_d^l(\mathcal{T}_h)]^{m_\sigma}, \qquad U_h := [\mathbb{P}_d^k(\mathcal{T}_h)]^{m_u}, \qquad V_h := \Sigma_h \times U_h.$$

We assume that
$$l \in \{k-1, k\}.$$

7.4.2.1 Design of the dG Bilinear Form

The dG bilinear form a_h is still such that $a_h = a_h^{\text{cf}} + s_h$ with a_h^{cf} defined by (7.31), but the stabilization bilinear form s_h is now taken in the form

$$s_h(z^u, y_h^u) = \sum_{F \in \mathcal{F}_h^b} (S_F^b z^u, y_h^u)_{L_u, F} + \sum_{F \in \mathcal{F}_h^i} (S_F^i [\![z^u]\!], [\![y_h^u]\!])_{L_u, F},$$

where the fields S_F^b and S_F^i are now $\mathbb{R}^{m_u, m_u}$-valued; in other words, the stabilization only acts on the u-component (compare with (7.35)). In the present setting, the design conditions (7.36)–(7.38) are modified as

$$\text{For the exact solution } z \text{ and for all } F \in \mathcal{F}_h^b,\ S_F^b z^u = 0, \tag{7.49a}$$

$$S_F^b \text{ and } S_F^i \text{ are symmetric and nonnegative,} \tag{7.49b}$$

$$\alpha_1 h_F^{-1} (\mathcal{D}_F^{u\sigma} \mathcal{D}_F^{\sigma u})^{1/2} \leq S_F^b \leq \alpha_2 \mathbb{1}_{m_u}, \tag{7.49c}$$

$$\alpha_3 h_F^{-1} (\mathcal{D}_F^{u\sigma} \mathcal{D}_F^{\sigma u})^{1/2} \leq S_F^i \leq \alpha_4 \mathbb{1}_{m_u}. \tag{7.49d}$$

Here, $\alpha_1, \ldots, \alpha_4$ are user-dependent positive parameters, and the inequalities in (7.49c) and (7.49d) are meant as usual on the associated quadratic forms. The matrices $\mathcal{D}_F^{u\sigma}$ and $\mathcal{D}_F^{\sigma u}$ are defined as $\mathcal{D}^{u\sigma}$ and $\mathcal{D}^{\sigma u}$, respectively, with n replaced by n_F. We observe that $\mathcal{D}_F^{u\sigma} \mathcal{D}_F^{\sigma u}$ is by construction symmetric and nonnegative, and it is actually positive definite since $\mathcal{D}^{\sigma u}$ is injective; hence, $(\mathcal{D}_F^{u\sigma} \mathcal{D}_F^{\sigma u})^{1/2}$ is well-defined. Finally, as in Chap. 4 (cf. Definition 4.5), in dimension $d \geq 2$, h_F denotes the diameter of the face F (other local length scales can be chosen), while in dimension 1, we set $h_F := \min(h_{T_1}, h_{T_2})$ if $F \in \mathcal{F}_h^i$ with $F = \partial T_1 \cap \partial T_2$ and $h_F := h_T$ if $F \in \mathcal{F}_h^b$ with $F = \partial T \cap \partial \Omega$.

Example 7.21 (Choices for S_F^b and S_F^i). For the div-grad problem, we can take $S_F^b = S_F^i = \eta h_F^{-1}$ (as in the SIP method presented in Sect. 4.2), for the linear elasticity problem, we can take $S_F^b = S_F^i = \eta h_F^{-1} \mathbb{1}_d$, and for the curl-curl problem, we can take $S_F^b = S_F^i = \eta h_F^{-1} T_F (T_F)^t$. In all cases, η is a positive user-dependent parameter.

7.4.2.2 The Discrete Problem

We consider the discrete problem:

$$\text{Find } z_h \in V_h \text{ s.t. } a_h(z_h, y_h) = (f, y_h)_L \text{ for all } y_h \in V_h. \tag{7.50}$$

Accounting for the two-field structure yields the expression

$$\begin{aligned}
a_h(z, y_h) = \sum_{T \in \mathcal{T}_h} & \left\{ (\mathcal{K}^{\sigma\sigma} z^\sigma, y_h^\sigma)_{L_\sigma, T} + (\mathcal{K}^{uu} z^u, y_h^u)_{L_u, T} \right. \\
& \left. + (Bz^u, y_h^\sigma)_{L_\sigma, T} + (\hat{B}z^\sigma, y_h^u)_{L_u, T} \right\} - \sum_{F \in \mathcal{F}_h^b} (\mathcal{D}_F^{\sigma u} z^u, y_h^\sigma)_{L_\sigma, F} \\
& - \sum_{F \in \mathcal{F}_h^i} \left\{ (\mathcal{D}_F^{\sigma u} [\![z^u]\!], \{\!\{y_h^\sigma\}\!\})_{L_\sigma, F} + (\mathcal{D}_F^{u\sigma} [\![z^\sigma]\!], \{\!\{y_h^u\}\!\})_{L_u, F} \right\} \\
& + \sum_{F \in \mathcal{F}_h^b} (S_F^b z^u, y_h^u)_{L_u, F} + \sum_{F \in \mathcal{F}_h^i} (S_F^i [\![z^u]\!], [\![y_h^u]\!])_{L_u, F}. \quad (7.51)
\end{aligned}$$

Condition (7.49a) ensures that the discrete bilinear form a_h is consistent. Condition (7.49b) combined with (7.33) (the seminorm $|\cdot|_\mathcal{M}$ now vanishes since $\mathcal{M}$ is skew-symmetric) yields

$$a_h(y_h, y_h) \geq a_h^{\text{cf}}(y_h, y_h) \geq \mu_0 \|y_h\|_L^2.$$

Hence, owing to the Lax–Milgram Lemma, the discrete problem (7.50) is well-posed.

7.4.2.3 Local Elimination of the σ-Component

As in the continuous setting, the σ-component can be eliminated from the discrete problem (7.50). Let $l \in \{k-1, k\}$ be the polynomial degree used to build the discrete space Σ_h and let $F \in \mathcal{F}_h$. We define the lifting operator $\mathrm{r}_F^l : [L^2(F)]^{m_u} \to [\mathbb{P}_d^l(\mathcal{T}_h)]^{m_\sigma}$ such that, for all $\varphi \in [L^2(F)]^{m_u}$,

$$(\mathrm{r}_F^l(\varphi), q_h)_{L_\sigma} = (\varphi, \mathcal{D}_F^{u\sigma} \{\!\{q_h\}\!\})_{L_u, F} \qquad \forall q_h \in [\mathbb{P}_d^l(\mathcal{T}_h)]^{m_\sigma}.$$

The support of r_F^l consists of the one or two mesh elements of which F is a face. For all $z_h^u \in U_h$ and for all $y_h^\sigma \in \Sigma_h$, we obtain

$$\sum_{F \in \mathcal{F}_h^b} (\mathcal{D}_F^{\sigma u} z_h^u, y_h^\sigma)_{L_\sigma, F} + \sum_{F \in \mathcal{F}_h^i} (\mathcal{D}_F^{\sigma u} [\![z_h^u]\!], \{\!\{y_h^\sigma\}\!\})_{L_\sigma, F} = \sum_{F \in \mathcal{F}_h} (\mathrm{r}_F^l([\![z_h^u]\!]), y_h^\sigma)_{L_\sigma}.$$

Assuming for simplicity $f^\sigma = 0$ in the discrete problem (7.50) and letting

$$\mathrm{R}_h^l([\![z_h^u]\!]) := \sum_{F \in \mathcal{F}_h} \mathrm{r}_F^l([\![z_h^u]\!]),$$

we deduce from (7.51) the reconstruction formula (compare with (7.48))

$$z_h^\sigma = -(\mathcal{K}^{\sigma\sigma})^{-1}(B_h z_h^u - \mathrm{R}_h^l([\![z_h^u]\!])), \qquad (7.52)$$

where B_h denotes the broken version of the differential operator B, that is, for all $T \in \mathcal{T}_h$, $B_h z_h^u|_T = B(z_h^u|_T)$. The reconstruction formula (7.52) is an

7.4. Two-Field Systems

extension to the present setting of the discrete gradient introduced in Sect. 4.3.2 in the context of the Poisson problem. Owing to (7.52), the discrete bilinear form (7.51) can be rewritten in the form

$$a_h(z_h, y_h) = (\mathcal{K}^{\sigma\sigma}(B_h z_h^u - \mathrm{R}_h^l(\llbracket z_h^u \rrbracket)), B_h y_h^u - \mathrm{R}_h^l(\llbracket y_h^u \rrbracket))_{L_\sigma} + (\mathcal{K}^{uu} z_h^u, y_h^u)_{L_u}$$
$$+ \sum_{F \in \mathcal{F}_h^b}(S_F^b z_h^u, y_h^u)_{L_u,F} + \sum_{F \in \mathcal{F}_h^i}(S_F^i \llbracket z_h^u \rrbracket, \llbracket y_h^u \rrbracket)_{L_u,F}$$
$$:= a_h^u(z_h^u, y_h^u). \tag{7.53}$$

Remark 7.22 (Stencil reduction). Similarly to the discussion in Sect. 4.4.2 for LDG methods, the stabilization bilinear form s_h can be further modified so as to reduce the stencil associated with the bilinear form a_h^u defined by (7.53). This entails here subtracting the term $(\mathcal{K}^{\sigma\sigma} \mathrm{R}_h^l(\llbracket z_h^u \rrbracket), \mathrm{R}_h^l(\llbracket y_h^u \rrbracket))_{L_\sigma}$ and taking the penalty parameters in S_F^b and S_F^i large enough to control this term.

7.4.2.4 Convergence Analysis

The convergence analysis of the discrete problem (7.50) can be performed by considering the discrete bilinear form a_h^u defined by (7.53) and proceeding as in Sect. 4.2 for the Poisson problem. In what follows, we consider instead the mixed formulation based on the discrete bilinear form a_h defined by (7.51). We keep the spaces V_* and V_{*h} defined in Assumption 7.15.

The design conditions (7.49c) and (7.49d) are motivated by the following discrete inf-sup stability result in terms of the norm defined on V_{*h} as

$$\|y\|^2 := \|y\|_L^2 + |y^u|_S^2 + \sum_{T \in \mathcal{T}_h} \|By^u\|_{L_\sigma,T}^2, \tag{7.54}$$

where $|y^u|_S^2 = s_h(y^u, y^u)$. We observe that, for all $y_h \in V_h$,

$$a_h(y_h, y_h) \geq \mu_0 \|y_h\|_L^2 + |y_h^u|_S^2. \tag{7.55}$$

Lemma 7.23 (Discrete inf-sup stability). *Let a_h be defined by (7.51). Then, there holds, for all $z_h \in V_h$,*

$$\|z_h\| \lesssim \sup_{y_h \in V_h \setminus \{0\}} \frac{a_h(z_h, y_h)}{\|y_h\|}.$$

Proof. Let $z_h \in V_h$ and set $\mathcal{S} = \sup_{y_h \in V_h \setminus \{0\}} \frac{a_h(z_h, y_h)}{\|y_h\|}$. Owing to (7.55),

$$\|z_h\|_L^2 + |z_h^u|_S^2 \lesssim \mathcal{S}\|z_h\|.$$

Moreover, the discrete function $y_h = (Bz_h^u, 0)$ is in V_h since $l \geq k-1$, and

$$\|y_h\|^2 = \|y_h\|_L^2 = \|y_h^\sigma\|_{L_\sigma}^2 = \sum_{T \in \mathcal{T}_h} \|Bz_h^u\|_{L_\sigma,T}^2.$$

A direct calculation shows that

$$\sum_{T\in\mathcal{T}_h} \|Bz_h^u\|_{L_\sigma,T}^2 = a_h(z_h, y_h) - \sum_{T\in\mathcal{T}_h} (\mathcal{K}^{\sigma\sigma} z_h^\sigma, y_h^\sigma) - \sum_{F\in\mathcal{F}_h^b} (\mathcal{D}_F^{\sigma u} z_h^u, y_h^\sigma)_{L_\sigma,F}$$

$$- \sum_{F\in\mathcal{F}_h^i} [(\mathcal{D}_F^{\sigma u}[\![z_h^u]\!], \{\!\{y_h^\sigma\}\!\})_{L_\sigma,F} = \mathfrak{T}_1 + \ldots + \mathfrak{T}_4,$$

and we bound the four terms on the right-hand side. It is clear that $|\mathfrak{T}_1| \leq \mathcal{S}\|y_h\|$ and $|\mathfrak{T}_2| \lesssim \|z_h\|_L \|y_h\|_L$. To bound $\mathfrak{T}_3$ and $\mathfrak{T}_4$, we observe that, for $x \in \mathbb{R}^{m_u}$ and $y \in \mathbb{R}^{m_\sigma}$, the Cauchy–Schwarz inequality yields $y^t \mathcal{D}^{\sigma u} x \leq \|\mathcal{D}^{\sigma u} x\|_{\mathbb{R}^{m_\sigma}} \|y\|_{\mathbb{R}^{m_\sigma}}$ where $\|\cdot\|_{\mathbb{R}^{m_\sigma}}$ denotes the Euclidean norm in $\mathbb{R}^{m_\sigma}$. Moreover,

$$\|\mathcal{D}^{\sigma u} x\|_{\mathbb{R}^{m_\sigma}}^2 = (\mathcal{D}^{\sigma u} x)^t (\mathcal{D}^{\sigma u} x) = x^t (\mathcal{D}^{u\sigma} \mathcal{D}^{\sigma u}) x \lesssim x^t (\mathcal{D}^{u\sigma} \mathcal{D}^{\sigma u})^{1/2} x.$$

Hence, using the lower bounds in (7.49c) and (7.49d) together with the discrete trace inequality (1.37), we infer

$$|\mathfrak{T}_3| + |\mathfrak{T}_4| \lesssim |z_h^u|_S \|y_h^\sigma\|_{L_\sigma}.$$

Collecting the above bounds and using Young's inequality yields (7.55). □

Our goal is now to estimate using Theorem 1.35 the approximation error $(z - z_h)$ in the $\|\cdot\|$-norm defined by (7.54). Discrete stability and consistency have already been proven; thus, it only remains to address the boundedness of the bilinear form a_h. It is convenient to integrate by parts the two terms involving the differential operators B and $\hat{B}$. This yields

$$a_h(z, y_h) = \sum_{T\in\mathcal{T}_h} \left\{ (\mathcal{K}^{\sigma\sigma} z^\sigma, y_h^\sigma)_{L_\sigma,T} + (\mathcal{K}^{uu} z^u, y_h^u)_{L_u,T} \right.$$

$$\left. - (z^u, \hat{B} y_h^\sigma)_{L_\sigma,T} - (z^\sigma, B y_h^u)_{L_u,T} \right\} + \sum_{F\in\mathcal{F}_h^b} (\mathcal{D}_F^{u\sigma} z^\sigma, y_h^u)_{L_\sigma,F}$$

$$+ \sum_{F\in\mathcal{F}_h^i} \left\{ (\mathcal{D}_F^{\sigma u}\{\!\{z^u\}\!\}, [\![y_h^\sigma]\!])_{L_\sigma,F} + (\mathcal{D}_F^{u\sigma}\{\!\{z^\sigma\}\!\}, [\![y_h^u]\!])_{L_u,F} \right\}$$

$$+ \sum_{F\in\mathcal{F}_h^b} (S_F^b z^u, y_h^u)_{L_u,F} + \sum_{F\in\mathcal{F}_h^i} (S_F^i [\![z^u]\!], [\![y_h^u]\!])_{L_u,F}. \quad (7.56)$$

We introduce the following norm on V_{*h}:

$$\|y\|_*^2 := \|y\|^2 + \sum_{T\in\mathcal{T}_h} \left\{ h_T^{-2} \|y^u\|_{L_u,T}^2 + h_T^{-1} \|y^u\|_{L_u,\partial T}^2 + h_T \|y^\sigma\|_{L_\sigma,\partial T}^2 \right\}.$$

Lemma 7.24 (Boundedness). *For all* $(w, y_h) \in V_{*h} \times V_h$, *there holds*

$$a_h(w, y_h) \lesssim \|w\|_* \|y_h\|. \quad (7.57)$$

7.4. Two-Field Systems

Proof. Let $(w, y_h) \in V_{*h} \times V_h$. Let $\mathfrak{T}_1, \ldots, \mathfrak{T}_9$ be the terms on the right-hand side of (7.56). We observe that

$$|\mathfrak{T}_1| + |\mathfrak{T}_2| + |\mathfrak{T}_4| + |\mathfrak{T}_8| + |\mathfrak{T}_9| \lesssim \|w\| \, \|y_h\| \leq \|w\|_* \, \|y_h\|.$$

Using the inverse inequality (1.36), $\mathfrak{T}_3$ is bounded as

$$|\mathfrak{T}_3| \lesssim \left(\sum_{T \in \mathcal{T}_h} h_T^{-2} \|w^u\|_{L_u,T}^2 \right)^{1/2} \|y_h\|_L \leq \|w\|_* \, \|y_h\|.$$

Using the lower bound in (7.49c) and (7.49d) leads to

$$|\mathfrak{T}_5| + |\mathfrak{T}_7| \lesssim \left(\sum_{T \in \mathcal{T}_h} h_T \|w^\sigma\|_{L_\sigma,\partial T}^2 \right)^{1/2} |y_h^u|_S \leq \|w\|_* \, \|y_h\|.$$

Finally, using the discrete trace inequality (1.37) and the mesh regularity yields

$$|\mathfrak{T}_6| \lesssim \left(\sum_{T \in \mathcal{T}_h} h_T^{-1} \|w^u\|_{L_u,\partial T}^2 \right)^{1/2} \|y_h\|_L \leq \|w\|_* \, \|y_h\|.$$

Collecting the above bounds, we infer (7.57). □

Theorem 7.25 (Error estimate and convergence rate for smooth solutions). *Let z solve (7.17) and assume that the Friedrichs' system features the two-field structure described in Sect. 7.4.1. Let z_h solve (7.50) with a_h defined by (7.51). Then, there holds*

$$\|z - z_h\| \lesssim \inf_{y_h \in V_h} \|z - y_h\|_*.$$

Moreover, if $z \in [H^k(\Omega)]^{m_\sigma} \times [H^{k+1}(\Omega)]^{m_u}$, there holds

$$\|z - z_h\| \leq C_z h^k, \tag{7.58}$$

where $C_z = C(\|z^\sigma\|_{[H^k(\Omega)]^{m_\sigma}} + \|z^u\|_{[H^{k+1}(\Omega)]^{m_u}})$ with C independent of h.

The error estimate (7.58) is optimal for the broken graph norm of the u-component. It is also optimal for the L_σ-norm of the σ-component if $l = k-1$. Furthermore, the estimate is suboptimal by one order for the L_u-norm. A sharper estimate can be derived using a duality argument assuming elliptic regularity. The duality argument presented for the Poisson problem in Sect. 4.2.4 can be extended to the setting of two-field Friedrichs' systems [143].

7.4.3 Partial Coercivity

For the div-grad problem and the linear elasticity equations, a slightly more general setting modifying assumption (7.2c) can be considered so as to avoid the need for a positive zero-order term in the u-component. Specifically, assumption (7.2c) is now replaced by

$$\exists \mu_0 > 0, \quad \forall (z^\sigma, z^u) \in L, \quad (\mathcal{A}^0 z, z)_L \geq \mu_0 \|z^\sigma\|_{L_\sigma}^2 \quad \text{a.e. in } \Omega. \tag{7.59}$$

In view of the block-decomposition (7.47), this means that the matrix $\mathcal{K}^{\sigma\sigma}$ is uniformly positive definite and that the matrix $\mathcal{K}^{uu}$ is only nonnegative. This situation is encountered for instance when considering the mixed formulation of the Poisson problem (7.9) leading to $\mathcal{K}^{uu} = 0$ or when considering the linear elasticity equations (7.11) with the coefficient α set to zero.

At the continuous level, the well-posedness of the Friedrichs' system in its weak formulation (7.16) can be established as in [144] provided the following Poincaré-like inequality holds true:

$$\exists \alpha_0 > 0, \quad \forall z \in V \cup V^*, \quad \alpha_0 \|z^u\|_{L_u} \leq a(z,z)^{1/2} + \|Bz^u\|_{L_\sigma}. \tag{7.60}$$

For the Poisson problem with homogeneous Dirichlet boundary conditions, $V = V^* = H_0^1(\Omega)$ and $Bz^u = \nabla z^u$, so that (7.60) results from the usual Poincaré inequality. Robin/Neumann boundary conditions can also be treated as detailed in [144]. For the linear elasticity equations with homogeneous Dirichlet boundary conditions, $V = V^* = [H_0^1(\Omega)]^d$ and $Bz^u = \frac{1}{2}(\nabla z^u + (\nabla z^u)^t)$, so that (7.60) results from Korn's First Inequality (see, e.g., Ciarlet [92, p. 24]).

At the discrete level, the dG bilinear form a_h defined by (7.51) can be used, provided the following discrete counterpart of (7.60) holds true:

$$\exists \alpha_1 > 0, \quad \forall z_h \in V_h, \quad \alpha_1 \|z_h^u\|_{L_u} \leq a_h(z_h, z_h)^{1/2} + \|B_h z_h^u\|_{L_\sigma}, \tag{7.61}$$

with α_1 independent of h. For the Poisson problem, (7.61) results from the discrete coercivity (7.55) and the discrete Poincaré inequality (4.20) stating that, for all $z_h^u \in U_h$,

$$\alpha_1 \|z_h^u\|_{L^2(\Omega)}^2 \leq |z_h^u|_S^2 + \|\nabla_h z_h^u\|_{[L^2(\Omega)]^d}^2,$$

since the penalty parameters in S_F^b and S_F^i scale as h_F^{-1}. For the linear elasticity equations, (7.61) results from a discrete Korn inequality; see Duarte, do Carmo, and Rochinha [137] and Brenner [52]. It is worthwhile to notice that the resulting dG method is not robust with respect to the compressibility parameter λ in (7.11), leading to the so-called locking phenomenon. The origin of the problem stems from the elimination of the stress tensor since the matrix $\mathcal{K}^{\sigma\sigma}$ is no longer invertible if $\lambda = 0$. A robust approximation in the incompressible limit can be designed by working on the mixed formulation, as observed by Franca and Stenberg [162] in the context of Galerkin/Least-Squares approximations and by Brezzi and Fortin [57, Chap. VI] and by Schwab and Suri [276] in the context of conforming mixed finite element methods. In the context of dG methods, one possibility analyzed by Ern and Guermond [144] consists in penalizing the boundary values and interface jumps of the trace of the stress tensor together with those of the displacement vector. Alternatively, an additional scalar variable (the pressure) can be introduced. Indeed, a direct calculation shows that (compare with (7.12))

$$\sigma^t (\mathcal{A}^0)^{\sigma\sigma} \sigma = \tfrac{\lambda}{(d+\lambda)^2} |\operatorname{tr}(\sigma)|^2 + |\sigma - \tfrac{1}{d+\lambda} \operatorname{tr}(\sigma) \mathbb{1}_d|_{\ell^2}^2,$$

so that in the limit $\lambda \to 0$, only the control on $\text{tr}(\sigma)$ is lost. This second approach leads to the dG methods analyzed in Sect. 6.1.2 in the context of steady Stokes flows.

7.5 The Unsteady Case

In this section, we consider unsteady Friedrichs' systems. Space discretization is achieved using the dG schemes analyzed in Sects. 7.3 and 7.4. Two approaches are considered for time discretization. For general Friedrichs' systems, we employ explicit Runge–Kutta schemes and show how the ideas presented in Sect. 3.1 for the unsteady advection-reaction problem (i.e., energy-based stability estimates) extend in a straightforward manner to unsteady Friedrichs' systems. Then, focusing on linear wave propagation problems in mixed form, that is, on unsteady Friedrichs' systems with the two-field structure identified in Sect. 7.4, we study explicit leap-frog schemes to march in time. The attractive property is that the use of centered fluxes leads to a certain energy conservation under a CFL condition. The approach can also accommodate local time stepping.

7.5.1 Explicit Runge–Kutta Schemes

In this section, following the ideas of Burman, Ern, and Fernández [66], we extend to Friedrichs' systems the explicit RK2 and RK3 schemes (3.23) and (3.27) analyzed in Chap. 3 in the context of the unsteady advection-reaction equation.

7.5.1.1 The Continuous Setting

We are interested in the time evolution problem

$$\partial_t z + Az = f \quad \text{in } \Omega \times (0, t_{\text{F}}), \tag{7.62a}$$

$$(\mathcal{M} - \mathcal{D})z = 0 \quad \text{on } \partial\Omega \times (0, t_{\text{F}}), \tag{7.62b}$$

$$z(\cdot, t = 0) = z_0 \quad \text{in } \Omega, \tag{7.62c}$$

with operator A defined by (7.3) and (7.4), source term f, initial datum z_0, and finite time $t_{\text{F}} > 0$. The field $\mathcal{D}$ is defined by (7.6) and the field $\mathcal{M}$ satisfies (7.7); cf. Sect. 7.1.2 for examples in the steady case. We assume that the fields $\mathcal{A}^0, \mathcal{A}^1, \ldots, \mathcal{A}^d$ defining the operator A are *time-independent* and that they satisfy assumptions (7.2a) and (7.2b). However, letting $\mathcal{X} = \sum_{k=1}^{d} \partial_k \mathcal{A}^k$, the field

$$\Lambda := \frac{1}{2}\left(\mathcal{A}^0 + (\mathcal{A}^0)^t\right) - \frac{1}{2}\mathcal{X} \in [L^\infty(\Omega)]^{m,m},$$

is no longer required to satisfy the positivity assumption (7.2c).

To fix the ideas, we assume that the model problem (7.62) has been normalized so that all the components of z are nondimensional. Then, the components of the fields $\mathcal{A}^1, \ldots, \mathcal{A}^d$ (and also those of $\mathcal{D}$ and $\mathcal{M}$) scale as velocities, while those of $\mathcal{A}^0$, $\mathcal{X}$, and Λ scale as the reciprocal of a time. We consider the reference velocity

$$\beta_c := \max(\|\mathcal{A}^1\|_{[L^\infty(\Omega)]^{m,m}}, \ldots, \|\mathcal{A}^d\|_{[L^\infty(\Omega)]^{m,m}}, \|\mathcal{M}\|_{[L^\infty(\Omega)]^{m,m}}).$$

We assume for simplicity that all the components of the fields $\mathcal{A}^1, \ldots, \mathcal{A}^d$ are Lipschitz continuous, and we denote by L_A the maximum Lipschitz module for all the components of all the fields. Then, similarly to the advection-reaction problem, we define the reference time

$$\tau_c := \{\max(\|\mathcal{A}^0\|_{[L^\infty(\Omega)]^{m,m}}, L_A)\}^{-1}.$$

We notice that it is possible that $\|\mathcal{A}^0\|_{[L^\infty(\Omega)]^{m,m}} = L_A = 0$ (e.g., no zero-order term and constant fields $\mathcal{A}^1, \ldots, \mathcal{A}^d$), in which case $\tau_c = \infty$. It is therefore convenient to introduce an additional reference time

$$\tau_* := \min(t_F, \tau_c),$$

which is always finite.

Using the notation of Sect. 3.1.1 for space-time functional spaces, we assume that $f \in C^0(L)$. Moreover, we are interested in *smooth solutions*, specifically,

$$z \in C^0([H^1(\Omega)]^m) \cap C^1(L).$$

(At this stage, we can replace $[H^1(\Omega)]^m$ by the graph space V defined by (7.22).) This implies that the initial datum z_0 is in $[H^1(\Omega)]^m$ and that we can formulate the time evolution problem (7.62) as

$$d_t z(t) + A z(t) = f(t) \qquad \forall t \in [0, t_F]. \tag{7.63}$$

The following basic energy estimate is proven by proceeding as in the proof of Lemma 3.2 for the unsteady advection-reaction equation.

Lemma 7.26 (Energy estimate). *Let $z \in C^0([H^1(\Omega)]^m) \cap C^1(L)$ solve (7.63). Then, introducing the time scale $\varsigma := (t_F^{-1} + 2\|\Lambda\|_{[L^\infty(\Omega)]^{m,m}})^{-1}$, there holds*

$$\|z(t)\|_L^2 \le e^{t/\varsigma}(\|z_0\|_L^2 + \varsigma t_F \|f\|_{C^0(L)}^2) \qquad \forall t \in [0, t_F].$$

7.5.1.2 Discretization

As in Sect. 7.3 for the steady case, the discrete space V_h is taken to be $[\mathbb{P}_d^k(\mathcal{T}_h)]^m$ with polynomial degree $k \ge 0$ and $\mathcal{T}_h$ belonging to an admissible mesh sequence. We set $V_{*h} = [H^1(\Omega)]^m + V_h$. For simplicity, as for the unsteady advection-reaction equation in Sect. 3.1.2, we consider quasi-uniform mesh sequences, meaning that the ratio of the largest to the smallest element diameter of a given mesh is uniformly bounded.

We define the discrete operators $A_h^{\mathrm{cf}} : V_{*h} \to V_h$ and $S_h : V_{*h} \to V_h$ such that, for all $(y, w_h) \in V_{*h} \times V_h$,

$$(A_h^{\mathrm{cf}} y, w_h)_L := a_h^{\mathrm{cf}}(y, w_h), \qquad (S_h y, w_h)_L := s_h(y, w_h),$$

7.5. The Unsteady Case

with the centered flux dG bilinear form a_h^{cf} defined by (7.31) and the stabilization bilinear form s_h defined by (7.35). Letting $a_h = a_h^{\text{cf}} + s_h$ as in (7.34), we also define the discrete operator $A_h : V_{*h} \to V_h$ such that

$$A_h := A_h^{\text{cf}} + S_h.$$

The discrete operator A_h can be used to formulate the space semidiscrete problem (notice that the zero-order term is incorporated in the definition of A_h)

$$d_t z_h(t) + A_h z_h(t) = f_h(t) \qquad \forall t \in [0, t_F],$$

with initial condition $z_h(0) = \pi_h z_0$ and source term $f_h(t) = \pi_h f(t)$ for all $t \in [0, t_F]$. Here, π_h denotes the L-orthogonal projection onto V_h.

The $\mathbb{R}^{m,m}$-valued fields S_F^b and S_F^i in the definition of s_h are assumed to satisfy the design conditions (7.36)–(7.38). Conditions (7.38) are slightly modified to account for the scaling of S_F^b and S_F^i in terms of the reference velocity β_c, yielding

$$S_F^b \le \alpha_1 \beta_c \mathbb{1}_m, \qquad \alpha_2 |\mathcal{D}_F| \le S_F^i \le \alpha_3 \beta_c \mathbb{1}_m,$$

and, for all $y, z \in [L^2(F)]^m$,

$$|((\mathcal{M} - \mathcal{D})y, z)_{L,F}| \le \alpha_4 \beta_c^{1/2} ((S_F^b + \mathcal{M})y, y)_{L,F}^{1/2} \|z\|_{L,F},$$

$$|((\mathcal{M} + \mathcal{D})y, z)_{L,F}| \le \alpha_5 \beta_c^{1/2} \|y\|_{L,F} ((S_F^b + \mathcal{M})z, z)_{L,F}^{1/2},$$

with user-dependent positive parameters $\alpha_1, \ldots, \alpha_5$. Owing to the adopted scalings, these parameters are nondimensional.

Let δt be the time step, taken to be constant for simplicity and such that $t_F = N \delta t$ where N is an integer. For $0 \le n \le N$, a superscript n indicates the value of a function at the discrete time $t^n := n\delta t$. We consider the explicit RK2 scheme

$$w_h^n = z_h^n - \delta t A_h z_h^n + \delta t f_h^n,$$
$$z_h^{n+1} = \tfrac{1}{2}(z_h^n + w_h^n) - \tfrac{1}{2}\delta t A_h w_h^n + \tfrac{1}{2}\delta t f_h^{n+1},$$

and the explicit RK3 scheme

$$w_h^n = z_h^n - \delta t A_h z_h^n + \delta t f_h^n,$$
$$y_h^n = \tfrac{1}{2}(z_h^n + w_h^n) - \tfrac{1}{2}\delta t A_h w_h^n + \tfrac{1}{2}\delta t (f_h^n + \delta t d_t f_h^n),$$
$$z_h^{n+1} = \tfrac{1}{3}(z_h^n + w_h^n + y_h^n) - \tfrac{1}{3}\delta t A_h y_h^n + \tfrac{1}{3}\delta t f_h^{n+1},$$

both with the initial condition $z_h^0 = \pi_h z_0$. Variants in handling the source term can be considered.

7.5.1.3 Convergence Results

We only quote the convergence results and refer the reader to [66] for the proofs. We abbreviate as $a \lesssim b$ the inequality $a \le Cb$ with positive C independent of h, δt, and the problem data (that is, f, $\mathcal{A}^0, \mathcal{A}^1, \ldots, \mathcal{A}^d$, and $\mathcal{M}$).

Theorem 7.27 (Convergence for RK2). *Assume $z \in C^3(L) \cap C^0([H^1(\Omega)]^m)$ and $f \in C^2(L)$.*

(i) *In the case $k \geq 2$, assume the strengthened 4/3-CFL condition $\delta t \leq \varrho' \tau_*^{-1/3}(h/\beta_c)^{4/3}$ for some positive real number ϱ';*

(ii) *In the piecewise affine case ($k = 1$), assume the CFL condition $\delta t \leq \varrho \beta_c^{-1} h$ with $\varrho \leq \varrho^{\mathrm{RK2}}$ for a suitable threshold ϱ^{RK2} independent of the h, δt, and the problem data.*

Assume $d_t^s z \in C^s([H^{k+1-s}(\Omega)]^m)$ for $s \in \{0,1\}$. Then,

$$\|z^N - z_h^N\|_L + \left(\sum_{m=0}^{N-1} \delta t |z^m - z_h^m|_{MS}^2\right)^{1/2} \lesssim \chi\left(\delta t^2 + h^{k+1/2}\right),$$

where χ depends on bounded norms of f and z. Here, for $y \in V_{*h}$,

$$|y|_{MS}^2 := \frac{1}{2}(\mathcal{M}y, y)_{L,\partial\Omega} + s_h(y,y),$$

Theorem 7.28 (Convergence for RK3). *Assume $z \in C^4(L) \cap C^0([H^1(\Omega)]^m)$ and $f \in C^3(L)$. Assume the CFL condition $\delta t \leq \varrho \beta_c^{-1} h$ with $\varrho \leq \varrho^{\mathrm{RK3}}$ for a suitable threshold ϱ^{RK3} independent of h, δt, and the problem data. Assume $d_t^s z \in C^s([H^{k+1-s}(\Omega)]^m)$ for $s \in \{0,1,2\}$. Then,*

$$\|z^N - z_h^N\|_L + \left(\sum_{m=0}^{N-1} \delta t |z^m - z_h^m|_{MS}^2\right)^{1/2} \lesssim \chi\left(\delta t^3 + h^{k+1/2}\right),$$

where χ depends on bounded norms of f and z.

7.5.2 Explicit Leap-Frog Schemes for Two-Field Systems

In this section, we consider the unsteady version of the two-field Friedrichs' systems examined in Sect. 7.4. Specifically, we are interested in linear wave propagation problems without zero-order terms ($\mathcal{A}^0 = 0$) in the autonomous case ($f = 0$). Under suitable boundary conditions (the reflecting boundary conditions of Definition 7.29 below), these problems conserve an energy. In many applications, e.g., for resonance computations in cavities, it is important to conserve this energy at the discrete level as much as possible. To avoid excessive dissipation in space, the stabilization bilinear form s_h is discarded, that is, the fields S_F^b and S_F^i are set to zero. This amounts to the use of centered fluxes in the dG scheme. We examine explicit leap-frog schemes to march in time combined with centered dG fluxes. This approach ensures energy-stability under a CFL condition and lends itself to the use of local time stepping.

7.5.2.1 The Continuous Setting

The fields $\mathcal{A}^1, \ldots, \mathcal{A}^d$ have the block-structure identified by (7.47), namely

$$\mathcal{A}^k = \left[\begin{array}{c|c} \mathbb{0}_{m_\sigma,m_\sigma} & \mathcal{B}^k \\ \hline (\mathcal{B}^k)^t & \mathbb{0}_{m_u,m_u} \end{array}\right] \qquad \forall k \in \{1,\ldots,d\},$$

and we suppose for simplicity that the $\mathbb{R}^{m_\sigma,m_u}$-valued fields $\mathcal{B}^k$ are constant in Ω. As before, we consider the differential operators $B = \sum_{k=1}^d \mathcal{B}^k \partial_k$ and $\hat{B} = \sum_{k=1}^d (\mathcal{B}^k)^t \partial_k$. The time evolution problem is (for simplicity, we write σ instead of z^σ and u instead of z^u)

$$\partial_t \sigma + Bu = 0 \qquad \text{in } \Omega \times (0, t_F), \tag{7.64a}$$
$$\partial_t u + \hat{B}\sigma = 0 \qquad \text{in } \Omega \times (0, t_F), \tag{7.64b}$$
$$(\mathcal{M} - \mathcal{D})(\sigma, u)^t = 0 \qquad \text{on } \partial\Omega \times (0, t_F), \tag{7.64c}$$
$$(\sigma, u)(\cdot, t=0) = (\sigma_0, u_0) \qquad \text{in } \Omega, \tag{7.64d}$$

with initial data $(\sigma_0, u_0) \in L_\sigma \times L_u$. We observe that (7.64a) and (7.64b) imply

$$\partial_{tt} u = \hat{B}Bu.$$

We are interested in *smooth solutions* such that $z = (\sigma, u)^t \in C^0([H^1(\Omega)]^m) \cap C^1(L)$. The boundary field $\mathcal{D}$ admits the block-structure

$$\mathcal{D} = \left[\begin{array}{c|c} \mathbb{0}_{m_\sigma,m_\sigma} & \mathcal{D}^{\sigma u} \\ \hline \mathcal{D}^{u\sigma} & \mathbb{0}_{m_u,m_u} \end{array}\right],$$

where $\mathcal{D}^{\sigma u} = \sum_{k=1}^d \mathcal{B}^k n_k$ is $\mathbb{R}^{m_\sigma,m_u}$-valued and $\mathcal{D}^{u\sigma} = (\mathcal{D}^{\sigma u})^t$ is $\mathbb{R}^{m_u,m_\sigma}$-valued. As in Sect. 7.4, we assume that $\mathcal{D}^{\sigma u}$ is injective.

Definition 7.29 (Reflecting boundary condition). We say that the boundary condition (7.64c) is *reflecting (or non-dissipative)* if the field $\mathcal{M}$ takes skew-symmetric values.

The interest in reflecting boundary conditions is motivated by the following energy conservation property.

Lemma 7.30 (Energy conservation). *Assume reflecting boundary conditions. Let $z = (\sigma, u)^t \in C^0([H^1(\Omega)]^m) \cap C^1(L)$ solve the time evolution problem (7.64). Then, defining the energy as*

$$\mathcal{E}(t) := \|\sigma\|_{L_\sigma}^2 + \|u\|_{L_u}^2 \qquad \forall t \in [0, t_F],$$

there holds

$$\mathcal{E}(t) = \mathcal{E}(0) \qquad \forall t \in [0, t_F].$$

Proof. We multiply (7.64a) by σ, (7.64b) by u, and integrate over Ω to infer
$$\frac{1}{2}d_t\mathcal{E}(t) + (\sigma, Bu)_{L_\sigma} + (u, \hat{B}\sigma)_{L_u} = 0.$$
Moreover,
$$(\sigma, Bu)_{L_\sigma} + (u, \hat{B}\sigma)_{L_u} = (\sigma, \mathcal{D}^{\sigma u} u)_{L_\sigma, \partial\Omega} = \tfrac{1}{2}(z, \mathcal{D}z)_{L,\partial\Omega} = \tfrac{1}{2}(z, \mathcal{M}z)_{L,\partial\Omega} = 0,$$
since the boundary condition is reflecting. Hence, $d_t\mathcal{E}(t) = 0$. $\square$

Two examples of reflecting boundary conditions are obtained by taking
$$\mathcal{M} = \begin{bmatrix} \mathbb{0}_{m_\sigma,m_\sigma} & -\gamma\mathcal{D}^{\sigma u} \\ \hline \gamma\mathcal{D}^{u\sigma} & \mathbb{0}_{m_u,m_u} \end{bmatrix},$$
with parameter $\gamma \in \{-1, 1\}$. The choice $\gamma = 1$ leads to $\mathcal{D}^{\sigma u} u = 0$ and hence to $u = 0$ since $\mathcal{D}^{\sigma u}$ is injective, i.e., a homogeneous Dirichlet boundary condition is enforced. The choice $\gamma = -1$ yields $\mathcal{D}^{u\sigma}\sigma = 0$, which can be interpreted as a homogeneous Neumann boundary condition.

7.5.2.2 Examples

The two classical examples for (7.64a) and (7.64b) are the unsteady div-grad problem associated with acoustic waves and the unsteady curl-curl problem associated with electromagnetic waves.

For the unsteady div-grad problem, let c_0 be a reference velocity, and consider the PDE system
$$\partial_t q + \nabla p = 0,$$
$$c_0^{-2}\partial_t p + \nabla \cdot q = 0,$$
where p is the pressure and q the momentum per unit volume. Setting $u = c_0^{-1}p$ and $\sigma = q$ yields (7.64a) and (7.64b) with the differential operators
$$B = c_0\nabla, \qquad \hat{B} = c_0\nabla\cdot.$$
Reflecting boundary conditions enforce p or the normal component of q to be zero on $\partial\Omega$.

For the unsteady curl-curl problem, let μ, ϵ be positive constants and let $c_0 = (\mu\epsilon)^{1/2}$ be the reference velocity. Consider the PDE system
$$\mu\partial_t H + \nabla\times E = 0,$$
$$\epsilon\partial_t E - \nabla\times H = 0,$$
where H is the magnetic field and E is the electric field. Setting $u = \epsilon^{1/2}E$ and $\sigma = \mu^{1/2}H$ yields (7.64a) and (7.64b) with the differential operators
$$B = c_0\nabla\times, \qquad \hat{B} = -c_0\nabla\times.$$
Reflecting boundary conditions enforce the tangential component of either H or E to be zero on $\partial\Omega$.

7.5. The Unsteady Case

7.5.2.3 Leap-Frog Scheme in Time and Centered Fluxes in Space

To discretize in space, we consider equal-order approximation for the two components u and σ, namely we set

$$U_h = [\mathbb{P}_d^k(\mathcal{T}_h)]^{m_u}, \qquad \Sigma_h = [\mathbb{P}_d^k(\mathcal{T}_h)]^{m_\sigma},$$

with polynomial degree $k \geq 0$ (observe that $k = 0$ is allowed here) and $\mathcal{T}_h$ belonging to an admissible mesh sequence. We define the following bilinear form on $\Sigma_h \times U_h$:

$$b_h(\tau_h, v_h) := \sum_{T \in \mathcal{T}_h} (Bv_h, \tau_h)_{L_\sigma, T} - \frac{\gamma+1}{2} \sum_{F \in \mathcal{F}_h^b} (\mathcal{D}_F^{\sigma u} v_h, \tau_h)_{L_\sigma, F}$$
$$- \sum_{F \in \mathcal{F}_h^i} (\mathcal{D}_F^{\sigma u} [\![v_h]\!], \{\!\{\tau_h\}\!\})_{L_\sigma, F}.$$

The discrete bilinear form b_h can be extended to larger functional spaces, but this extension is not needed in what follows. To allow for a more compact notation, we define the discrete differential operator $B_h : U_h \to \Sigma_h$ such that, for all $(\tau_h, v_h) \in \Sigma_h \times U_h$,

$$(B_h v_h, \tau_h)_{L_\sigma} := b_h(\tau_h, v_h).$$

We also define the discrete differential operator $\hat{B}_h : \Sigma_h \to U_h$ such that, for all $(\tau_h, v_h) \in \Sigma_h \times U_h$,

$$(\hat{B}_h \tau_h, v_h)_{L_u} := -b_h(\tau_h, v_h) = -(B_h v_h, \tau_h)_{L_\sigma}.$$

Integrating by parts on each mesh element yields

$$(\hat{B}_h \tau_h, v_h)_{L_u} = \sum_{T \in \mathcal{T}_h} (v_h, \hat{B} \tau_h)_{L_u, T} + \frac{\gamma-1}{2} \sum_{F \in \mathcal{F}_h^b} (\mathcal{D}_F^{u\sigma} \tau_h, v_h)_{L_u, F}$$
$$- \sum_{F \in \mathcal{F}_h^i} (\mathcal{D}_F^{u\sigma} [\![\tau_h]\!], \{\!\{v_h\}\!\})_{L_u, F}.$$

Remark 7.31 (Centered fluxes). The link between the discrete bilinear form b_h and the discrete bilinear form a_h defined by (7.51) and considered in Sect. 7.4.2 for homogeneous Dirichlet boundary conditions on u is that, for $\gamma = 1$ and $\mathcal{K} = 0$, taking $S_F^b = 0$ and $S_F^i = 0$ yields

$$a_h((\sigma_h, u_h), (\tau_h, v_h)) = b_h(\tau_h, u_h) - b_h(\sigma_h, v_h).$$

Hence, the discrete bilinear form b_h is associated with the use of centered fluxes for both components.

For future use, we state the following boundedness result, which is a consequence of the definition of B_h, the inverse inequality (1.36), and the discrete trace inequality (1.37).

Lemma 7.32 (Bound on B_h). *There is C_B, independent of h and the problem data, such that, for all $v_h \in U_h$,*
$$\|B_h v_h\|_{L_\sigma} \leq C_B \beta_c h^{-1} \|v_h\|_{L_u},$$
with the reference velocity $\beta_c := \max_{k \in \{1,\ldots,d\}} \|\mathcal{B}^k\|_{\mathbb{R}^{m_\sigma,m_u}}$.

We consider the following leap-frog scheme: Given $u_h^{-1/2}$ and σ_h^0, compute, for all $n \geq 0$,
$$u_h^{n+1/2} - u_h^{n-1/2} + \delta t \hat{B}_h \sigma_h^n = 0, \tag{7.65a}$$
$$\sigma_h^{n+1} - \sigma_h^n + \delta t B_h u_h^{n+1/2} = 0. \tag{7.65b}$$

This scheme is fully explicit.

Lemma 7.33 (Pseudo-energy conservation). *The leap-frog scheme (7.65) conserves the discrete pseudo-energy defined, for all $n \in \{0, \ldots, N\}$, as*
$$\tilde{\mathcal{E}}_h^n := \|\sigma_h^n\|_{L_\sigma}^2 + (u_h^{n+1/2}, u_h^{n-1/2})_{L_u},$$
that is, for all $n \geq 0$, $\tilde{\mathcal{E}}_h^n = \tilde{\mathcal{E}}_h^0$.

Proof. We multiply (7.65b) by $\bar{\sigma}_h^{n+1/2} = \frac{1}{2}(\sigma_h^n + \sigma_h^{n+1})$ and integrate over Ω to infer
$$\|\sigma_h^{n+1}\|_{L_\sigma}^2 - \|\sigma_h^n\|_{L_\sigma}^2 + 2\delta t (\bar{\sigma}_h^{n+1/2}, B_h u_h^{n+1/2})_{L_\sigma} = 0.$$
Furthermore, we form the average of (7.65a) at n and $(n+1)$, multiply by $u_h^{n+1/2}$, and integrate over Ω to obtain
$$(u_h^{n+3/2}, u_h^{n+1/2})_{L_u} - (u_h^{n+1/2}, u_h^{n-1/2})_{L_u} + 2\delta t (u_h^{n+1/2}, \hat{B}_h \bar{\sigma}_h^{n+1/2})_{L_u} = 0.$$
Summing the two above equations yields
$$\tilde{\mathcal{E}}_h^{n+1} - \tilde{\mathcal{E}}_h^n + 2\delta t \left((\bar{\sigma}_h^{n+1/2}, B_h u_h^{n+1/2})_{L_\sigma} + (u_h^{n+1/2}, \hat{B}_h \bar{\sigma}_h^{n+1/2})_{L_u} \right) = 0,$$
and hence, $\tilde{\mathcal{E}}_h^{n+1} - \tilde{\mathcal{E}}_h^n = 0$ owing to the definitions of B_h and $\hat{B}_h$. □

The discrete pseudo-energy $\tilde{\mathcal{E}}_h^n$ is a quadratic form on the discrete unknowns σ_h^n and u_h^n, but is not positive definite. To remedy this, a CFL condition on the time step can be invoked.

Theorem 7.34 (Energy estimate). *Under the CFL condition*
$$\delta t \leq C_B^{-1} \frac{h}{\beta_c}, \tag{7.66}$$
there holds, for all $n \in \{1, \ldots, N\}$,
$$\|\sigma_h^n\|_{L_\sigma}^2 + \|u_h^{n-1/2}\|_{L_u}^2 \leq 2\tilde{\mathcal{E}}_h^0.$$

7.5. The Unsteady Case

Proof. We observe that

$$\begin{aligned}
\tilde{\mathcal{E}}_h^n &= \|\sigma_h^n\|_{L_\sigma}^2 + (u_h^{n+1/2}, u_h^{n-1/2})_{L_u} \\
&= \|\sigma_h^n\|_{L_\sigma}^2 + \|u_h^{n-1/2}\|_{L_u}^2 + (u_h^{n+1/2} - u_h^{n-1/2}, u_h^{n-1/2})_{L_u} \\
&= \|\sigma_h^n\|_{L_\sigma}^2 + \|u_h^{n-1/2}\|_{L_u}^2 - \delta t (u_h^{n-1/2}, \hat{B}_h \sigma_h^n)_{L_u} \\
&= \|\sigma_h^n\|_{L_\sigma}^2 + \|u_h^{n-1/2}\|_{L_u}^2 + \delta t (B_h u_h^{n-1/2}, \sigma_h^n)_{L_\sigma} \\
&\geq \|\sigma_h^n\|_{L_\sigma}^2 + \|u_h^{n-1/2}\|_{L_u}^2 - (\delta t C_B \beta_c h^{-1}) \|u_h^{n-1/2}\|_{L_u} \|\sigma_h^n\|_{L_\sigma},
\end{aligned}$$

where we have used the Cauchy–Schwarz inequality together with Lemma 7.32. Hence, using the CFL condition (7.66) and Young's inequality yields

$$\tilde{\mathcal{E}}_h^n \geq \frac{1}{2}\left(\|\sigma_h^n\|_{L_\sigma}^2 + \|u_h^{n-1/2}\|_{L_u}^2\right),$$

whence the assertion owing to Lemma 7.33. □

7.5.2.4 Verlet Schemes and Local Time Stepping

Owing to the CFL condition (7.66), the time step is restricted by the diameter of the smallest mesh element. When modeling systems with small geometrical details (e.g., devices in emerging technologies related to optical waveguides or furtivity), this condition leads to very small time steps and, hence, to prohibitive computational costs. A possible work-around is to use local time stepping, meaning that mesh elements are advanced with different local time steps, and that the size of the local time step is adapted locally to the meshsize. It turns out that the particular context of linear wave propagation problems approximated by centered dG schemes is one of the rare cases where local time stepping can be proven rigorously to be stable, at least in the two-scale situation (corresponding to the algorithm $V^2(\delta t)$ described below). For the sake of brevity, we only state here the main ideas. We refer the reader to Piperno [259] for further insight and to Hardy, Okunbor, and Skeel [186] for the theory of symplectic integrators for dynamical Hamiltonian systems.

The elementary brick to formulate the time stepping scheme is based on the Verlet method (equivalent to the leap-frog scheme), which can be written as

$$u_h^{n+1/2} - u_h^n + \tfrac{1}{2}\delta t \hat{B}_h \sigma_h^n = 0, \tag{7.67a}$$

$$\sigma_h^{n+1} - \sigma_h^n + \delta t B_h u_h^{n+1/2} = 0, \tag{7.67b}$$

$$u_h^{n+1} - u_h^{n+1/2} + \tfrac{1}{2}\delta t \hat{B}_h \sigma_h^{n+1} = 0. \tag{7.67c}$$

A recursive Verlet method can be used to derive a multiscale algorithm with local time stepping. To this purpose, we assume that:

(a) The mesh elements have been partitioned into N classes (this partition is performed once and for all before starting the simulation and is based for instance on geometrical or physical criteria).

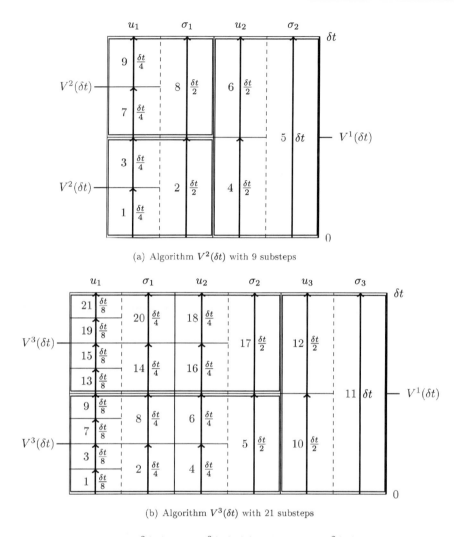

Fig. 7.1: Algorithms $V^2(\delta t)$ and $V^3(\delta t)$ (**a**) Algorithm $V^2(\delta t)$ with 9 substeps (**b**) Algorithm $V^3(\delta t)$ with 21 substeps

(b) The global time step for the algorithm is δt.

(c) The mesh elements of class $k \in \{1, \ldots, N\}$ are time-advanced using the Verlet method (7.67) with local time step $2^{k-N}\delta t$. In particular, the larger elements in the computational domain belong to class N and the smallest to class 1.

7.5. The Unsteady Case

We denote by $V^N(\delta t)$ the algorithm for time-advancing N classes over the time interval δt. This algorithm is defined recursively. The algorithm $V^1(\delta t)$ with only one class is the Verlet method (7.67). For $N \geq 1$, $V^{N+1}(\delta t)$ is defined as follows:

1. Advance all the elements with class $k \leq N$ with the algorithm $V^N(\frac{1}{2}\delta t)$.

2. Advance all the elements with class $k = N + 1$ with the Verlet method $V^1(\delta t)$.

3. Advance all the elements with class $k \leq N$ with the algorithm $V^N(\frac{1}{2}\delta t)$.

Figure 7.1 illustrates algorithms $V^2(\delta t)$ and $V^3(\delta t)$. It is clear that these algorithms are fully explicit. For all the substeps, updated values for unknowns in neighboring elements are used whenever available.

The main stability result for algorithm $V^2(\delta t)$ is the following; we refer the reader to Piperno [259] for the proof.

Lemma 7.35 (Stability for $V^2(\delta t)$). *Algorithm $V^2(\delta t)$ preserves a quadratic form of the unknowns. For δt small enough, this quadratic form is positive definite.*

Determining an explicit bound on the time step that yields a sufficient condition for stability is a difficult task. In actual computations when the time step and the mesh size vary at the same time, it seems that the scheme becomes unstable if the usual local CFL condition is considered. It is conjectured that a sufficient stability condition involves an extended non-local CFL condition, where the class and size of neighbors of an element is taken into account. In general, the scheme behaves well if the user exerts some caution, i.e., if the local time step is driven not only by the local size of the elements, but also by the possible smaller time steps of the neighboring elements.

Appendix A: Implementation

In this appendix, we discuss two relevant issues in the implementation of dG methods, namely matrix assembly and the selection of basis functions in broken polynomial spaces. In particular, we emphasize that the choice of basis functions can have a substantial impact on both computational accuracy and efficiency in practical implementations of dG methods.

A.1 Matrix Assembly

Matrix assembly for dG methods differs from that for conforming FE methods because (1) the degrees of freedom associated with each mesh element are decoupled from those associated with the remaining elements; (2) terms involving integrals on interfaces are generally present. To illustrate the first point, we consider the mass matrix, which is particularly relevant in the context of time-marching schemes. To illustrate the second point, we consider the stiffness matrix resulting from the discrete bilinear form associated with a purely diffusive model problem.

A.1.1 Basic Notation

At the discrete level, we consider an approximation space V_h. A simple example is the broken polynomial space $\mathbb{P}_d^k(\mathcal{T}_h)$ defined by (1.15).

We exploit the fact that the restriction of a function $v_h \in V_h$ to a given mesh element $T \in \mathcal{T}_h$ can be chosen independently of its restriction to other elements, whereby we restrict the support of basis functions to a single mesh element. This is generally a good idea, since it reduces communications between mesh elements and, therefore, yields the minimal stencil (cf. Definition 2.26). In this situation, it is natural to assume that the global enumeration of the degrees of freedom is such that the local degrees of freedom are numbered contiguously for each mesh element. This leads to a basis for V_h of the form

$$\Phi := \{\{\varphi_i^T\}_{i \in D_T}\}_{T \in \mathcal{T}_h},$$

where the set $D_T = \{1, \ldots, N_{\mathrm{dof}}^T\}$ collects the local indices of the N_{dof}^T degrees of freedom for the mesh element T and where

$$\mathrm{supp}(\varphi_i^T) = \overline{T}, \qquad \forall T \in \mathcal{T}_h, \ \forall i \in D_T.$$

The dimension of V_h is therefore equal to $\sum_{T \in \mathcal{T}_h} N_{\mathrm{dof}}^T$. Several choices for the local basis $\{\varphi_i^T\}_{i \in D_T}$ are discussed in Sect. A.2.

The number of degrees of freedom N_{dof}^T can vary from element to element. In the simple case where $V_h = \mathbb{P}_d^k(\mathcal{T}_h)$, this number is constant and equal to $\dim(\mathbb{P}_d^k) = \frac{(k+d)!}{k!d!}$ (cf. (1.14)). An example where this number varies is a broken polynomial space built on a mesh containing both triangles and quadrangles for which a Lagrangian basis at element nodes has been selected.

A.1.2 Mass Matrix

The global mass matrix is associated with the bilinear form

$$m(v_h, w_h) = \int_\Omega v_h w_h, \qquad \forall v_h, w_h \in V_h.$$

Exploiting the localization of basis functions to single mesh elements, the *global mass matrix* can be block-partitioned in the form

$$\mathbf{M} = \begin{bmatrix} \mathbf{M}_{T_1 T_1} & 0 & \cdots & 0 \\ 0 & \mathbf{M}_{T_2 T_2} & \ddots & \vdots \\ \vdots & \ddots & \ddots & 0 \\ 0 & \cdots & 0 & \mathbf{M}_{T_N T_N} \end{bmatrix},$$

where $N := \mathrm{card}(\mathcal{T}_h)$. The *local mass matrix* corresponding to a generic element $T \in \mathcal{T}_h$ is

$$\mathbf{M}_{TT} = \left[m(\varphi_j^T, \varphi_i^T) \right] \in \mathbb{R}^{N_{\mathrm{dof}}^T, N_{\mathrm{dof}}^T},$$

and this matrix is obviously symmetric positive definite.

The matrix $\mathbf{M}$ is relatively easy to invert owing to its block diagonal structure. A typical situation where this inverse is needed is when computing the L^2-orthogonal projection $\pi_h v$ of a given function $v \in L^2(\Omega)$ onto the approximation space V_h (cf. (1.29)). The discrete function $\pi_h v$ is decomposed in the basis Φ in such a way that, for all $T \in \mathcal{T}_h$,

$$\pi_h v|_T = \sum_{j \in D_T} X_j^T \varphi_j^T,$$

leading to the following independent linear systems: For all $T \in \mathcal{T}_h$,

$$\mathbf{M}_{TT} \mathbf{X}^T = \mathbf{V}^T,$$

where $\mathbf{X}^T = [X_j^T] \in \mathbb{R}^{N_{\mathrm{dof}}^T}$ and $\mathbf{V}^T = [\int_T v \varphi_j^T] \in \mathbb{R}^{N_{\mathrm{dof}}^T}$.

A.1.3 Stiffness Matrix

Focusing on the purely diffusive model problem (4.14) with the discrete bilinear form a_h^{sip} defined by (4.12), the *global stiffness matrix* can be block-partitioned

A.1. Matrix Assembly

in the form

$$\mathbf{A} = \begin{bmatrix} \mathbf{A}_{T_1 T_1} & \mathbf{A}_{T_1 T_2} & \cdots & \mathbf{A}_{T_1 T_N} \\ \mathbf{A}_{T_2 T_1} & \mathbf{A}_{T_2 T_2} & \cdots & \mathbf{A}_{T_2 T_N} \\ \vdots & \vdots & \ddots & \vdots \\ \mathbf{A}_{T_N T_1} & \mathbf{A}_{T_N T_2} & \cdots & \mathbf{A}_{T_N T_N} \end{bmatrix},$$

where, for all $T_k, T_l \in \mathcal{T}_h$,

$$\mathbf{A}_{T_k T_l} = [a_h^{\text{sip}}(\varphi_j^{T_l}, \varphi_i^{T_k})] \in \mathbb{R}^{N_{\text{dof}}^{T_k}, N_{\text{dof}}^{T_l}}.$$

In view of the elementary stencil associated with the SIP bilinear form (cf. Remark 4.10), the block $\mathbf{A}_{T_k T_l}$ is nonzero only if $T_k = T_l$ (diagonal block) or if T_k and T_l share a common interface.

We split the discrete bilinear form a_h^{sip} into volume, interface, and boundary face contributions as follows: For all $v_h, w_h \in V_h$,

$$a_h^{\text{sip}}(v_h, w_h) = a_h^{\text{v}}(v_h, w_h) + a_h^{\text{if}}(v_h, w_h) + a_h^{\text{bf}}(v_h, w_h),$$

where

$$a_h^{\text{v}}(v_h, w_h) = \sum_{T \in \mathcal{T}_h} \int_T \nabla v_h \cdot \nabla w_h,$$

$$a_h^{\text{if}}(v_h, w_h) = \sum_{F \in \mathcal{F}_h^i} \int_F \left(\{\!\!\{\nabla_h v_h\}\!\!\} \cdot \mathbf{n}_F [\![w_h]\!] + [\![v_h]\!] \{\!\!\{\nabla_h w_h\}\!\!\} \cdot \mathbf{n}_F + \frac{\eta}{h_F} [\![v_h]\!] [\![w_h]\!] \right),$$

$$a_h^{\text{bf}}(v_h, w_h) = \sum_{F \in \mathcal{F}_h^b} \int_F \left(\nabla_h v_h \cdot \mathbf{n} w_h + v_h \nabla_h w_h \cdot \mathbf{n} + \frac{\eta}{h_F} v_h w_h \right).$$

Each summation yields a loop over the corresponding mesh entities to assemble local contributions into the global stiffness matrix. The *local stiffness matrix* stemming from the volume contribution of a generic mesh element $T \in \mathcal{T}_h$ is

$$\mathbf{A}^T = [a_h^{\text{v}}(\varphi_j^T, \varphi_i^T)] \in \mathbb{R}^{N_{\text{dof}}^T, N_{\text{dof}}^T},$$

and it contributes to the diagonal block $\mathbf{A}_{TT}$ of the global stiffness matrix $\mathbf{A}$. An interface $F \in \mathcal{F}_h^i$ contributes to four blocks of the global stiffness matrix $\mathbf{A}$, and the local stiffness matrix stemming from the interface contribution can be block-partitioned in the form

$$\mathbf{A}^F = \begin{bmatrix} \mathbf{A}_{T_1 T_1}^F & \mathbf{A}_{T_1 T_2}^F \\ \mathbf{A}_{T_2 T_1}^F & \mathbf{A}_{T_2 T_2}^F \end{bmatrix},$$

where $F = \partial T_1 \cap \partial T_2$,

$$\mathbf{A}_{T_m T_n}^F = \left[a_h^{\text{if}}(\varphi_j^{T_{n'}}, \varphi_i^{T_{m'}}) \right] \in \mathbb{R}^{N_{\text{dof}}^{T_{m'}}, N_{\text{dof}}^{T_{n'}}}, \qquad \forall m, n \in \{1, 2\},$$

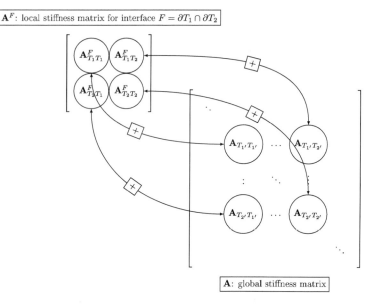

Fig. A.1: Assembly of interface contributions

and $1', 2' \in \{1, \ldots, N\}$ are the indices of T_1, T_2 in the global enumeration of mesh elements. Finally, a boundary face $F \in \mathcal{F}_h^b$ contributes through the local stiffness matrix

$$\mathbf{A}^F = [a_h^{\mathrm{bf}}(\varphi_j^T, \varphi_i^T)] \in \mathbb{R}^{N_{\mathrm{dof}}^T, N_{\mathrm{dof}}^T},$$

where $F = \partial T \cap \partial \Omega$. The procedure for global matrix assembly is summarized in Algorithm 1. The assembly of interface terms is depicted in Fig. A.1.

A common operation with the global stiffness matrix is to multiply it by a given component vector $\mathbf{U}$ associated with a discrete function $u_h \in V_h$ in such a way that, for all $T \in \mathcal{T}_h$,

$$u_h|_T = \sum_{j \in D_T} U_j^T \varphi_j^T, \qquad \mathbf{U}_T = [U_j^T] \in \mathbb{R}^{N_{\mathrm{dof}}^T}.$$

Concerning the volume contributions, we obtain, for all $T \in \mathcal{T}_h$ and all $i \in D_T$,

$$a_h^{\mathrm{v}}(u_h, \varphi_i^T) = \mathbf{A}^T(i, :)\mathbf{U}_T,$$

Algorithm 1 Global matrix assembly

1: **for** $T \in \mathcal{T}_h$ **do** {Loop over elements}
2:    $\mathbf{A}_{TT} \leftarrow \mathbf{A}_{TT} + \mathbf{A}^T$
3: **end for**
4: **for** $F \in \mathcal{F}_h^i$ s.t. $F = \partial T_1 \cap \partial T_2$ **do** {Loop over interfaces}
5:    **for** $m = 1$ to 2 **do**
6:      **for** $n = 1$ to 2 **do**
7:        $\mathbf{A}_{T_{m'}T_{n'}} \leftarrow \mathbf{A}_{T_{m'}T_{n'}} + \mathbf{A}^F_{T_m T_n}$
8:      **end for**
9:    **end for**
10: **end for**
11: **for** $F \in \mathcal{F}_h^b$ s.t. $F = \partial T \cap \partial \Omega$ **do** {Loop over boundary faces}
12:    $\mathbf{A}_{TT} \leftarrow \mathbf{A}_{TT} + \mathbf{A}^F$
13: **end for**

where $\mathbf{A}^T(i,:)$ denotes the ith line of the local stiffness matrix $\mathbf{A}^T$. For the interface contributions, we obtain, for all $F = \partial T_1 \cap \partial T_2 \in \mathcal{F}_h^i$, all $l \in \{1, 2\}$, and all $i \in D_{T_{l'}}$,

$$a_h^{\text{if}}(u_h, \varphi_i^{T_{l'}}) = \mathbf{A}^F_{T_l T_1}(i,:)\mathbf{U}^{T_{1'}} + \mathbf{A}^F_{T_l T_2}(i,:)\mathbf{U}^{T_{2'}}.$$

Finally, for the boundary face contributions, we obtain, for all $F = \partial T \cap \partial \Omega$ and all $i \in D_T$,

$$a_h^{\text{bf}}(u_h, \varphi_i^T) = \mathbf{A}^F(i,:)\mathbf{U}^T.$$

A.2 Choice of Basis Functions

The choice of basis functions can have a substantial impact on the performance (including accuracy) of a dG code. This is, in particular, the case when using polynomials with high degree. In this section, we discuss some criteria to select basis functions, and we present two main families of such functions, namely nodal and modal basis functions. A brief section recalling some basic facts concerning Legendre and Jacobi polynomials is also included for completeness.

A.2.1 Requirements

Various criteria can be adopted to select basis functions. Orthogonality and hierarchism are often considered in the context of modal basis functions, while ease of integral computation is often considered in the context of nodal basis functions.

A.2.1.1 Orthogonality and Hierarchism

A first possible criterion to select a high-degree polynomial basis is that it be orthogonal or nearly orthogonal with respect to an appropriate inner product.

When the inner product corresponds to local matrices contributing to the global system, this ensures that their condition number does not increase too much as the polynomial degree k grows. In the case of unsteady problems discretized with an explicit time integration method, an appropriate choice may be one yielding a (quasi-)diagonal mass matrix (and, hence, a diagonal global system). Using well-conditioned mass matrices also mitigates time step restrictions when explicit schemes are applied to advective problems (see, e.g., Karniadakis and Spencer [209, p. 187]).

Improving the condition number of local mass matrices is particularly relevant for high-order approximations: for instance, when using Lagrange polynomials of degree k associated with equispaced nodes in d space dimensions, the condition number of elemental mass matrices is known to grow as 4^{kd} (see Olsen and Douglas [252]). Also, the condition number of the global matrix has an important role in determining the numerical error in the solution (see, e.g., Quarteroni, Sacco, and Saleri [264, Theorem 3.1]).

A second important criterion which is often associated with orthogonality is that the basis be hierarchical, that is to say, the basis for a given polynomial degree k includes the bases for polynomial degrees less than k. This point is particularly relevant when one wishes to locally adapt (in space and/or in time) the polynomial degree according to some a posteriori regularity estimates, following the paradigm that, when the exact solution is locally smooth enough, increasing the polynomial degree generally pays off more in terms of error reduction than reducing the meshsize.

A.2.1.2 Ease of Integral Computation

A different criterion for selecting basis functions is to make the computation of volume and face integrals as efficient as possible. While for linear problems with constant coefficients, exact volume and face integrations can be performed, quadratures are required in general. The principle of a quadrature is to approximate the integral of a generic function f over a domain D by a linear combination of the values of f at a set of nodes $\{\xi_n\}_{0 \leq n \leq N}$ of D, i.e.,

$$\int_D f(t) \simeq \sum_{n=0}^{N} \omega_n f(\xi_n).$$

The points $\{\xi_n\}_{0 \leq n \leq N}$ are called the *quadrature nodes*, and the real numbers $\{\omega_n\}_{0 \leq n \leq N}$ the *quadrature weights*. Thus, another reasonable criterion is to select basis functions whose values at a specific set of quadrature points can be easily computed. A simple example (whenever possible) is that of Lagrange polynomials associated with these quadrature nodes; cf. the discussion in Sect. A.2.3.

It is also worthwhile to try to alleviate the computation of integrals over mesh faces as much as possible. As an example, we consider the element $T = (0, 1)$ in one space dimension along with the basis of $\mathbb{P}_1^k(T)$ given by $\{\varphi_j(x) = x^j\}_{0 \leq j \leq k}$. The trace at $x = 0$ of a function $v \in \mathbb{P}_1^k(T)$ expressed in terms of the selected

A.2. Choice of Basis Functions

basis as

$$v(x) = \sum_{j=0}^{k} \alpha_j \varphi_j(x)$$

can simply be evaluated as $v|_{x=0} = \alpha_0$, so that it solely depends on one degree of freedom. On the other hand, the trace $v|_{x=1} = \sum_{j=0}^{k} \alpha_j$ depends on all the degrees of freedom and is thus more computationally expensive to evaluate.

On a d-simplex T, while the dimension of $\mathbb{P}_d^k(T)$ is

$$S_d^k := \frac{(k+d)!}{k!d!},$$

the dimension of the space $\mathbb{P}_{d-1}^k(F)$, spanned by the traces of functions in $\mathbb{P}_d^k(T)$ over a face $F \in \mathcal{F}_T$ is

$$S_{d-1}^k = \frac{(k+d-1)!}{k!(d-1)!}.$$

Thus, the dimension of $\mathbb{P}_{d-1}^k(F)$ is smaller than that of $\mathbb{P}_d^k(T)$ by a factor $(k+d)/d$. This fact can be exploited by localizing the degrees of freedom with exactly S_{d-1}^k nonzero basis functions on each face $F \in \mathcal{F}_T$; cf. Sect. A.2.3 for further discussion.

Remark 7.36 (Applicability to general elements). When handling general elements, the localization of the degrees of freedom can become difficult since no simple reference element is available. For the same reason, numerical integration requires adapted techniques. In this situation, ease of integral computation is often renounced in favour of orthogonality and hierarchism.

A.2.2 Jacobi and Legendre Polynomials

We briefly present here some basic facts concerning Jacobi and Legendre orthogonal polynomial families, which are used in the next sections to construct nodal and modal bases on the reference element.

For real numbers $\alpha > -1$ and $\beta > -1$ and an integer $k \geq 0$, the Jacobi polynomials $\mathcal{J}_{\alpha,\beta}^k$ are the solutions of the singular Sturm–Liouville problem

$$(1-x)(1+x)\frac{\mathrm{d}^2 y(x)}{\mathrm{d}x^2} + [\beta - \alpha - (\alpha + \beta + 2)x]\frac{\mathrm{d}y(x)}{\mathrm{d}y} = -\lambda_k y(x),$$

with $\lambda_k = k(k+\alpha+\beta+1)$. For any $k \geq 0$, the Jacobi polynomial $\mathcal{J}_{\alpha,\beta}^k$ on the reference interval $[-1,1]$ is of degree k and can be computed by the following recursive relation (see Abramowitz and Stegun [2, Table 22.7]):

$$\mathcal{J}_{\alpha,\beta}^0(x) = 1,$$

$$\mathcal{J}_{\alpha,\beta}^1(x) = \frac{1}{2}\left[\alpha - \beta + (\alpha + \beta + 2)x\right],$$

$$a_k^1 \mathcal{J}_{\alpha,\beta}^{k+1}(x) = (a_k^2 + a_k^3 x)\mathcal{J}_{\alpha,\beta}^k(x) - a_k^4 \mathcal{J}_{\alpha,\beta}^{k-1}(x),$$

with
$$a_k^1 = 2(k+1)(k+\alpha+\beta+1)(2k+\alpha+\beta),$$
$$a_k^2 = (2k+\alpha+\beta+1)(\alpha^2-\beta^2),$$
$$a_k^3 = (2k+\alpha+\beta+2)(2k+\alpha+\beta+1)(2k+\alpha+\beta),$$
$$a_k^4 = 2(k+\alpha)(k+\beta)(2k+\alpha+\beta+2).$$

The Jacobi polynomials satisfy the orthogonality property
$$\int_{-1}^{1} (1-t)^\alpha (1+t)^\beta \mathsf{J}_{\alpha,\beta}^k(t) \mathsf{J}_{\alpha,\beta}^l(t) \, dt = c_{k,\alpha,\beta} \delta_{lk},$$

with constant
$$c_{k,\alpha,\beta} := \frac{2^{\alpha+\beta+1}}{2k+\alpha+\beta+1} \frac{\Gamma(k+\alpha+1)\Gamma(k+\beta+1)}{k!\Gamma(k+\alpha+\beta+1)},$$

where the so-called Γ function is such that, for any positive integer k, $\Gamma(k) := (k-1)!$. More details on Jacobi polynomials can be found in the books by Abramowitz and Stegun [2, Chap. 22] or by Karniadakis and Spencer [209, p. 350]. Jacobi polynomials with $\alpha = \beta = 0$ are also termed Legendre polynomials. For convenience, we define the Legendre polynomials over the reference interval $[0,1]$ by setting
$$\mathbb{L}^k(t) = \mathsf{J}_{0,0}^k(2t-1) \qquad \forall t \in [0,1]. \tag{A.1}$$

The Legendre polynomial $\mathbb{L}^k$ is of degree k. The orthogonality relation for the Legendre polynomials simplifies to
$$\int_0^1 \mathbb{L}^k(t) \mathbb{L}^l(t) \, dt = \frac{1}{2k+1} \delta_{lk}. \tag{A.2}$$

The first four Legendre polynomials are represented in the left panel of Fig. A.2.

A.2.3 Nodal Basis Functions

In this section, we discuss the use of Lagrange polynomials in the context of nodal basis functions.

A.2.3.1 Lagrange Polynomials

Lagrange polynomials are the interpolating polynomials for a given set of distinct data points $\{x_i\}_{0 \leq i \leq k}$. In one space dimension, the Lagrange polynomial equal to 1 at x_i and to 0 at x_j, $j \neq i$, can be expressed as
$$\mathbb{I}_i^k(x) = \prod_{j \in \{0,\ldots,k\} \setminus \{i\}} \frac{(x-x_j)}{(x_i-x_j)} \qquad \forall i \in \{0,\ldots,k\}.$$

The set $\{\mathbb{I}_i^k\}_{0 \leq i \leq k}$ forms a basis of the polynomial space $\mathbb{P}_1^k$. When $d > 1$, in order for the set of Lagrange polynomials to form a basis for the polynomial

A.2. Choice of Basis Functions

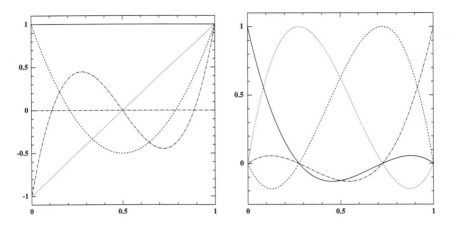

Fig. A.2: *Left*: Legendre polynomials of degree ≤ 3; *Right*: Lagrange polynomials of degree ≤ 3 associated with Gauß-Lobatto nodes

space $\mathbb{P}_d^k$, a set containing S_d^k unisolvent points must be provided. Examples on simplices can be found in Ern and Guermond [141, p. 23], Braess [49, p. 60], and Brenner and Scott [54, p. 69].

A.2.3.2 Simplicial Meshes

A basis of $\mathbb{P}_d^k(T)$ on a simplex T in $\mathbb{R}^d$ meeting the localization requirement can be easily built using Lagrange polynomials. Indeed, it suffices to choose $(k+d)!/(k!d!)$ nodes inside the simplex, with exactly $(k+d-1)!/(k!(d-1)!)$ nodes on each face of the simplex. The basis is then obtained building the Lagrange polynomials corresponding to the selected set of points. The nodes can be chosen to optimize the condition number of the local mass matrix.

When working on a simplex T in $\mathbb{R}^d$ and the polynomial space $\mathbb{P}_d^k$ with k large, it is possible to define sets of quadrature points with nearly optimal interpolation properties: the so-called *Fekete points*; see, e.g., Chen and Babuška [87] and Taylor, Wingate, and Vincent [290]. When using Lagrange interpolation polynomials associated with these points as local basis functions, the condition number of the mass matrix is also nearly optimal. Fekete points in one space dimension coincide with the so-called Gauß-Lobatto quadrature nodes. For an integer $k \geq 2$, the Gauß-Lobatto quadrature nodes are defined as the endpoints of the interval $[0,1]$ together with the $(k-1)$ roots of the polynomial $(\mathbb{L}^k)'$, where $\mathbb{L}^k$ denotes the Legendre polynomial of degree k defined by (A.1). The resulting quadrature is exact for polynomials up to degree $(2k-1)$. Lagrange polynomials associated with the Gauß-Lobatto quadrature nodes for $k=3$ are plotted in the right panel of Fig. A.2. The Fekete points on edges turn out to be Gauß-Lobatto points on each edge of the triangle. In three space dimensions, the theory is less complete. One possibility is to use Lagrange interpolation nodes such that, on each face of the tetrahedron, the nodes on that face are the two-dimensional Fekete points.

A.2.4 Modal Basis Functions

In this section, we discuss modal basis functions, first on simplicial meshes and then on general meshes.

A.2.4.1 Simplicial Meshes

Instead of taking Lagrange polynomials as local basis functions, it can be more convenient to consider non-nodal polynomials for which the mass matrix is nearly diagonal, diagonal, or simply equal to the identity matrix. In one space dimension, this can be achieved by taking the basis functions on the reference element $[0,1]$ to be the Legendre polynomials defined by (A.1). Owing to (A.2), the elemental mass matrix is diagonal, and its condition number grows moderately as $(2k+1)$. Extensions of such polynomials to two or three space dimensions are possible using nonlinear transformations that map the simplex into the d-cube where tensor-products of one-dimensional modal basis functions can be used; see Dubiner [138]. For example, in three space dimensions, the following polynomials $\psi_{lmn} \in \mathbb{P}_d^k$ are L^2-orthogonal on a tetrahedron:

$$\psi_{lmn} = c_{lmn}\, J_{0,0}^l(2r-1)\left(\frac{1-s}{2}\right)^l J_{2l+1,0}^m(2s-1)\left(\frac{1-t}{2}\right)^{l+m} J_{2l+2m+2,0}^m(2t-1),$$

with integer parameters l, m, $n \geq 0$ such that $l+m+n = k$ and

$$r = \frac{\lambda_2 - \lambda_1}{\lambda_2 + \lambda_1}, \qquad s = \frac{\lambda_3 - \lambda_2 - \lambda_1}{\lambda_3 + \lambda_2 + \lambda_1}, \qquad t = \lambda_4 - \lambda_3 - \lambda_2 - \lambda_1,$$

the family $\{\lambda_i\}_{1 \leq i \leq 4}$ denoting the barycentric coordinates in the tetrahedron; cf. Definition 5.23. Although the functions ψ_{lmn} lead to diagonal mass matrices, using them requires accurate pre-computation and storage of other matrices (for instance, if derivatives are needed) and pointwise values if quadratures are used. We refer to Hesthaven and Warburton [191, 192] for further insight. Finally, one nice feature of the above bases is that they are hierarchical.

A.2.4.2 General Meshes

The above approach mapping tetrahedra onto d-cubes can be extended to elements with some other shapes (e.g., prisms, pyramids). However, when elements featuring more general shapes are used, other ideas must be devised. Following Bassi, Botti, Colombo, Di Pietro, and Tesini [31], a hierarchical orthogonal basis can be constructed on the physical elements. Since the notion of reference element may fail to exist, the procedure needs to be repeated for each mesh element. Let T denote a general polyhedral mesh element of $\mathbb{R}^d$. Since quadratures are usually not available on general elements, numerical integration on T has to be performed by decomposing T into simpler elements (e.g., simplices) for which quadratures are available. Let $\widehat{\Phi} := \{\widehat{\varphi}_j\}_{j \in \{1,\ldots,S_d^k\}}$ denote a starting set of linearly independent basis functions for $\mathbb{P}_d^k(T)$. We apply the Gram–Schmidt orthonormalization procedure,

A.2. Choice of Basis Functions

1: $\varphi_1 \leftarrow \widehat{\varphi}_1 / \|\widehat{\varphi}_1\|_{L^2(T)}$ {Initialize}
2: **for** $i = 2$ to S_d^k **do**
3: $\quad \widetilde{\varphi}_i \leftarrow \widehat{\varphi}_i$
4: $\quad$ **for** $j = 1$ to $i - 1$ **do**
5: $\quad\quad \widetilde{\varphi}_i \leftarrow \widetilde{\varphi}_i - (\widehat{\varphi}_i, \varphi_j)_{L^2(T)} \varphi_j$ {Remove the projection of $\widehat{\varphi}_i$ onto φ_j}
6: $\quad$ **end for**
7: $\quad \varphi_i \leftarrow \widetilde{\varphi}_i / \|\widetilde{\varphi}_i\|_{L^2(T)}$ {Normalize}
8: **end for**

By construction, the basis $\Phi := \{\varphi_j\}_{j \in \{1, \ldots, S_d^k\}}$ obtained via the above Gram–Schmidt procedure is orthonormal. Moreover, its elements satisfy

$$\varphi_i = a_{ii} \widehat{\varphi}_i - \sum_{j=1}^{i-1} a_{ij} \varphi_j \qquad \forall i \in \{1, \ldots, S_d^k\}, \tag{A.3}$$

with

$$\frac{1}{a_{ii}} = \left(\|\widehat{\varphi}_i\|_{L^2(T)}^2 - \sum_{j=1}^{i-1} (\widehat{\varphi}_i, \varphi_j)_{L^2(T)}^2 \right)^{1/2} \qquad \forall i \in \{1, \ldots, S_d^k\},$$

$$\frac{a_{ij}}{a_{ii}} = (\widehat{\varphi}_i, \varphi_j)_{L^2(T)} \qquad \forall j \in \{1, \ldots, i-1\}.$$

It is clear from the above result that adding basis functions to increase the polynomial degree leaves unaltered the equations of the form (A.3) corresponding to lower degrees. In practice, it may be desirable to repeat the orthogonalization process more than once as the results can be affected by round-off errors. Giraud, Langou, and Rozloznik [169] suggest that performing orthonormalization twice yields in most cases a set of basis functions orthogonal up to machine precision.

It only remains to select an appropriate initial basis $\widehat{\Phi}$. This choice is relevant since working in finite precision can affect the quality of the final orthonormal basis Φ. Following again [31], one possibility is to introduce the frame ξ centered at the barycenter of T whose axes are aligned with the principal axes of inertia of T. Then, if we define the set of monomials $\{\mathsf{M}_d^\alpha\}_{\alpha \in A_d^k}$ with A_d^k given by (1.8) (recall that $S_d^k = \text{card}(A_d^k)$) such that, for all $\alpha \in A_d^k$,

$$\mathsf{M}_d^\alpha(\xi) := \prod_{i=1}^d \xi_i^{\alpha_i}, \qquad \xi = (\xi_1, \ldots, \xi_d) \in \mathbb{R}^d,$$

a numerically satisfactory choice for the initial basis is to set

$$\widehat{\Phi} := \{\mathsf{M}_d^\alpha / \|\mathsf{M}_d^\alpha\|_{L^2(T)}\}_{\alpha \in A_d^k}.$$

Some comments are in order. The procedure outlined in this section does not require the identification of the degrees of freedom on a reference element to construct the basis. Furthermore, while this technique allows for elements of general shape, its major drawback is the computational cost, since a basis has to be computed for each element in the discretization. Nevertheless, tessellations using general polyhedral elements usually contain fewer elements than the corresponding tetrahedral meshes, and this fact can partially compensate for the additional cost. Numerical integration can be expensive as well, especially if a decomposition into simpler subelements is required to perform quadratures in each mesh element. Possible strategies to reduce the cost of integration are presented in [31, Sect. 3.3]. Finally, we observe that the technique presented in this section is particularly appropriate in the case of agglomeration h-multigrid methods, where the additional cost to obtain the basis is balanced by the fact that the agglomerate mesh contains significantly fewer elements than the fine mesh.

Bibliography

1. R. Abgrall. Méthodes variationnelles pour les problèmes hyperboliques et paraboliques, application à la mécanique des fluides. Notes from Matmeca course, University of Bordeaux I, www.math.u-bordeaux1.fr/~abgrall/Cours/cours.ps, 2010.

2. M. Abramowitz and I. Stegun. *Handbook of Mathematical Functions*. Dover, 9th edition, 1972.

3. Y. Achdou, C. Bernardi, and F. Coquel. A priori and a posteriori analysis of finite volume discretizations of Darcy's equations. *Numer. Math.*, 96(1):17–42, 2003.

4. R. A. Adams. *Sobolev Spaces*, volume 65 of *Pure and Applied Mathematics*. Academic Press, 1975.

5. L. Agélas, D. A. Di Pietro, and J. Droniou. The G method for heterogeneous anisotropic diffusion on general meshes. *M2AN Math. Model. Numer. Anal.*, 44(4):597–625, 2010.

6. L. Agélas, D. A. Di Pietro, R. Eymard, and R. Masson. An abstract analysis framework for nonconforming approximations of diffusion problems on general meshes. *Int. J. Finite Volumes*, 7(1):1–29, 2010.

7. M. Ainsworth. A posteriori error estimation for discontinuous Galerkin finite element approximation. *SIAM J. Numer. Anal.*, 45(4):1777–1798, 2007.

8. M. Ainsworth and J. T. Oden. *A posteriori error estimation in finite element analysis*. Pure and Applied Mathematics (New York). Wiley-Interscience [John Wiley & Sons], New York, 2000.

9. M. Ainsworth and R. Rankin. Fully computable error bounds for discontinuous Galerkin finite element approximations on meshes with an arbitrary number of levels of hanging nodes. *SIAM J. Numer. Anal.*, 47(6):4112–4141, 2010.

10. L. Ambrosio. Transport equation and Cauchy problem for BV vector fields. *Invent. Math.*, 158(2):227–260, 2004.

11. C. Amrouche and V. Girault. On the existence and regularity of the solution of Stokes problem in arbitrary dimension. *Proc. Japan. Acad.*, 67:171–175, 1991.

12. N. Antonić and K. Burazin. On equivalent descriptions of boundary conditions for Friedrichs systems. To appear in Mathematic Montisnigri, www.mathos.hr/~kburazin/papers/antonic_burazin.pdf, 2011.

13. T. Arbogast and Z. Chen. On the implementation of mixed methods as nonconforming methods for second-order elliptic problems. *Math. Comp.*, 64(211):943–972, 1995.

14. D. N. Arnold. An interior penalty finite element method with discontinuous elements. *SIAM J. Numer. Anal.*, 19:742–760, 1982.

15. D. N. Arnold and F. Brezzi. Mixed and nonconforming finite element methods: implementation, postprocessing and error estimates. *RAIRO Modél. Math. Anal. Numér.*, 19(1):7–32, 1985.

16. D. N. Arnold, F. Brezzi, B. Cockburn, and L. D. Marini. Unified analysis of discontinuous Galerkin methods for elliptic problems. *SIAM J. Numer. Anal.*, 39(5):1749–1779, 2002.

17. J.-P. Aubin. *Analyse fonctionnelle appliquée.* Presses Universitaires de France, Paris, 1987.

18. J.-P. Aubin. *Applied Functional Analysis.* Pure and Applied Mathematics. Wiley-Interscience, New York, second edition, 2000.

19. I. Babuška. The finite element method with lagrangian multipliers. *Numer. Math.*, 20:179–192, 1973.

20. I. Babuška. The finite element method with penalty. *Math. Comp.*, 27:221–228, 1973.

21. I. Babuška and A. K. Aziz. Survey lectures on the mathematical foundations of the finite element method. In *The mathematical foundations of the finite element method with applications to partial differential equations (Proc. Sympos., Univ. Maryland, Baltimore, Md., 1972)*, pages 1–359. Academic Press, New York, 1972.

22. I. Babuška and M. R. Dorr. Error estimates for the combined h and p versions of the finite element method. *Numer. Math.*, 37(2):257–277, 1981.

23. I. Babuška, B. A. Szabo, and I. N. Katz. The p-version of the finite element method. *SIAM J. Numer. Anal.*, 18(3):515–545, 1981.

24. I. Babuška and M. Zlámal. Nonconforming elements in the finite element method with penalty. *SIAM J. Numer. Anal.*, 10(5):863–875, 1973.

25. G. A. Baker. Finite element methods for elliptic equations using nonconforming elements. *Math. Comp.*, 31(137):45–49, 1977.

26. S. Balay, K. Buschelman, W. D. Gropp, D. Kaushik, M. G. Knepley, L. C. McInnes, B. F. Smith, and H. Zhang. PETSc Web page. http://www.mcs.anl.gov/petsc.

27. W. Bangerth, R. Hartmann, and G. Kanschat. deal.II — a general-purpose object-oriented finite element library. *ACM Trans. Math. Softw.*, 33(4), 2007.

28. C. Bardos. Problèmes aux limites pour les équations aux dérivées partielles du premier ordre à coefficients réels; théorèmes d'approximation; application à l'équation de transport. *Ann. Sci. École Norm. Sup. (4)*, 3:185–233, 1970.

29. C. Bardos, A. Y. Le Roux, and J.-C. Nédélec. First order quasilinear equations with boundary conditions. *Comm. Partial Differential Equations*, 4(9):1017–1034, 1979.

30. A. J.-C. Barré de Saint Venant. Note à joindre au Mémoire sur la dynamique des fluides, présenté le 14 avril 1834. *Compt. Rend. Acad. Sci., Paris*, 17:1240–1243, 1843.

31. F. Bassi, L. Botti, A. Colombo, D. A. Di Pietro, and P. Tesini. On the flexibility of agglomeration based physical space discontinuous Galerkin discretizations. *J. Comput. Phys.* published online. DOI 10.1016/j.jcp.2011.08.018, 2011.

32. F. Bassi, A. Crivellini, D. A. Di Pietro, and S. Rebay. An artificial compressibility flux for the discontinuous Galerkin solution of the incompressible Navier-Stokes equations. *J. Comput. Phys.*, 218(2):794–815, 2006.

33. F. Bassi, A. Crivellini, D. A. Di Pietro, and S. Rebay. An implicit high-order discontinuous Galerkin method for steady and unsteady incompressible flows. *Comp. & Fl.*, 36(10):1529–1546, 2007.

34. F. Bassi and S. Rebay. A high-order accurate discontinuous finite element method for the numerical solution of the compressible Navier-Stokes equations. *J. Comput. Phys.*, 131(2):267–279, 1997.

35. F. Bassi, S. Rebay, G. Mariotti, S. Pedinotti, and M. Savini. A high-order accurate discontinuous finite element method for inviscid and viscous turbomachinery flows. In R. Decuypere and G. Dibelius, editors, *Proceedings of the 2nd European Conference on Turbomachinery Fluid Dynamics and Thermodynamics*, pages 99–109, 1997.

36. P. Bastian, M. Droske, C. Engwer, R. Klöfkorn, T. Neubauer, M. Ohlberger, and M. Rumpf. Towards a unified framework for scientific computing. In *Proc. of the 15th international conference on domain decomposition methods*, 2005.

37. P. Bastian and B. Rivière. Superconvergence and $H(\text{div})$ projection for discontinuous Galerkin methods. *Internat. J. Numer. Methods Fluids*, 42(10):1043–1057, 2003.

38. G. K. Batchelor. *An introduction to fluid dynamics*. Cambridge University Press, 1967.

39. M. Bebendorf. A note on the Poincaré inequality for convex domains. *Z. Anal. Anwendungen*, 22(4):751–756, 2003.

40. R. Becker, P. Hansbo, and M. G. Larson. Energy norm a posteriori error estimation for discontinuous Galerkin methods. *Comput. Methods Appl. Mech. Engrg.*, 192(5-6):723–733, 2003.

41. R. Becker, P. Hansbo, and R. Stenberg. A finite element method for domain decomposition with non-matching grids. *M2AN Math. Model. Numer. Anal.*, 37(2):209–225, 2003.

42. C. Bernardi and Y. Maday. Spectral methods. In *Handbook of numerical analysis, Vol. V*, Handb. Numer. Anal., V, pages 209–485. North-Holland, Amsterdam, 1997.

43. R. Biswas, K. D. Devine, and J. E. Flaherty. Parallel, adaptive finite element methods for conservation laws. *App. Num. Math.*, 14:255–283, 1994.

44. D. Boffi. Fortin operator and discrete compactness for edge elements. *Numer. Math.*, 87(2):229–246, 2000.

45. M. Bogovskiĭ. *Theory of cubature formulas and the application of functional analysis to problems of mathematical physics*, volume 149(1) of *Trudy Sem. S. L. Soboleva*, chapter Solutions of some problems of vector analysis associated with the operators div and grad, pages 5–40. Akad. Nauk SSSR Sibirsk. Otdel. Inst. Mat., Novosibirsk, Russia, 1980.

46. A. Bonito and R. H. Nochetto. Quasi-optimal convergence rate of an adaptive discontinuous Galerkin method. *SIAM J. Numer. Anal.*, 48(2):734–771, 2010.

47. L. Botti and D. A. Di Pietro. A pressure-correction scheme for convection-dominated incompressible flows with discontinuous velocity and continuous pressure. *J. Comput. Phys.*, 230(3):572–585, 2011.

48. F. Boyer. Trace theorems and spatial continuity properties for the solutions of the transport equation. *Differential Integral Equations*, 18(8):891–934, 2005.

49. D. Braess. *Finite elements*. Cambridge University Press, Cambridge, third edition, 2007. Theory, fast solvers, and applications in elasticity theory, Translated from the German by Larry L. Schumaker.

50. D. Braess and J. Schöberl. Equilibrated residual error estimator for edge elements. *Math. Comp.*, 77(262):651–672, 2008.

51. S. C. Brenner. Poincaré-Friedrichs inequalities for piecewise H^1 functions. *SIAM J. Numer. Anal.*, 41(1):306–324, 2003.

52. S. C. Brenner. Korn's inequalities for piecewise H^1 vector fields. *Math. Comp.*, 73(247):1067–1087 (electronic), 2004.

53. S. C. Brenner and L. Owens. A weakly over-penalized non-symmetric interior penalty method. *JNAIAM J. Numer. Anal. Ind. Appl. Math.*, 2(1-2):35–48, 2007.

54. S. C. Brenner and L. R. Scott. *The mathematical theory of finite element methods*, volume 15 of *Texts in Applied Mathematics*. Springer, New York, third edition, 2008.

55. H. Brézis. *Analyse fonctionnelle. Théorie et applications. [Functional Analysis. Theory and applications]*. Applied Mathematics Series for the Master's Degree. Masson, Paris, France, 1983.

56. F. Brezzi. On the existence, uniqueness and approximation of saddle-point problems arising from Lagrange multipliers. *RAIRO Anal. Numér.*, pages 129–151, 1974.

57. F. Brezzi and M. Fortin. *Mixed and hybrid finite element methods*, volume 15 of *Springer Series in Computational Mathematics*. Springer-Verlag, New York, 1991.

58. F. Brezzi, G. Manzini, L. D. Marini, P. Pietra, and A. Russo. Discontinuous finite elements for diffusion problems. In *Atti del convegno in onore di F. Brioschi (Milano 1997)*, pages 197–217. Istituto Lombardo, Accademia di Scienze e di Lettere, 1999.

59. F. Brezzi, G. Manzini, L. D. Marini, P. Pietra, and A. Russo. Discontinuous Galerkin approximations for elliptic problems. *Numer. Methods Partial Differential Equations*, 16(4):365–378, 2000.

60. F. Brezzi, L. D. Marini, and E. Süli. Discontinuous Galerkin methods for first-order hyperbolic problems. *Math. Models Methods Appl. Sci.*, 14(12):1893–1903, 2004.

61. A. Buffa and C. Ortner. Compact embeddings of broken Sobolev spaces and applications. *IMA J. Numer. Anal.*, 4(29):827–855, 2009.

62. A. Buffa and I. Perugia. Discontinuous Galerkin approximation of the Maxwell eigenproblem. *SINUM*, 44:2198–2226, 2006.

63. A. Buffa, I. Perugia, and T. Warburton. The mortar-discontinuous Galerkin method for the 2D Maxwell eigenproblem. *J. Sci. Comput.*, 40:86–114, 2009.

64. E. Burman and A. Ern. Continuous interior penalty hp-finite element methods for advection and advection-diffusion equations. *Math. Comp.*, 76(259):1119–1140, 2007.

65. E. Burman and A. Ern. Discontinuous Galerkin approximation with discrete variational principle for the nonlinear Laplacian. *C. R. Math. Acad. Sci. Paris*, 346(17–18):1013–1016, 2008.

66. E. Burman, A. Ern, and M. A. Fernández. Explicit Runge-Kutta schemes and finite elements with symmetric stabilization for first-order linear PDE systems. *SIAM J. Numer. Anal.*, 48(6):2019–2042, 2010.

67. E. Burman, A. Ern, I. Mozolevski, and B. Stamm. The symmetric discontinuous Galerkin method does not need stabilization in 1D for polynomial orders $p \geq 2$. *C. R. Math. Acad. Sci. Paris*, 345(10):599–602, 2007.

68. E. Burman, A. Quarteroni, and B. Stamm. Interior penalty continuous and discontinuous finite element approximations of hyperbolic equations. *J. Sci. Comput.*, 43(3):293–312, 2010.

69. E. Burman and B. Stamm. Minimal stabilization for discontinuous Galerkin finite element methods for hyperbolic problems. *J. Sci. Comput.*, 33:183–208, 2007.

70. E. Burman and B. Stamm. Low order discontinuous Galerkin methods for second order elliptic problems. *SIAM J. Numer. Anal.*, 47(1):508–533, 2008/09.

71. E. Burman and B. Stamm. Local discontinuous Galerkin method with reduced stabilization for diffusion equations. *Communications in Computational Physics*, 5:498–524, 2009.

72. E. Burman and B. Stamm. Bubble stabilized discontinuous Galerkin method for Stokes' problem. *Math. Models Methods Appl. Sci.*, 20(2):297–313, 2010.

73. E. Burman and P. Zunino. A domain decomposition method for partial differential equations with non-negative form based on interior penalties. *SIAM J. Numer. Anal.*, 44:1612–1638, 2006.

74. J. Butcher. *The numerical analysis of ordinary differential equations: Runge-Kutta and general linear methods*. Wiley, Chichester, 1987.

75. C. Canuto and A. Quarteroni. Approximation results for orthogonal polynomials in Sobolev spaces. *Math. Comp.*, 38(157):67–86, 1982.

76. C. Carstensen. Quasi-interpolation and a posteriori error analysis in finite element methods. *M2AN Math. Model. Numer. Anal.*, 33(6):1187–1202, 1999.

77. C. Carstensen and S. A. Funken. Constants in Clément-interpolation error and residual based a posteriori error estimates in finite element methods. *East-West J. Numer. Math.*, 8(3):153–175, 2000.

78. C. Carstensen and D. Peterseim. personal communication. 2011.

79. P. Castillo. Stencil reduction algorithms for the local discontinuous Galerkin method. *Int. J. Numer. Meth. Engng.*, 81:1475–1491, 2010.

80. P. Castillo, B. Cockburn, I. Perugia, and D. Schötzau. An a priori error analysis of the local discontinuous Galerkin method for elliptic problems. *SIAM J. Numer. Anal.*, 38(5):1676–1706 (electronic), 2000.

81. P. Castillo, B. Cockburn, D. Schötzau, and C. Schwab. Optimal a priori error estimates for the hp-version of the local discontinuous Galerkin method for convection-diffusion problems. *Math. Comp.*, 71(238):455–478, 2002.

82. L. Cattabriga. Su un problema al contorno relativo al sistema di equazioni di Stokes. *Rend. Sem. Mat. Univ. Padova*, 31:308–340, 1961.
83. P. Causin and R. Sacco. A discontinuous Petrov-Galerkin method with Lagrangian multipliers for second order elliptic problems. *SIAM J. Numer. Anal.*, 43(1):280–302 (electronic), 2005.
84. P. Caussignac and R. Touzani. Solution of three-dimensional boundary layer equations by a discontinuous finite element method. I. Numerical analysis of a linear model problem. *Comput. Methods Appl. Mech. Engrg.*, 78(3):249–271, 1990.
85. P. Caussignac and R. Touzani. Solution of three-dimensional boundary layer equations by a discontinuous finite element method. II. Implementation and numerical results. *Comput. Methods Appl. Mech. Engrg.*, 79(1):1–20, 1990.
86. G. Chavent and B. Cockburn. The local projection $P^0 P^1$-discontinuous-Galerkin finite element method for scalar conservation laws. *RAIRO Modél. Math. Anal. Numér.*, 23(4):565–592, 1989.
87. Q. Chen and I. Babuška. Approximate optimal points for polynomial interpolation of real functions in an interval and in a triangle. *Comput. Methods Appl. Mech. Engrg.*, 128(3-4):405–417, 1995.
88. Z. Chen. Analysis of mixed methods using conforming and nonconforming finite element methods. *RAIRO Modél. Math. Anal. Numér.*, 27(1):9–34, 1993.
89. Z. Chen and H. Chen. Pointwise error estimates of discontinuous Galerkin methods with penalty for second-order elliptic problems. *SIAM J. Numer. Anal.*, 42(3):1146–1166 (electronic), 2004.
90. A. J. Chorin. Numerical solution of the Navier–Stokes equations. *Math. Comp.*, 22:745–762, 1968.
91. A. J. Chorin. On the convergence of discrete approximations to the unsteady Navier–Stokes equations. *Math. Comp.*, 23:341–353, 1969.
92. P. G. Ciarlet. *The finite element method for elliptic problems*, volume 40 of *Classics in Applied Mathematics*. Society for Industrial and Applied Mathematics (SIAM), Philadelphia, PA, 2002. Reprint of the 1978 original [North-Holland, Amsterdam; MR0520174 (58 #25001)].
93. S. Cochez-Dhondt and S. Nicaise. Equilibrated error estimators for discontinuous Galerkin methods. *Numer. Methods Partial Differential Equations*, 24(5):1236–1252, 2008.
94. B. Cockburn. *Advanced numerical approximation of nonlinear hyperbolic equations*, chapter An introduction to the discontinuous Galerkin method for convection-dominated problems, pages 151–269. Lecture Notes in Mathematics. Springer, 1997.
95. B. Cockburn, B. Dong, and J. Guzmán. A superconvergent LDG-hybridizable Galerkin method for second-order elliptic problems. *Math. Comp.*, 77(264):1887–1916, 2008.
96. B. Cockburn, B. Dong, J. Guzmán, M. Restelli, and R. Sacco. A hybridizable discontinuous Galerkin method for steady-state convection-diffusion-reaction problems. *SIAM J. Sci. Comput.*, 31(5):3827–3846, 2009.
97. B. Cockburn, J. Gopalakrishnan, and R. Lazarov. Unified hybridization of discontinuous Galerkin, mixed, and continuous Galerkin methods for second order elliptic problems. *SIAM J. Numer. Anal.*, 47(2):1319–1365, 2009.

98. B. Cockburn, J. Guzmán, and H. Wang. Superconvergent discontinuous Galerkin methods for second-order elliptic problems. *Math. Comp.*, 78(265):1–24, 2009.

99. B. Cockburn, S. Hou, and C.-W. Shu. The Runge-Kutta local projection discontinuous Galerkin finite element method for conservation laws. IV. The multidimensional case. *Math. Comp.*, 54(190):545–581, 1990.

100. B. Cockburn, G. Kanschat, I. Perugia, and D. Schötzau. Superconvergence of the local discontinuous Galerkin method for elliptic problems on Cartesian grids. *SIAM J. Numer. Anal.*, 39(1):264–285 (electronic), 2001.

101. B. Cockburn, G. Kanschat, and D. Schötzau. The local discontinuous Galerkin method for the Oseen equations. *Math. Comp.*, 73:569–593, 2004.

102. B. Cockburn, G. Kanschat, and D. Schötzau. A locally conservative LDG method for the incompressible Navier-Stokes equations. *Math. Comp.*, 74:1067–1095, 2005.

103. B. Cockburn, G. Kanschat, and D. Schötzau. A note on discontinuous Galerkin divergence-free solutions of the Navier-Stokes equations. *J. Sci. Comput.*, 31(1-2):61–73, 2007.

104. B. Cockburn, G. Kanschat, and D. Schötzau. An equal-order DG method for the incompressible Navier-Stokes equations. *J. Sci. Comput.*, 40(1-3):188–210, 2009.

105. B. Cockburn, G. Kanschat, D. Schötzau, and C. Schwab. Local Discontinuous Galerkin methods for the Stokes system. *SIAM J. Numer. Anal.*, 40(1):319–343 (electronic), 2002.

106. B. Cockburn, G. E. Karniadakis, and C.-W. Shu. *Discontinuous Galerkin Methods - Theory, Computation and Applications*, volume 11 of *Lecture Notes in Computer Science and Engineering*. Springer, 2000.

107. B. Cockburn, F. Li, and C.-W. Shu. Locally divergence-free discontinuous Galerkin methods for the Maxwell equations. *J. Comput. Phys.*, 194:588–610, 2004.

108. B. Cockburn, S. Lin, and C.-W. Shu. TVB Runge-Kutta local projection discontinuous Galerkin finite element method for conservation laws. III. One-dimensional systems. *J. Comput. Phys.*, 84(1):90–113, 1989.

109. B. Cockburn, D. Schötzau, and Jing Wang. Discontinuous Galerkin methods for incompressible elastic materials. *Comput. Methods Appl. Mech. Engrg.*, 195(25-28):3184–3204, 2006.

110. B. Cockburn and C.-W. Shu. TVB Runge-Kutta local projection discontinuous Galerkin finite element method for conservation laws. II. General framework. *Math. Comp.*, 52(186):411–435, 1989.

111. B. Cockburn and C.-W. Shu. The Runge-Kutta local projection P^1-discontinuous-Galerkin finite element method for scalar conservation laws. *RAIRO Modél. Math. Anal. Numér.*, 25(3):337–361, 1991.

112. B. Cockburn and C.-W. Shu. The local discontinuous Galerkin finite element method for convection-diffusion systems. *SIAM J. Numer. Anal.*, 35:2440–2463, 1998.

113. B. Cockburn and C.-W. Shu. The Runge-Kutta discontinuous Galerkin method for conservation laws. V. Multidimensional systems. *J. Comput. Phys.*, 141(2):199–224, 1998.

114. B. Cockburn and C.-W. Shu. Foreword [Proceedings of the First International Symposium on DG Methods]. *J. Sci. Comput.*, 22/23:1–3, 2005. Held in Newport, RI, May 24–26, 1999.

115. G. Cohen, X. Ferrieres, and S. Pernet. A spatial high-order hexahedral discontinuous Galerkin method to solve Maxwell's equations in time domain. *J. Comput. Phys.*, 217(2):340–363, 2006.

116. E. Creusé and S. Nicaise. Discrete compactness for a discontinuous Galerkin approximation of Maxwell's system. *M2AN Math. Model. Numer. Anal.*, 40(2): 413–430, 2006.

117. G. Crippa. *The flow associated to weakly differentiable vector fields*, volume 12 of *Tesi. Scuola Normale Superiore di Pisa (Nuova Series) [Theses of Scuola Normale Superiore di Pisa (New Series)]*. Edizioni della Normale, Pisa, http://cvgmt.sns.it/papers/cri07/crippa_phd_web.pdf, 2009.

118. I. Danaila, F. Hecht, and O. Pironneau. *Simulation numérique en C++*. Dunod, Paris, 2003. http://www.freefem.org.

119. M. Dauge. *Elliptic boundary value problems on corner domains*, volume 1341 of *Lecture Notes in Mathematics*. Springer-Verlag, Berlin, 1988. Smoothness and asymptotics of solutions.

120. C. Dawson. Foreword. *Comput. Methods Appl. Mech. Eng.*, 195(25–28):3183, 2006.

121. C. Dawson, S. Sun, and M. F. Wheeler. Compatible algorithms for coupled flow and transport. *Comp. Meth. Appl. Mech. Eng.*, 193:2565–2580, 2004.

122. K. Deimling. *Nonlinear functional analysis*. Springer-Verlag, Berlin, 1985.

123. J. Deny and J.-L. Lions. Les espaces du type de Beppo Levi. *Ann. Inst. Fourier, Grenoble*, 5:305–370, 1955.

124. B. Després. Convergence of non-linear finite volume schemes for linear transport. In *Notes from the XIth Jacques-Louis Lions Hispano-French School on Numerical Simulation in Physics and Engineering (Spanish)*, pages 219–239. Grupo Anal. Teor. Numer. Modelos Cienc. Exp. Univ. Cádiz, Cádiz, 2004.

125. B. Després. Lax theorem and finite volume schemes. *Math. Comp.*, 73(247):1203–1234 (electronic), 2004.

126. P. Destuynder and B. Métivet. Explicit error bounds in a conforming finite element method. *Math. Comp.*, 68(228):1379–1396, 1999.

127. D. A. Di Pietro. Analysis of a discontinuous Galerkin approximation of the Stokes problem based on an artificial compressibility flux. *Int. J. Numer. Methods Fluids*, 55:793–813, 2007.

128. D. A. Di Pietro. Cell centered Galerkin methods. *C. R. Math. Acad. Sci. Paris*, 348:31–34, 2010.

129. D. A. Di Pietro. Cell centered Galerkin methods for diffusive problems. *M2AN Math. Model. Numer. Anal.*, 46(1):111–144, 2012.

130. D. A. Di Pietro. A compact cell-centered Galerkin method with subgrid stabilization. *C. R. Math. Acad. Sci. Paris*, 349(1–2):93–98, 2011.

131. D. A. Di Pietro and A Ern. Discrete functional analysis tools for discontinuous Galerkin methods with application to the incompressible Navier-Stokes equations. *Math. Comp.*, 79(271):1303–1330, 2010.

132. D. A. Di Pietro and A Ern. Analysis of a discontinuous Galerkin method for heterogeneous diffusion problems with low-regularity solutions. *Numer. Methods Partial Differential Equations*, published online. DOI 10.1012/num.20675, 2011.

133. D. A. Di Pietro, A. Ern, and J.-L. Guermond. Discontinuous Galerkin methods for anisotropic semi-definite diffusion with advection. *SIAM J. Numer. Anal.*, 46(2):805–831, 2008.

134. J. Douglas Jr. and T. Dupont. *Lecture Notes in Physics*, volume 58, chapter Interior penalty procedures for elliptic and parabolic Galerkin methods. Springer-Verlag, Berlin, 1976.

135. J. Droniou and R. Eymard. A mixed finite volume scheme for anisotropic diffusion problems on any grid. *Num. Math.*, 105(1):35–71, 2006.

136. M. Dryja. On discontinuous Galerkin methods for elliptic problems with discontinuous coefficients. *Comput. Methods Appl. Math.*, 3(1):76–85, 2003.

137. A. V. C. Duarte, E. G. D. do Carmo, and F. A. Rochinha. Consistent discontinuous finite elements in elastodynamics. *Comput. Methods Appl. Mech. Engrg.*, 190(1-2):193–223, 2000.

138. M. Dubiner. Spectral methods on triangles and other domains. *J. Sci. Comput.*, 6(4):345–390, 1991.

139. R. G. Durán and M. A. Muschietti. An explicit right inverse of the divergence operator which is continuous in weighted norms. *Studia Math.*, 148(3):207–219, 2001.

140. L. El Alaoui and A. Ern. Residual and hierarchical a posteriori error estimates for nonconforming mixed finite element methods. *M2AN Math. Model. Numer. Anal.*, 38(6):903–929, 2004.

141. A. Ern and J.-L. Guermond. *Theory and Practice of Finite Elements*, volume 159 of *Applied Mathematical Sciences*. Springer-Verlag, New York, NY, 2004.

142. A. Ern and J.-L. Guermond. Discontinuous Galerkin methods for Friedrichs' systems. I. General theory. *SIAM J. Numer. Anal.*, 44(2):753–778, 2006.

143. A. Ern and J.-L. Guermond. Discontinuous Galerkin methods for Friedrichs' systems. II. Second-order elliptic PDEs. *SIAM J. Numer. Anal.*, 44(6):2363–2388, 2006.

144. A. Ern and J.-L. Guermond. Discontinuous Galerkin methods for Friedrichs' systems. III. Multi-field theories with partial coercivity. *SIAM J. Numer. Anal.*, 46(2):776–804, 2008.

145. A. Ern, J.-L. Guermond, and G. Caplain. An intrinsic criterion for the bijectivity of Hilbert operators related to Friedrichs' systems. *Comm. Partial Differ. Eq.*, 32:317–341, 2007.

146. A. Ern, I. Mozolevski, and L. Schuh. Accurate velocity reconstruction for Discontinuous Galerkin approximations of two-phase porous media flows. *C. R. Acad. Sci. Paris, Ser. I*, 347:551–554, 2009.

147. A. Ern, S. Nicaise, and M. Vohralík. An accurate **H**(div) flux reconstruction for discontinuous Galerkin approximations of elliptic problems. *C. R. Math. Acad. Sci. Paris*, 345(12):709–712, 2007.

148. A. Ern and A. F. Stephansen. A posteriori energy-norm error estimates for advection-diffusion equations approximated by weighted interior penalty methods. *J. Comp. Math.*, 26(4):488–510, 2008.

149. A. Ern, A. F. Stephansen, and M. Vohralík. Guaranteed and robust discontinuous Galerkin a posteriori error estimates for convection–diffusion–reaction problems. *J. Comp. Appl. Math.*, 234(1):114–130, 2010.

150. A. Ern, A. F. Stephansen, and P. Zunino. A discontinuous Galerkin method with weighted averages for advection-diffusion equations with locally small and anisotropic diffusivity. *IMA J. Numer. Anal.*, 29(2):235–256, 2009.

151. A. Ern and M. Vohralík. Flux reconstruction and a posteriori error estimation for discontinuous Galerkin methods on general nonmatching grids. *C. R. Math. Acad. Sci. Paris*, 347:441–444, 2009.

152. C. R. Ethier and D. A. Steinman. Exact fully 3D Navier–Stokes solutions for benchmarking. *Int. J. Num. Meth. Fluids*, 19(5):369–375, 1994.

153. L. C. Evans. *Partial differential equations*, volume 19 of *Graduate Studies in Mathematics*. American Mathematical Society, Providence, RI, 1998.

154. R. E. Ewing, Junping Wang, and Y. Yang. A stabilized discontinuous finite element method for elliptic problems. *Numer. Linear Algebra Appl.*, 10(1-2):83–104, 2003. Dedicated to the 60th birthday of Raytcho Lazarov.

155. R. Eymard, T. Gallouët, M. Ghilani, and R. Herbin. Error estimates for the approximate solutions of a nonlinear hyperbolic equation given by finite volume schemes. *IMA J. Numer. Anal.*, 18(4):563–594, 1998.

156. R. Eymard, T. Gallouët, and R. Herbin. Finite volume methods. In *Handbook of numerical analysis, Vol. VII*, Handb. Numer. Anal., VII, pages 713–1020. North-Holland, Amsterdam, 2000.

157. R. Eymard, T. Gallouët, and R. Herbin. Finite volume approximation of elliptic problems and convergence of an approximate gradient. *Appl. Numer. Math.*, 37:31–53, 2001.

158. R. Eymard, T. Gallouët, and R. Herbin. A finite volume scheme for anisotropic diffusion problems. *Compt. Rend. Acad. Sci., Paris*, 339(4):527–548, 2004.

159. R. Eymard, T. Gallouët, and R. Herbin. Discretization of heterogeneous and anisotropic diffusion problems on general nonconforming meshes SUSHI: a scheme using stabilization and hybrid interfaces. *IMA J. Numer. Anal.*, 30(4):1009–1043, 2010.

160. R. Eymard, R. Herbin, and J.-C. Latché. Convergence analysis of a colocated finite volume scheme for the incompressible Navier–Stokes equations on general 2D or 3D meshes. *SIAM J. Numer. Anal.*, 45(1):1–36, 2007.

161. R. Eymard, D. Hilhorst, and M. Vohralík. A combined finite volume nonconforming/mixed-hybrid finite element scheme for degenerate parabolic problems. *Num. Math.*, 105(1):73–131, 2006.

162. L. P. Franca and R. Stenberg. Error analysis of Galerkin least squares methods for the elasticity equations. *SIAM J. Numer. Anal.*, 28(6):1680–1697, 1991.

163. K. O. Friedrichs. Symmetric positive linear differential equations. *Comm. Pure and Appl. Math.*, 11:333–418, 1958.

164. G. J. Gassner, F. Lörcher, C.-D. Munz, and J. S. Hesthaven. Polymorphic nodal elements and their application in discontinuous Galerkin methods. *J. Comput. Phys.*, 228(5):1573–1590, 2009.

165. F. Gastaldi and A. Quarteroni. On the coupling of hyperbolic and parabolic systems: Analytical and numerical approach. *Appl. Numer. Math.*, 6:3–31, 1989.

166. E. H. Georgoulis. *hp*-version interior penalty discontinuous Galerkin finite element methods on anisotropic meshes. *Int. J. Numer. Anal. Model.*, 3(1):52–79, 2006.

167. E. H. Georgoulis, E. Hall, and P. Houston. Discontinuous Galerkin methods for advection-diffusion-reaction problems on anisotropically refined meshes. *SIAM J. Sci. Comput.*, 30(1):246–271, 2007/08.

168. U. Ghia, K. N. Ghia, and C. T. Shin. High-Re solutions for incompressible flow using the Navier–Stokes equations and a multigrid method. *J. Comput. Phys.*, 12:387–411, 1982.

169. L. Giraud, J. Langou, and M. Rozloznik. On the loss of orthogonality in the Gram-Schmidt orthogonalization process. Technical report TR/PA/03/25, CERFACS, 2003.

170. V. Girault and P.-A. Raviart. *Finite element methods for Navier-Stokes equations*, volume 5 of *Springer Series in Computational Mathematics*. Springer-Verlag, Berlin, 1986. Theory and algorithms.

171. V. Girault, B. Rivière, and M. F. Wheeler. A discontinuous Galerkin method with nonoverlapping domain decomposition for the Stokes and Navier-Stokes problems. *Math. Comp.*, 74(249):53–84 (electronic), 2004.

172. E. Godlewski and P.-A. Raviart. *Hyperbolic systems of conservation laws*. Number 3/4 in Mathématiques & Applications. Ellipses, Paris, 1991.

173. E. Godlewski and P.-A. Raviart. *Numerical approximation of systems of conservation laws*, volume 118 of *Applied Mathematical Sciences*. Springer-Verlag, New York, Inc., 1996.

174. D. Gottlieb and C.-W. Shu. On the Gibbs phenomenon and its resolution. *SIAM Rev.*, 39(4):644–668, 1997.

175. S. Gottlieb and C.-W. Shu. Total variation diminishing Runge-Kutta schemes. *Math. Comp.*, 67(221):73–85, 1998.

176. S. Gottlieb, C.-W. Shu, and E. Tadmor. Strong stability-preserving high-order time discretization methods. *SIAM Rev.*, 43(1):89–112 (electronic), 2001.

177. P. Grisvard. *Singularities in Boundary Value Problems*. Masson, Paris, 1992.

178. M. J. Grote, A. Schneebeli, and D. Schötzau. Interior penalty discontinuous Galerkin method for Maxwell's equations: Energy norm error estimates. *Journal of Computational and Applied Mathematics*, 204:375–386, 2007.

179. J.-L. Guermond, P. Minev, and J. Shen. Error analysis of pressure-correction schemes with open boundary conditinos. *SIAM J. Numer. Anal.*, 43(1):239–258, 2005.

180. J.-L. Guermond and L. Quartapelle. On the approximation of the unsteady Navier–Stokes equations by finite element projection methods. *Numer. Math.*, 80:207–238, 1998.

181. J. Guzmán. Pointwise error estimates for discontinuous Galerkin methods with lifting operators for elliptic problems. *Math. Comp.*, 75(255):1067–1085 (electronic), 2006.

182. A. Hansbo and P. Hansbo. An unfitted finite element method, based on Nitsche's method, for elliptic interface problems. *Comput. Methods Appl. Mech. Engrg.*, 191(47-48):5537–5552, 2002.

183. P. Hansbo and M. G. Larson. Discontinuous Galerkin methods for incompressible and nearly incompressible elasticity by Nitsche's method. *Comput. Methods Appl. Mech. Engrg.*, 191(17-18):1895–1908, 2002.

184. P. Hansbo and M. G. Larson. Discontinuous Galerkin and the Crouzeix-Raviart element: application to elasticity. *M2AN Math. Model. Numer. Anal.*, 37(1):63–72, 2003.

185. P. Hansbo and M. G. Larson. Piecewise divergence-free discontinuous Galerkin methods for Stokes flow. *Commun. Numer. Meth. Engng*, 24:355–366, 2008.

186. D. J. Hardy, D. I. Okunbor, and R. D. Skeel. Symplectic variable step size integration for n-body problems. *Appl. Numer. Math.*, 29:19–30, 1999.

187. J. Haslinger and I. Hlaváček. Convergence of a finite element method based on the dual variational formulation. *Apl. Mat.*, 21(1):43–65, 1976.

188. B. Heinrich and S. Nicaise. The Nitsche mortar finite-element method for transmission problems with singularities. *IMA J. Numer. Anal.*, 23(2):331–358, 2003.

189. B. Heinrich and K. Pietsch. Nitsche type mortaring for some elliptic problem with corner singularities. *Computing*, 68(3):217–238, 2002.

190. B. Heinrich and K. Pönitz. Nitsche type mortaring for singularly perturbed reaction-diffusion problems. *Computing*, 75(4):257–279, 2005.

191. J. S. Hesthaven and T. Warburton. Nodal high-order methods on unstructured grids. i: Time-domain solution of maxwell's equations. *J. Comput. Phys.*, 181(1):186–221, 2002.

192. J. S. Hesthaven and T. Warburton. *Nodal discontinuous Galerkin methods*, volume 54 of *Texts in Applied Mathematics*. Springer, New York, 2008. Algorithms, analysis, and applications.

193. I. Hlaváček, J. Haslinger, J. Nečas, and J. Lovíšek. *Solution of variational inequalities in mechanics*, volume 66 of *Applied Mathematical Sciences*. Springer-Verlag, New York, 1988. Translated from the Slovak by J. Jarník.

194. R. H. W. Hoppe, G. Kanschat, and T. Warburton. Convergence analysis of an adaptive interior penalty discontinuous Galerkin method. *SIAM J. Numer. Anal.*, 47(1):534–550, 2008/09.

195. H. Hoteit, P. Ackerer, R. Mosé, J. Erhel, and B. Philippe. New two-dimensional slope limiters for discontinuous Galerkin methods on arbitrary meshes. *Internat. J. Numer. Methods Engrg.*, 61(14):2566–2593, 2004.

196. T. Y. Hou and P. G. LeFloch. Why nonconservative schemes converge to wrong solutions: error analysis. *Math. Comp.*, 62(206):497–530, 1994.

197. P. Houston, I. Perugia, A. Schneebeli, and D. Schötzau. Interior penalty method for the indefinite time-harmonic Maxwell equations. *Numer. Math.*, 100:485–518, 2005.

198. P. Houston, I. Perugia, A. Schneebeli, and D. Schötzau. Mixed discontinuous Galerkin approximation of the Maxwell operator: the indefinite case. *M2AN Math. Model. Numer. Anal.*, 39(4):727–753, 2005.

199. P. Houston, D. Schötzau, and T. P. Wihler. Energy norm a posteriori error estimation of hp-adaptive discontinuous Galerkin methods for elliptic problems. *Math. Models Methods Appl. Sci.*, 17(1):33–62, 2007.

200. J. Jaffré, C. Johnson, and A. Szepessy. Convergence of the discontinuous Galerkin finite element method for hyperbolic conservation laws. *M3AS*, 5(3):367–386, 1995.

201. M. Jensen. *Discontinuous Galerkin Methods for Friedrichs Systems with Irregular Solutions*. PhD thesis, University of Oxford, www.comlab.ox.ac.uk/research/na/thesis/thesisjensen.pdf, 2004.

202. G. S. Jiang and C.-W. Shu. On a cell entropy inequality for discontinuous Galerkin methods. *Math. Comp.*, 62(206):531–538, 1994.

203. C. Johnson, U. Nävert, and J. Pitkäranta. Finite element methods for linear hyperbolic equations. *Comput. Methods Appl. Mech. Engrg.*, 45:285–312, 1984.

204. C. Johnson and J. Pitkäranta. An analysis of the discontinuous Galerkin method for a scalar hyperbolic equation. *Math. Comp.*, 46(173):1–26, 1986.

205. G. Kanschat. *Discontinuous Galerkin methods for viscous incompressible flow*. Advances in Numerical Mathematics. Vieweg-Teubner, Wiesbaden, Germany, 2008.

206. O. A. Karakashian and W. N. Jureidini. A nonconforming finite element method for the stationary Navier-Stokes equations. *SIAM J. Numer. Anal.*, 35(1):93–120 (electronic), 1998.

207. O. A. Karakashian and F. Pascal. A posteriori error estimates for a discontinuous Galerkin approximation of second-order elliptic problems. *SIAM J. Numer. Anal.*, 41(6):2374–2399, 2003.

208. O. A. Karakashian and F. Pascal. Convergence of adaptive discontinuous Galerkin approximations of second-order elliptic problems. *SIAM J. Numer. Anal.*, 45(2):641–665, 2007.

209. G. E. Karniadakis and J. S. Spencer. *Spectral/hp Element Methods for CFD*. Numerical Mathematics and Scientific Computation. Oxford University Press, New York, NY, 1999.

210. R. B. Kellogg. On the Poisson equation with intersecting interfaces. *Applicable Anal.*, 4:101–129, 1974/75. Collection of articles dedicated to Nikolai Ivanovich Muskhelishvili.

211. F. Kikuchi. On a discrete compactness property for the Nédélec finite elements. *J. Fac. Sci. Univ. Tokyo Sect. IA Math.*, 36(3):479–490, 1989.

212. K. Y. Kim. A posteriori error estimators for locally conservative methods of nonlinear elliptic problems. *Appl. Numer. Math.*, 57(9):1065–1080, 2007.

213. B. Kirk, J. W. Peterson, R. H. Stogner, and G. F. Carey. libmesh: A C++ library for parallel adaptive mesh refinement/coarsening simulation. *Engineering with Computers*, 22(3–4):237–254, 2006.

214. L. S. G. Kovasznay. Laminar flow behind a two-dimensional grid. *Proc. Camb. Philos. Soc.*, 44:58–62, 1948.

215. L. Krivodonova. Limiters for high-order discontinuous Galerkin methods. *J. Comput. Phys.*, 226(1):879–896, 2007.

216. L. Krivodonova, J. Xin, J.-F. Remacle, N. Chevaugeon, and J. E. Flaherty. Shock detection and limiting with discontinuous Galerkin methods for hyperbolic conservation laws. *Appl. Numer. Math.*, 48(3-4):323–338, 2004. Workshop on Innovative Time Integrators for PDEs.

217. D. Kuzmin. A vertex-based hierarchical slope limiter for p-adaptive discontinuous Galerkin methods. *J. Comput. Appl. Math.*, 233(12):3077–3085, 2010.

218. P. Ladevèze. *Comparaison de modèles de milieux continus.* Ph.D. thesis, Université Pierre et Marie Curie (Paris 6), 1975.

219. P. Ladevèze and D. Leguillon. Error estimate procedure in the finite element method and applications. *SIAM J. Numer. Anal.*, 20(3):485–509, 1983.

220. M. G. Larson and A. J. Niklasson. Analysis of a nonsymmetric discontinuous Galerkin method for elliptic problems: stability and energy error estimates. *SIAM J. Numer. Anal.*, 42(1):252–264 (electronic), 2004.

221. A. Lasis and E. Süli. Poincaré-type inequalities for broken Sobolev spaces. Technical Report NI03067-CPD, http://www.newton.ac.uk/preprints2003.html, Isaac Newton Institute for Mathematical Sciences, Oxford, England, 2003.

222. P. D. Lax and A. N. Milgram. Parabolic equations. In *Contributions to the theory of partial differential equations*, Annals of Mathematics Studies, no. 33, pages 167–190. Princeton University Press, Princeton, N. J., 1954.

223. P. D. Lax and R. S. Phillips. Local boundary conditions for dissipative symmetric linear differential operators. *Comm. Pure Appl. Math.*, 13:427–455, 1960.

224. R. Lazarov, S. Repin, and S. Tomar. Functional a posteriori error estimates for discontinuous Galerkin approximations of elliptic problems. *Numer. Methods Partial Differential Equations*, 25(4):952–971, 2009.

225. P. G. LeFloch. *Hyperbolic systems of conservation laws: the theory of classical and nonclassical shock waves.* Lectures in Mathematics. ETH Zürich. Birkhäuser, 2002.

226. T. Leicht and R. Hartmann. Anisotropic mesh refinement for discontinuous Galerkin methods in two-dimensional aerodynamic flow simulations. *Internat. J. Numer. Methods Fluids*, 56(11):2111–2138, 2008.

227. P. Lesaint. Finite element methods for symmetric hyperbolic equations. *Numer. Math.*, 21:244–255, 1973/74.

228. P. Lesaint. *Sur la résolution des systèmes hyperboliques du premier ordre par des méthodes d'éléments finis.* PhD thesis, University of Paris VI, 1975.

229. P. Lesaint and P.-A. Raviart. On a finite element method for solving the neutron transport equation. In *Mathematical Aspects of Finite Elements in Partial Differential Equations*, pages 89–123. Publication No. 33. Math. Res. Center, Univ. of Wisconsin-Madison, Academic Press, New York, 1974.

230. R. J. LeVeque. *Finite volume methods for hyperbolic problems.* Cambridge Texts in Applied Mathematics. Cambridge University Press, Cambridge, 2002.

231. D. Levy and E. Tadmor. From semidiscrete to fully discrete: stability of Runge-Kutta schemes by the energy method. *SIAM Rev.*, 40(1):40–73 (electronic), 1998.

232. A. Lew, P. Neff, D. Sulsky, and M. Ortiz. Optimal BV estimates for a discontinuous Galerkin method for linear elasticity. *AMRX Appl. Math. Res. Express*, 3:73–106, 2004.

233. R. Luce and B. I. Wohlmuth. A local a posteriori error estimator based on equilibrated fluxes. *SIAM J. Numer. Anal.*, 42(4):1394–1414, 2004.

234. H. Luo, J. D. Baum, and R. Löhner. A Hermite WENO-based limiter for discontinuous Galerkin method on unstructured grids. *J. Comput. Phys.*, 225(1):686–713, 2007.

235. J. M. Melenk and B. I. Wohlmuth. On residual-based a posteriori error estimation in hp-FEM. A posteriori error estimation and adaptive computational methods. *Adv. Comput. Math.*, 15(1-4):311–331 (2002), 2001.

236. B. Merlet and J. Vovelle. Error estimate for finite volume scheme. *Numer. Math.*, 106(1):129–155, 2007.

237. P. Monk and L. Demkowicz. Discrete compactness and the approximation of Maxwell's equations in $\mathbb{R}^3$. *Math. Comp.*, 70(234):507–523, 2001.

238. P. Monk and E. Süli. The adaptive computation of far-field patterns by a posteriori error estimation of linear functionals. *SIAM J. Numer. Anal.*, 36(1):251–274, 1999.

239. A. Montlaur, S. Fernandez-Mendez, and A. Huerta. Discontinuous Galerkin methods for the Stokes equations using divergence-free approximations. *Int. J. Numer. Meth. Fluids*, 57:1071–1092, 2008.

240. P. Morin, R. H. Nochetto, and K. G. Siebert. Convergence of adaptive finite element methods. *SIAM Rev.*, 44(4):631–658 (electronic) (2003), 2002. Revised reprint of "Data oscillation and convergence of adaptive FEM" [SIAM J. Numer. Anal. 38 (2000), no. 2, 466–488 (electronic)].

241. C.-L. M. H. Navier. Mémoire sur les lois de l'équilibre et du mouvement des corps solides élastiques [read on 14 May, 1821]. *Mém. Acad. R. Sci.*, 7:375–393, 1827.

242. J.-C. Nédélec. Mixed finite elements in $\mathbb{R}^3$. *Numer. Math.*, 35(3):315–341, 1980.

243. J. Nečas. Sur une méthode pour résoudre les équations aux dérivées partielles de type elliptique, voisine de la variationnelle. *Ann. Scuola Norm. Sup. Pisa*, 16:305–326, 1962.

244. J. Nečas. *Equations aux Dérivées Partielles*. Presses de l'Université de Montréal, Montréal, Canada, 1965.

245. N. C. Nguyen, J. Peraire, and B. Cockburn. An implicit high-order hybridizable discontinuous Galerkin method for linear convection-diffusion equations. *J. Comput. Phys.*, 228(9):3232–3254, 2009.

246. S. Nicaise and S. Cochez-Dhondt. Adaptive finite element methods for elliptic problems: abstract framework and applications. *M2AN Math. Model. Numer. Anal.*, 44(3):485–508, 2010.

247. S. Nicaise and A.-M. Sändig. General interface problems. I, II. *Math. Methods Appl. Sci.*, 17(6):395–429, 431–450, 1994.

248. J. Nitsche. Über ein Variationsprinzip zur Lösung von Dirichlet-Problemen bei Verwendung von Teilräumen, die keinen Randbedingungen unterworfen sind. *Abh. Math. Sem. Univ. Hamburg*, 36:9–15, 1971. Collection of articles dedicated to Lothar Collatz on his sixtieth birthday.

249. J. Nitsche. On Dirichlet problems using subspaces with nearly zero boundary conditions. In *The mathematical foundations of the finite element method with applications to partial differential equations (Proc. Sympos., Univ. Maryland, Baltimore, Md., 1972)*, pages 603–627. Academic Press, New York, 1972.

250. J. T. Oden, I. Babuška, and C. E. Baumann. A discontinuous hp finite element method for diffusion problems. *J. Comput. Phys.*, 146(2):491–519, 1998.

251. K. B. Ølgaard, A. Logg, and G. N. Wells. Automated code generation for discontinuous Galerkin methods. *SIAM J. Sci. Comput.*, 31(2):849–864, 2008.

252. E. T. Olsen and J. Douglas Jr. Bounds on spectral condition numbers of matrices arising in the p-version of the finite element method. *Num. Math.*, 69(3):333–352, 1995.

253. S. Osher. Riemann solvers, the entropy condition, and difference approximations. *SIAM J. Numer. Anal.*, 21(2):217–235, 1984.

254. F. Otto. Initial-boundary value problem for a scalar conservation law. *C. R. Acad. Sci. Paris Sér. I Math.*, 322(8):729–734, 1996.

255. L. E. Payne and H. F. Weinberger. An optimal Poincaré inequality for convex domains. *Arch. Rational Mech. Anal.*, 5:286–292, 1960.

256. J. Peraire and P.-O. Persson. The compact discontinuous Galerkin (CDG) method for elliptic problems. *SIAM J. Sci. Comput.*, 30(4):1806–1824, 2008.

257. I. Perugia and D. Schötzau. The hp-local discontinuous Galerkin method for low-frequency time-harmonic Maxwell equations. *Math. Comp.*, 72(243):1179–1214, 2003.

258. T. E. Peterson. A note on the convergence of the discontinuous Galerkin method for a scalar hyperbolic equation. *SIAM J. Numer. Anal.*, 28(1):133–140, 1991.

259. S. Piperno. Symplectic local time-stepping in non-dissipative DGTD methods applied to wave propagation problems. *ESAIM: M2AN*, 405(5):815–841, 2006.

260. S. D. Poisson. Mémoire sur les équations générales de l'équilibre et du mouvement des corps solides élastiques et des fluides. *J. École Polytech.*, 13:1–174, 1831.

261. W. Prager and J. L. Synge. Approximations in elasticity based on the concept of function space. *Quart. Appl. Math.*, 5:241–269, 1947.

262. C. Prud'homme. A domain specific embedded language in C++ for automatic differentiation, projection, integration and variational formulations. *Scientific Programming*, 14(2):81–110, 2006.

263. L. Quartapelle. *Numerical solution of the incompressible Navier-Stokes equations*, volume 113 of *International Series of Numerical Mathematics*. Birkhäuser Verlag, Basel, 1993.

264. A. Quarteroni, R. Sacco, and F. Saleri. *Numerical Mathematics*. Springer Verlag, Berlin, 2000.

265. J. Rauch. Symmetric positive systems with boundary characteristic of constant multiplicity. *Trans. Amer. Math. Soc.*, 291(1):167–187, 1985.

266. J. Rauch. Boundary value problems with nonuniformly characteristic boundary. *J. Math. Pures Appl. (9)*, 73(4):347–353, 1994.

267. P.-A. Raviart and J.-M. Thomas. A mixed finite element method for 2nd order elliptic problems. In *Mathematical aspects of finite element methods (Proc. Conf., Consiglio Naz. delle Ricerche (C.N.R.), Rome, 1975)*, pages 292–315. Lecture Notes in Math., Vol. 606. Springer, Berlin, 1977.

268. W. H. Reed and T. R. Hill. Triangular mesh methods for the neutron transport equation. Technical Report LA-UR-73-0479, http://lib-www.lanl.gov/cgi-bin/getfile?00354107.pdf, Los Alamos Scientific Laboratory, Los Alamos, NM, 1973.

269. B. Rivière. *Discontinuous Galerkin methods for solving elliptic and parabolic equations: Theory and implementation*. Frontiers in Mathematics. SIAM, Philadelphia, 2008.

270. B. Rivière and M. F. Wheeler. A posteriori error estimates for a discontinuous Galerkin method applied to elliptic problems. *Comput. Math. Appl.*, 46(1):141–163, 2003.

271. B. Rivière, M. F. Wheeler, and V. Girault. Improved energy estimates for interior penalty, constrained and discontinuous Galerkin methods for elliptic problems. I. *Computat. Geosci.*, 8:337–360, 1999.

272. B. Rivière, M. F. Wheeler, and V. Girault. A priori error estimates for finite element methods based on discontinuous approximation spaces for elliptic problems. *SIAM J. Numer. Anal.*, 39(3):902–931 (electronic), 2001.

273. J. E. Roberts and J.-M. Thomas. Mixed and hybrid methods. In *Handbook of Numerical Analysis, Vol. II*, pages 523–639. North-Holland, Amsterdam, 1991.

274. H.-G. Roos, M. Stynes, and L. Tobiska. *Robust numerical methods for singularly perturbed differential equations. Convection-diffusion-reaction and flow problems*, volume 24 of *Springer Series in Computational Mathematics*. Springer-Verlag, Berlin, second edition, 2008.

275. C. Schwab. *p- and hp-Finite Element Methods*. Numerical Mathematics and Scientific Computation. The Clarendon Press Oxford University Press, New York, NY, 1998.

276. C. Schwab and M. Suri. Mixed hp finite element methods for Stokes and non-Newtonian flow. *Comput. Methods Appl. Mech. Engrg.*, 175(3-4):217–241, 1999.

277. S. J. Sherwin, R. M. Kirby, J. Peiró, R. L. Taylor, and O. C. Zienkiewicz. On 2D elliptic discontinuous Galerkin methods. *Int. J. Numer. Meth. Engng.*, 65:752–784, 2006.

278. C.-W. Shu. TVB boundary treatment for numerical solutions of conservation laws. *Math. Comp.*, 49(179):123–134, 1987.

279. C.-W. Shu and S. Osher. Efficient implementation of essentially nonoscillatory shock-capturing schemes. *J. Comput. Phys.*, 77(2):439–471, 1988.

280. V. A. Solonnikov. l^p-estimates for solutions of the heat equation in a dihedral angle. *Rend. Mat. Appl.*, 21:1–15, 2001.

281. R. J. Spiteri and S. J. Ruuth. A new class of optimal high-order strong-stability-preserving time discretization methods. *SIAM J. Numer. Anal.*, 40(2):469–491 (electronic), 2002.

282. R. Stenberg. Mortaring by a method of J.A. Nitsche. In Idelsohn S.R., Oñate E., and Dvorkin E.N., editors, *Computational Mechanics: New trends and applications*, pages 1–6, Barcelona, Spain, 1998. Centro Internacional de Métodos Numéricos en Ingeniería.

283. A. F. Stephansen. *Méthodes de Galerkine discontinues et analyse d'erreur a posteriori pour les problèmes de diffusion hétérogène.* Ph.D. thesis, Ecole Nationale des Ponts et Chaussées, http://pastel.paristech.org/3419/, 2007.

284. G. G. Stokes. On the theories of the internal friction of fluids in motion, and of the equilibrium and motion of elastic solids. *Trans. Cambridge Phil. Soc.*, 8:287–305, 1845.

285. G. Strang. Variational crimes in the finite element method. In A.K. Aziz, editor, *The Mathematical Foundations of the Finite Element Method with Applications to Partial Differential Equations*, New York, NY, 1972. Academic Press.

286. F. Stummel. Basic compactness properties of nonconforming and hybrid finite element spaces. *RAIRO Anal. Numér.*, 14(1):81–115, 1980.

287. S. Sun and M. F. Wheeler. Anisotropic and dynamic mesh adaptation for discontinuous Galerkin methods applied to reactive transport. *Comput. Methods Appl. Mech. Engrg.*, 195(25-28):3382–3405, 2006.

288. B. A. Szabó and I. Babuška. *Finite element analysis.* A Wiley-Interscience Publication. John Wiley & Sons Inc., New York, 1991.

289. E. Tadmor. From semidiscrete to fully discrete: stability of Runge-Kutta schemes by the energy method. II. *SIAM Proceedings in Applied Mathematics*, 109:25–49, 2002.

290. M. A. Taylor, B. A. Wingate, and R. E. Vincent. An algorithm for computing Fekete points in the triangle. *SIAM J. Numer. Anal.*, 38(5):1707–1720 (electronic), 2000.

291. R. Temam. Une méthode d'approximation de la solution des équations de Navier–Stokes. *Bull. Soc. Math. France*, 98:115–152, 1968.

292. R. Temam. *Navier-Stokes Equations*, volume 2 of *Studies in Mathematics and its Applications.* North-Holland Publishing Co., Amsterdam, revised edition, 1979. Theory and numerical analysis, With an appendix by F. Thomasset.

293. A. ten Eyck and A. Lew. Discontinuous Galerkin methods for non-linear elasticity. *Internat. J. Numer. Methods Engrg.*, 67(9):1204–1243, 2006.

294. V. Thomée. *Galerkin finite element methods for parabolic problems*, volume 25 of *Springer Series in Computational Mathematics.* Springer-Verlag, Berlin, second edition, 2006.

295. E. F. Toro. *Riemann solvers and numerical methods for fluid dynamics. A practical introduction.* Springer-Verlag, Berlin, second edition, 1999.

296. A. Toselli. hp-finite element discontinuous Galerkin approximations for the Stokes problem. *M3AS*, 12(11):1565–1616, 2002.

297. J. J. W. van der Vegt and H. van der Ven. Discontinuous Galerkin finite element method with anisotropic local grid refinement for inviscid compressible flows. *J. Comput. Phys.*, 141(1):47–77, 1998.

298. R. Verfürth. *A review of a posteriori error estimation and adaptive mesh-refinement techniques.* Teubner-Wiley, Stuttgart, 1996.

299. R. Verfürth. personal communication. 2009.

300. M. Vohralík. On the discrete Poincaré–Friedrichs inequalities for nonconforming approximations of the Sobolev space H^1. *Numer. Funct. Anal. Optim.*, 26(7–8):925–952, 2005.

301. M. Vohralík. A posteriori error estimates for lowest-order mixed finite element discretizations of convection-diffusion-reaction equations. *SIAM J. Numer. Anal.*, 45(4):1570–1599 (electronic), 2007.

302. M. Vohralík. Unified primal formulation-based a priori and a posteriori error analysis of mixed finite element methods. *Math. Comp.*, 79(272):2001–2032, 2010.

303. J. Vovelle. Convergence of finite volume monotone schemes for scalar conservation laws on bounded domains. *Numer. Math.*, 90(3):563–596, 2002.

304. T. Warburton and J. S. Hesthaven. On the constants in hp-finite element trace inverse inequalities. *Comput. Methods Appl. Mech. Engrg.*, 192(25):2765–2773, 2003.

305. M. F. Wheeler. A priori L_2 error estimates for Galerkin approximations to parabolic partial differential equations. *SIAM J. Numer. Anal.*, 10:723–759, 1973.

306. M. F. Wheeler. An elliptic collocation-finite element method with interior penalties. *SIAM J. Numer. Anal.*, 15:152–161, 1978.

307. T. P. Wihler. Locking-free adaptive discontinuous Galerkin FEM for linear elasticity problems. *Math. Comp.*, 75(255):1087–1102, 2006.

308. T. P. Wihler and B. Rivière. Discontinuous Galerkin method for second-order elliptic PDE with low-regularity solutions. *J. Sci. Comput.*, 46(12):151–165, 2011.

309. L. Yuan and C.-W. Shu. Discontinuous Galerkin method based on non-polynomial approximation spaces. *J. Comput. Phys.*, 218(1):295–323, 2006.

310. Q. Zhang and C.-W. Shu. Error estimates to smooth solutions of Runge-Kutta discontinuous Galerkin methods for scalar conservation laws. *SIAM J. Numer. Anal.*, 42(2):641–666 (electronic), 2004.

311. Q. Zhang and C.-W. Shu. Stability analysis and a priori error estimates of the third order explicit Runge-Kutta discontinuous Galerkin method for scalar conservation laws. *SIAM J. Numer. Anal.*, 48(3):1038–1063, 2010.

Author Index

Abgrall, R. 110
Abramowitz, M. 349, 350
Achdou, Y. 214
Ackerer, P. 114
Adams, R. A. 4
Agélas, L. VII, 141, 201, 207
Ainsworth, M. 131, 216, 229, 230, 236
Ambrosio, L. 70
Amrouche, C. 258
Antonić, N. 306, 307
Arbogast, T. 226
Arnold, D. N. VI, IX, 119, 120, 122, 125, 130, 134, 147, 150, 226
Aubin, J.-P. 133, 258, 303
Aziz, A. K. 3

Babuška, I. VI, VIII, 3, 125, 199, 252, 351
Baker, G. A. VI, 125
Balay, S. 128
Bangerth, W. V
Bardos, C. 38, 99
Barré de Saint Venant, A. J.-C. 241
Bassi, F. VI, 120, 137, 138, 143, 203, 254, 352–354
Bastian, P. V, 219
Batchelor, G. K. 241
Baum, J. D. 114
Baumann, C. E. 199
Bebendorf, M. 232
Becker, R. 230
Bernardi, C. 28, 214
Biswas, R. 113
Boffi, D. 193
Bogovskiĭ, M. 246
Bonito, A. 131, 216, 229
Botti, L. 283, 287, 352–354
Boyer, F. 70

Braess, D. X, 141, 236, 351
Brenner, S. C. X, 7, 24, 26, 33, 38, 130, 192, 199, 245, 330, 351
Brézis, H. 4, 5, 121, 135, 190, 193
Brezzi, F. VI, 28, 56, 120, 134, 137, 140, 147, 150, 218, 226, 252, 330
Buffa, A. 24, 141, 188, 301
Burazin, K. 306, 307
Burman, E. VII, 13, 58, 68, 78, 81, 82, 130, 141, 155, 214, 266, 331, 333
Buschelman, K. 128
Butcher, J. 76

Canuto, C. 28
Caplain, G. 293, 301, 304–307, 312
Carey, G. F. V
Carstensen, C. 17, 28, 234
Castillo, P. 147
Cattabriga, L. 258
Causin, P. 120, 148
Caussignac, P. V
Chavent, G. V
Chen, H. 134
Chen, Q. 351
Chen, Z. 134, 226
Chevaugeon, N. 114
Chorin, A. J. 283
Ciarlet, P. G. X, 7, 24, 245, 330
Cochez-Dhondt, S. 229, 236
Cockburn, B. V, VI, 68, 110, 112–114, 120, 134, 143, 147, 148, 150, 241, 242, 249, 271, 282, 300, 301
Cohen, G. 301
Colombo, A. 352–354
Coquel, F. 214
Creusé, E. 193

Crippa, G. 70
Crivellini, A. 254

Danaila, I. V
Dauge, M. 135
Dawson, C. V, 142, 199
Deimling, K. 275
Demkowicz, L. 193
Deny, J. 33
Després, B. 78
Destuynder, P. 236
Devine, K. D. 113
Di Pietro, D. A. VII, 119, 135, 141, 155, 160, 163, 168, 172, 173, 188, 195, 201, 207, 242, 254, 282, 283, 287, 352–354
do Carmo, E. G. D. 245, 330
Dong, B. 150
Dorr, M. R. VIII
Douglas Jr., J. VI, 125, 348
Droniou, J. VII, 120, 148
Droske, M. V
Dryja, M. 150, 155
Duarte, A. V. C. 245, 330
Dubiner, M. 352
Dupont, T. VI, 125
Durán, R. G. 246

El Alaoui, L. 214
Engwer, C. V
Erhel, J. 114
Ern, A. VI, VII, X, 2–4, 7, 18, 24, 26, 28, 33, 38, 41, 61, 68, 78, 81, 82, 130, 141, 155, 163, 168, 171–173, 211, 214, 218, 222, 230, 234, 236, 243, 246, 247, 249, 293, 301, 304–307, 312, 323, 329–331, 333, 351
Ethier, C. R. 288
Evans, L. C. 4, 5, 121, 135, 190, 193
Ewing, R. E. 148
Eymard, R. VII, 120, 139, 141, 142, 148, 188, 191, 193, 201, 207–210, 234, 275

Fernández, M. A. VII, 68, 78, 81, 82, 331, 333
Fernandez-Mendez, S. 242
Ferrieres, X. 301
Flaherty, J. E. 113, 114
Fortin, M. 28, 218, 330
Franca, L. P. 330

Friedrichs, K. O. X, 293, 295, 306
Funken, S. A. 28

Gallouët, T. VII, 188, 191, 193, 207–210, 234, 275
Gassner, G. J. 13
Gastaldi, F. 173
Georgoulis, E. H. 24
Ghia, K. N. 266
Ghia, U. 266
Ghilani, M. 275
Giraud, L. 353
Girault, V. 34, 199, 246, 258, 266, 269–271
Godlewski, E. 68, 110
Gopalakrishnan, J. 120, 148
Gottlieb, D. 108
Gottlieb, S. VIII, 68, 107
Grisvard, P. 122, 133
Gropp, W. D. 128
Grote, M. J. 301
Guermond, J.-L. VI, X, 2–4, 7, 18, 26, 28, 33, 38, 41, 61, 155, 163, 168, 172, 173, 218, 243, 246, 247, 249, 283, 287, 293, 301, 304–307, 312, 323, 329, 330, 351
Guzmán, J. 134, 150

Hall, E. 24
Hansbo, A. 155
Hansbo, P. 130, 155, 230, 242, 266, 300
Hardy, D. J. 339
Hartmann, R. V, 24
Haslinger, J. 236
Hecht, F. V
Heinrich, B. 155
Herbin, R. VII, 139, 188, 191, 193, 207–210, 234, 275
Hesthaven, J. S. VII, VIII, X, 13, 28, 352
Hilhorst, D. 142
Hill, T. R. V
Hlaváček, I. 236
Hoppe, R. H. W. 229
Hoteit, H. 114
Hou, S. VI, 68, 114
Hou, T. Y. 101
Houston, P. 24, 230, 301
Huerta, A. 242

Jaffré, J. VI
Jensen, M. 44, 293, 305

Author Index

Jiang, G. S. 103
Johnson, C. V, VI, 57, 61, 293
Jureidini, W. N. 271

Kanschat, G. V, X, 147, 229, 241, 242, 249, 271, 282
Karakashian, O. A. 214, 216, 229, 230, 271
Karniadakis, G. E. V, 348, 350
Katz, I. N. VIII
Kaushik, D. 128
Kellogg, R. B. 152
Kikuchi, F. 193
Kim, K. Y. 218, 230, 236
Kirby, R. M. 147
Kirk, B. V
Klöfkorn, R. V
Knepley, M. G. 128
Kovasznay, L. S. G. 289
Krivodonova, L. 114
Kuzmin, D. 114

Ladevèze, P. 236
Langou, J. 353
Larson, M. G. 130, 199, 230, 242, 266, 300
Lasis, A. 192
Latché, J.-C. 275
Lax, P. D. 3, 306
Lazarov, R. 120, 148, 236
Le Roux, A. Y. 99
LeFloch, P. G. 68, 101
Leguillon, D. 236
Leicht, T. 24
Lesaint, P. V, 293
LeVeque, R. J. 68, 103, 105
Levy, D. VIII, 67, 85
Lew, A. 141, 190
Li, F. 301
Lin, S. VI, 68
Lions, J.-L. 33
Logg, A. V
Löhner, R. 114
Lörcher, F. 13
Lovíšek, J. 236
Luce, R. 236
Luo, H. 114

Maday, Y. 28
Manzini, G. 137, 140

Marini, L. D. VI, 56, 120, 134, 137, 140, 147
Mariotti, G. 120, 137, 138, 143, 203
Masson, R. VII, 141, 201, 207
McInnes, L. C. 128
Melenk, J. M. 28
Merlet, B. 78
Métivet, B. 236
Milgram, A. N. 3
Minev, P. 287
Monk, P. 28, 193
Montlaur, A. 242
Morin, P. 229
Mosé, R. 114
Mozolevski, I. 130, 211
Munz, C.-D. 13
Muschietti, M. A. 246

Nävert, U. 293
Navier, C.-L. M. H. 241
Nédélec, J.-C. 99, 218
Neff, P. 141, 190
Neubauer, T. V
Nečas, J. 3, 236, 247
Nguyen, N. C. 150
Nicaise, S. VII, 152, 155, 193, 218, 229, 236
Niklasson, A. J. 199
Nitsche, J. VI, 125
Nochetto, R. H. 131, 216, 229

Oden, J. T. 199, 229, 236
Ohlberger, M. V
Okunbor, D. I. 339
Ølgaard, K. B. V
Olsen, E. T. 348
Ortiz, M. 141, 190
Ortner, C. 24, 141, 188
Osher, S. 76, 102, 107
Otto, F. 99
Owens, L. 199

Pascal, F. 214, 216, 229, 230
Payne, L. E. 232
Pedinotti, S. 120, 137, 138, 143, 203
Peiró, J. 147
Peraire, J. 147, 150
Pernet, S. 301
Persson, P.-O. 147
Perugia, I. 137, 147, 301

Peterseim, D. 17
Peterson, J. W. V
Peterson, T. E. 61
Philippe, B. 114
Phillips, R. S. 306
Pietra, P. 137, 140
Pietsch, K. 155
Piperno, S. 339, 341
Pironneau, O. V
Pitkäranta, J. 293
Poisson, S. D. 241
Pönitz, K. 155
Prager, W. 236
Prud'homme, C. V

Quartapelle, L. 241, 283
Quarteroni, A. 28, 58, 173, 348

Rankin, R. 131, 216
Rauch, J. 305
Raviart, P.-A. V, 34, 68, 110, 218, 246, 269, 270, 293
Rebay, S. VI, 120, 137, 138, 143, 203, 254
Reed, W. H. V
Remacle, J.-F. 114
Repin, S. 236
Restelli, M. 150
Rivière, B. 135, 199, 219, 230, 266, 271
Roberts, J. E. 218
Rochinha, F. A. 245, 330
Roos, H.-G. 166
Rozloznik, M. 353
Rumpf, M. V
Russo, A. 137, 140
Ruuth, S. J. 68

Sacco, R. 120, 148, 150, 348
Saleri, F. 348
Sändig, A.-M. 152
Savini, M. 120, 137, 138, 143, 203
Schneebeli, A. 301
Schöberl, J. 236
Schötzau, D. 301
Schuh, L. 211
Schwab, C. VIII, 28, 147, 241, 249, 330
Scott, L. R. X, 7, 26, 33, 38, 351
Shen, J. 287
Sherwin, S. J. 147
Shin, C. T. 266

Shu, C.-W. V, VI, VIII, 12, 68, 76, 78, 103, 107, 108, 113, 114, 120, 143, 301
Siebert, K. G. 229
Skeel, R. D. 339
Smith, B. F. 128
Solonnikov, V. A. 246
Spencer, J. S. 348, 350
Spiteri, R. J. 68
Stamm, B. 13, 58, 130, 266
Stegun, I. 349, 350
Steinman, D. A. 288
Stenberg, R. 155, 230, 330
Stephansen, A. F. VII, 28, 155, 163, 168, 171, 230, 234, 236
Stogner, R. H. V
Stokes, G. G. 241
Strang, G. 18, 141
Stummel, F. 193
Stynes, M. 166
Süli, E. 28, 56, 192
Sulsky, D. 141, 190
Sun, S. 24, 142, 199
Suri, M. 330
Synge, J. L. 236
Szabó, B. A. VIII
Szepessy, A. VI

Tadmor, E. VIII, 67, 68, 81, 85, 107
Taylor, M. A. 351
Taylor, R. L. 147
Temam, R. 242, 271, 283
ten Eyck, A. 141
Tesini, P. 352–354
Thomas, J.-M. 218
Thomée, V. VII, 67
Tobiska, L. 166
Tomar, S. 236
Toro, E. F. 68, 105
Toselli, A. 13, 266
Touzani, R. V

van der Vegt, J. J. W. 24
van der Ven, H. 24
Verfürth, R. 29, 212, 229
Vincent, R. E. 351
Vohralík, M. VII, 24, 142, 204, 218, 222, 226, 227, 230, 234, 236
Vovelle, J. 78, 99, 102

Wang, H. 150
Wang, Jing 300

Author Index

Wang, Junping 148
Warburton, T. VII, VIII, X, 28, 229, 301, 352
Weinberger, H. F. 232
Wells, G. N. V
Wheeler, M. F. VI, 24, 125, 142, 184, 199, 230, 266, 271
Wihler, T. P. 135, 230, 300
Wingate, B. A. 351
Wohlmuth, B. I. 28, 236

Xin, J. 114

Yang, Y. 148
Yuan, L. 12

Zhang, H. 128
Zhang, Q. 78, 108
Zienkiewicz, O. C. 147
Zlámal, M. VI, 125
Zunino, P. 155, 163, 168, 171

Index

A

Admissible states   99
Advection-reaction
   steady   37
   unsteady   67
Approximation error   21, 30
Asymptotic consistency   196
Average   10
   weighted   154
Averaging operator   214

B

Backward Euler scheme   177
Banach–Nečas–Babuška Theorem
   2
Barycentric coordinates   208
BDF2 scheme   182
Boundedness   21
   on orthogonal subscales   59
BRMPS   203
Broken divergence   16
Broken gradient   14
Broken polynomial spaces   12
Broken Sobolev spaces   14
Butcher's array   75
BV-norm   188

C

Cattabriga's regularity   258
Coercivity   3
Consistency   21
   asymptotic   196
   asymptotic adjoint   201
   strong on smooth functions   208
Consistency term   125
Contact-regularity   24
Curl-curl problem   300

D

Diffusion-advection-reaction   164
Diffusive flux   121
   discrete   143
   reconstruction by local solves   224
   reconstruction by prescription   219
Discrete gradient   140
Discrete Rellich–Kondrachov Theorem   193
Discrete Sobolev embeddings   190
Discrete stability   20
Div-grad problem   297

E

E-flux   102
Elliptic regularity   133
Error estimate
   abstract   22
   optimal   31
   quasi-optimal   31
   suboptimal   31
Explicit Euler scheme   74
Explicit RK2 scheme   78
Explicit RK3 scheme   78

F

Face normal   11
$\mathfrak{F}_T, \mathfrak{F}_F$ 23
$\mathcal{F}_h, \mathcal{F}_h^i, \mathcal{F}_h^b$ 10
Flux function   99
Forward Euler scheme   74, 75
   improved   75
Friedrichs' operator   303
$\mathcal{F}_T$ 10

Function spaces
   $H(\mathrm{div};\Omega)$ 16
   $H(\mathrm{div};\mathcal{T}_h)$ 16
   $H^m(\Omega)$ 6
   $H^m(\mathcal{T}_h)$ 14
   $L^p(\Omega)$ 5
   $\mathbb{P}_d^k$ 12
   $\mathbb{P}_d^k(\mathcal{T}_h)$ 12
   $\mathbb{Q}_d^k$ 13
   $W^{m,p}(\Omega)$ 6
   $W^{m,p}(\mathcal{T}_h)$ 14

G
Galerkin orthogonality   21
Gelfand triple   40
Graph space   39
Gronwall's Lemma   71

H
Harmonic mean   155
Harten's Lemma   110
HDG   120, 148
Heat equation   176
Heterogeneous diffusion   151
Heun scheme
   three-stage   76
   two-stage   75

I
IIP   199
Implicit Euler scheme   177
Inflow   37
Interface   9
Inverse inequality   26

J
Jump   11

L
L^2-orthogonal projection   19
   approximation   31
Lax–Milgram Lemma   4
LBB condition   252
LDG   120, 146
Lebesgue spaces   5
Liftings   138

Linear elasticity   298
Local conservation
   advection   55, 65
   Poisson   142
   Stokes   254
Local time stepping   339

M
mass matrix
   global   344
   local   344
Matching simplicial submesh   23
Maxwell, diffusive regime   300
Mesh
   general   9
   matching simplicial   23
   meshsize   9
   simplicial   8
Mesh compatibility   153
Mesh element   9
   diameter   9
   outward normal   9
Mesh face   9
   boundary face   9
   diameter   25
   interface   9
   length scale   125
Mesh regularity parameters   24
Mesh sequence
   admissible   31
   admissible in 1d   34
   contact-regular   24
   finitely shaped   33
   shape-regular   24
   star-shaped   33
Mixed dG methods   143
Monotone numerical flux   103

N
Navier–Stokes problem
   steady   267
   unsteady   284
N_∂ 10
n_F 11
NIP   199

Index

Norms
    $\|\cdot\|_{BV}$ 188
    $\|\cdot\|_{dG,p}$ 188
Norms for advection
    $\|\cdot\|_{cf}$ 51
    $\|\cdot\|_{cf,*}$ 53
    $\|\cdot\|_{uw\flat}$ 57
    $\|\cdot\|_{uw\flat,*}$ 59
    $\|\cdot\|_{uw\sharp}$ 61
    $\|\cdot\|_{uw\sharp,*}$ 64
Norms for diffusion
    $\|\cdot\|_{brmps}$ 203
    $\|\cdot\|_{sip}$ 128
    $\|\cdot\|_{sip,*}$ 130, 135
    $\|\cdot\|_{swip}$ 157
    $\|\cdot\|_{swip,*}$ 159, 160
Norms for diffusion-advection
    $\|\cdot\|_{da\flat}$ 167
    $\|\cdot\|_{da\flat,*}$ 168
    $\|\cdot\|_{da\sharp}$ 169
    $\|\cdot\|_{da\sharp,*}$ 170
Norms for Navier–Stokes flow
    $\|\cdot\|_{ns}$ 274
Norms for Stokes flow
    $\|\cdot\|_P$ 244
    $\|\cdot\|_{sto}$ 253
    $\|\cdot\|_{sto,*}$ 257
    $\|\cdot\|_U$ 244
    $\|\cdot\|_{vel}$ 249
    $\|\cdot\|_X$ 244
n_T 9
Numerical flux for conservation laws 100
    consistent 101
    E-flux 102
    generalized Lax–Friedrichs 105
    Godunov 105
    local Lax–Friedrichs 106
    monotone 103
    Roe 106
    Rusanov 105
Numerical fluxes
    advection, centered 55
    advection, upwind 65
    Friedrichs 320
    Poisson 142
    Stokes 254

O
Oswald interpolator 214
Outflow 41

P
Penalty term 125
Poincaré inequality
    broken 130
    continuous 121
    discrete 192
Poisson problem 119
    mixed formulation 121
Polyhedron 7
Polynomial approximation
    in mesh elements 31
    on mesh faces 32
    optimal 31
Potential 121
    discrete 143
    reconstruction 215, 216
Projection method 284

R
Raviart–Thomas–Nédélec elements 218
Reflecting boundary condition 335
Regularity of exact solution
    advection-reaction 48
    Friedrichs 313
    Poisson 122, 134
    Stokes 255
Riemann problem 99, 105
Runge scheme 75
Runge–Kutta scheme 74
    SSP 107

S
Seminorms
    $|\cdot|_\beta$ 167
Seminorms for diffusion
    $|\cdot|_J$ 128
    $|\cdot|_{J,\kappa}$ 157
Seminorms for Stokes flow
    $|\cdot|_J$ 250
    $|\cdot|_p$ 251
Shape-regularity 24
Simplex 8
    face 8

SIP   119
Sobolev spaces   6
   fractional exponent   7
Star-shaped   32
Stencil   56
   advection-reaction   56
   BRMPS   205
   LDG   146
   SIP   126
stiffness matrix
   global   344
   local   345
Stokes problem   243
Subelements   23
Subfaces   23
Submesh   23
Symmetry term   125

T

Temam's device
   continuous   271
   discrete   272

Trace inequality
   continuous   7, 28
   discrete   27
Trilinear form
   continuous   268
   discrete   272
$\mathfrak{S}_T$ 23
TVDM scheme   109

V

Vanishing diffusion   171
Velocity lifting   246
Verlet scheme   339

W

Weighted averages   154
Well-prepared initial datum   70